Practical Fourier Transform Infrared Spectroscopy

Practical Fourier Transform Infrared Spectroscopy

Industrial and Laboratory Chemical Analysis

Edited by

John R. Ferraro
Consultant, Argonne National Laboratories
Argonne, Illinois
(Emeritus Searle Professor of Chemistry
Loyola University
Chicago, Illinois)

K. Krishnan
BioRad Digilab Division
Cambridge, Massachusetts

Academic Press, Inc.
Harcourt Brace Jovanovich, Publishers
San Diego New York
Berkeley Boston London Sydney Tokyo Toronto

Academic Press, Inc.
San Diego, California 92101

United Kingdom Edition published by
Academic Press Limited
24–28 Oval Road, London NW1 7DX

Library of Congress Cataloging-in-Publication Data

Practical Fourier transform infrared spectroscopy : industrial and laboratory chemical analysis / [edited by] John R. Ferraro and K. Krishnan.
p. cm.
Includes index.
ISBN 0-12-254125-1 (alk. paper)
1. Infrared spectrosocpy. 2. Fourier transform spectroscopy. I. Ferraro, John R., Date. II. Krishnan, K.
QD96.I5P73 1989
543'.08583--dc20 89-31673
CIP

Printed in the United States of America
89 90 91 92 9 8 7 6 5 4 3 2 1

Contents

Contributors

Numbers in parentheses indicate the pages on which the authors' contributions begin.

John R. Ferraro (1, 41), Argonne National Laboratory, 9700 South Cass Avenue, Argonne, Illinois 60439

David M. Haaland (395), Sandia National Laboratories, Division 1823, Albuquerque, New Mexico 87185

S. L. Hill (103), BioRad Digilab Division, 237 Putnam Avenue, Cambridge, Massachusetts 02139

H. Ishida (351), Material Science Laboratories, Toray Research Center, Incorporated, Otsu, Shiga 520, Japan

A. Ishitani (351), Material Science Laboratories, Toray Research Center, Incorporated, Otsu, Shiga 520, Japan

Timothy A. Keiderling (203), Department of Chemistry, Box 4348, University of Illinois at Chicago, Chicago, Illinois 60680

K. Krishnan (103, 285), BioRad Digilab Division, 237 Putnam Avenue, Cambridge, Massachusetts 02139

Victor A. Maroni (1), Argonne National Laboratory, 9700 South Cass Avenue, Argonne, Illinois 60439

Ryujiro Namba (469), Shiseido Toxicological and Analytical Research Center, 1050 Nippa-cho Kohoku-ku, Yokohama, Japan

B. Schrader (167), Institut für Physikalische und Theoretische Chemie, Universität Essen, D 4300 Essen 1, Federal Republic of Germany

P. J. Stout (285), BioRad Digilab Division, 237 Putnam Avenue, Cambridge, Massachusetts 02139

Masaharu Watanabe (285), Toshiba ULSI Research Center, Kawasaki, Japan

Jack M. Williams (41), Argonne National Laboratory, 9700 South Cass Avenue, Argonne, Illinois 60439

Preface

The four volumes of the work entitled, *Fourier Transform Infrared Spectroscopy: Applications to Chemical Systems* (Academic Press, Inc.) have been found to be extremely useful. They have served the scientific community with state-of-the-art topics, written by experts and accompanied by ample bibliographies, and in some cases, provided directions for further research. A recent *Analytical Chemistry* survey has listed Fourier Transform Infrared Spectroscopy (FT-IR) as one of the fastest growing analytical techniques. The continued interest in FT-IR and its utility in solving problems has encouraged us to provide a new, related volume. This volume attempts to indicate the value of the tool for solving chemical laboratory problems and emphasizes industrial applications. Please note that the title is somewhat different, and we have chosen not to number this volume. Additionally, we have an international group of authors.

Chapter 1 deals with the potentially revolutionary, high-T_c (critical temperature) inorganic superconductors of the type LBCO (lanthanum barium copper oxide) and YBCO (yttrium barium copper oxide). Chapter 2 addresses the less well-known, charge-transfer organic superconductors. These materials have low Tc values [$(ET)_2Cu(SCN)_2$ has the highest T_c of $\approx$10K, where ET is bis(ethylenedithio)tetrathiafulvalene] and are tremendously interesting because they lend themselves to molecular engineering in their synthetic methods, and their chemistry is of considerable interest. Chapter 3 discusses the present status of FT-IR microsampling techniques which have become extremely useful in solving practical problems, difficult or insoluble by other methods.

Chapter 4 presents an update on the status of FT-Raman spectroscopy, a potentially valuable addition to the arsenal of instrumentation available to analytical scientists. Chapter 5 is devoted to vibrational circular dichroism with practical applications provided in the area of biochemistry. Chapter 6 presents the characterization of impurities such as oxygen, carbon, nitrogen, and hydrogen in semiconductor silicon by FT-IR, and also the determination of epitaxial film thicknesses on silicon.

Chapter 7 considers the solution of industrial problems using FT-IR and compares the technique to other analytical techniques. Both surface and bulk analyses are discussed regarding impurities on silicon and gallium arsenide wafers. Chapter 8 deals with the multivariate statistical methods used in modern quantitative analysis by FT-IR, an area of considerable importance relative to mathematical manipulation of spectroscopic data. Chapter 9 discusses the solution of industrial problems using GC/FT-IR. Some of the applications relate to the food and pesticide industries, the environment, carbohydrates, fragrant compounds, polymers, and forensic (drug) areas.

In some instances, the chapters present new and only peripherally researched material. The editors have encouraged such inclusions, as it allows the reader to formulate directions of research he or she may want to pursue.

It is the hope of the publisher and editors that this volume will serve the analytical scientific community as effectively as have the previous four related volumes.

Acknowledgments

With the appearance of commercial FT-IR instrumentation over the past two decades, we have observed a mushrooming of interest in, and applications of, the FT-IR technique in solving chemical analyses problems. We believe that much of this increased attention has been due to the availability of FT-IR instrumentation and accessory equipment to every laboratory—commercial and academic. It is our intention to acknowledge the contribution of the FT-IR instrumentation manufacturers and those providing FT-IR sampling accessories to solve the various analytical problems facing the scientific community. The following companies are saluted for their contribution to the popularity of FT-IR:

Analect Instruments, BioRad Digilab Division, Bomem Incorporated, Bruker Analytische Messtechnik GmbH, Buck Scientific, Incorporated, Foxboro Wilks, Harrick Scientific Corporation, Hewlett-Packard Company, IBM Instruments, Infrared Analysis, Incorporated, Janos Technology, JASCO, JEOL, Mattson Instruments, McCarthy Scientific Corporation, Midac Corporation, Nicolet Instruments Corporation, Perkin-Elmer Corporation, Phillips Instruments, Shimadzu Scientific Instruments, Specac Limited–ARIES, and Spectra-Tech Incorporated. We acknowledge their endeavors and dedicate this book to them.

John R. Ferraro
K. Krishnan

1

The Use of Vibrational Spectroscopy in the Characterization of High-Critical-Temperature Ceramic Superconductors

Victor A. Maroni
John R. Ferraro*
Argonne National Laboratory
Argonne, Illinois

*Consultant to Argonne National Laboratory and Searle Professor of Chemistry (Emeritus), Loyola University, Chicago, Illinois.

The finding, less than two years ago, of superconductivity in La-Ba-Cu-O compounds at ~40 K (Bednorz and Müller, 1986) and the ensuing discovery by Chu *et al.* (1987) of other copper oxide-based ceramics with superconducting critical temperatures (T_c) exceeding the boiling point of liquid nitrogen (77 K) has inspired an enormous body of scientific research aimed at the synthesis, characterization, and application of these novel materials. Vibrational spectroscopies (mainly infrared and Raman) have played a key role in the characterization of the new high-T_c materials, and there is already a collection of literature on just this aspect of the characterization process that is well in excess of one hundred articles. However, because so much information of the same general type emerged so quickly, there is an appreciable level of both redundancy and discrepancy in the findings from these vibrational spectroscopic investigations. Determining the consistency of results and resolving discrepancies is important in this instance because the vibrational spectroscopic data may hold one of the keys to elucidation of the high-T_c mechanism. In fact, most of the study of vibrational spectra of high-T_c materials has been devoted to searches for evidence of electron–phonon coupling and low-lying energy gaps that mediate the superconductivity process.

In this chapter, an attempt is made to sort through the rapidly evolving literature addressing the vibrational spectroscopy of high-T_c ceramics. The emphasis is on infrared and Raman spectroscopy because most of the spectroscopic studies have been of this type. Both techniques are used to about the same extent, and the results from each have tended to be complementary rather than duplicative. In addition to discussing the key results and the important conclusions that have come from this body of work, mention is made of the techniques employed in sample preparation and spectral measurement, although it is worthy of note, even at the outset of this review, that very few of the published papers cover these experimental details in a clear and comprehensive manner. Also, because several research groups have published their findings widely and in some cases with changing views of how the results should be interpreted, only the more recent publications from these groups are included in the review. Many papers that contained poorly presented spectral results (often not truly supportive of the conclusions drawn from them) and vague discussions are not covered in this review. Finally, the review is limited to the two most thoroughly studied phases, $La_{2-x}(Sr\text{ or }Ba)_xCuO_4$ (also referred to herein as LCO) and $ABa_2Cu_3O_{7-x}$ with A = e.g., Y, Sm, Eu, Gd, Ho (also referred to herein as 123).

I. STRUCTURAL CONSIDERATIONS AND GROUP THEORY

In the LCO series of high-T_c materials, the optimum superconducting properties are achieved at a composition near $La_{2-x}M_xCuO_4$ with M = Ba or Sr and $0.1 \leq x \leq 0.2$. At this stoichiometry, the structural phase is a tetragonal perovskite (see Fig. 1) having the space group I4/mmm (D_{4h}^{17}) with one formula unit (7 atoms) per unit cell. The representation of the zone center phonons for this structure is given in Fig. 2. La_2CuO_4 and numerous other related phases (Singh *et al.*, 1984) have the orthorhombic structure Cmca (D_{2h}^{18}) with 14 atoms per Bravais unit cell. For purposes of comparison and because it is germane to subsequent discussions, the representation of the zone center phonons for this structure is also given in Fig. 2.

In the case of the 123 materials, the phase $ABa_2Cu_3O_{7-x}$

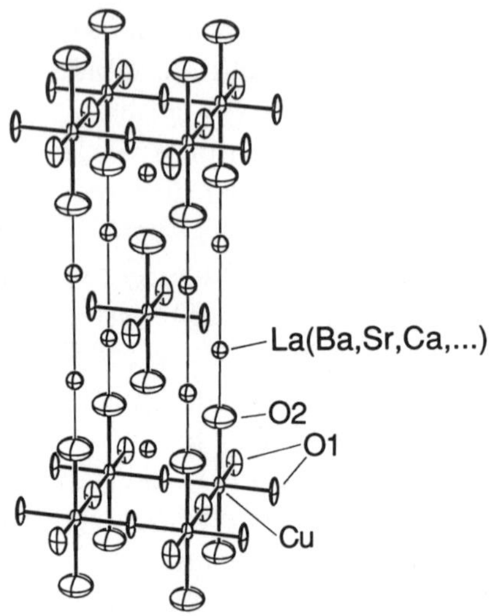

Fig. 1. Structure diagram for tetragonal $La_{2-x}M_xCuO_4$.

La_2CuO_4 $(Cmca = D_{2h}^{18})$

$$\Gamma_{TOT} = 5A_g + 4B_{1g} + 3B_{2g} + 6B_{3g} + 4A_u + 8B_{1u} + 7B_{2u} + 5B_{3u}$$
$$\Gamma_{Acou} = B_{1u} + B_{2u} + B_{3u}$$
$$\Gamma_R = 5A_g + 4B_{1g} + 3B_{2g} + 6B_{3g}$$
$$\Gamma_{IR} = 7B_{1u} + 6B_{2u} + 4B_{3u}$$

$La_{2-x}Sr_xCuO_4$ $(I4/mmm = D_{4h}^{17})$

$$\Gamma_{TOT} = 2A_{1g} + 2E_g + 4A_{2u} + B_{2u} + 5E_u$$
$$\Gamma_{Acou} = A_{2u} + E_u$$
$$\Gamma_R = 2A_{1g} + 2E_g$$
$$\Gamma_{IR} = 3A_{2u} + 4E_u$$
$$\Gamma_{TOT}^{T'} = A_{1g} + B_{1g} + 2E_g + 4A_{2u} + B_{2u} + 5E_u$$

$YBa_2Cu_3O_7$ $(Pmmm = D_{2h}^{1})$

$$\Gamma_{TOT} = 5A_g + 5B_{2g} + 5B_{3g} + 8B_{1u} + 8B_{2u} + 8B_{3u}$$
$$\Gamma_{Acou} = B_{1u} + B_{2u} + B_{3u}$$
$$\Gamma_R = 5A_g + 5B_{2g} + 5B_{3g}$$
$$\Gamma_{IR} = 7B_{1u} + 7B_{2u} + 7B_{3u}$$

$YBa_2Cu_3O_6$ $(P4/mmm = D_{4h}^{1})$

$$\Gamma_{TOT} = 4A_{1g} + B_{1g} + 5E_g + 6A_{2u} + B_{2u} + 7E_u$$
$$\Gamma_{Acou} = A_{2u} + E_u$$
$$\Gamma_R = 4A_{1g} + B_{1g} + 5E_g$$
$$\Gamma_{IR} = 5A_{2u} + 6E_u$$

Fig. 2. Vibrational representations of High-T_C and related structures. There are two tetragonal forms of $La_{2-x}Sr_xCuO_4$, T and T′; Γ_{TOT} for both is given.

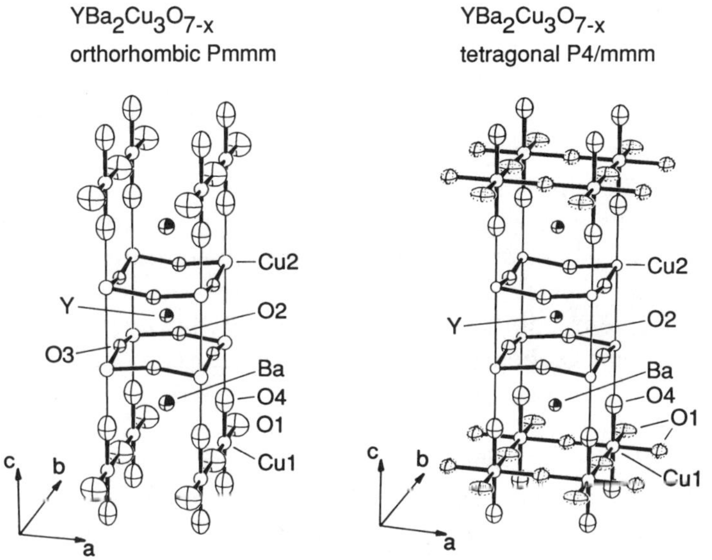

Fig. 3. Structure diagrams for orthorhombic $YBa_2Cu_3O_7$ and tetragonal $YBa_2Cu_3O_6$.

approaches optimum superconducting properties as $x \rightarrow 0$. This phase is an orthorhombic perovskite (see Fig. 3) belonging to the space group Pmmm (D^1_{2h}) with 13 atoms per Bravais unit cell. The representation of the degrees of freedom of this structure is included in Fig. 2, together with the representation for the tetragonal phase $ABa_2Cu_3O_6$ (i.e., $ABa_2Cu_3O_{7-x}$ with $x = 1$, as depicted in Fig. 3).

II. METHODS FOR INFRARED AND RAMAN MEASUREMENTS

A variety of sample preparation and handling techniques have been used to record infrared and Raman spectra of LCO, 123, and related phases. Most of the infrared measurements have been made using Fourier transform interferometric infrared

(FT-IR) instruments, but some have been made with dispersive spectrometers. In the remaining discussions of infrared measurements and data interpretation, no further distinction will be made with respect to the kind of instrumentation used; the reader is referred to the referenced literature for these types of details.

Because most of the infrared measurements have been made on solid samples that are highly opaque (mostly black) and either electrically conducting or semiconducting under measurement conditions (i.e., many of the spectra are recorded near or below T_c), it has been convenient to use reflectivity methods (e.g., Bonn *et al.*, 1988; Genzel *et al.*, 1987); Ose *et al.*, 1988; Degiorgi *et al.*, 1987; Gervais *et al.*, 1988). Typically, these infrared measurements are made on pellets in the as-pressed, as-annealed, or surface-polished condition. Where polishing is done, the usual method has been to use submicron-size alumina slurried in benzene, kerosene, or some other solvent that does not react with the high-T_c phase. (Methanol has been used by some investigators, but is not recommended, because it slowly degrades some high-T_c materials if it is not completely anhydrous.) The types of materials used to produce reference reflectivity spectra have included gold, aluminum, lead, stainless steel, and brass. Another means of standardizing the reflectivity data has been to coat the surface of the high-T_c sample with ~1000 Å of a reflective material (e.g., gold, lead, aluminum) and to record a reflectivity spectrum under the same instrumental and geometric conditions used to obtain the spectrum of the uncoated sample. This procedure helps to compensate for the incoherent scattering from rough spots on surfaces or from pores on polished high-T_c compacts, which typically show pitted areas making up 10 to 20% of the total area within the polished region. A Kramers-Kronig transformation or some other type of optical/conductivity analysis is normally used for thorough assessment of the reflectivity data.

In spite of the opaque nature of the high-T_c phases, some success has been achieved in detecting their infrared spectra by transmission techniques. Powdered samples of the high-T_c

material have been slurried in a volatile liquid and applied to infrared transmissive wafers made of, e.g., silicon (Stavola *et al.*, 1987a). Evaporation of the solvent then leaves a film of the high-T_c specimen on the wafer substrate. Powdered samples have also been (1) pressed into KBr (Sawada *et al.*, 1987; Ogita *et al.*, 1987), CsI (Taliani *et al.*, 1987; Oh-ishi *et al.*, 1987), and polyethylene disks (Kuroda *et al.*, 1987), (2) pressed between thin polyethylene windows (Nagasaka *et al.*, 1987), and (3) mulled into vacuum grease and mounted on infrared transmitting substrates (Schlesinger *et al.*, 1987b; Vuong *et al.*, 1987). Transmission infrared spectra have been recorded for sputter-deposited films of high-T_c material on substrates such as $SrTiO_3$ (Bozovic *et al.*, 1987). Attempts have even been made to record the infrared spectra of small single-crystal specimens of high-T_c phases (Kamarás *et al.*, 1988a).

Raman spectra have been obtained from sintered pellets (polished and unpolished), thin films, uncompacted powders, and single crystals of high-T_c phases. Usually the 488- or 514-nm line of an Ar^+ laser is used at power levels in the range 10–200 mW. Some defocusing, often to a line image, is done to reduce the power density delivered to the sample surface. Weber *et al.* (1988) have shown that focused laser radiation at 300 mW for 20 sec will convert $YBa_2Cu_3O_{7-x}$ to Y_2O_3 and Y_2BaCuO_5. Copič *et al.* (1987) claim that better spectra are obtained with the shorter wavelength exciting line. We have found no references reporting the use of pulsed or chopped lasers with synchronous detection systems or two-photon-type systems.

III. VIBRATIONAL SPECTROSCOPY OF $La_{2-x}M_xCuO_4$

Raman scattering and infrared (absorption and reflectance) frequencies reported for tetragonal $La_{1.85}Sr_{0.15}CuO_4$ over the past year and a half since the discovery of high-temperature superconductivity in this ceramic oxide are summarized in Table I. To consolidate the observations made by the various teams of investigations, values that fall in a frequency range of $\pm 10\ cm^{-1}$ are grouped together for simplicity. This is a reasonable way to present such data since one is dealing with solid-state spectroscopic measurements wherein effects due to oxygen stoichiometry variations, crystalline disorder, mixed phases, attainable signal-to-noise, instrument calibration errors, and variations in technique can easily lead to $\pm$ 10 cm^{-1} variability in individual peak frequency determinations.

What emerges from this assessment in the case of both the Raman and infrared data is that there is reasonable agreement with regard to the observation and assignment of at least three of the Raman active vibrational phonons and two of the infrared active phonons for $La_{1.85}Sr_{0.15}CuO_4$. Raman bands at 422 $\pm$ 10, 218 $\pm$ 8, and 145 $\pm$ 7 cm^{-1} are seen with good accord by several teams of investigators (Brun *et al.*, 1987; Copič *et al.*, 1987; Goshchitskii *et al.*, 1987; Maroni *et al.*, 1989). Results of polarization studies give evidence that the "422"-and "145"-cm^{-1} bands are *c*-axis phonons attributable to the two A_{1g} modes. The polarization characteristics of the "218"-cm^{-1} band are consistent with its assignment as one of

Table I Summary of Literature Data on Observed Raman and Infrared Phonons for $La_{1.85}Sr_{0.15}CuO_4$

Raman frequencies (cm^{-1})	Number of reporting references	Assignment	Infrared frequencies (cm^{-1})	Number of reporting references	Assignment
422 ± 10	6/7	A_{1g}	675 ± 5	2/12	?
373 ± 7	4/7	—[a]	560	1/12	?
218 ± 8	3/7	E_g	500 ± 10	11/12	A_{2u}
180	1/7	—[a]	445	1/12	?
145 ± 7	4/7	A_{1g}	346 ± 10	2/12	?
97 ± 3	2/7	E_g	248 ± 8	6/12	E_u
			117	1/12	?

[a]See the discussion of oxygen stoichiometry effects for $La_{2-x}M_xCuO_4$ given in the text.

the two expected E_g phonons (Copič *et al.*, 1987). There have been at least two observations of a fourth Raman band near 100 cm^{-1} (97 ± 3 cm^{-1}) (Blumenroeder *et al.*, 1987; Maroni *et al.*, 1989). Further confirmation of phonon activity at frequencies near those of the two lower energy Raman active modes comes from an inelastic neutron scattering study of LCO by Böni *et al.* (1988), who find peaks at 147 and 88 cm^{-1}. The Raman spectrum of a freshly prepared polished pellet of high-purity $La_{1.85}Sr_{0.15}CuO_4$, containing evidence of all four of the aforementioned Raman active bands, is shown in Fig. 4 (Maroni *et al.*, 1989). One can deduce from the review of the Raman studies of LCO that the "373"-cm^{-1} band does not appear for carefully prepared samples.

It must be noted that one of the two studies reporting the "97"-cm^{-1} Raman band (Blumenroeder *et al.*, 1987) is peculiar in that the other observed Raman bands are not at the

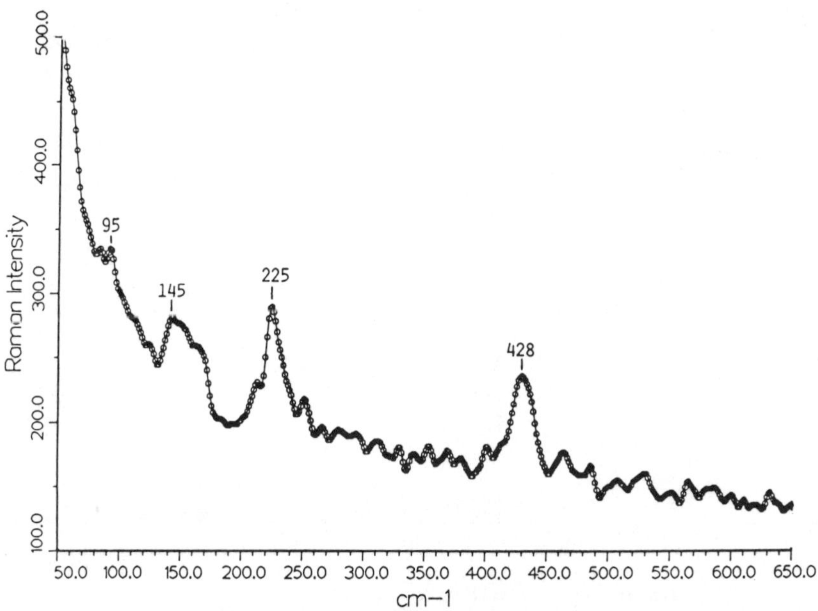

Fig. 4. Backscatter Raman spectrum of a polished sintered pellet of $La_{1.85}Sr_{0.15}CuO_4$. (From Maroni *et al.*, 1989.)

frequencies listed in Table I. Instead, their highest frequency mode is at 380 cm^{-1}; another weak feature is seen near 180 cm^{-1}; and the band they observe at 100 cm^{-1} is the most intense of the three. If their spectrum is indeed due to LCO, it may be that they produced an oxygen-deficient phase that shows the same type of behavior discussed in the next section of this review for $ABa_2Cu_3O_{7-x}$ as x goes from $0 \rightarrow 1$,i.e., a decrease in the highest frequency totally symmetric phonon. (The possibility that their sample is oxygen deficient is actually mentioned by Blumenroeder *et al.*) Unfortunately, no detailed studies have been found that probe the effect of oxygen stoichiometry on the vibrational modes of LCO.

To add further confirmation to the proposed assignments of "422" and "145" cm^{-1} to the A_{1g} phonons and of "218" and "97" cm^{-1} to the E_g phonons of LCO, the correlation chart shown in Fig. 5 was prepared. In this chart, the frequency of each observed phonon is ratioed to the highest frequency A_{1g} phonon and a plot is made of this relative frequency for both LCO and its isostructural analogy, Sr_2TiO_4, for which phonon frequencies and assignments are known

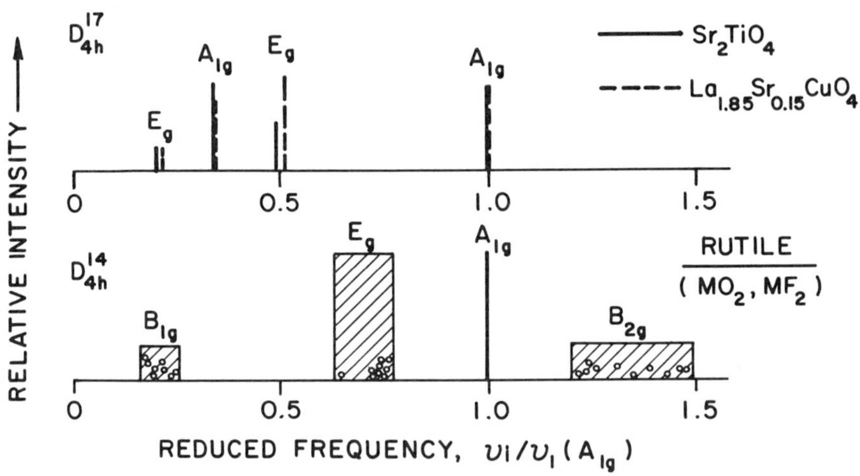

Fig. 5. Frequency correlation diagram for the Raman-active phonons of two types of tetragonal structures. (From Maroni *et al.* 1989.)

(Burns *et al.*, 1988a). A similar correlation diagram based on data for 10 rutile compounds (oxides and fluorides; see Maroni, 1988) is also presented in Fig. 5. Such diagrams for these tetragonal crystal systems can be expected to provide a representative correlation of *gerade* phonons (which include the Raman-active ones) because the kinetic energy terms of each phonon are either exclusively (in the case of rutile) or almost exclusively (in the case of LCO) determined by the mass of the light atom (oxygen or fluorine). The other assumption inherent to these diagrams is that the character of the force field and the proportioning of the corresponding potential constants are roughly the same from one isostructural compound to the next. As can be seen in Fig. 5, the correlation of LCO and Sr_2TiO_4 frequencies is excellent.

In the case of the infrared spectroscopy of LCO (Herr *et al.*, 1987; Ogita *et al.*, 1987; Oh-ishi *et al.*, 1987; Sawada *et al.*, 1987; Stavola *et al.*, 1987a; Sugai, 1987a; Sugai *et al.*, 1987; Sulewski *et al.*, 1987), matters are considerably less clear at this time. There is consistent observation (see Table I) of a sharp infrared transition at 500 ± 10 cm^{-1} and a broader band at 248 ± 8 cm^{-1}. Although several other bands are reported in the 330- to 360-cm^{-1} region and at frequencies from 550 to 700 cm^{-1}, there is little firm evidence that these are zone center phonons of LCO, and the most carefully prepared samples do not as a rule give much indication of such bands.

One revealing observation that has been made in the infrared spectra of tetragonal $La_{2-x}M_xCuO_4$ and orthorhombic La_2CuO_4 is the appearance of a strong infrared absorption near 680 cm^{-1} for the orthorhombic phase that does not appear for the tetragonal phase (both phases show the "500"-cm^{-1} infrared band). This effect, actually first seen by Singh *et al.* (1984) for LCO-type perovskites, has been reproduced in several laboratories for La_2CuO_4 and the tetragonal analog produced by doping with barium, strontium, or calcium (Ohbayashi *et al.*, 1987; Ogita *et al.*, 1987; Oh-ishi *et al.*,1987; Sawada *et al.*, 1987; Maeno *et al.*, 1987; Stavola *et al.*, 1987a). The most logical deduction to be made here is that the

680 cm^{-1} band is a zone center phonon of the orthorhombic structure that becomes a zone boundary phonon when the symmetry is raised to tetragonal. The possibility that this mode becomes the inactive B_{2u} phonon of the tetragonal phase seems unlikely, as Prade *et al.* (1987) have calculated a value near 200 cm^{-1} for B_{2u} of LCO. Alternatively, the 680-cm^{-1} mode may be an *ab*-plane phonon of the orthorhombic (O) structure that is also infrared active in the tetragonal (T) phase, but is not seen because of enhanced metal-like character in the Cu—O planes following transformation of O → T. It would follow from this argument that the "500"-cm^{-1} band is a *c*-axis phonon since it is seen for both O and T forms. Normal mode calculations for LCO discussed later in this chapter suggest the existence of a high-frequency mode (⩾600 cm^{-1}) of E_u symmetry that is mainly due to Cu—O stretch in the *ab*-plane.

An effect similar to the one described in the preceding paragraph has been observed in the Raman spectra of orthorhombic La_2CuO_4 and LCO (Kourouklis *et al.*, 1987a). The O form shows Raman bands at 526 and "422" cm^{-1}, whereas only the latter is present in the T form. Polarization studies by Kourouklis *et al.* (1987a) place the eigenvector for the 526-cm^{-1} band in the *ab*-plane (i.e., a symmetric stretching mode involving the short Cu—O bonds). They contend that this mode is a zone center phonon of the O form, but a zone boundary phonon of the T form; thus, it appears in the Raman spectrum of the O form but not of the T form. Maksimow *et al.* (1988), however, find this mode (527 cm^{-1} in their work) to be a *c*-axis A_g phonon of La_2CuO_4. They also report the appearance of the "218"-cm^{-1} phonon with *c*-axis A_g symmetry in a single crystal of the orthorhombic phase.

Batlogg *et al.* (1987a) have studied the effect of isotopic substitution (^{18}O for ^{16}O) in tetragonal LCO. They find a 14.5-cm^{-1} downward shift of the "422"-cm^{-1} band upon 52% ^{16}O replacement by ^{18}O and an 18-cm^{-1} shift for 73% ^{16}O replacement. The predicted shift at 422 cm^{-1} for 100% ^{18}O substitution should approach ~24 cm^{-1}, i.e., 422 cm^{-1} × [1 − (16/

$18)^{\frac{1}{2}}] \simeq 24\ cm^{-1}$. They see a corresponding distinct lowering of T_c that is proportional to the amount of ^{18}O substitution. These observations provide two significant pieces of information. First, they provide strong evidence that the superconducting pairing in tetragonal LCO is at least partly caused by an electron–phonon interaction; second, they prove that the "422"-cm^{-1} band originates from an eigenvector that involves mainly oxygen motion. Since the correlation in Fig. 5 indicates that the "422"-cm^{-1} band is an A_{1g} mode of the tetragonal phase of LCO, the corresponding oxygen motion must be along the *c*-axis. That such coupling of the motion of a light atom like oxygen to the conduction electrons should lead to a higher T_c has been predicted by Weber (1987). It is more likely, however, that motions of oxygens in the *ab*-plane are the ones of significance for high-T_c in LCO.

Numerous investigators have used infrared and Raman techniques to search for energy gaps in the spectra of tetragonal LCO and related phases, both superconducting and nonsuperconducting. Bonn *et al.* (1987) detected (by infrared reflectivity) a low-energy gap similar to the one first illustrated by Joyce and Richards (1970) for lead in the superconducting state. Their value for the gap parameter, $2\Delta/k_BT_c$, ranged from 2.9 to 4.5 based on a comparison of the frequency-dependent conductivity of LCO at temperatures above (70 K) and below (2 K) the critical temperature. This range of values brackets the Bardeen-Cooper-Schrieffer (BCS) prediction that $2\Delta_{BCS} = 3.5k_BT_c$. Nagasaka *et al.* (1987), Schlesinger *et al.* (1987a,b), Sulewski *et al.* (1987), and Thomas *et al.* (1987a) have obtained similar results, but it is clear that the interpretation of reflectivity data in terms of the gap parameter is sensitive to where one determines T_c to be, the amount of nonsuperconducting phase present, grain structure of the sample, and the usual uncertainties attendant to studies of weak features in the far-infrared. This subject will be addressed again later when the gap studies for the 123 phase are discussed.

Other types of optical transitions have been reported for LCO and its nonsuperconducting analogs. These transitions, which generally lie in the range 0.3 to 1.3 eV (2400 to 10,400 cm^{-1}), depending on the research group, have been attached to exciton effects that may (or may not) be connected with the high-T_c mechanism. Etemad *et al.* (1988a,b) observed a dopant-induced absorption band in the $La_{2-x}Sr_xCuO_{4-\delta}$ system at 0.5 eV (4000 cm^{-1}), which they claim is strongly correlated with an electronically driven superconductivity mechanism. Herr *et al.* (1987) detected similar features at 0.44 and 1.3 eV, which were interpreted within a model of strongly correlated electrons interacting with vibrational modes. Orenstein *et al.* (1987) saw a similar gaplike transition around 0.9 eV, which they reproduced with a modified Drude model. As will be discussed later for the 123 phase, the exact origin and significance of these broad mid-to-near-infrared features is not yet fully resolved.

Using the existing vibrational spectroscopic data as a guide, several groups have carried out lattice dynamics analyses at zero wave vector and/or phonon-density-of-state calculations. Brun *et al.* (1987) calculated phonon dispersion curves for LCO (based on their own Raman and infrared data), using a simple valence-bond-type force field. The observed Raman bands in this initial analysis were not assigned according to the recommended manner in Table I; therefore, a second more detailed vibrational analysis (at zero wave vector) was undertaken (Maroni *et al.*, 1989) to revise and augment the earlier work. The results, summarized in Table II, show that the four observed Raman-active phonon frequencies can be reproduced with a simple three-parameter force field that includes a La—Cu interaction plus appropriately scaled values for the various Cu—O and La—O bonds. The potential energy distribution reveals that it is not strictly correct to refer to the "422"-cm^{-1} Raman-active band as being the *c*-axis Cu—O stretching mode; some La—O stretch is also involved. In fact, all of the calculated phonons exhibit a distributed potential energy behavior. In a recent paper, Gervais *et al.* (1988) analyzed the infrared reflectivity spectrum of La_2CuO_4 in the tetragonal approximation. Their "best-fit" values for the A_{2u}

Table II Calculated Frequencies and Potential Energy Distributions for Tetragonal La_2CuO_4[a]

Symmetry	Observed frequency (cm^{-1})	Calculated frequency (cm^{-1})	Calculated potential energy distribution (%)		
			Cu—O	La—O	Cu—La
A_{1g}	422	431	69[c]	31	0
	145	141	6[c]	49	45
E_g	218	218	0	97	3
	97	103	0	29	71
A_{2u}	500	493	79[c]	16	5
	?	259	1	65	34
	?	219	0	72	28
E_u	?	—[b]	99[d]	0	1
	248	245	0	86	14
	?	216	0	87	13
	?	189	0	27	73

[a]Lattice parameters used in the calculation (from Cava *et al.*, 1987) were as follows: a = 3.779 Å; c = 13.226 Å; ec(La) = 0.360; ec(O_{II}) = 0.182, where ec is the fractional coordinate along the c axis. The iteratively determined force constants were as follows: Cu − O_I[b]; Cu − O_{II} = 1.21 mdyne/Å; La^c − O_{II}^c = 0.49 mdyne/Å; La − O_I = 0.24 mdyne/Å; La^c − $O_{II}^{c'}$ = 0.20 mdyne/Å; Cu^c − $La^{c'}$ = 0.55 mdyne/Å.

[b]The calculated Cu—O_I force constant is closely correlated with the value of the highest frequency E_u phonon, E_u^1 (not observed), and is not constrained by any of the observed frequencies. Frequencies ⩾600 cm^{-1} can be used as input for E_u^1 with very little effect on the other force constants, frequencies, or eigenvectors.

[c]All from Cu—O_{II}.

[d]All from Cu—O_I.

and E_u transverse optical modes are in reasonably good accord with the results in Table II. Specifically, they find that the two A_{2u} modes are at 501 and 240 cm^{-1} (with certain caveats concerning the lower frequency mode), and that the E_u modes are at 671, 363, 220, and 162 cm^{-1}. At least the latter two of these E_u phonons are in the range of the calculated values in Table II.

Prade *et al.* (1987) have also calculated phonon dispersion curves for LCO, but they, too, did not use the recommended assignments as a basis. Their work, however, shows the same type of distributed potential energy effects summarized in Table II. Mase *et al.* (1988) calculated phonon dispersion curves for LCO based on force constants deduced from acoustic measurements. Because of certain restrictions due to the nature of the sound velocity data, they conclude that there are no phonons for LCO above ~360 cm^{-1}. This finding is not in agreement with infrared and Raman studies, as can be seen in Table I. Although they did not make specific calculations, Burns *et al.* (1988a) give a discussion of the eigenvectors for LCO based on a prior normal mode analysis of the isostructural K_2ZnF_4 crystal. They suggest that the highest frequency Raman mode ("422" cm^{-1}) is dominated by La—O stretching along the *c*-axis. This is contrary to the results in Table II, which imply that the Cu—O bond force constants and their contributions to the potential energy distribution for the higher frequency modes are larger than those for the La—O bonds. Prade *et al.* (1987) also find that the Cu—O force constants are at least twice the value of the La—O force constants. Weber (1987), too, has suggested that the Cu—O forces are important to the high-frequency modes. In fact, his calculated vibrational density of states for LCO shows a surprisingly good correlation with the calculated frequencies in Table II.

There have been several attempts to predict the value of the highest frequency E_u phonon for LCO at $k = 0$. Using a correlation approach, Burns *et al.* (1988a) place it at 672 cm^{-1}. On the basis of the potential induced breathing model, Cohen *et al.* (1988) give a value of 964 cm^{-1}. Prade *et al.* (1987) calculate that this mode is near 547 cm^{-1}, by application of a shell model. Fu and Freeman (1987) performed a frozen phonon calculation that gave a value of ~880 cm^{-1} for the Cu—O breathing mode in the *ab*-plane. To our knowledge, no conclusive experimental observations of a high-frequency E_u phonon attributable to LCO have been made to date.

The vibrational spectroscopic study of LCO and related phases got off to a good start after the discovery of high-T_c superconductivity in LCO by Bednorz and Müller (1986). But when the tide of research endeavors shifted to the 123 materials, important remaining work on LCO was left undone.

Hopefully, when things settle down in the high-T_c arena, there will be a return of attention to the LCO-type phases to resolve lingering discrepancies and complete the undone tasks. Detailed single-crystal studies of both the orthorhombic and tetragonal phases would be a significant contribution in this regard. An investigation of oxygen stoichiometry effects on the phonon frequencies and intensities is also needed.

IV. VIBRATIONAL SPECTROSCOPY OF $ABa_2Cu_3O_{7-x}$

In the annals of modern science, there are few materials that have been so extensively studied in such a short time as the phase $ABa_2Cu_3O_{7-x}$. This phase exists in what has proved to be a complex ternary system formed by oxide mixes of the type A_2O_3–BaO–CuO, where A can be yttrium or any one of a number of rare earth metals. The numerous other phases that tend to form in this system have complicated the early spectroscopic study of the 123 phase, and only recently has there emerged some semblance of agreement concerning the phonons and other spectral features (exitons) that are attributable to the orthorhombic and tetragonal structures of $ABa_2Cu_3O_{7-x}$ within the stoichiometry range $0 \leq x \leq 1$. A summary of the observed infrared and Raman spectral data for 123 is given in Table III, in the same format as was used in Table I for LCO. Three types of studies have contributed greatly to the interpretation of these data—Raman measurements on single crystals, Raman and infrared studies of oxygen stoichiometry effects, and comprehensive investigations of the Raman spectra of starting materials and stable phases (other than the 123 phase) that tend to form during the synthesis of 123.

First to be discussed here will be the Raman studies of starting materials and second phases. Bhadra *et al.* (1988), Mascarenhas *et al.* (1988), Morioka *et al.* (1987), Udagawa *et al.* (1987), Popovic *et al.* (1988), and Rosen *et al.* (1987, 1988) have investigated various compounds, including Y_2O_3, CuO,

Table III Summary of Literature Data for Observed Raman and Infrared Phonons for $ABa_2Cu_3O_{7-x}$

Raman frequencies (cm^{-1})	Number of reporting references[a]	Assignment[b]	Infrared frequencies (cm^{-1})	Number of reporting references[c]	Assignment
639 ± 5	10/24	123(T) $BaCuO_2$	642 ± 6	3/14	123(T)
595 ± 7	15/24	211	590 ± 3	6/14	123(T) 211
503 ± 4	21/24	123(O,A_g)	572 ± 3	7/14	123(O)
480 ± 5	—	123(T,A_{1g})	521 ± 11	3/14	123(O) 211
452 ± 10	4/24	123(T,A_{1g})	477 ± 3	6/14	123(O) 211
436 ± 6	17/24	123(O,A_g)	310 ± 7[d]	9/14	123(O,T)
380 ± 10	3/24	211	276 ± 8	8/14	123(O)
335 ± 5	19/24	123(O,A_g) 123(T,B_{1g})	241 ± 8	5/14	?
300 ± 10	4/24	211	213 ± 4	4/14	?
215 ± 10	7/24	211	193 ± 3	8/14	123(O,T)
146 ± 6	13/24	123(O,A_g) 123(T,A_{1g})	169 ± 5	3/14	?
112 ± 6[e]	6/24	123(O,A_g)	148 ± 7	11/14	123(O,T)
			110 ± 10	6/14	123(T)

p. 18

[a]Number of references reporting a particular Raman frequency out of the 24 that were surveyed.

[b]T = tetragonal ABa_2CuO_{7-x} ($x \rightarrow 1$); O = orthorhombic $ABa_2Cu_3O_{7-\delta}$ ($x \rightarrow 0$).

[c]Number of references reporting a particular infrared frequency out of the 14 that were surveyed.

[d]Most mid-infrared studies do not go below 400 cm^{-1}; several far-infrared studies terminate at 400–500 cm^{-1}.

[e]Many of the Raman studies did not show data below 140 cm^{-1}.

$BaCO_3$, Y_2BaCuO_5 (211), $BaCuO_2$, and $Y_2Cu_2O_5$, and found numerous bands that have repeatedly shown up in reported Raman spectra of 123. In particular, a band often seen at around 640 cm^{-1} (namely, "639" cm^{-1} in Table III) is probably best ascribed to the most intense phonon of $BaCuO_2$ in many if not all cases. Similarly, a Raman band around 600 cm^{-1} (namely, "595" cm^{-1} in Table III), which is often accompanied by a band in the vicinity of 380 cm^{-1}, is very likely due to the 211 phase, which is green in color and may actually be resonantly enhanced by 514-nm excitation. The key implication of these studies is that second-phase concentrations below the detection level of X-ray diffraction (e.g., in the 1 to 2% range) can give Raman bands that are comparable in intensity to the 123 phonons. This effect is nicely illustrated by Rosen *et al.* (1988), who show Raman spectra of 123 + 10% $BaCuO_2$ mixtures wherein the 640-cm^{-1} $BaCuO_2$ peak is more than twice the intensity of the strongest 123 phonon at "503" cm^{-1}. One can conclude from other spectral data given in this latter report by Rosen *et al.* that a mixture of 123 + 1% 211 would give a spectrum wherein the "600"-cm^{-1} band of 211 is about equal in intensity to the "503"-cm^{-1} band of 123. Clearly, the 123 phase has a much lower Raman scattering cross section than most of the second phases that are commonly present as impurities in preparations of 123. The Raman spectrum of a 123 sample containing <1% $BaCuO_2$ is displayed in Fig. 6.

Popovic *et al.* (1988) recorded infrared reflectance spectra of 211 samples and reported frequencies for 14 phonons of 211. Kamarás *et al.* (1988a) list five infrared frequencies for 211, but no spectra are displayed and no information is given on relative band strengths. Nonetheless, the results from these two studies have influenced the infrared band assignments in Table III, although it must be noted that these assignments are considered to be tentative ones. A detailed study of the infrared spectra of starting materials and second-phase compounds pertinent to the 123 system (e.g., $BaCuO_2$ and 211) would be a worthwhile contribution to the field.

Further definitive information that has contributed to the Raman band assignments in Table III comes from investigations of single-crystal samples of 123. With good quality single crystals, it should be possible to greatly reduce the relative concentration of impurities and to perform the type of polarization studies that define the symmetry species of observed Raman bands. Results of such studies on oriented orthorhombic 123 single crystals (123/O) have been reported by Bhadra *et al.* (1988), Cooper *et al.* (1988), Denisov *et al.* (1988), and

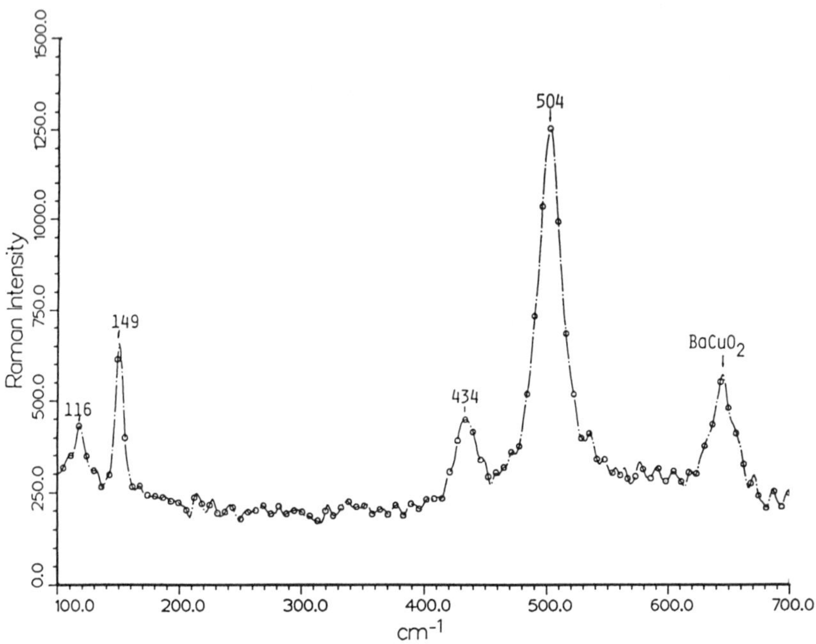

Fig. 6. Raman spectrum of single-crystal $YBa_2Cu_3O_{7-x}$ oriented for detection of A_g modes. ∘ = ZZ pol., (100), 100K.

Hadjiev and Iliev (1988). All four groups are in agreement that the five A_g modes of orthorhombic 123 are nominally at "503," "436," "335," "146," and "112" cm^{-1}, although there is not complete accord with regard to the assignment of these bands to specific eigenvectors. Liu *et al.* (1988a) and Krol *et al.* (1987) are in concurrence with most of these observations. Hemley and Mao (1987) used Raman microprobe techniques to examine crystallites in a mixed phase Y-Ba-Cu-O preparation. They obtained spectra of both 123 and 211 crystals. The 123 in their work was a tetragonal crystal (123/T), which is known to have a Raman spectrum very much like that of the orthorhombic form (Liu *et al.*, 1988a). The interesting and predictable difference between the two is that the "335"-cm^{-1} band, which is A_g for 123/O, becomes B_{1g} for 123/T, whereas the other four A_g modes of 123/O go to A_{1g} for

123/T (Liu *et al.*, 1988a). A correlation diagram for the addition of an oxygen atom to a D_{2h} site in tetragonal $ABa_2Cu_3O_6$ to produce orthorhombic $ABa_2Cu_3O_7$ is shown in Fig. 7. The Raman results of Yamanaka *et al.* (1987) for a single crystal of 123/T are consistent with the above, but only three modes are observed. Burns *et al.* (1988b) attempted to record Raman spectra of 123/O and 123/T single crystals, but both spectra resemble 123/T-type results.

As was mentioned earlier, the origins of the five phonons attributed to the A_g modes of 123/O have been a matter of controversy. More recent work by Stavola *et al.* (1989) and Morioka *et al.* (1988) has given a consistent set of assignments for the five A_g phonons of 123/O, which can be summarized as follows (referring to Fig. 3):

503 cm^{-1}	axial motion of the O(4) atoms
436 cm^{-1}	Cu(2)—O(2) and —O(3) bond bending, with the O(2) and O(3) atoms moving in phase
335 cm^{-1}	Cu(2)—O(2) and —O(3) bond bending, with the O(2) and O(3) atoms moving out of phase
146 cm^{-1}	axial stretching of the Cu(2) atoms
112 cm^{-1}	axial stretching of the Ba atoms

Other interesting observations from the single-crystal studies have been that (1) as expected there is less evidence of the troublesome second-phase bands, particularly those

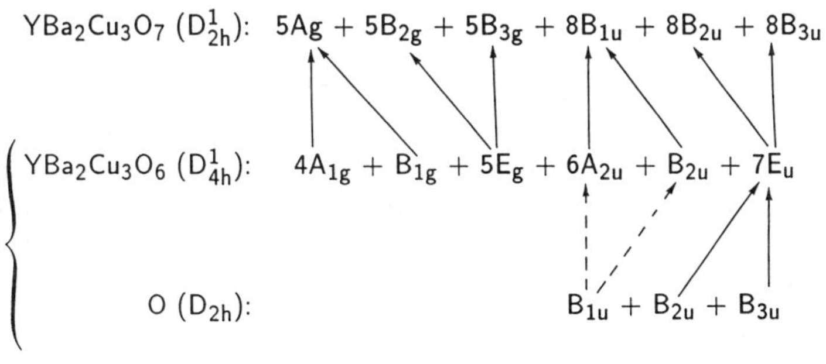

Fig. 7. Phonon correlation diagram for $y = 6$ to $y = 7$ restructuring of $YBa_2Cu_3O_y$.

around 640 and 600 cm^{-1} due to $BaCuO_2$ and 211, respectively, and (2) only the centrosymmetric *c*-axis phonons of 123/O (the A_g modes) and 123/T (the A_{1g} and B_{1g} modes) are unambiguously present in the Raman spectra. Hadjiev and Iliev (1988) do report a band at 59 cm^{-1} with B_{2g} (B_{3g}) character, but even so they acknowledge the possible existence of a strong optical anisotropy in 123 crystals resulting from the much higher conductivity of the *ab*-plane relative to the *ac*- and *bc*-planes. It seems that the 123 phases (O and T) exhibit metal-like scattering behavior in the *ab*-plane but ceramic-like behavior along the *c*-axis. If this is indeed the case, then Raman spectroscopy may not be a very informative tool for probing the most significant type of electron–phonon coupling in 123, i.e., that in the *ab*-plane.

Apparently, the origins of Raman scattering in the 550- to 700-cm^{-1} region are not just limited to trace levels of impurity phases. Stavola *et al.* (1987b) and McCarty *et al.* (1988) have observed that spectral features begin to appear in this frequency region when polycrystalline or single-crystal 123/O ($x \rightarrow 0$) is annealed under conditions that cause loss of oxygen from the lattice, i.e., allowing $x \rightarrow 1$. These features are attributed to oxygen defect-induced activity of *ungerade* phonons (made active by the loss of inversion symmetry in the disordered lattice), but they may also be manifestations of the noncentrosymmetric *gerade* modes (e.g., B_{2g} and B_{3g} of 123/O), which perhaps gain intensity as a consequence of the way oxygen vacancy disorder affects the polarizability tensor matrix elements.

Having clarified with reasonable certainty the Raman scattering features that are attributable to 123/O, it is now appropriate to consider the effects of oxygen stoichiometry on the Raman spectrum of this phase. A number of groups (Bhadra *et al.*, 1988; Hangyo *et al.*, (1988; Kourouklis *et al.*, 1987b; Mascarenhas *et al.*, 1988; Stavola *et al.*, 1987b; Macfarlane *et al.*, 1988; Thomsen *et al.*, 1988a) have addressed this subject with very much the same result. The findings of these studies are best represented by the plots in Fig. 8 taken from Thomsen *et al.* (1988a), which show how the frequencies of four

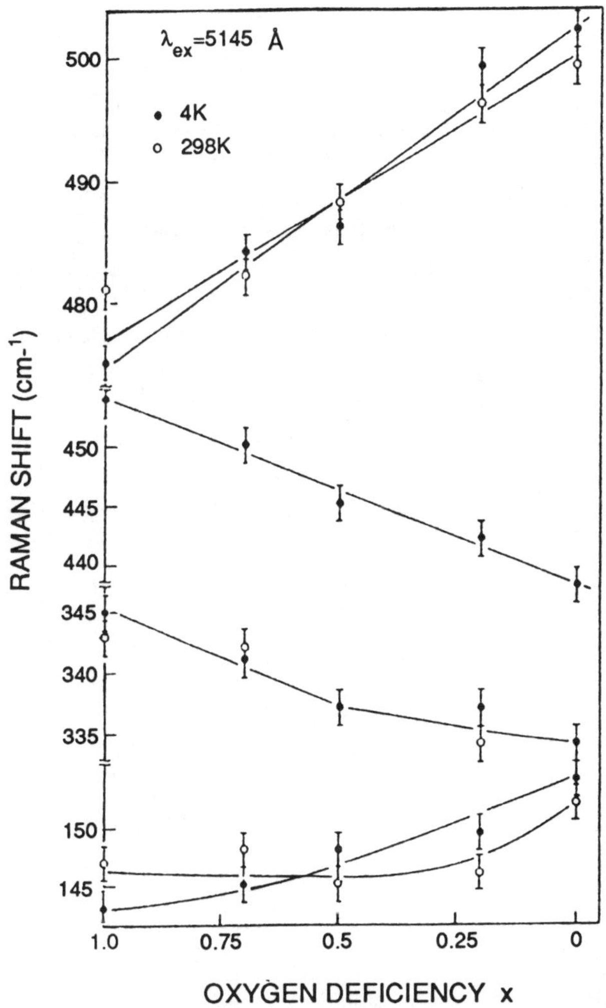

Fig. 8. Dependence of various Raman peak frequencies on oxygen concentration at 4 K and 298 K. (From Thomsen *et al.*, 1988a.)

of the interrelated modes of 123/O and 123/T vary as x in $ABa_2Cu_3O_{7-x}$ goes from $0 \rightarrow 1$. What is not evident in this figure is how the intensities change with increasing x. Most notable in this regard is that the intensity of the "335"-cm^{-1} phonon seems to increase considerably as the lattice mode character transforms from A_g of 123/O to B_{1g} of 123/T (see, for example, Weber *et al.*, 1988). Figure 9 shows the Raman spectrum of a compacted, polished polycrystalline sample of high-purity 123 with $x \cong 0.3$. The effect of oxygen stoichiometry can be at least partly appreciated by comparing the intensity of the 336 cm^{-1} band in Fig. 9 with that in Fig. 6, which was obtained for a single-crystal sample with $x \rightarrow 0$. Two factors can be contributing to this effect. First of all, the A_g mode of 123/O that transforms to B_{1g} of 123/T has polarization properties different from those of the other four A_g phonons of 123/O, so the effect can be orientational (Feile *et al.*, 1988). Second, because its polarization tensor has *a*/*b* asymmetry, the question arises (Feile *et al.*, 1988; Thomsen *et al.*, 1988d) whether this mode can mediate electron–phonon interaction in the 123/O phase. If it can, then its intensity in 123/O may be diminished by the *ab*-plane conductivity effect described earlier.

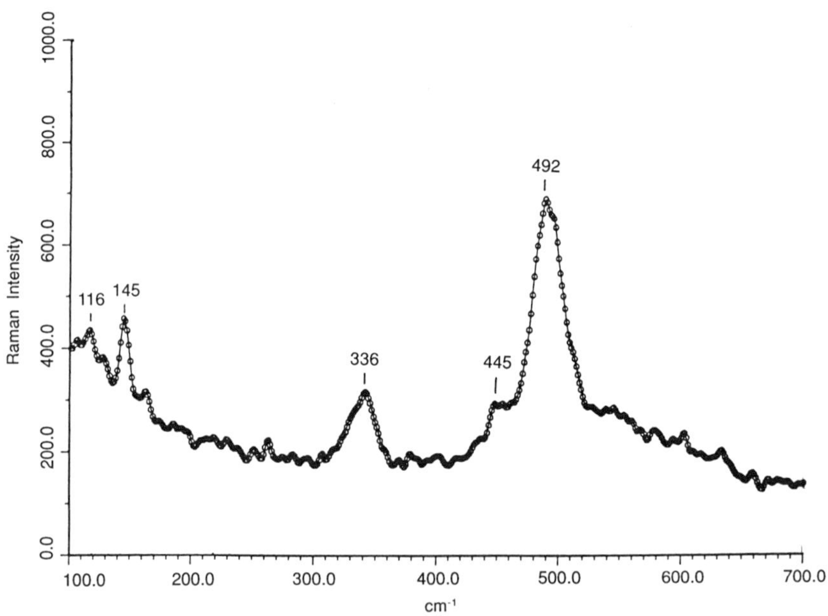

Fig. 9. Backscatter Raman spectrum of a polished sintered pellet of $YBa_2Cu_3O_{7-x}$.

Infrared investigations have also been made on 123 as a function of oxygen stoichiometry, but here the findings and their interpretation are not as clear-cut as for the Raman data. Results from several studies (Burns *et al.*, 1988c; Onari *et al.*, 1988; Saito *et al.*, 1987; Stavola *et al.*, 1987b; Sugai, 1987b; Thomsen *et al.*, 1988a,b) seem to indicate that bands in the 550- to 650-cm^{-1} region (presumably Cu—O stretching vibrations) are more intense in the tetragonal phase than in the orthorhombic phase. Similar observations are made for bands in the 350- to 400-cm^{-1} region and around 150 cm^{-1}.

In the course of studies of oxygen stoichiometry effects on the $ABa_2Cu_3O_{7-x}$ system as x varies from $0 \rightarrow 1$, Liu *et al.* (1988b) have detected Cu_2O in samples having x very close to 1.0 (123/T). The amount of Cu_2O is believed to be at trace levels, but it still yields relatively intense bands, particularly when the 488-nm Ar^+ line is used for Raman excitation. This effect is presumably due to resonance Raman scattering induced by the blue exciton of Cu_2O. No evidence of Cu_2O has been seen in 123/O samples.

There have been several zero wave vector and phonon-density-of-states calculations on orthorhombic and tetragonal 123. Bates and Eldridge (1987) made a normal coordinate analysis of 123/O using estimated force constants. Their calculated frequencies for the five A_g modes are in fair agreement with the recommended values in Table III, despite the fact that they made no attempt to refine their force constants. Like LCO, 123 gave results that indicate considerable mixing of the internal displacement coordinates for most calculated phonons, and this mixing leads to a highly distributed potential energy behavior. Also like the high-frequency modes of LCO, the highest frequency modes of 123 tend to have mainly Cu—O stretching character whereas the lower frequency modes show higher percentages of contributions from forces between the barium atoms and the Cu—O planes or chains. Liu *et al.* (1988a) and Thomsen *et al.* (1988b) made zero wave vector calculations for 123/O and 123/T, respectively, using estimated force constants. They present eigenvector maps for both structures and give frequencies for the totally symmetric modes that are in good accord with the recommended values in Table III. McMullan *et al.* (1988) carried out a similar analysis, but their results for the A_g modes are in lesser accord with the values in Table III. Stavola *et al.* (1987b) and Kress *et al.* (1988) performed phonon-density-of-states calculations for 123/O with similar findings; however, in neither case do the results completely agree with the recommended values for the A_g modes given in Table III. A force-constant-refinement calculation (similar to the one summarized in Table II), which uses a minimum number of adjustable parameters (e.g., Cu—O, Ba—O, and Y—O stretch with scaling relationships to accommodate differences in internuclear distance) and is constrained by the A_g modes of 123/O and the A_{1g} + B_{1g} modes of 123/T, should provide assignments for observed infrared modes and improved estimates for the yet-to-be-observed phonons of 123/O and 123/T. Gupta (1988) has analyzed the zone center phonons of 123/O using an angular force model.

Other properties of the 123 structures that have been probed by infrared and Raman spectroscopy are the effect of changing the A cation in $ABa_2Cu_3O_{7-x}$ and the effect of temperature. Some of the more comprehensive studies of this type are ones by Cardona *et al.* (1987, 1988), Thomsen *et al.* (1988c), and Wittlin *et al.* (1987). A plot of the correlation between selected Raman and infrared frequencies and ionic radius of A, taken from their work, is presented in Fig. 10. Figure 11, also from Cardona *et al.* (1988), shows the temperature dependence of these phonons for several different A cations. The observed monotonic increases of most modes with increasing cationic radius has not been fully explained, but they may be a result of how changes in the *c*-axis parameter affect modes that have all or part of their eigenvector directed along the *c*-axis. An intriguing feature in these types of studies has been the tendency of bands in the 280- to 340-cm^{-1} region to show anomalous softening starting near T_c. This softening effect has also been seen by Wrobel *et al.* (1987) and Macfarlane

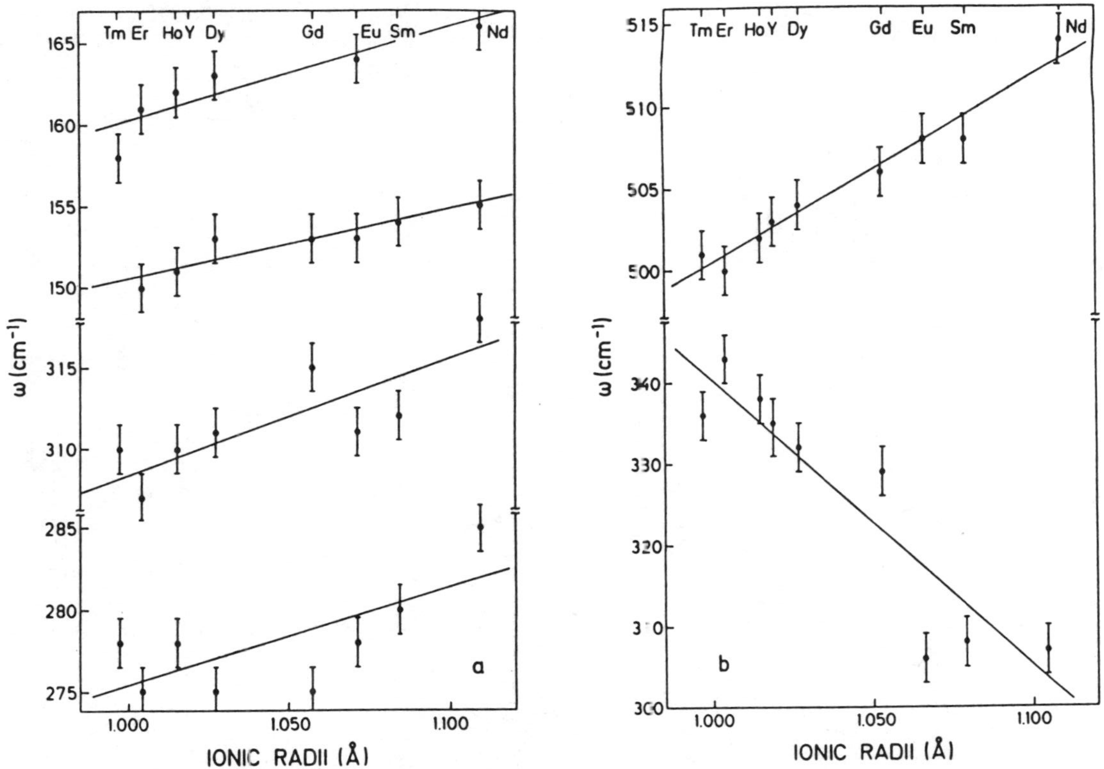

Fig. 10. Frequency dependence of selected Raman (a) and infrared (b) bands for $ABa_2Cu_3O_{7-x}$ on the ionic radius of the A^{3+} ion. (From Cardona *et al.*, 1988.)

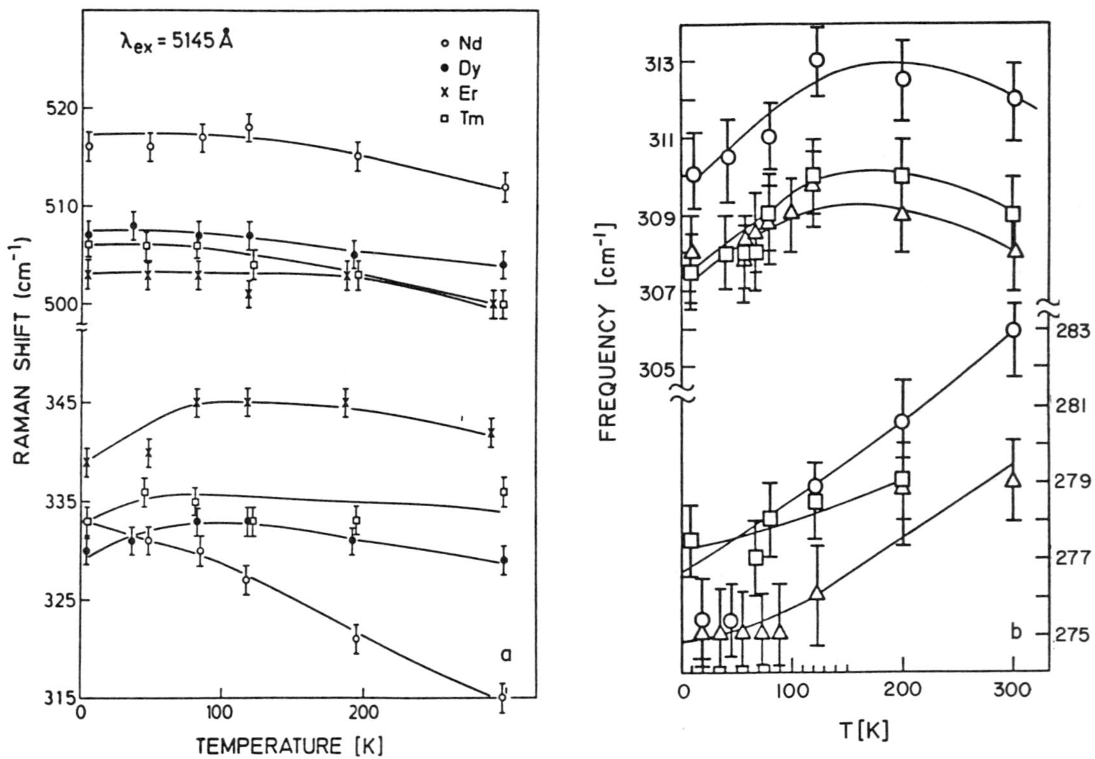

Fig. 11. Frequency dependence of selected Raman (a) and infrared (b) bands for $ABa_2Cu_3O_{7-x}$ (A = Nd, Dy, Er, Tm) on temperature. (From Cardona *et al.*, 1988.)

et al. (1987). Raman spectra of $ABa_2Cu_3O_{7-x}$ compounds reported by Choudhury *et al.* (1988) are replete with bands not seen by other investigators.

The effect of isotopic substitution of ^{18}O for ^{16}O in 123/O has been investigated by Batlogg *et al.* (1987b), who found an 18-cm^{-1} downward shift of the "503"-cm^{-1} Raman band and a 13-cm^{-1} shift of the "436"-cm^{-1} band following 74 ± 7% replacement of ^{16}O by ^{18}O. Their spectra also contained the high-frequency $BaCuO_2$ bands (between 550 and 650 cm^{-1}), which shifted downward as well upon ^{18}O substitution. (This latter observation, if nothing else, shows that the phase chemistry of 123 and its common second-phase impurities are in roughly the same kinetic and thermodynamic equilibria.) Unlike the reports on LCO, however, conflicting reports (Morris *et al.*, 1988) regarding the magnitude of the shift in T_c following ^{18}O substitution in 123 have been published. Whether or not T_c for 123/O is perturbed by substitution of ^{18}O, the amount of the shift is clearly smaller than that observed for LCO, and it is much less than the shift expected for strongly coupled phonon-mediated superconductivity (Loye *et al.*, 1987).

A solid confirmation of the origin of the "503"-cm^{-1} Raman band of 123/O has been attained through (1) the numerous studies that show its oxygen sensitivity, including especially the afore mentioned isotopic work by Batlogg *et al.* (1987b), and (2) the work of Krol *et al.* (1988), which reveals its sensitivity to the Cu(1)—Cu(2) distance in a series of ABa_2Cu_3-O_{7-x} phases having different *c*-axis parameters. This result (shown in Fig. 12), the ^{18}O substitution work, and the oriented single-crystal studies leave little doubt that the "503"-cm^{-1} Raman band is due to vibration along the *c*-axis of the oxygen atoms at (0,0, *z*) positions.

The superconducting gap in orthorhombic 123 has been studied extensively using both infrared and Raman spectroscopy. Collins *et al.* (1987) applied far-infrared reflectivity to (1) an epitaxial film of 123/O with its crystallographic *c*-axis oriented perpendicular to the film surface and (2) several bulk

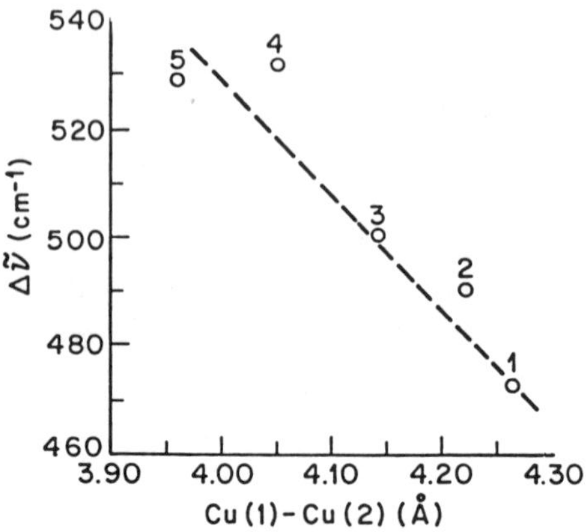

Fig. 12. Position of the "503"-cm^{-1} Raman band of 123/O ("480" cm^{-1} for 123/T) versus the Cu(1) — Cu(2) distance. Data are for (1) $YBa_2Cu_3O_6$, (2) $YBa_2Cu_3O_{6.5}$, (3) $YBa_2Cu_3O_7$, (4) $La_{1.5}Ba_{1.5}Cu_3O_{7+x}$, (5) $YSr_2Cu_3O_x$. (From Krol *et al.*, 1988.)

polycrystalline samples (both single and multiphase). Their estimate of the energy gap for the film sample was $2\Delta = (4.7 \pm 1.2)k_BT_c$, a value consistent with strong-coupling superconductivity. The reflectivity from bulk samples, which was more difficult to interpret, tended to suggest a considerably smaller superconducting energy gap. These results are in reasonable agreement with earlier tunneling studies by the same group (Kirtley *et al.*, 1987), which gave a value of $2\Delta = (4.7 \pm 1.0)$ k_BT_c. Noh *et al.* (1988) performed far-infrared reflectivity studies on bulk (sintered) samples, films, and single crystals, and reported values of $2\Delta/k_BT_c = 2.6$, 6.4, and 7.7, respectively. Lyons *et al.* (1987) used an iodine absorption cell to facilitate low-frequency Raman studies on a thin film of 123/O and extracted from their data the value $2\Delta = (3.4 \pm 1.5)k_BT_c$. Vuong *et al.* (1987), Wittlin *et al.* (1988), Thomas *et al.* (1987b), Ose *et al.* (1988), and Kuroda *et al.* (1987) all searched for the energy gap in powdered and sintered samples of 123/O by far-infrared reflectance methods. A range of values for $2\Delta/k_BT_c$ ($0.7 \rightarrow 4.6$) emerged from these studies. When viewed collectively, the results of all the superconducting gap studies on 123 leave the impression that the true value of the gap parameter is not yet fully resolved. As an interesting sidelight to this discussion, Cooper *et al.* (1988) have observed a broad diffuse scattering in the Raman spectrum of 123/O single crystals at temperatures well below T_c. They attribute this scattering, centered near 470 cm^{-1}, to a continuum of electronic states inside the gap; furthermore, they cite evidence in their spectra for a coupling of the continuum to two discrete phonons associated with 123/O. In spite of the uncertainties, evidence is building to support the view that electron–phonon coupling is at least partly responsible for high-T_c superconductivity in the $ABa_2Cu_3O_{7-x}$ system.

Another area of controversy for 123/O concerns the existence and origin of excitonic absorption in the mid-infrared region. Kamarás *et al.* (1987, 1988a) report the observation of two such bands for polycrystalline 123/O pellets at 0.37 eV (3000 cm^{-1}) and 2.5 eV (~20,000 cm^{-1}), which they interpret as being due to charge-transfer processes. The lower energy transition (0.37 eV) is extremely intense in the case of pelletized 123/O, but it is not observed for 123/T or for a single crystal of 123/O oriented with its *ab*-plane facing the incident radiation. From this, it is posited by Kamarás *et al.* (1988a) that the 0.37-eV feature may be polarized along the crystallographic *c*-axis. Little *et al.* (1988) offer confirming evidence of the existence of a reflectivity maximum in the 0.1- to 0.5-eV (800- to 4000-cm^{-1}) region for a 123/O pellet; and they further show that this maximum all but disappears when the 123/O phase is reduced completely to the 123/T phase (i.e., x goes from 0 to 1). Chui *et al.* (1988) made an *ab initio* computation of the optical conductivity of 123/O and obtained a value of 0.37 eV for an interband transition from bridging oxygens that connect the Cu—O planes to the Cu—O chains. But Bozovic *et al.* (1987) do not find an absorption band in the 0.1- to 0.4-eV range for magnetron-sputtered thin films of 123/O on $SrTiO_3$ substrates using either reflectance or transmission techniques. In fact, both infrared and Raman backgrounds obtained with these thin films were featureless in the 0.1- to

0.5-eV region from 300 K down to 10 K. Orenstein and Rapkine (1988) have offered an explanation for the discrepancies among the previously described results that is based on the properties of wave propagation in isotropic and anisotropic crystalline media. The extent to which Drude absorption is involved in the appearance of the so-called exciton bands has been debated by Kamarás *et al.* (1988b). Mihailović *et al.* (1987) report an unexpectedly intense Stokes-shifted spectrum in light scattering studies on polished, polycrystalline samples of $YBa_2Cu_3O_{7-x}$. This scattering, which typically covered the 100- to 4000-cm^{-1} region, was attributed to fluctuating quasi-particle states above the superconducting ground state.

V. CONCLUDING REMARKS

The study of high-T_c superconducting ceramics by Raman and infrared spectroscopy has been extensive. Although there was considerable discrepancy among the earliest reports of spectral data, more recent studies have tended to show a steadily improving concurrence regarding observations and interpretations. Raman studies of oriented single crystals, infrared reflectance measurements analyzed by the Kramers-Kronig formalism, comprehensive investigations of the effects of oxygen stoichiometry and metal atom substitution, and determinations of impurity-related spectral features have provided the most useful sets of data from which these interpretations have been made. An unequivocal connection between lattice phonons of the $ABa_2Cu_3O_7$ structure and its superconducting properties has not been made to date, but it also may be the case that the phonons most likely to be involved in the superconducting mechanism, i.e., those in the Cu—O planes and chains, have thus far defied detection. Nonetheless, vibrational spectroscopy (Raman, in particular) has proved to be a useful indicator of oxygen stoichiometry variations, changes in lattice parameters, and the presence of parts-per-hundred levels of impurity phases. The fact that Raman and infrared techniques tend to probe only the near-surface region of bulk high-T_c samples gives cause for concern with respect to interfacial composition gradients and the extent to which surface and bulk properties vary from sample to sample. Standardized sample preparation/handling and spectral measurement procedures should help to alleviate much of this type of concern.

ACKNOWLEDGMENT

The authors are grateful to M. Grimsditch and J. D. Jorgensen for providing several of the figures and for helpful commentary on this chapter. The work reported in this chapter was sponsored by the Division of Materials Sciences, Office of Basic Energy Sciences, U.S. Department of Energy, under contract W-31-109-ENG-38.

REFERENCES

Bates, F. E., and Eldridge, J. E. (1987). *Solid State Commun.* **64,** 1435.

Batlogg, B., Kourouklis, G., Weber, W., Cava, R. J., Jayaraman, A., White, A. E., Short, K. T., Rupp, L. W., and Rietman, E. A. (1987a). *Phys. Rev. Lett.* **59,** 912.

Batlogg, B., Cava, R. J., Jarayaman, A., van Dover, R. B., Kourouklis, G. A., Sunshine, S., Murphy, D. W., Rupp, L. W., Chen, H. S., White, A., Short, K. T., Mujsce, A. M., and Rietman, E. A. (1987b). *Phys. Rev. Lett.* **58,** 2333.

Bednorz, J. G., and Müller, K. A. (1986). *Z. Phys. B* **64,** 189.

Bhadra, R., Brun, T. O., Beno, M. A., Dabrowksi, B., Hinks, D. G., Liu, J. Z., Jorgensen, J. D., Nowicki, L. J., Paulikas, A. P., Schuller, I. K., Segre, C. U., Soderholm, L., Veal, B., Wang, H. H., Williams, J. M., Zhang, K., and Grimsditch, M. (1988). *Phys. Rev. B: Condens. Matter* **37,** 5142.

Blumenroeder, S. Zirngiebl, E., Thompson, J. D., Killough, P., Smith, J. L., and Fisk, Z. (1987). *Phys. Rev. B: Condens. Matter* **35,** 8840.

Böni, P., Axe, J. D., Shirane, G., Birgeneau, R. J., Gabbe, D. R., Jenssen, H. P., Kastner, M. A., Peters, C. J., Picone, P. J., and Thurston, T. R. (1988). *Phys. Rev. B: Condens. Matter* **38,** 185.

Bonn, D. A., Greedan, J. E., Stager, C. V., Timusk, T., Doss, M. G., Herr, S. L., Kamarás, K., Porter, C. D., Tanner, B. D., Tarascon, J. M., McKinnon, W. R., and Greene, L. H. (1987). *Phys. Rev. B: Condens. Matter* **35,** 8843.

Bonn, D. A., O'Reilly, A. H., Greedan, J. E., Stager, C. V., Timusk, T., Kamarás, K., and Tanner, D. B. (1988). *Phys. Rev. B: Condens. Matter* **37,** 1574.

Bozovic, I., Kirillov, D., Kapitulnik, A., Char, K., Hahn, M. R., Beasley, M. R., Geballe, T. H., Kim, Y. H., and Heeger, A. J. (1987). *Phys. Rev. Lett.* **59,** 2219.

Brun, T., Grimsditch, M., Gray, K. E., Bhadra, R., Maroni, V. A., and Loong, C.-K. (1987). *Phys. Rev. B: Condens. Matter* **35,** 8837.

Burns, G., Dacol, F. H., Kliche, G., König, W., and Shafer, M. W. (1988a). *Phys. Rev. B: Condens. Matter* **37,** 3381.

Burns, G., Dacol, F. H., Holtzberg, F., and Kaiser, D. L. (1988b). *Solid State Commun.* **66,** 217.

Burns, G., Dacol, F. H., Freitas, P. P., König, W., and Plaskett, T. S. (1988c). *Phys. Rev. B: Condens. Matter* **37,** 5171.

Cardona, M., Genzel, L., Liu, R., Wittlin, A., Mattauch, H., García-Alvarado, F., and García-Gonzalez, E. (1987). *Solid State Commun.* **64,** 727.

Cardona, M., Liu, R., Thomsen, C., Bauer, M., Genzel, L., König, W., Wittlin, A., Amador, U., Barahona, M., Fernandez, F., Otero, C., and Saez, R. (1988). *Solid State Commun.* **65,** 71.

Cava, R. J., Santoro, A., Johnson, D. W., Jr., and Rhodes, W. W. (1987). *Phys. Rev. B Condens. Matter* **35,** 6716.

Choudhury, P., Sukla, A. K., Mavi, H. S., Jain, K. P., Abbi, S. C., Mandal, P., Poddar, A., Das, A. N., and Ghosh, B. (1988). *J. Phys. C* **21,** L41.

Chu, C. W., Hor, P. H., Mong, R. L., Gao, L., Huang, Z. J., and Wang, Y. Q. (1987). *Phys. Rev. Lett.* **58,** 405.

Chui, S. T., Kasowski, R. V., and Hsu, W. Y. (1988). *Phys. Rev. Lett.* **61,** 885.

Cohen, R. E., Pickett, W. E., Krakauer, H., and Boyer, L. L. (1988). *Physica B (Amsterdam)* **150,** 61.

Collins, R. T., Schlesinger, Z., Koch, R. H., Laibowitz, R. B., Plaskett, T. S., Freitas, P., Gallagher, W. J., Sandstrom, R. L., and Dinger, T. R. (1987). *Phys. Rev. Lett.* **59,** 704.

Cooper, S. L., Klein, M. V., Pazol, B. G., Rice, J. P., and Ginsberg, D. M. (1988). *Phys. Rev. B: Condens. Matter* **37,** 5920.

Copič, M., Mihailóvic, D., Zgonik, M., Prester, M., Biljaković, K., Orel, B., and Brničević, N. (1987). *Solid State Commun.* **64,** 297.

Degiorgi, L., Kaldis, E., and Wachter, P. (1987). *Solid State Commun.* **64,** 873.

Denisov, V. N., Mavrin, B. N., Podobedov, V. B., Alexandrov, I. V., Bykov, A. B., Goncharov, A. F., and Mel'nikov, O. K. (1988). *Phys. Lett. A* **130,** 411.

Etemad, S., Aspnes, D. E., Kelley, M. K., Thompson, R., Tarascon, J.-M., and Hull, G. W. (1988a). *Phys. Rev. B: Condens. Matter* **37,** 3396.

Etemad, S., Aspnes, D. E., Barboux, P., Hull, G. W., Kelley, M. K., Tarascon, J.-M., Thompson, R., Herr, S. L., Kamarás, K., Porter, C. D., and Tanner, D. B. (1988b). *Proc. Mater. Res. Soc. Symp.* **99,** 135.

Feile, R., Kowalewski, J., and Assmus, W. (1988). *Physica C (Amsterdam)* **152,** 491.

Fu, C. L., and Freeman, A. J. (1987). *Phys. Rev. B: Condens. Matter* **35,** 8861.

Genzel, L., Wittlin, A., Kuhl, J., Mattausch, H., Bauhofer, W., and Simon, A. (1987). *Solid State Commun.* **63,** 843.

Gervais, F., Echegut, P., Bassat, J. M., and Odier, P. (1988). *Phys. Rev. B: Condens. Matter* **37,** 9364.

Goshchitskii, B. N., Davydov, S. A., Zemlyanov, M. G., Karkin, A. E., Kozhevnikov, V. L., Mirmelshtein, A. V., Panova, G. II., Parshin, P. P., Ponosov, Y. S., Fotiev, A. A., Khlopkin, M. N., Chernoplyokov, N. A., Cheshnitskii, S. M., and Shikov, A. A. (1987). *Fiz. Met. Metalloved.* **64,** 188.

Gupta, H. C. (1988). *Solid State Commun.* **65,** 495.

Hadjiev, V. G., and Iliev, M. N. (1988). *Solid State Commun.* **66,** 451.

Hangyo, M., Nakashima, S., Mizoguchi, K., Fujii, A., Mitsuishi, A., and Yotsuya, T. (1988). *Solid State Commun.* **65,** 835.

Hemley, R. J., and Mao, H. K. (1987). *Phys. Rev. Lett.* **58,** 2340.

Herr, S. L., Kamarás, K., Porter, C. D., Doss, M. G., Tanner, D. B., Bonn, D. A., Greedan, J. E., Stager, C. V., and Timusk, T. (1987). *Phys. Rev. B: Condens. Matter* **36,** 733.

Joyce, R. R., and Richards, P. L. (1970). *Phys. Rev. Lett.* **24,** 1007.

Kamarás, K., Porter, C. D., Doss, M. G., Herr, S. L., Tanner, D. B., Bonn, D. A., Greedan, J. E., O'Reilly, A. H., Stager, C. V., and Timusk, T. (1987). *Phys. Rev. Lett.* **59,** 919.

Kamarás, K., Doss, M. G., Herr, S. L., Kim, J. S., Porter, C. D., Stewart, G. R., Tanner, D. B., Bonn, D. A., Greedan, J. E., O'Reilly, A. H., Stager, C. V., Timusk, T., Keszei, B., Pekker, S., Hutiray, G., and Mihaly, L. (1988a). *Proc. Mater. Res. Soc. Symp.* **99,** 777.

Kamarás, K., Porter, C. D., Doss, M. G., Herr, S. L., Tanner, D. B., Bonn, D. A., Greedan, J. E., O'Reilly, A. H., Stager, C. V., and Timusk, T. (1988b). *Phys. Rev. Lett.* **60,** 969.

Kirtley, J. R., Collins, R. T., Schlesinger, Z., Gallagher, W. J., Sandstrom, R. L., Dinger, T. R., and Chance, D. A. (1987). *Phys. Rev. B: Condens. Matter* **35,** 8846.

Kourouklis, G. A., Jayaraman, A., Weber, W., Remeika, J. P., Espinosa, G. P., Cooper, A. S., and Maines, R. G., Sr. (1987a). *Phys. Rev. B: Condens. Matter* **36,** 7218.

Kourouklis, G. A., Jayaraman, A., Batlogg, B., Cava, R. J., Stavola, M., Krol, D. M., Rietman, E. A., and Schneemeyer, L. F. (1987b). *Phys. Rev. B: Condens. Matter* **36,** 8320.

Kress, W., Schröder, U., Prade, J., Kulkarni, A. D., de Wette, F. W. (1988). *Phys. Rev. B: Condens. Matter* **38,** 2906.

Krol, D. M., Stavola, M., Weber, W., Schneemeyer, L. F., Waszczak, J. V., Zahurak, S. M., and Kosinski, S. G. (1987). *Phys. Rev. B: Condens. Matter* **36,** 8325.

Krol, D. M., Stavola, M., Schneemeyer, L. F., Waszczak, J. V., and Weber, W. (1988). *Proc. Mater. Res. Soc. Symp.* **99,** 781.

Kuroda, N., Chida, F., Sasaki, Y. Nishina, Y., Kikuchi, M., Tokiwa, A., and Syono, Y. (1987). *J. Phys. Soc. Jpn.* **56,** 3797.

Little, W. A., Collman, J. P., Yee, G. T., Holcomb, M. J., McDevitt, J. T., and Brown, G. E. (1988). *J. Am. Chem. Soc.* **110,** 1301.

Liu, R., Thomsen, C., Kress, W., Cardona, M., Gegenheimer, B., de Wette, F. W., Prade, J., Kulkarni, A. D., and Schröder, U. (1988a). *Phys. Rev. B: Condens. Matter* **37,** 7971.

Liu, R., Thomsen, C., Cardona, M., and Mattausch, H. (1988b). *Solid State Commun.* **65,** 67.

Loye, H.-C., Leary, K. J., Keller, S. W., Ham, W. K., Faltens, T. A., Michaels, J. N., Stacy, A. M. (1987). *Science* **238,** 1558.

Lyons, K. B., Liou, S. H., Hong, M., Chen, H. S., Kwo, J., and Negran, T. J. (1987). *Phys. Rev. B: Condens. Matter* **36,** 5592.

McCarty, K. F., Hamilton, J. C., Shelton, R. N., and Ginley, D. S. (1988). *Phys. Rev. B: Condens. Matter* **38,** 2914.

McMullan, W. G., Gygax, S., and Irwin, J. C. (1988). *Solid State Commun.* **66,** 165.

Macfarlane, R. M., Rosen, H., and Seki, H. (1987). *Solid State Commun.* **63,** 831.

Macfarlane, R. M., Rosen, H. J., Engler, E. M., Jacowitz, R. D., and Lee, V. Y. (1988). *Phys. Rev. B: Condens. Matter* **38,** 284.

Maeno, Y., Kato, M., and Fujita, T. (1987). *Jpn. J. Appl. Phys.* **26,** L329.

Maksimow, A. I., Misochko, O. V., Tartakovsky, I. T., Timofeev, V. B., Remeika, J. P., Cooper, A. S., and Fisk, Z. (1988). *Solid State Commun.* **66,** 1077.

Maroni, V. A. (1988). *J. Phys. Chem. Solids* **49,** 307.

Maroni, V. A., Brun, T. O., Grimsditch, M., and Loong, C.-K. (1989). *Phys. Rev. B: Condens. Matter,* **39,** 4127.

Mascarenhas, A., Geller, S., Xu, L. C., Katayama-Yoshida, H., Pankove, J. I., and Deb, S. K. (1988). *Appl. Phys. Lett.* **52,** 242.

Mase, S., Yasuda, T., Horie, Y., and Fukami, T. (1988). *Solid State Commun.* **65,** 477.
Mihailovíc, D., Zgonik, M., Copič, M., and Hrovat, M. (1987). *Phys. Rev. B: Condens. Matter* **36,** 3997.
Morioka, Y., Kikuchi, M., and Syono, Y. (1987). *Jpn. J. Appl. Phys.* **26,** L1499.
Morioka, Y., Tokiwa, A., Kikuchi, M., and Syono, Y. (1988). *Solid State Commun.* **67,** 267.
Morris, D. E., Kuroda, R. M., Markelz, A. G., Nickel, J. H., and Wei, J. Y. T. (1988). *Phys. Rev. B: Condens. Matter* **37,** 5936.
Nagasaka, K., Soto, M., Ihara, H., Tokumoto, M., Hirabayashi, M., Terada, N., Senzaki, K., and Kimura, Y. (1987). *Jpn. J. Appl. Phys.* **26,** L479.
Noh, T. W., Sulewski, P. E., Kaplan, S. G., Sievers, A. J., Lathrop, D. K., and Buhrman, R. A. (1988). *Proc. Mater. Res. Soc. Symp.* **99,** 435.
Ogita, N., Ohbayashi, K., Udagawa, M., Aoki, Y., Maeno, Y., and Fujita, T. (1987). *Jpn. J. Appl. Phys.* **26,** L415.
Ohbayashi, K., Ogita, N., Udagawa, M., Aoki, Y., Maeno, Y., and Fujita, T. (1987). *Jpn. J. Appl. Phys.* **26,** L423.
Oh-ishi, K., Kikuchi, M., Syono, Y., Hiraga, K., and Morioka, Y. (1987). *Jpn. J. Appl. Phys.* **26,** L484.
Onari, S., Hiroaki, T., Ohshima, K., Honma, H., and Arai, T. (1988). *Solid State Commun.* **66,** 303.
Orenstein, J., and Rapkine, D. H. (1988). *Phys. Rev. Lett.* **60,** 968.
Orenstein, J., Thomas, G. A., Rapkine, D. H., Bethea, C. G., Levine, B. F., Cava, R. J., Rietman, E. A., and Johnson, D. W. (1987). *Phys. Rev. B: Condens. Matter* **36,** 729.
Ose, W., Obermayer, P. E., Otto, H. H., Zetterer, T., Lengfellner, H., Keller, J., and Renk, K. F. (1988). *Z. Phys. B* **70,** 307.
Popovic, Z. V., Thomsen, C., Cardona, M., Liu, R., Stanisic, G., and König, W. (1988). *Solid State Commun.* **66,** 43.
Prade, J., Kulkarni, A. D., de Wette, F. W., Kress, W., Cardona, M., Reiger, R., and Schröder, U. (1987). *Solid State Commun.* **64,** 1267.
Rosen, H., Engler, E. M., Strand, T. C., Lee, V. Y., and Bethune, D. (1987). *Phys. Rev. B: Condens. Matter* **36,** 726.
Rosen, H., Macfarlane, R. M., Engler, E. M., Lee, V. Y., and Jacowitz, R. D. (1988). *Phys. Rev. B: Condens. Matter* **38,** 2460.
Saito, Y., Sawada, H., Iwazumi, T., Abe, Y., Ikeda, H., and Yoshizaki, R. (1987). *Solid State Commun.* **64,** 1047.
Sawada, H., Saito, Y., Iwazumi, T., Yoshizaki, R., Abe, Y., and Matsuura, E. (1987). *Jpn. J. Appl. Phys.* **26,** L426.
Schlesinger, Z., Green, R. L., Bednorz, J. G., and Müller, K. A. (1987a). *Phys. Rev. B: Condens. Matter* **35,** 5334.

Schlesinger, Z., Collins, R. T., and Shafer, M. W. (1987b). *Phys. Rev. B: Condens. Matter* **35,** 7232.

Singh, K. K., Ganguly, P., and Goodenough, J. B. (1984). *J. Solid State Chem.* **52,** 254.

Stavola, M., Cava, R. J., and Rietman, E. (1987a). *Phys. Rev. Lett.* **58,** 1571.

Stavola, M., Krol, D. M., Weber, W., Sunshine, S. A., Jayaraman, A., Kourouklis, G. A., Cava, R. J., and Rietman, E. A. (1987b). *Phys. Rev. B: Condens. Matter* **36,** 850.

Stavola, M., Krol, D. M., Schneemeyer, L. F., Sunshine, S. A., Waszcak, J. V., and Kosinski, S. G. (1989). *Phys. Rev. B: Condens. Matter* **39,** 287.

Sugai, S. (1987a). *Jpn. J. Appl. Phys.* **26,** L1517.

Sugai, S. (1987b). *Phys. Rev. B: Condens. Matter* **36,** 7133.

Sugai, S., Uchida, S., Takagi, H., Kitazawa, K., and Tanaka, S. (1987). *Jpn. J. Appl. Phys.* **26,** L879.

Sulewski, P. E., Noh, T. W., McWhirter, J. T., and Sievers, A. J. (1987). *Phys. Rev. B: Condens. Matter* **36,** 5735.

Taliani, C., Zamboni, R., and Licci, F. (1987). *Solid State Commun.* **64,** 911.

Thomas, G. A., Millis, A. J., Bhatt, R. N., Cava, R. J., and Rietman, E. A. (1987a). *Phys. Rev. B: Condens. Matter* **36,** 736.

Thomas, G. A., Ng, H. K., Millis, A. J., Bhatt, R. N., Cava, R. J., Rietman, E. A., Johnson, D. W., Jr., Espinosa, G. P., and Vandenberg, J. M. (1987b). *Phys. Rev. B: Condens. Matter* **36,** 846.

Thomsen, C., Liu, R., Bauer, M., Wittlin, A., Genzel, L., Cardona, M., Schönherr, E., Bauhofer, W., and König, W. (1988a). *Solid State Commun.* **65,** 55.

Thomsen, C., Cardona, M., Kress, W., Liu, R., Genzel, L., Bauer, M., Schönherr, E., and Schröder, U. (1988b). *Solid State Commun.* **65,** 1139.

Thomsen, C., Liu, R., Wittlin, A., Genzel, L., Cardona, M., König, W., Cabañas, M. V., and García, E. (1988c). *Solid State Commun.* **65,** 219.

Thomsen, C., Liu, R., Cardona, M., Amador, U., and Moran, E. (1988d). *Solid State Commun.* **67,** 271.

Udagawa, M., Ogita, N., Fukumoto, A., Utsunomiya, Y., and Ohbayashi, K. (1987). *Jpn. J. Appl. Phys.* **26,** L858.

Vuong, T. H. H., Tsui, D. C., Goldman, V. J., Hor, P. H., Meng, R. L., and Chu, C. W. (1987). *Solid State Commun.* **63,** 525.

Weber, W. (1987). *Phys. Rev. Lett.* **58,** 1371.

Weber, W. H., Logothetis, E. M., Soltis, R. E., Graham, G. W., Chen, J. T., and Wenger, L. E. (1988). *Proc. Mater. Res. Soc. Symp.* **99,** 631.

Wittlin, A., Liu, R., Cardona, M., Genzel, L., König, W., Bauhofer, W., Mattausch, H., Simon, A., and García-Alvarado, F. (1987). *Solid State Commun.* **64,** 477.

Wittlin, A., Genzel, L., Cardona, M., Bauer, M., König, W., García, E., Barahona, M., and Cabañas, M. V. (1988). *Phys. Rev. B: Condens. Matter* **37,** 652.

Wrobel, J. M., Wang, S., Gygax, S., Clayman, B. P., and Peterson, L. K. (1987). *Phys. Rev. B: Condens. Matter* **36,** 2368.

Yamanaka, A., Minami, F., Watanabe, K., Inoue, K., Takekawa, S., and Iyi, N. (1987). *Jpn. J. Appl. Phys.* **26,** L1404.

2

The Use of Vibrational Spectroscopy in the Characterization of Synthetic Organic Electrical Conductors and Superconductors

John R. Ferraro*
Jack M. Williams
Argonne National Laboratory
Argonne, Illinois

*Consultant to Argonne National Laboratory and Searle Professor of Chemistry (Emeritus) Loyola University, Chicago, Illinois.

Practical Fourier Transform Infrared Spectroscopy

This chapter is devoted to a description of the role vibrational spectroscopy is playing in the characterization of synthetic electrical conductors and superconductors. Both Raman and infrared spectroscopies are included. Most of the infrared data cited have been obtained with FT-IR instrumentation. Earlier dispersive data are also included to provide a consistent overview of the subject matter. The various synthetic electrical materials to be discussed include the organic charge-transfer compounds, polymer–salt complexes, transition element–macrocyclic ligand complexes, and organic polymers,* and range in electrical conductivity from semiconductors through conductors to superconductors. This chapter

*Abbreviations used in text, tables, and figures: BEDT-TTF, or ET, bis(ethylenedithio) tetrathiafulvalene; bipy, bipyridyl; bpz, 2,2[1]-bipyrazine; bqd, benzoquinonedioxime; cod, cyclooctadiene; DBTTF, dibenzotetrathiafulvalene; dipyam, di(α,α^1-pyridyl)amine; DMTTF, dimethyltetrathiafulvalene; dpg, diphenylglyoxime; DPTTF, diphenyltetrathiafulvalene; EBTTF, tetrahydrodithinotetrathiafulvalene; ET, see BEDT-TTF; HMDS-DTF, hexamethylenediselenadithiofulvalene; HMTSF, hexamethylenetetraselenafulvalene; HMTTF, hexamethylenetetrathiafulvalene; mdb, nor-

is not to be construed as an exhaustive presentation. Additionally, it is not meant to be a complete review. To this end, we have chosen to cite the published work that best illustrates our topic.

I. ORGANIC CHARGE-TRANSFER CONDUCTORS AND SUPERCONDUCTORS

Figure 1 illustrates structures of several donor molecules, acceptor molecules, and anions that form charge-transfer conductors and are pertinent to the discussion that follows.

The primary spectroscopic interest in the organic charge-transfer conductors and superconductors has centered in three major areas:

1. The use of vibrational spectroscopy to determine the extent of charge-transfer occurring in charge-transfer complexes (e.g., TTF-TCNQ).

---

bordiene, Me, methyl; M(OEP), metal salt of OEP; M(OMTBP), metal salt of OMTBP; MTCNQ, metal salt of TCNQ; MTPP, methyltriphenylphosphine; NMP, *N*-methylphenylazine; OEP, 1,2,3,4,5,6,7,8-octaethyl-porphyrin; OMTBP, 1,4,5,8,9,12,13,16-octamethyltetrabenzoporphyrin; PBPA, poly(*N*-benzylpropargylamine) acetylene; Pc, phthalocyanine; PEO, poly (ethylene oxide); Per, perylene; POHP, poly(2-propyn-1-ol) acetylene; PPA, poly(phenylacetylene); PPO, poly(propylene oxide); PPP, poly(paraphenylene); PTF, bis(propylene dithio)tetrathiafulvalene; 2 PTF's; PVP, poly(vinylpyrrolidone); Qn, quinoline; RRS, resonance Raman Spectroscopy; TAA, dihydrobenzo [b, i](1,4,8,11) tetra-azocyclotetradecine; TAAB, tetrabenzo [b,q,f,j,m] (1,5,9,13) tetra-azocylohexadecine; TBP, tetrabenzoporphyrin; TCA, tetracyanoethylene; TCNQ, tetracyanoquinodimethane; TEA, triethylammonium; TMTSF, tetramethylenetetraselenafulvalene; TMTTA, tetramethyl complex of TAA; TMTTF, tetramethyltetrathiafulvalene; TNAP, 11,11,12,12-tetracyanonaphtho-2,6-quinodimethane; TSF, tetraselenafulvalene; TSeF, tetraselenatetracene; TTF, tetrathiafulvalene; TTFX, tetrathiafulvalene salt, where X = anion.

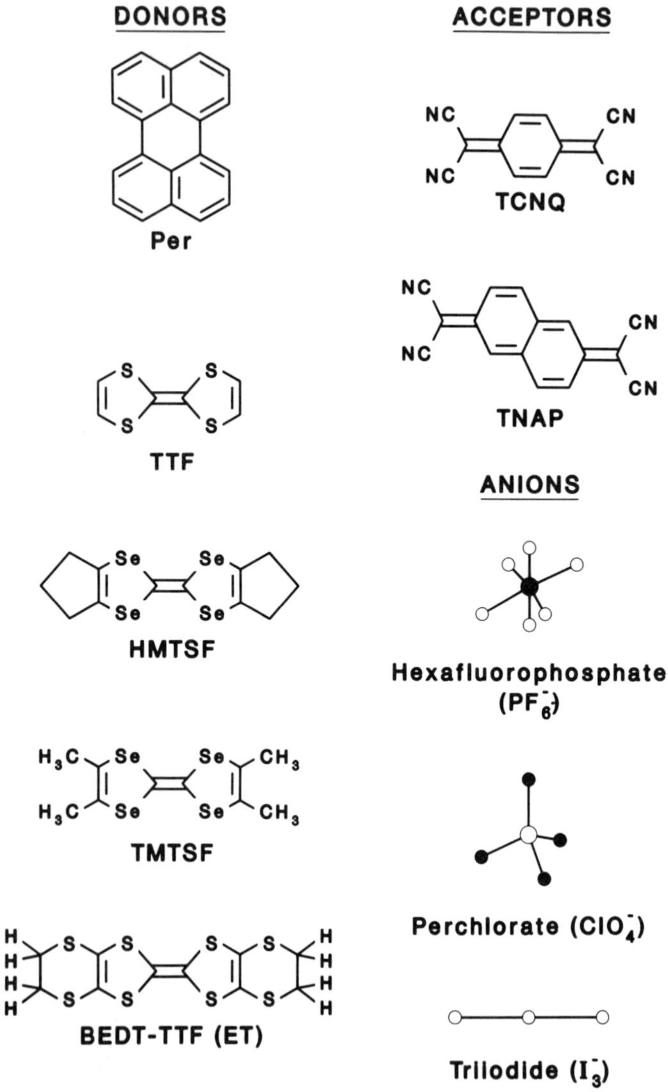

Fig. 1. Electron-donor, electron-acceptor, and anionic species that make up selected charge-transfer compounds. (From Ferraro and Williams, 1987.)

2. Obtaining polarized reflectance spectra to ascertain the extent of anisotropy, electronic structure, plasma frequencies, existence of an optical band gap, calculation of the optical conductivity values, determination of an electron–phonon coupling constant, and the dimensionality of the complex.
3. Assignments of vibrations in neat molecules (e.g., TCNQ, TTF, BEDT-TTF) and in their complexes [e.g., $(BEDT\text{-}TTF)_2I_3$, TTF-TCNQ].

A. The Use of Vibrational Spectroscopy to Determine the Degree of Charge-Transfer Occurring in Charge-Transfer Complexes

The discussion in this section deals with TTF and TCNQ variants for which the extent of charge-transfer between donor and acceptor is considered to be less than unity. Torrance (1979) arranged various cations (electron donors) used with TCNQ (acceptor) in terms of four classes. Table I summarizes some of these data. Complete charge-transfer leads to poor conductivity, whereas incomplete charge-transfer leads to improved conductivity. It is important to be able to determine the extent of charge-transfer in these salts, and vibrational

Table I Charge-Transfer Status for Several Charge-Transfer Salts

Class	Example	Conductivity (σ)	Energy	Activation CT status
1	1:1 TCNQ inorganic salts	Insulators	Large	Complete
2	1:1 TCNQ organic salts	σ high	Low	Incomplete
3	1:2 salts	$\sigma<10$ (ohm cm)$^{-1}$	High	Complete
4	1:2 salts	$\sigma>50$ (ohm cm)$^{-1}$	Low	Incomplete

spectroscopy has played a major role in making such determinations.

Van Duyne *et al.* (1986) used resonance Raman spectroscopy (RRS) to study partially oxidized derivatives of TTF and TCNQ. The ν_3 vibrational mode in TTF was used as a monitor, because it has a large shift with oxidation of the donor molecule. For example, it shifts from 1512 cm^{-1} in neat TTF to 1427 cm^{-1} in TTF^+. A second band (ν_6) at 468 cm^{-1} in TTF and 510 cm^{-1} in TTF^+ may also be used. For TCNQ, the ν_4 band, which is sensitive to reduction, can be used. A linear relationship was derived between ν_3 and ρ (degree of charge-transfer) in TTF^+ and between ν_4 and ρ in $TCNQ^-$. Both ν_3

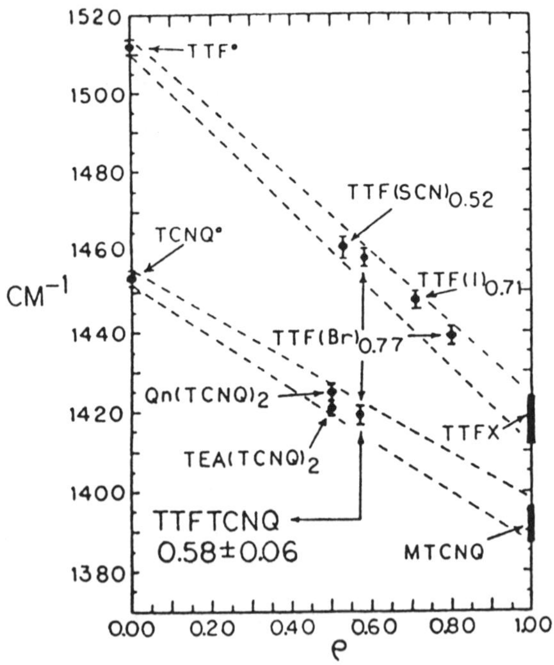

Fig. 2. Frequency dependence of ν_3 for TTF derivatives (top) and ν_4 for TCNQ compounds (bottom) on ρ (degree of charge-transfer). (From Van Duyne *et al.*, 1986.)

for TTF and ν_4 for TCNQ have been assigned as the C═C stretching vibrations in their respective molecules. Figure 2 shows the interdependence of ν_3 and ρ for TTF and related derivatives, and for ν_4 for TCNQ compounds. Values for ρ range from ~0.5 to ~0.8. Similar studies were made for DB-TTF salts [e.g., $(DBTTF)_2(Cu_2Br_6)$] by Tanaka *et al.* (1986). A linear dependency of the totally symmetrical vibrational mode of the C═C bond (ν_5) at 1540 cm^{-1} in neat DBTTF versus ρ was found (see Tanaka *et al.*, 1986). Values of ρ ranged from ~0.5 to 0.75. The extent of charge-transfer in molecules of the TTF-TCNQ family was determined by Chappell *et al.* (1981), using the C—N stretching frequency. The results are shown in Fig. 3.

Studies of TTF derivatives and TCNQ have also been made by Matsuzaki *et al.* (1980). Jurgenson and Drickamer

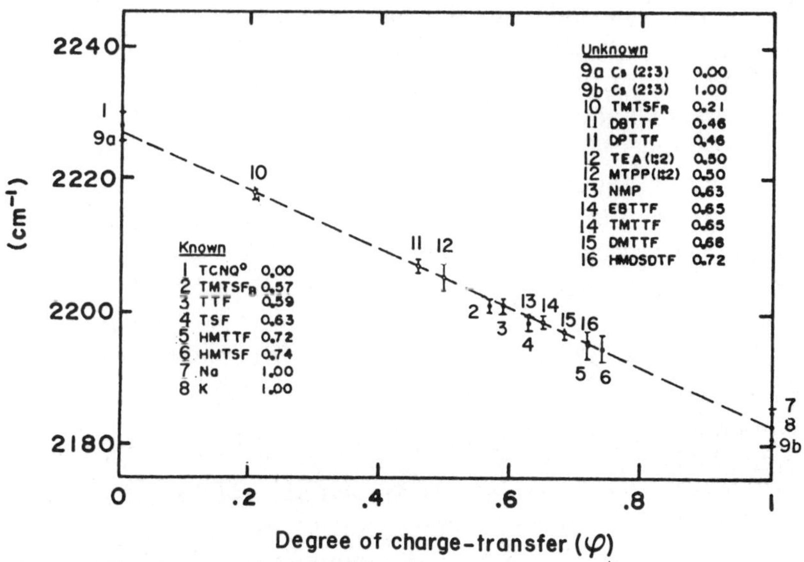

Fig. 3. The C═N strectching frequency ν versus ρ for TCNQ salts. (●) ρ known from independent measurements; (○) ρ previously unknown. (From Chappell *et al.*, 1981.)

(1986) showed that pressures of ~7.8 kbar* could shift the equilibrium in the charge-transfer compounds perylene-tetracyanoethylene and TTF-chloranil to the ionic state. Visible and infrared spectroscopies were used to monitor the transition. Tokura *et al.* (1986) studied the pressure-induced neutral-ionic transition in TTF-chloranil using the infrared C=O stretching mode as monitor. Hanfland *et al.* (1988) demonstrated the transition to the ionic state of TTF-chloranil using Raman spectroscopy by following the shift of four A_g modes with pressure.

B. Polarized Reflectance Spectra of the $(ET)_2X$ Salts

The polarized reflectance spectra for several salts of the type $(ET)_2X$, where X is a monovalent anion and ET is an acronym for BEDT-TTF (see Fig. 1), have been measured. Table II lists the various salts and the range of the experimental measurements. All of the materials have reflectance spectra that resemble one another. They all show a broad band increasing in reflectance with decreasing frequency and sharp bands appearing at frequencies less than 1000 cm^{-1}. Jacobsen *et al.* (1985) studied the polarized reflectance spectra as a function of temperature and observed a transition from a semiconducting material at 300 K to a metallic-like substance at 40 K for β-$(ET)_2I_3$. Plasma edges were observed in two directions in the crystals at low temperature, and it was concluded that the band structure was isotropic in two dimensions. Table III shows the plasma frequency (ω_p) obtained from reflectance data for several ET salts. Figure 4 illustrates the $R(\omega)$ (the measured reflectance) curves for several $(ET)_2X$ salts.

The polarized reflectance measurements indicate that all of these charge-transfer compounds possess anisotropic

---

*1 kbar = 10^8 Nm^{-2} = 10^8 Pa, where Nm is Newton meter.

Table II Polarized Reflectance Measurements for Several ET Salts[a]

Salt	Polarized reflectance range $(cm^{-1}) \times 10^{-3}$	References
α-$(ET)_2I_3$	>15	Koch *et al.* (1985)
	>5	Meneghetti *et al.* (1986)
	>25	Sugano *et al.* (1985)
	~8	Jacobsen *et al.* (1985)
	>15	Kaplunov *et al.* (1985)
β-$(ET)_2I_3$	>15	Kuroda *et al.* (1985)
	>25	Sugano *et al.* (1985)
	>25	Tajima *et al.* (1985)
	>25	Tajima *et al.* (1986)
β-$(ET)_2PF_6$	>25	Kuroda *et al.* (1985)
	>25	Tajima *et al.* (1985)
$(ET)_2ClO_4(C_2H_3Cl_3)_{0.5}$	~25	Kuroda *et al.* (1985)
		Tajima *et al.* (1984)
β-$(ET)_2IBr_2$	>25	Sugano and Saito (1986)
β-$(ET)_2I_2Br$	~8	Jacobsen *et al.* (1987)
β-$(ET)_2AuI_2$	~8	Jacobsen *et al.* (1987)
β''-$(ET)_2IAuBr$	~20	Ugawa *et al.* (1986)
$(ET)_3(ClO_4)_2$	>25	Kuroda *et al.* (1985)

[a] α Structures belong to the $P\bar{1}$ space group with $z = 2$ and cell volume equal to ~1700 Å^3. α^1-structures belong to the P2/n space group with z = 4 and cell volume equal to ~1600 Å^3. β, β', and β'' structures are similar and belong to the $P\bar{1}$ space group with z = 1 and cell volumes equal to ~825–850 Å^3.

Table III Plasma Frequencies Obtained from Reflectance Data

Compound	Conductivity at 300 K $(ohm\ cm^{-1})$	$\omega_p(cm^{-1}) \times 10^3$	Metal-insulator transition (K)
$(ET)_2PF_6$	10	5.1[a]	297
		5.3	
$(ET)_2ClO_4(TCE)_{0.5}$	20	8.8	None
		4.7	
$(ET)_2ClO_4$	50	10.5	170
		4.4	
β-$(ET)_2I_3$	30	8.5	None
		5.3	
α-$(ET)_2I_3$	—	4.4	135
β-$(ET)_2I_2Br$	—	9.3	None
		5.3	
β-$(ET)_2AuI_2$	—	10.3	None
		5.8	

[a] The first number applies to $\sigma(\omega)$ parallel to the *a*-axis; the second, parallel to the *c*-axis.

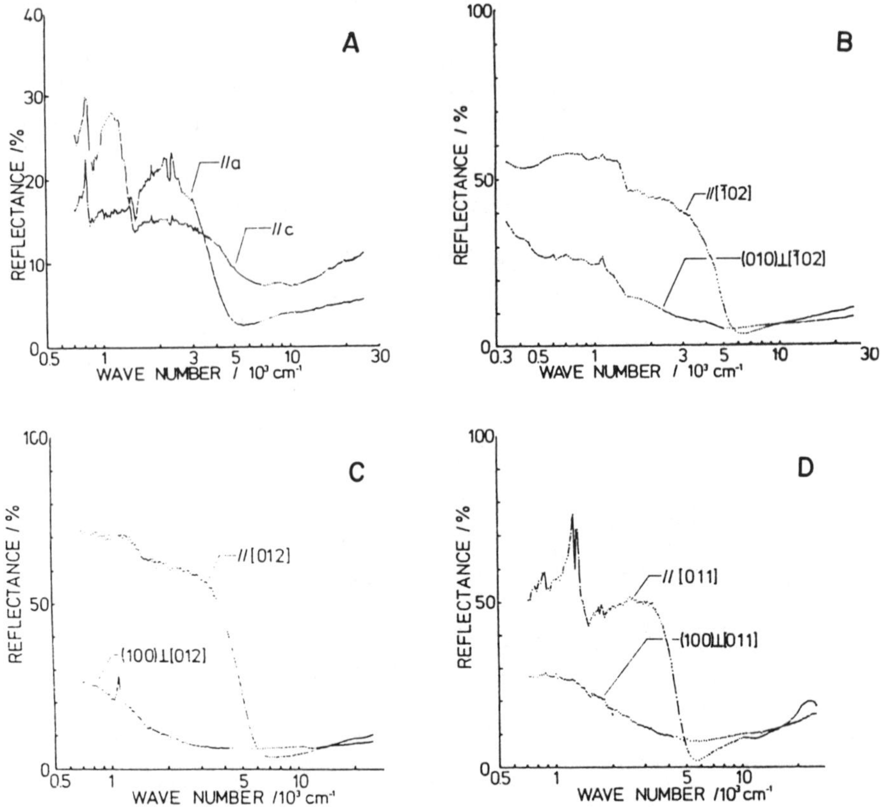

Fig. 4. Polarized reflectance spectra for several ET salts. (A) Polarized reflectance spectrum of β-$(ET)_2PF_6$. (B) Polarized reflectance spectrum of β-$(ET)_2ClO_4(C_2H_3Cl_3)_{0.5}$. (C) Polarized reflectance spectrum of β-$(ET)_3(ClO_4)_2$. (D) Polarized reflectance spectrum of β-$(ET)_2I_3$. (From Kuroda *et al.*, 1985.)

physical properties. The degree of anisotropy varies among the salts, but generally the ET salts appear to be less anisotropic than the TMTSF salts.

From these reflectance spectra values for R(ω), one can obtain values of σ(ω) (the optical conductivities) through Kramers-Kronig analysis. All of σ(ω) curves resemble one another, as do the R(ω) curves (see Fig. 5). They all have a broad band with a maximum, and superimposed sharp bands also appear.

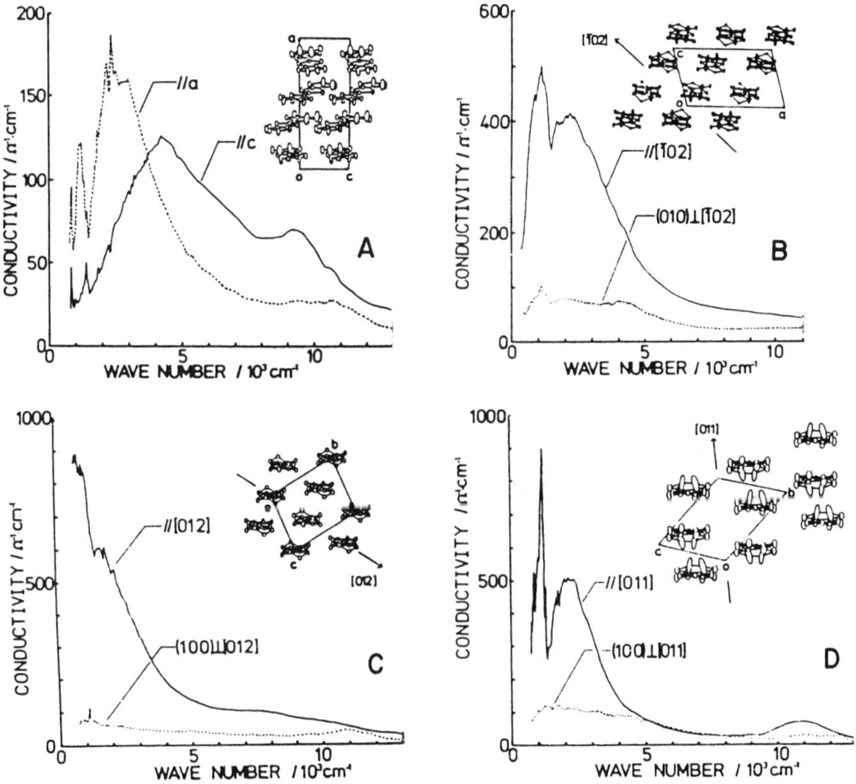

Fig. 5. Optical conductivity spectra for several ET salts. (A) β-$(ET)_2PF_6$. (B) β-$(ET)_2ClO_4(C_2H_3Cl_3)_{0.5}$. (C) β-$(ET)_3(ClO_4)_2$. (D) β-$(ET)_2I_3$. (From Kuroda *et al.*, 1985.)

Unfortunately, the measurements were not made to sufficiently low ω values, thus precluding their use to determine whether the substance was a metal, a semiconductor, or a semimetal [see Williams *et al.* (1987) for details of this method].

The spectra and σ(ω) were analyzed by use of the Rice model (see Rice, 1976; Rice *et al.*, 1977). The broad band is assigned to the charge-transfer between the donor and acceptor molecules. The superimposed fine structure is associated with an electron–phonon coupled transition (Williams *et al.*, 1987).

C. Vibrational Assignments Made for Various Neat Donor and Acceptor Species and Their Salts

We present several studies of the BEDT-TTF, TMTSF, TTF, and TCNQ families to illustrate the use of vibrational spectroscopy in the study of synthetic electrical conductors.

1. Vibrational Spectra of Neat Donor Molecules and Acceptors

The vibrational spectra of TTF, TTF^+, TTF-d_4, and TTF-d_4^+ have been reported (Bozio *et al.*, 1977, 1979). Data were collected by use of polarized infrared spectra of single crystals and by Raman spectroscopy. Table IV shows the symmetry classes and selection rules for the TTF molecule. Table V tabulates the assignments for TTF, TTF^+, TTF-d_4, and TTF-d_4^+. The vibrations of the C═C stretching mode, corresponding to the ν_2, ν_3 (A_g type), and ν_{14} (B_{1u} type) species, and the C—S stretching vibration (ν_{16}, B_{1u}) show significant shifts

Table IV Symmetry Classes and Selection Rules for TTF[a]

Molecular space group D_{2h}			Factor group C_{2h}	
Degree of freedom	Species	Site group C_i	Species	Degrees of freedom
7	a_g			
$2+R_z$	b_{1g}		A_g	21 (3 torsional)
$3+R_y$	b_{2g}	a_g		
$6+R_x$	b_{3g}		B_g	21 (3 torsional)
3	a_u			
$6+T_z$	b_{1u}		A_u	20 (2 trans.) + T_b
$6+T_y$	b_{2u}	a_u		
$3+T_x$	b_{2u}		B_u	19 (1 trans.) + $T_{a,c}$

[a]The choice of the molecular axes is *z*, in-plane short axis; *y*, in-plane short axis; and *x*, axis normal to the molecular plane. (Taken in part from Bozio *et al.* 1979.)

Table V Vibrational Assignments of TTF, TTF^+, TTF-d_4, and TTF-d_4^+

Species	Mode	Observed frequencies (cm^{-1})[a]			
		TTF	TTF^+	TTF-d_4	TTF-d_4^+
a_g	ν_1	3083		2280	
	ν_2	1555	1505	1544	1482
	ν_3	1518	1420	1504	1420
	ν_4	1094	1078	787	775
	ν_5	735	758	715	741
	ν_6	474	501	470	499
	ν_7	244	265	242	265
b_{3g}	ν_{28}	3072		(2280)	
	ν_{29}	1258		1057	
	ν_{30}	944		975	
	ν_{31}	800		(715)	
	ν_{32}	612		594	
	ν_{33}	308		305	
b_{1u}	ν_{13}	3108	3079	2337	2316
	ν_{14}	1530	1478	1508	1438
	ν_{15}	1090	1072	779	828
	ν_{16}	781	836	758	770
	ν_{17}	734	751	719	(731)
	ν_{18}	427	460	425	458
b_{2u}	ν_{22}	3073	3063	2285	2275
	ν_{23}	1254	1237	1040	1041
	ν_{24}	863		865	
	ν_{25}	794	825	703	731
	ν_{26}	(639)		603	614
	ν_{27}	(110)			
b_{3u}	ν_{34}	639	705	492	538
	ν_{35}	247*		246*	
	ν_{36}	110		108	

[a]All observed frequencies of neutral molecules are from solution samples except those marked with an asterisk, which are from monoclinic crystals (Bozio *et al.*, 1977). All of the observed frequencies of the cations are from powder samples of (TTF)Br and (TTF-d_4)Br. Values in parentheses refer to overlapped bands. (Taken in part from Bozio *et al.* 1979.)

upon the oxidation of TTF to TTF^+ and reflect the degree of charge-transfer occurring in TTF upon oxidation. Normal coordinate analyses (NCA) give calculated frequencies in reasonable agreement with the observed values.

A complete vibrational analysis of TCNQ was made by Girlando and Pecile (1973), Bozio *et al.* (1973, 1975), and Girlando *et al.* (1974). Polarized Raman spectra of TCNQ and TCNQ-d_4 single crystals were obtained by Girlando and Pecile (1973). Infrared data were obtained by Takenaka 1971. Table VI shows the correlation table and the selection rules for the TCNQ molecule (Girlando and Pecile, 1973). Table VII tabulates assignments made for TCNQ fundamentals. Figure 6 shows the spectra of TTF, TCNQ, and TTF-TCNQ. Normal coordinate analyses were performed on TCNQ and TCNQ-d_4 by Takenaka (1971) and Girlando and Pecile (1973). Calculated frequencies were found to be in agreement with observed values.

Table VIII lists the symmetry classes and selection rules for ET. Assignments of fundamental vibrations for ET and ET-d_8 have been made by Kozlov *et al.* (1987). Table IX tabulates the assignments made from infrared and Raman data.

Table VI Correlation Table Showing the Selection Rules for the Molecule and for the Crystal of TCNQ[a]

Molecular group D_{2h}			Factor group C_{2h}	
Degrees of freedom	Species and activity	Site group C_i	Species and activity	Degrees of freedom
10	$a_g(xx, yy, zz)$		$A_g(aa, bb, cc, ac)$	30
$3 + R_z$	$b_{1g}(xy)$			(3 torsional)
$5 + R_y$	$b_{2g}(xz)$	a_g	$B_g(ab, bc)$	30
$9 + R_x$	$b_{3g}(yz)$			(3 torsional)
4	a_u		$A_u(b)$	29
$9 + T_z$	$b_{1u}(z)$			(2 translational) + T_b
$9 + T_y$	$b_{2u}(y)$	a_u		28
$9 + T_x$	$b_{3u}(x)$		$B_u(a, c)$	(1 translational) + $T_{a,c}$

[a]Reprinted with permission of Pergamon Press, Inc. and authors, *Spectrochim. Acta, Part A* **29A**, 1859 (1973), A Girlando and C. Pecile.

Table VII Observed and Calculated In-plane Fundamentals of TCNQ and TCNQ-d_4[a]

Species	Mode[b]	Frequencies for TCNQ (cm^{-1}) Observed	Calculated	Frequencies for TCNQ-d_4 (cm^{-1}) Observed[c]	Calculated
a_g	ν_1	3048	3052	2291	2275
	ν_2	2229	2230	2232	2230
	ν_3	1602	1602	1564	1579
	ν_4	1454	1454	1449	1444
	ν_5	1207	1207	954	926
	ν_6	948	933	864	855
	ν_7	711	726	690	709
	ν_8	602	600	599	600
	ν_9	334	338	333	338
	ν_{10}	144	126	143	126
b_{3g}	ν_{41}	3030	3049	2265	2265
	ν_{42}	2223	2226	2223	2226
	ν_{43}	1451	1439	1413	1410
	ν_{44}	1323	1328	1205	1215
	ν_{45}	1187	1184	1025	1019
	ν_{46}	609	604	591	584
	ν_{47}	519	505	502	503
	ν_{48}	—	431	—	402
	ν_{49}	—	93	—	92
b_{1u}	ν_{18}	3065	3041	2259	2240
	ν_{19}	2228	2230	2229	2230
	ν_{20}	1545	1542	1533	1527
	ν_{21}	1405	1404	1246	1246
	ν_{22}	998	1013	1001	1002
	ν_{23}	962	961	801	804
	ν_{24}	600	600	602	600
	ν_{25}	540	526	536	512
	ν_{26}	146	160	142	160
b_{3u}	ν_{32}	3053	3050	2293	2270
	ν_{33}	2228	2226	2229	2226
	ν_{34}	1540	1530	1510	1510
	ν_{35}	1354	1360	1316	1310
	ν_{36}	—	1209	1156	1150
	ν_{37}	1125	1119	855	856
	ν_{38}	498	503	496	503
	ν_{39}	—	301	—	299
	ν_{40}	—	60	—	60

[a]Reprinted with permission of Pergamon Press, Inc. and authors, *Spectrochim. Acta, Part A* **29A,** 1859 (1973), A. Girlando and C. Pecile.
[b]From Mulliken (1955).

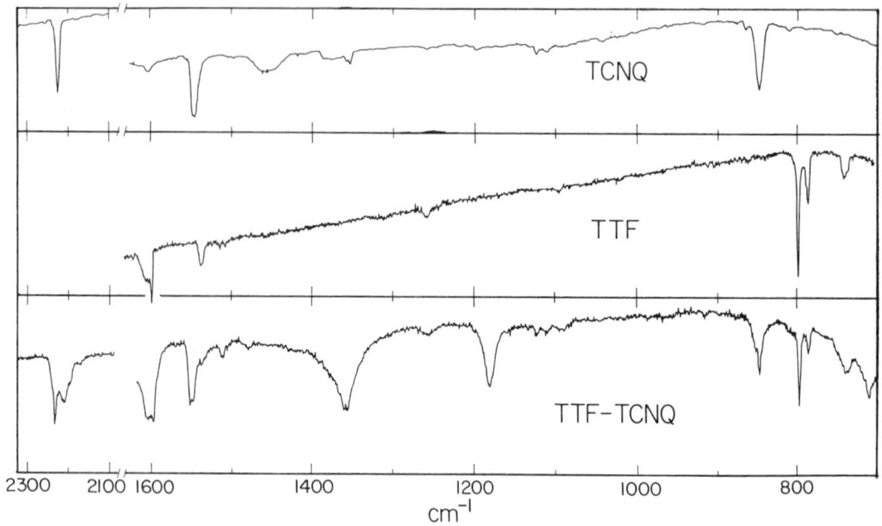

Fig. 6. Infrared spectra of TCNQ, TTF, and TTF-TCNQ in N_2 matrices at 10 K. (From Wozniak *et al.*, 1975.)

2. Vibrational Studies of Charge-Transfer Salts of TMTSF and ET

Infrared and Raman spectra of $(TMTSF)_2X$, where X = ReO_4^-, BF_4^- and ClO_4^- have been reported (Bozio *et al.*, 1982, 1985). Raman scattering measurements for TSeF, TMTSF-d_{12}, $(TMTSF)_2PF_6$, and $(TMTSF)_2AsF_6$ (Iwahuna *et al.*, 1982),

Table VIII Symmetry Classes and Selection Rules for ET[a]

$\Gamma(C_{2v})$ =	$12A_1$ Raman, ir			$6A_2$ Raman		$7B_1$ Raman, ir		$11B_2$ Raman, ir	
$\Gamma(D_{2h})$ =	$12A_g$	$1A_\mu$	$11B_{1\mu}$	$6B_{1g}$	$6A_\mu$	$7B_{2g}$	$7B_{3\mu}$	$11B_{3g}$	$11B_{2\mu}$
	Raman	ia	ir	Raman	ia	Raman	ir	Raman	ir

[a]Reprinted with permission of Pergamon Press, Inc. and authors, *Spectrochim. Acta, Part A* **43A,** 323 (1987), M. E. Kozlov *et al.*

and $(TMTSF)_2PF_6$ (Krauzman *et al.*, 1986) were made. Far-infrared studies for $(TMTSF)_2ClO_4$ (Challener and Richards, 1984), $(TMTSF)_2SbF_6$ (Ng *et al.*, 1984), $(TMTSF)_2AsF_6$ (Ng *et al.*, 1984), $(TMTSF)_2SbF_6$ (Eldridge and Bates, 1986) also have been performed. Challener and Richards (1984) identified a low-frequency feature at 28 cm^{-1}, which was interpreted in terms of an electron–phonon coupling interaction. Ng *et al.* (1984) identified a frequency at 180 cm^{-1} for $(TMTSF)_2$-SbF_6 and assigned this as a gap frequency but observed no such gap in $(TMTSF)_2AsF_6$.

Resonant Raman scattering in the superconductors β-$(ET)_2I_3$ and $(ET)_2IBr_2$ were made (Sugai and Saito, 1986). Figure 7 shows the low-frequency Raman spectra of several ET salts. Strong resonance effects were noted, and more than 10 overtones of the symmetrical stretching modes of the anions were observed. Raman spectra of α- and β-$(ET)_2I_3$ and α- and β-$(ET)_2IBr_2$ were obtained by Sugai and Saito (1986). Strong resonant effects were noted for the I_3^- compounds. Infrared spectra of various alkaline earth salts of TCNQ have been obtained by Bozio *et al.* (1975, 1977). Vibrational spectra have indicated that $Cu(TCNQ)_2$, like $Cu_2(TCNQ)_3$, has both neutral and negatively charged TCNQ units (see Girlando *et al.*, 1974). The effect of reduction of TCNQ to $TCNQ^-$ in the formation of alkaline salts of TCNQ resulted in frequency shifts of up to ~70 cm^{-1} and were reproduced by normal coordinate analysis (Bozio *et al.*, 1975). For further discussion on topics discussed in this section, see Bozio and Pecile (1980), Jacobsen (1986), and references therein.

II. POLYMER–SALT COMPLEXES

Polymer–salt complexes, known also as polymer–ion electrolytes or ionomers, are a relatively new class of electrical conductors. The electrical conductivities of these materials are close to those obtained for semiconductors, and polymers of

Table IX Infrared and Raman Spectra of BEDT-TTF and BEDT-TTF-d_8[a]

Frequencies for BEDT-TTF (cm^{-1})				Assignment for BEDT-TTF	Frequencies for BEDT-TTF-d_8 (cm^{-1})				Assignment for BEDT-TTF-d_8[d]
Infrared[b]		Raman[c]			Infrared[b]		Raman[c]		
ν_i	i	ν_i	i		ν_i	i	ν_i	i	
2958 w	66				2237 w	66			
					2225 w				
2958 w	26			νCH_2	2169 vw	26			νCD_2
					2141 vw				
2916 w	44								
		1552 m	2				1552 m	2	
1505 w	27	1511 m	27		1506 w	27	1511 m	27	
		1494 s	3				1494 s	3	
1420 w	28				1011 m	29			
					1002 vw	5			
1406 m	45			δCH_2	1041 m	46	1044 w	46	δCD_2
		1409 vw	56		1030 w	57	1029 w	57	
		1285 vw	5				794 w	7	$_{\nu}SCD_2$ + ωCD_2
1282 m	29				793 w	31			
1259 w	46			ωCH_2	990 vw	47			
1253 vw	57	1256 vw	57		984 vw	59	984 w	59	ωCD_2
1173 w	21	1175 vw	21		930 w	21	935 vw	21	
1132 sh	38	1132 vw	38	tCH_2			805 w	38	tCD_2
1125 w	67	1126 vw	67		806 w	67			
		1016 vw	59				1018 w	58	
996 w	30	1002 w	30	νCC	1110 m	28			$_{\nu}CC$ + ωCD_2
987 w	6	990 w	6						

938 vw	22			ρCH_2	741 w	22	741 w	22	ρCD_2
917 s	48	919 vw	60		827 w	50			
905 m	31	911 vw	7		905 m	30			
890 m	49	888 vw	49		879 m	49			
875 w	50	875 vw	50		879 m	48			
860 vw	61	860 vw	61						
772 s	32	765 w	32		772 s	32	784 vw	32	
687 w	68			} ρCH_2	693 m	68			} ρCD_2
		687 w	39		678 w	39	693 w	39	
653 w	33	653 m	8		634 w	33	635 m	8	
624 w	51	625 w	62		609 w	51	610 m	62	
499 m	34	486 m	9		500 m	34	486 m	9	
450 w	10	440 m	10		452 w	10	439 m	10	
							351 w		
390 m	35				388 m	35			
		348 w	63				339 w	63	
335 m	52	334 w	64		326 w	52	323 w	64	
		308 w	11				296 w	11	
278 m	36	272 vw	36		269 vw	36			
257 m	53	260 vw	53		257 m	53	255 w	53	
		159 s	65				155 s	65	
		151 s	12				147 s	12	
96 m	72	127 vw	43		88 w	72	125 w	43	
		52 vs					52 vs		
30 w	54	31 vs	54		30 w	54	30 vs	54	

[a]Relative intensities: vs, very strong; s, strong; m, medium; w, weak; vw, very weak sh, shoulder. Reprinted with permission of Pergamon Press, Inc. and authors, *Spectrochim. Acta, Part A* **43A,** 323 (1987), M. E. Kozlov *et al.*

[b]KBr pellets above 400 cm^{-1} and Nujol mull below.

[c]Monocrystalline sample. Exciting line 676.45 nm.

[d]The following symbols have been used for ethylene group vibrations: ν, stretching; δ, bending; ω, wagging; t, twisting; ρ, rocking.

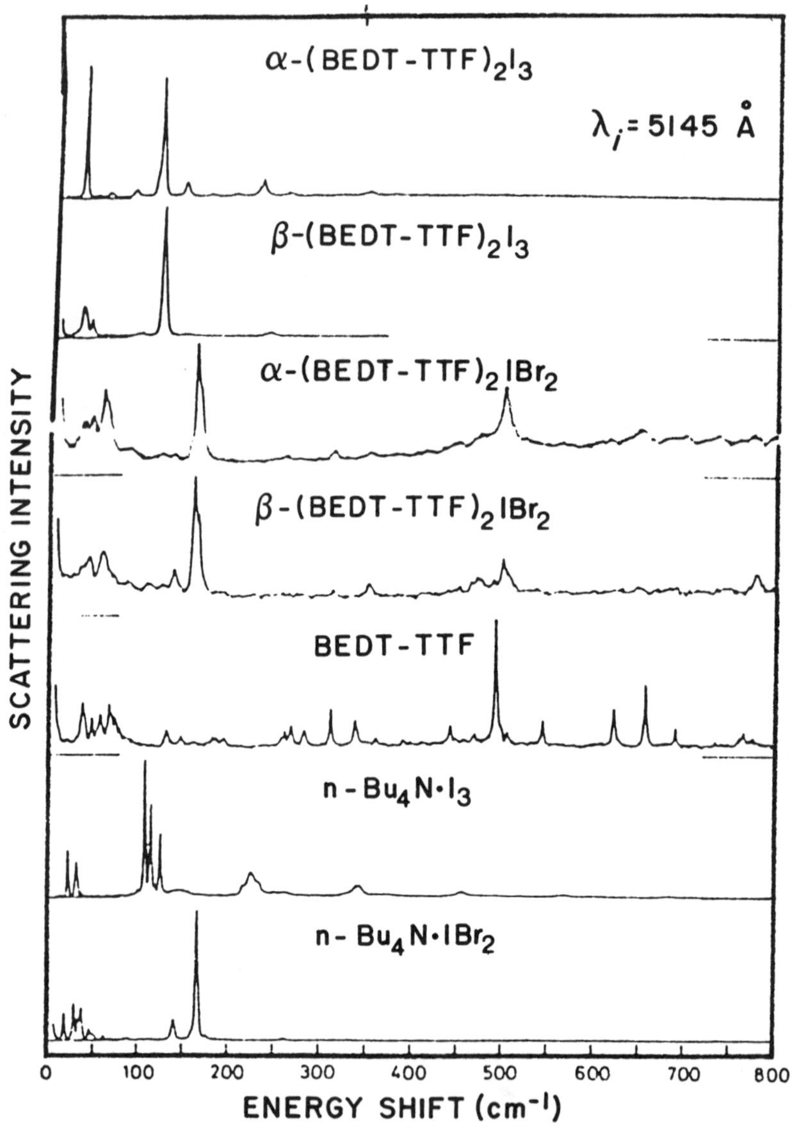

Fig. 7. Raman spectra $(BEDT\text{-}TTF)_2X$ [or $(ET)_2X$] at about 80 K, BEDT-TTF (or ET) at 31 K and *n*-$Bu_4N{\cdot}X$ at 32 K. [Reprinted with permission of Pergamon Press, Inc. and authors, *Solid State Commun.* **58**, 759 (1986), S. Sugai and G. Saito.]

this type have been known for some time. The materials originally were not obtained solvent free and were used as polyelectrolyte ion-exchange polymers. Solvent-free systems were obtained in 1978 when polymers of poly(ethylene oxide) complexed with alkali-metal salts were synthesized (Armand *et al.*, 1978, 1979, 1980; Wright, 1976; James *et al.*, 1979; Fenton *et al.*, 1973; Shriver *et al.*, 1981; Dupon *et al.*, 1982; Papke *et al.*, 1982a,b; Shriver and Farrington, 1985; Spindler and Shriver, 1986; Hardy and Shriver, 1986). For reviews, see Shriver and Farrington (1985), Ferraro and Williams (1987), Vincent (1987), and references therein. Figure 8 shows the backbone structures of various polymers that are used to form polymer–salt complexes and are discussed in this section.

Molecular spectroscopy has been utilized in studying these materials to elucidate:

1. The nature of the conformers in the crystalline PEO-ion polymers. (For infrared and Raman spectroscopic data, see Davidson, 1955; Kuroda and Kubo, 1957, 1959; Yoshihara *et al.*, 1964; Tadokoro *et al.*, 1964; Matsuara and Miyazawa, 1968, 1969; Lin and Parsons, 1969; Koenig and Angood, 1970; Maxfield and Shepherd, 1975.)
2. The configuration of the polymer in the polymer salts after salt formation. [For example, Papke *et al.* (1981, 1982a,b) provided evidence from infrared and Raman spectroscopy that the gauche configuration for the —CH_2—CH_2— group was retained in the PEO-alkali metal systems.]
3. The nature of the polymer–ion interactions in the salts. (These interactions have been studied by infrared and Raman spectroscopy; see Papke *et al.*, 1981, 1982a,b; Teeters and Frech, 1986; Spindler and Shriver, 1986; Hardy and Shriver, 1986.)
4. The ion–ion interactions in polymer–electrolytes. (These interactions have been studied by infrared and Raman spectroscopy; see Papke *et al.*, 1981, 1982a,b;

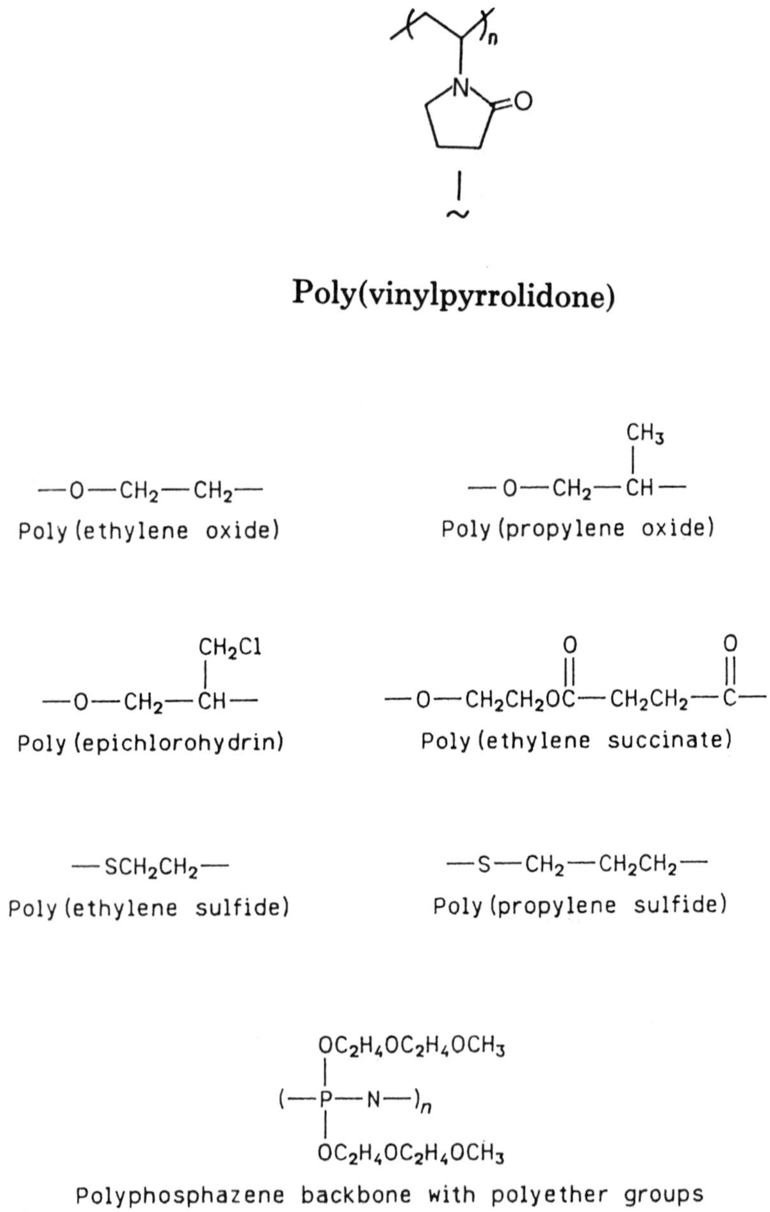

Fig. 8. Backbone structures of various polymers used to form polymer–salt complexes. (Taken in part from Ferraro and Williams, 1987.)

Shriver *et al.*, 1981; Shriver and Farrington, 1985; Teeters and Frech, 1986.)

X-ray diffraction studies combined with vibrational spectroscopy have been used to determine the structure of polymer–ion complexes. Poly(ethylene oxide) consists of an extended helix, having a fiber repeat distance of ~19.48 Å. An X-ray analysis of PEO-$HgCl_2$ was made by Takahashi and Tadokoro (1973) and by Iwamoto *et al.* (1968). X-ray studies with PEO and KSCN show shorter fiber repeat distances of ~8.1 Å (Fenton *et al.*, 1973).

The polymer–ion interactions have been studied by infrared and Raman spectroscopy. Papke *et al.* (1981, 1982a) have shown that in a series of PEO-alkali metal salts, the —CH_2—CH_2— group in PEO maintains a gauche configuration after salt formation. Several new bands appeared in the 1000- to 800-cm^{-1} region in the infrared spectrum. These were assigned to the ether oxygen atom vibrational modes affected by the alkali metal cations. A new band in the Raman spectrum in the ~865-cm^{-1} region was assigned to a metal–oxygen stretching mode, with the cation coordinated symmetrically to oxygen atoms. In the far-infrared region, Papke *et al.* (1981) observed bands that were assigned to ion–solvent cage vibrations. Teeters and Frech (1986), studying poly(propylene oxide)-based electrolytes, obtained a far-infrared band at ~240 cm^{-1}, which could be assigned as a bending or torsional mode of the polymer backbone.

The nature of the ion moiety in these polymer–salt complexes has been elucidated by molecular spectroscopy. The identification of ion-pairing processes, which lowers the electrical conductivity, has been made using Raman and infrared measurements. The use of the different selection rules, involved as one proceeds from ionic to ion pairing, allows their differentiation. In ionic, tetrahedral species, one has two infrared modes (ν_3 and ν_4), whereas when the symmetry is lowered, ν_1 and ν_2 may appear. Likewise, in ionic species with D_{3h} symmetry, ν_3, ν_2, and ν_4 are observed in the infrared and

ν_1 is only observed in the Raman effect. As the symmetry is lowered, ν_1 will begin to appear in the infrared region and the ν_3 and ν_4 vibrations split. The symmetry lowering may be ascribed to ion pairing occurring in the polymer–ion electrolytes. Table X shows some spectroscopic results obtained for several PEO polymer systems involving salts such as $LiBH_4$, $NaBH_4$, and $LiNO_3$. The results show that, for $LiBH_4$ and $LiNO_3$ polymer salts, extensive ion pairing occurs.

Since the cation is the mobile ion involved in the ionic conductivity in these materials, any ion pairing would lower the conductivity. Figure 9 illustrates the influence of ion pairing involving the $NaBH_4$ salt (Dupon *et al.*, 1982). For a one-dimensional ion transport in the helical tunnels, ion pairing can block the movement of the cation. For three-dimensional transport, the ion pairing does not block the ion transport, because pathways for ion transport remain.

Teeters and Frech (1986) found in the Raman spectra of PPO-NaSCN (7 : 1) electrolytes that the CN stretch (at ~2060 cm^{-1}) and the SC stretch (at ~750 cm^{-1}) were much higher than the values for the free ions of the salts in dimethylformamide, but close to the ion-pair values at 2066 and 754 cm^{-1}. The frequencies of these vibrations were found to be cation dependent for LiSCN-, NaSCN-, and KSCN-based electrolytes, a finding consistent with ion–ion interactions. Poly(ethylene oxide)-sodium polyiodide complexes were recently found to be electrical conductors of the semiconductor type (Hardy and Shriver, 1986). The methods of charge transport were found to be ionic and electronic (hole). Raman spectroscopy has indicated that polyiodide species occur in the salts. The nature of the iodine moiety was found to be dependent upon the salt concentration in the polymer, as well as the iodine to sodium ion ratio. For low PEO/NaI and low I_2/NaI concentrations, the I_3^- specie was present; this conclusion is based on a strong Raman band at 116 cm^{-1}. As these concentrations increase, new bands appear at ~170 cm^{-1}, an observation indicative of higher polyiodides greater than I_3^-; perhaps the I_4^{2-} specie (see Herbstein *et al.*, 1983). Figure 10

Table X Spectroscopic Results Obtained for Several Polymer Salts[a]

PEO-$LiBH_4$ (ion pairing) ($<T_d$)		$NaBH_4$ (ionic) Basic aqueous solutions (T_d)	PEO-$LiBF_4$ (ionic) (T_d)		$LiBF_4$ (ionic) (T_d)	PEO-$LiNO_3$ (ion pairing) ($<D_{3h}$)		Free NO_3^- (ionic) (D_{3h})
2347(vs)	[$\nu_2 + \nu_4$]		1065(m)	[ν_3]	1070(m)	1415(s)	[ν_4]	
								1380(s) [ν_3]
2295(m)	[ν_1]		775(m)	[ν_1]	777(m)	1363(m)		
2232(vs)	[ν_3]	2272(s)				1324(vs)	[ν_1]	
2178(s)						827(m)	[ν_2]	828(m) [ν_2]
2169(s)						729(vv)	[ν_3]	
	[$2\nu_4$]	2200(sh)						
						719(vvw)		716(m) [ν_4]
ir data			Raman data			ir data		
Extensive interaction, ion pairing			Slight interaction no ion pairing			Extensive interaction, ion pairing		

[a]Frequencies in cm^{-1}. See footnote to Table IX for meaning of abbreviations. (From Ferraro and Williams, 1987).

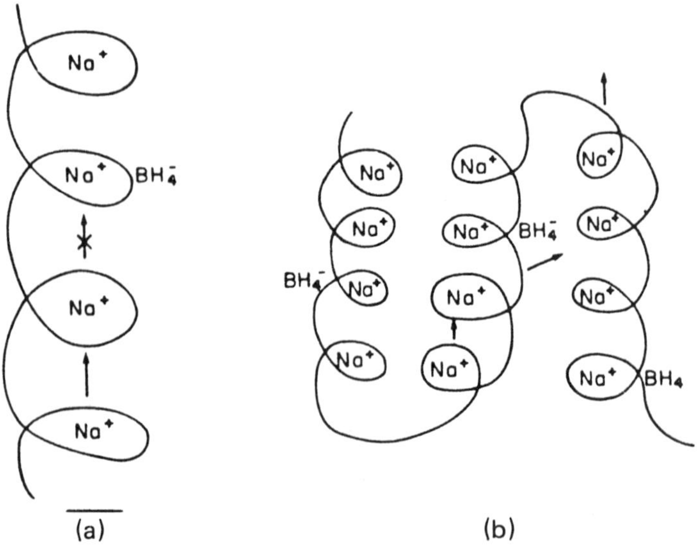

Fig. 9. Schematic illustration of the influence of ion pairing on conductivity. (a) One-dimensional ion transport in helical tunnels is blocked by ion pairing with BH_4^-; (b) three-dimensional ion transport is not blocked by ion pairs. (From Ferraro and Williams, 1987.)

shows the resonance Raman spectra for PEO-NaI_2 compounds; the spectra of the iodine moieties existing in the compounds are also shown. Spindler and Shriver (1986) have prepared poly(vinylpyrrolidone) complexes with lithium trifluoromethanesulfonate and on the basis of infrared results have concluded that strong interactions occur between the lithium cation and the carbonyl oxygen of PVP.

III. TRANSITION ELEMENT–MACROCYCLIC LIGAND COMPLEXES

In the mid-seventies interest in the metallomacrocyclic [M(mac)] molecules (M is a transition element and mac is a macrocyclic ligand) developed. When oxidized by I_2 or Br_2,

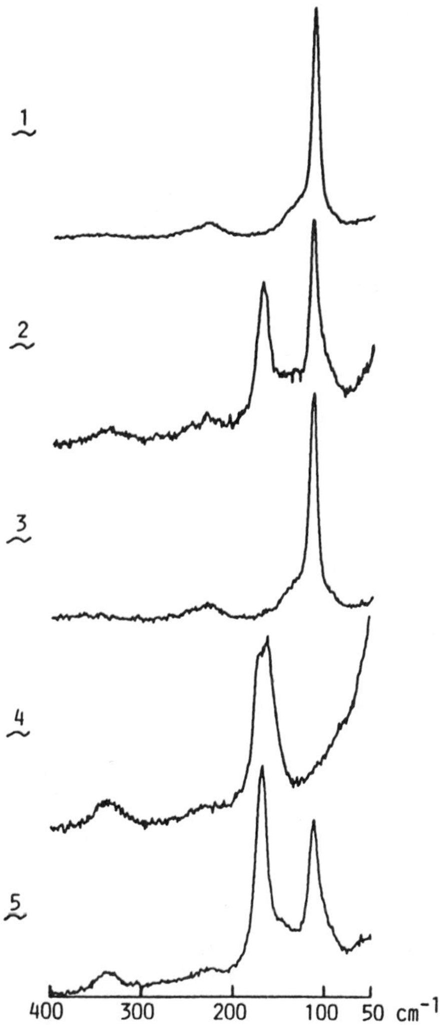

Fig. 10. Resonance Raman spectra of PEO-NaI_2 compounds (ν_0 = 488 nm). (1) $(PEO)_3NaI_{1.5}$, triiodide; (2) $(PEO)_3NaI_2$, triiodide plus polyiodide; (3) $(PEO)NaI_2$, triiodide; (4) $(PEO)_4NaI_3$, polyiodide; (5) $(PEO)_8NaI_3$, triodide plus polyiodide. (From Hardy and Shriver, 1986.)

the materials become simultaneously doped and electrically conductive. These doped materials produce lattices that contain one-dimensional arrays of planar donor molecules of M(mac) with fractionally occupied electronic-valence shells. In effect, a partial oxidation state of the metal cation-radical results, and chains of polyhalide ions, parallel to the M(mac) stacks, occur in the structure. Structures of several M(mac) materials are shown in Fig. 11. A variety of transition metals can be used to form these complexes, including Fe(II), Co(II), Ni(II), Cu(II), Zn(II), and Pt(II).

The complexes are structurally one-dimensional in nature and are subject to a Peierls distortion as one lowers the temperature; as a result, they become insulators. At room temperature they are, in general, semiconductors.

In order to determine the formal charge on the M(mac) cation following oxidation, various experimental methods have been used including Raman, Mössbauer (^{129}I), X-ray diffuse scattering, electron spin resonance, magnetic susceptibility, and charge transport measurements. Conspicuous by their absence are far-infrared data, which would complement the Raman data. This is in part due to the fact that the far-infrared data are more difficult to obtain, because only high-resolution interferometers can be used and these are considerably more expensive than bench-top interferometers. We will highlight the Raman data in this section to indicate the use of vibrational spectroscopy in determining the formal charge on the M(mac) cation, and to provide insight as to the nature of the conduction process.

The general oxidation occurring in these complexes may be considered to be

$$\mathrm{M(mac)} + {}^{1}\!/_{2}\mathrm{I}_2 \rightarrow \mathrm{M\ (mac)I} \tag{1}$$

No evidence for an I^- anion has been found; thus, the divalent metal has not been oxidized to a trivalent state. Experimental data, among these Raman spectroscopy, has indicated that the iodine- or bromine-containing moiety is present as a

Table XI Selection Rules for Linear Diatomic and Triatomic Molecules

Type	Symmetry	Vibrations expected
Linear diatomic		
X_2	$D_{\infty h}$	ν_1, sym. st.; Raman active (p)
Linear triatomic-sym.		
X_3	$D_{\infty h}$	ν_1, sym. st.; Raman active (p)
		ν_2, bend (degenerate); ir active
		ν_3, asym. st.; ir active

polarized, and of medium to strong intensity. The ν_3 vibration in the I_3^- portion of the polyiodide may become Raman active at ~145 cm^{-1}, albeit of weak intensity. The ν_2 vibration for I_3^- will appear in the infrared and cannot be seen in the Raman because of interference from the exciting line. Table XII tabulates data for molecules containing I_3^- and I_5^- moieties. Results for the oxidized M(mac) complexes are shown in Figs. 12–15.

Figure 12 shows Raman spectra of $Pd(dpg)_2I$ and Ni(dpg)I, which contain I_5^- moieties that have been substantiated by ^{129}I Mössbauer spectra (Marks and Kalina, 1982). Figure 13 shows the Raman spectra of $Pd(bqd)I_{0.5}$·0.52(*o*-dichlorobenzene) and $Ni(bqd)_2I_{0.52}$·0.32(toluene). The Raman data are typical for an I_3^- species (Marks and Kalina, 1982). Figure 14 illustrates the Raman spectra of $FePcI_{1.9}$, $CoPcI_{1.0}$, and $NiPcI_{1.0}$, which are typical for I_3^- species (Marks and Kalina, 1982). In general, for complexes $MPcI_x$, where $x < 3$, the chief polyiodide moiety is I_3^- (Scaringe *et al.*, 1980; Petersen *et al.*, 1977; Marks, 1976); but when $x > 3$, I_5^- species and other coordinated species have been detected. Figure 15 shows the resonance Raman spectra of $[Ni(4\text{-}Me)_4Pc]I_{1.24}$, $[Cu(3\text{-}Me)_4Pc]I_{1.56}$, and $[Cu(3\text{-}Me)_4Pc]I_{3.41}$, and results are consistent with an I_5^- specie (Stojakovic, 1977; D. W. Kalina, unpublished observations). For $[M(TAA)]I_x$ and $[(TMTTA)]I_x$, when $x = \leqslant 3$, the predominant species is I_3^-; and when $x \geqslant 3$, the

Table XII Summary of Metallomacrocyclic Complexes with Respect to Partial Oxidation State Formulation and the Polyhalide Moiety Present[a]

Complex moiety	Partial oxidation formulation	Polyhalide present
$Ni(dpg)_2I$	$[Ni(dpg)_2]^{+0.2}(I_5^-)_{0.2}$	I_5^-
$Pd(dpg)_2I$	$[Pd(dpg)_2]^{+0.2}(I_5^-)_{0.2}$	I_5^-
$Ni(dpg)_2Br$	$[Ni(dpg)_2]^{+0.2}(Br_5^-)_{0.2}$	Br_5^-
$Pd(dpg)_2Br_{1.1}$	$[Pd(dpg)_2]^{+0.22}(Br_5^-)_{0.22}$	Br_5^-
$Ni(bqd)_2I_{0.5}S$	$Ni(bqd)_2^{+0.17}(I_3^-)_{0.17}$	I_3^-
$Pd(bqd)_2I_{0.5}S$	$Pd(bqd)_2^{+0.17}(I_3^-)_{0.17}$	I_3^-
NiPcI	$[NiPc]^{+0.33}(I_3^-)_{0.33}$	I_3^-
$Ni[OMTBP]_{1.08}$	$[NiOMTBP]^{+0.36}(I_3^-)_{0.36}$	I_3^-
$[M(4,5\text{-}Me_2)_4Pc]I_x$	$[M(4,5\text{-}Me_2)_4Pc]^{+x/5}(I_5^-)_{x/5}$	I_5^-
$[MOE_2P_2]I_x$	$[MOEP]^{+x/5}(I_5^-)_{x/5}$	I_5^-
$Ni(TAA)I_{1.80}$	$[NiTAA]^{+0.6}(I_3^-)_{0.6}$	I_3^-
$Ni(TMTAA)I_{2.44}$	$[Ni(TMTAA)]^{+0.8}(I_3^-)_{0.8}$	I_3^-
$\{[Si(Pc)O]I_{0.50}\}_n$	$\{[Si(Pc)O]^{0.17}(I_3^-)_{0.17}\}_n$	I_3^-
$\{[Ge(Pc)O]I_{1.80}\}_n$	$\{[Ge(Pc)O]^{+0.6}(I_3^-)_{0.6}\}_n$	I_3^-
$\{[Sn(Pc)O]I_{1.2}\}_n$	$\{[Sn(Pc)O]^{+0.4}(I_3^-)_{0.4}\}_n$	I_3^-

[a]Me, Methyl; S, solvent; M, metal. Also, see footnote of abbreviations at the beginning of this chapter. (From Ferraro and Williams, 1987.)

species is I_5^- (McClure *et al.*, 1980; Lin *et al.*, 1980). In all cases, supplemental measurements were used to substantiate Raman results, e.g., X-ray diffraction, ^{129}I Mössbauer spectroscopy.

Table XII summarizes the results obtained primarily from Raman data (Teitelbaum *et al.*, 1979, 1980; Schoch *et al.*, 1979). Also included in the table are data from the iodine-doped polymer $[M(Pc)0]_n$. These materials have been synthesized from $M(Pc)Cl_2$ complexes where M = Si, Ge, Sn (Meyer and Wöhrle, 1974; Meyer *et al.*, 1975; Joyner and Kenney, 1960). These polymers are of practical interest because they can be processed into fibers by extruding acid solutions of the polymers, which are soluble in strong acids. In combination with the plastic Kevlar they form strong fibers (weight for

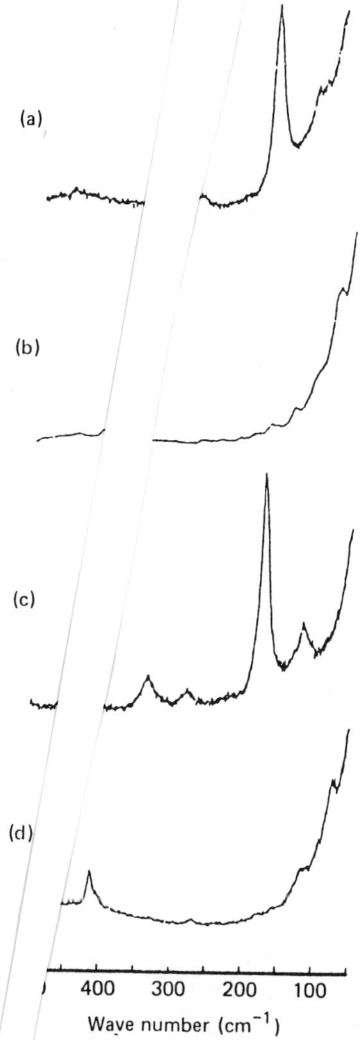

Fig. 12. Raman s[illegible]tra (5145 Å excitation, polycrystalline samples) of (a) Pd(dpg)[illegible] (b) $Pd(dpg)_2$; (c) $Ni(dpg)_2I$; and (d) $Ni(dpg)_2$. (From Marks [illegible] Kalina, 1982.)

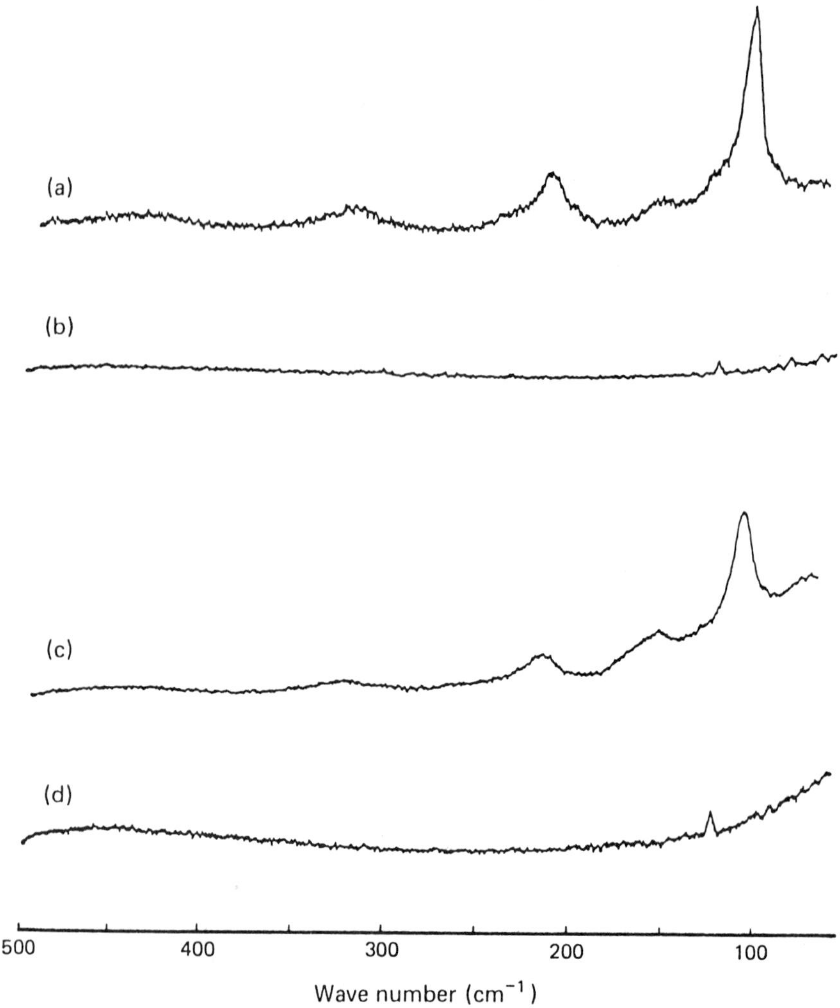

Fig. 13. Resonance Raman spectra (5145 Å excitation, polycrystalline samples) of (a) $Pd(bqd)_2I_{0.5}$·0.52(*o*-dichlorobenzene); (b) $Pd(bqd)_2$; (c) $Ni(bqd)_2I_{0.52}$·0.32(toluene); and (d) $Ni(bqd)_2$. Weak transitions in (b) and (d) at 117 and 77 cm^{-1} result from laser plasma emission. (From Marks and Kalina, 1982.)

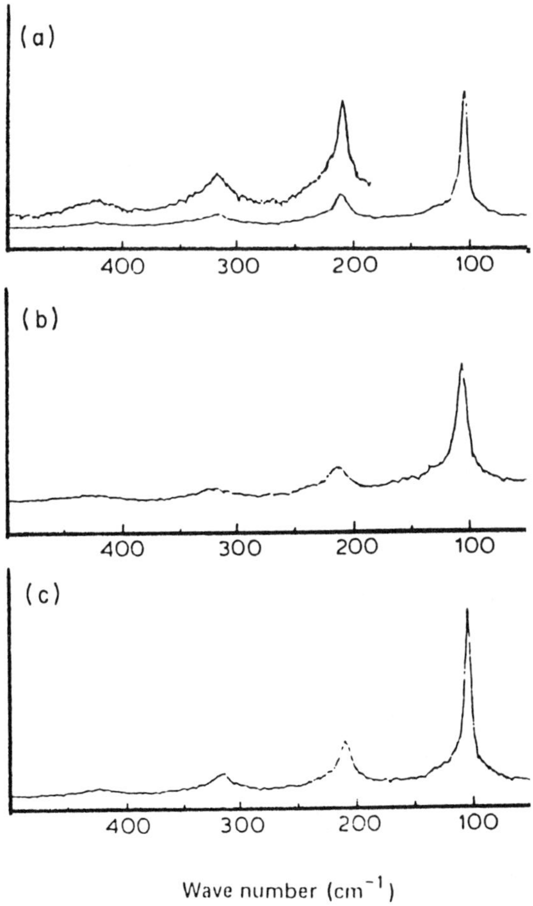

Fig. 14. Resonance Raman spectra of partially oxidized phthalocyanine complexes with 5145 Å excitation; (a) $FePcI_{1.9}$; (b) $CoPcI_{1.0}$; and (c) $NiPcI_{1.0}$. (From Marks and Kalina, 1982.)

weight six times the modulus of steel) (Inabe *et al.*, 1983a,b, 1984).

For reviews on the subject, see Ferraro (1982), Ferraro and Williams (1987), Marks (1978, 1980, 1985), Marks and Kalina (1982), Allcock (1985), and references therein.

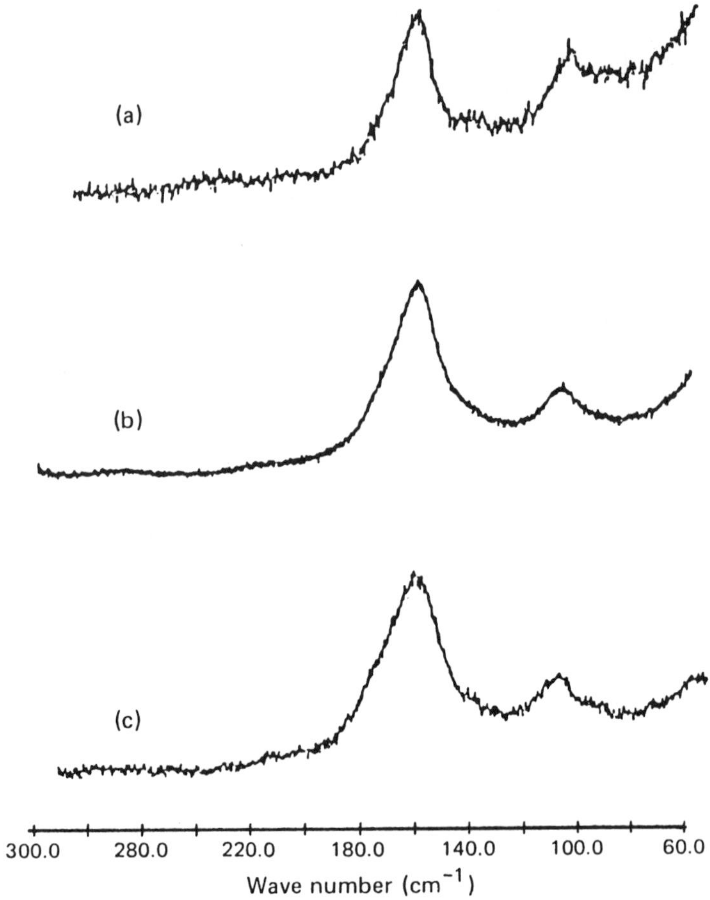

Fig. 15. Resonance Raman spectra of polycrystalline samples of (a) $[Ni(4\text{-}Me)_4Pc]I_{1.24}$; (b) $[Cu(3\text{-}Me)_4Pc]I_{1.56}$; and (c) $[Cu(3\text{-}Me)_4Pc]I_{3.41}$, ν_0 = 5145 Å. (From Marks and Kalina, 1982.)

IV. ORGANIC POLYMERS: POLYACETYLENE

Hatano *et al.* (1961) found that polyacetylene conducts electrical current. Shirakawa *et al.* (1973) and Ito *et al.* (1974) were the first to observe that polyacetylene [$(CH)_x$] could be prepared as films having metallic luster. Considerable interest in conducting organic polymers therefore followed in the mid-seventies. In 1977 it was found that the conductivity of polyacetylene could be increased by 13 orders of magnitude by doping with various donors or acceptors to give p-type or n-type semiconductors and conductors (Shirakawa *et al.*, 1977, 1978; Shirakawa and Kobayashi, 1983; Chiang *et al.*, 1977, 1978a–c; Park *et al., 1979;* Fincher *et al.*, 1978; for a summary of the various dopants used with polyacetylene, see Ferraro and Williams, 1987). Conductivities of 10^3 (ohm cm)$^{-1}$ were found in doped $(CH)_x$ (e.g., $[(CH)(AsF)_{0.1}]_x$). Recently, Naarmann (1987a,b) has synthesized $(CH)_x$ that demonstrates conductivities on a weight-to-weight basis of two times that of Cu metal. Furthermore, the polymer is stable in air for weeks.

Figure 16 shows the conductivity ratio and stability of the doped $(CH)_x$ produced by the Naarmann method. The $(CH)_x$ prepared by this method is more crystalline and has fewer amorphous areas, and smaller amounts of cross-linking between chains, than $(CH)_x$ prepared by previous methods. To date several derivatives of polyacetylene have been synthesized, but none have approached the high conductivity of $(CH)_x$. These polymers remain semiconductors, which does not detract from their interest in device applications.

A. Vibrational Studies of Pristine $(CH)_x$

Vibrational spectroscopy has been extensively utilized in the study of electrically conducting polymers. For example, for polyacetylene, the following areas have been examined:

1. Assignments of infrared modes for the cis and trans forms (Shirakawa and Ikeda, 1971; Inagaki *et al.*, 1975; Shirakawa *et al.*, 1973; Blanchet *et al.*, 1983; Riseman *et al.*, 1981; Rabolt *et al.*, 1979, 1982;

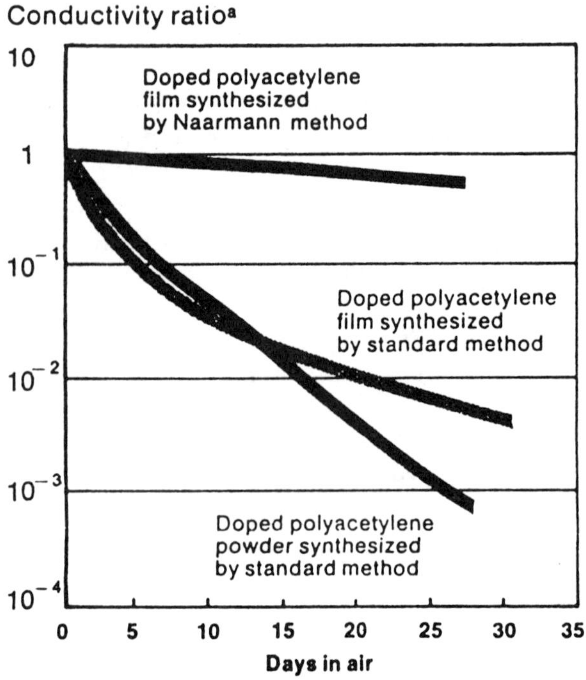

a Ratio of conductivity after air exposure to initial conductivity.

Fig. 16. Conductivity ratio and stability of the doped $(CH)_x$ by the Naarmann method. (From Naarmann, 1987a.)

Montaner *et al.,* 1981; Bechtold *et al.,* 1985; Horovitz, 1981; Galtier and Montaner, 1985).

2. Assignments of Raman modes for the cis and trans forms (Shirakawa *et al.,* 1973; Inagaki *et al.,* 1975; Harada *et al.,* 1978, 1980; Kuzmany 1980, 1981; Lichtmann and Fitcher, 1979, 1981; Schügerl and Kuzmany, 1981; Lichtmann, 1981; Lichtmann *et al.,* 1980, 1981; Lefrant *et al.,* 1979, 1982; Hsu *et al.,* 1978; Mele and Rice, 1980; Lefrant *et al.,* 1985; Mulazzi *et al.,* 1985a,b; Vardeny *et al.,* 1985; Ehrenfreund *et al.,* 1985; Kuzmany and Knoll, 1985; Fitchen, 1985).

3. Attempts to identify the existence of solitons in polyacetylene (Rabolt *et al.*, 1982; Mele and Rice, 1980; Horovitz, 1981; Etemad *et al.*, 1981; Zannoni and Zerbi, 1982a; Chien, 1984).
4. Identification of lattice modes of the cis and trans forms of $(CH)_x$ in the infrared region (Bechtold *et al.*, 1985).
5. Determination of the differences occurring in the vibrational spectrum of $(CH)_x$ upon doping (Fincher *et al.*, 1979; Tanaka *et al.*, 1980; McQuillan *et al.*, 1982; Mele and Rice, 1980; Zannoni and Zerbi, 1982a; Lefrant *et al.*, 1979, 1985; Kuzmany, 1980; Hsu *et al.*, 1978; Blanchet *et al.*, 1983; Riseman *et al.*, 1981, Kanicki *et al.*, 1981; Horovitz, 1981).
6. Determination of the dopant species moiety in doped polyacetylene [for a summary, see Ferraro and Williams, (1987)].
7. Cis–trans isomerization of polyacetylene followed by infrared spectroscopy (Montaner *et al.*, 1981; Mulazzi *et al.*, 1985a,b; Harada *et al.*, 1985, Gibson *et al.*, 1983) and by Raman spectroscopy (Lefrant *et al.*, 1979; Faulques *et al.*, 1984).
8. Zerbi and co-workers have utilized vibrational spectroscopy in various studies of polyacetylene. They have obtained the density of states of chemically and structurally disordered *trans*-$(CH)_x$ (Zannoni and Zerbi, 1982a, 1983d); the lattice dynamics of *cis*- and *trans*-$(CH)_x$ (Zerbi *et al.*, 1985; Zerbi, 1987; Zerbi and Zannoni, 1983; Zannoni and Zerbi, 1982a, 1983e, 1984); made normal coordinate calculations on a simple model of a cross-link in *trans*-polyacetylene (Zannoni and Zerbi, 1983a); used vibrational spectroscopy to obtain information on a possible Peierls distortion occurring in pristine $(CH)_x$ (Zannoni and Zerbi, 1983b); obtained charge distributions in $(CH)_x$ from infrared data (Castiglioni *et al.*, 1985; Gussoni *et al.*,

1985). Some of these studies have also been made for the doped polyacetylenes (Zannoni and Zerbi, 1982a,b, 1983b,c,f; Zerbi and Dellepiane, 1987; Zerbi and Zannoni, 1983; Eckhardt *et al.*, 1984).

Shirakawa and Ikeda (1971) have analyzed the infrared spectra of $(CH)_x$ and $(CD)_x$ in the cis and trans conformations and made normal mode assignments. The cis form of polyacetylene corresponds to a C_{2h} factor group (z = 2), and the vibrational modes for this symmetry are

$$\Gamma_{\text{cis}} = 4A_g + B_{1g} + 4B_{2g} + 3B_{1u} + B_{2u} + 3B_{3u}$$

For trans $(CH)_x$ corresponding to a D_{2h} symmetry (z = 1), the vibrational modes are

$$\Gamma_{\text{trans}} = 4A_g + B_g + A_u + 2B_u$$

Assignments made for *cis*-$(CH)_x$ and *trans*-$(CH)_x$ are tabulated in Table XIII. It was not possible to determine whether the cis

Table XIII Comparison of Infrared Data for *cis*- and *trans*-Polyacetylene (cm^{-1})

Observed frequencies for *cis* form (cm^{-1})	Assignments	Observed frequencies for *trans* form (cm^{-1})
3047	CH stretch	
3044	CH stretch	
	CH stretch	3013
1329	CH deformation in-plane	
	CH deformation in-plane	1292
1249	CH deformation in-plane	
1118	C—C stretch	
	CH deformation in-plane	1015
980		
940		
740	CH deformation out-of-plane	
446	CCC deformation	

configuration of polyacetylene had the cisoid or cis-transoid structure from the infrared data.

Raman spectra of $(CH)_x$ have been reported (see references in item 2 in Section IV,A). Two strong Raman bands are observed for the trans polymer, while the cis isomer shows three strong bands. The three bands in the cis isomer have been assigned to two C—C single bond stretches (A_g and B_{2g} at 908 and 1247 cm^{-1}, respectively, and one C═C stretch (A_g at 1541 cm^{-1}) (Litchmann *et al.*, 1980). For the cis-transoid configuration, one C═C stretch (A_g) and two C—C stretching modes (A_g, B_{2g}) should be observed. For the trans-cisoid configuration, two C═C stretching modes (A_g and B_{2g}) and one C—C stretching mode (A_g) should be observed. Thus, *cis*-$(CH)_x$ conforms to a cis-transoid configuration, and the conformation determination is possible from Raman data.

cis-transoid

trans-cisoid

B. Vibrational Studies of Doped Polyacetylene

When *cis*- or *trans*-polyacetylene are doped, new infrared bands appear in the spectra (Fincher *et al.*, 1979). In AsF_5-doped $(CH)_x$, a narrow and strong band appears at ~1370 cm^{-1}, and a broad band at ~900 cm^{-1} is also observed. Iodine-doped *cis*-polyacetylene shows the same bands, although a weak band is also observed at ~1280 cm^{-1}. In *n*-doped *cis*-polyacetylene (e.g., Na-doped), bands appear at ~1370 and ~900 cm^{-1}, as was found in AsF_5-doped $(CH)_x$. Since *p*- and *n*-doped polymers produce the same absorptions, the conclusion reached was that these bands are associated with charged solitons and not with neutral solitons. Similar absorptions were observed with BF_3-doped and HF-doped polyacetylene (see Tanaka *et al.*, 1980; McQuillan *et al.*, 1982). Figure 17 shows infrared spectra of HF-doped polyacetylene. The origin of these observed absorptions remains unknown, and several postulates have been advanced, including the assignment to charged solitons, which was made by Mele and Rice (1980) and by Etemad *et al.* (1981). Contrary conclusions were suggested by Zannoni and Zerbi (1982a), who proposed that the absorptions were due to a centrosymmetrical defect sandwiched between two dopant molecules. Rabolt *et al.* (1979) proposed that vibronic enhancement of the totally symmetric Raman-active A_g modes, which become activated in the infrared, caused the absorptions in doped polyacetylene. The activation of these infrared vibrations are due to an interaction between the intramolecular charge oscillations parallel to the chain axis and the skeleton stretching vibrations. The band at $\sim$1385 cm^{-1} was assigned to the Raman-active C—C and C=C stretching modes. The band at $\sim$900 cm^{-1} was attributed to a vibronically activated A_g mode, which is shifted to lower frequency by doping and has become infrared active (assigned as a CH out-of-plane bending mode). Etemad *et al.* (1981) has suggested an alternate explanation, i.e., that such frequency shifts should be dopant level dependent, and this is contrary to experimental observations. Results show that the infrared intensities are proportional to dopant concentrations, whereas frequency shifts are concentration independent. Rabolt *et al.* (1982), in experiments with AsF_5-doped $(CH)_x$, showed that the bands in infrared do shift with dopant concentration. Horovitz (1982) interpreted the infrared bands in doped-$(CH)_x$ as arising from Raman bands due to the introduction of charges in the polyacetylene upon doping. The controversy over these suggestions exists today.

Raman studies of doped $(CH)_x$ indicate a significant reduction for the resonance-enhanced C—C and C=C stretches, whereas the weak C—C stretch at $\sim$1291 cm^{-1} was not affected by doping (Kuzmany, 1980; Lefrant *et al.*, 1979).

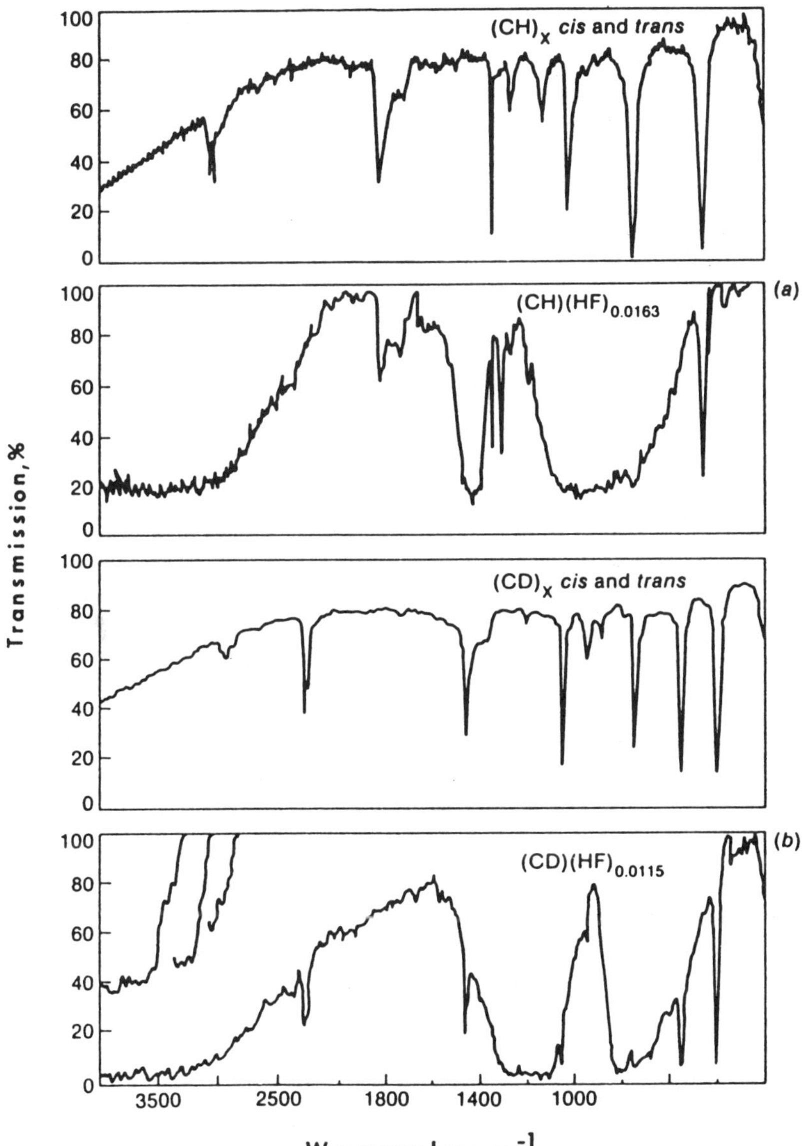

Fig. 17. Infrared spectra: (a) $[CH(HF)_{0.0163}]_x$; (b) $[CD(HF)_{0.0115}]_x$. [From McQuillan *et al.*, 1982. Reprinted with permission from the *Journal of Electronic Materials* **11**, 412 (1982), a publication of The Metallurgical Society, Warrendale, Pennsylvania.]

Bands characteristic of trans-$(CH)_x$ appeared during the doping process, a finding indicative of a cis→trans isomerization (Lefrant *et al.*, 1979; Faulques *et al.*, 1984).

Perhaps the greatest utilization of infrared spectroscopy has occurred in the far-infrared studies, where the identity of the doped species can be determined. Other instrumental techniques have been also used, such as Raman spectroscopy, Mössbauer, X-ray photoelectron spectroscopy (XPS) (Salaneck *et al.*, 1980), and mass spectroscopy (Allen *et al.*, 1979). In polyacetylene doped with iodine, resonance Raman and Mössbauer spectroscopy studies show that the iodine moiety contains I_3^- and I_5^- anions. Raman bands at 105–107 cm^{-1} were indicative of I_3^- and assigned to the symmetrical stretching of the I_3^- ion. Bands at 150–160 cm^{-1} were assigned to I_5^- (Hsu *et al.*, 1978; Lefrant *et al.*, 1979; Harada *et al.*, 1980). Hsu *et al.* (1978) determined the Raman spectrum of trans-rich polyacetylene and observed I_5^- species. Faulques and Lefrant (1983) showed that pumping on the iodine-doped polyacetylene causes the I_5^- band to disappear. Results with bromine-doped polyacetylene showed the presence of the Br_3^- and Br_5^- species (Faulques and Lefrant, 1983). Details of assignments for I_5^- and I_3^- anions are discussed in the section of M(mac) complexes. Pron *et al.* (1981, 1983) and Chen *et al.* (1987) have observed $FeCl_4^-$ in $FeCl_3$-doped polyacetylene.

The far-infrared data has also been examined for an absorption that could be assigned to a pinned translation mode of the charged soliton in $(CH)_x$. However, Rabolt *et al.* (1982) found no evidence for the predicted charge soliton pinning mode at any doping level in the region of 300–500 cm^{-1}.

C. Variants of Polyacetylene

A number of variants of polyacetylene have been synthesized and some of these are shown in Fig. 18. For these materials, vibrational spectroscopy has played a similar investigative

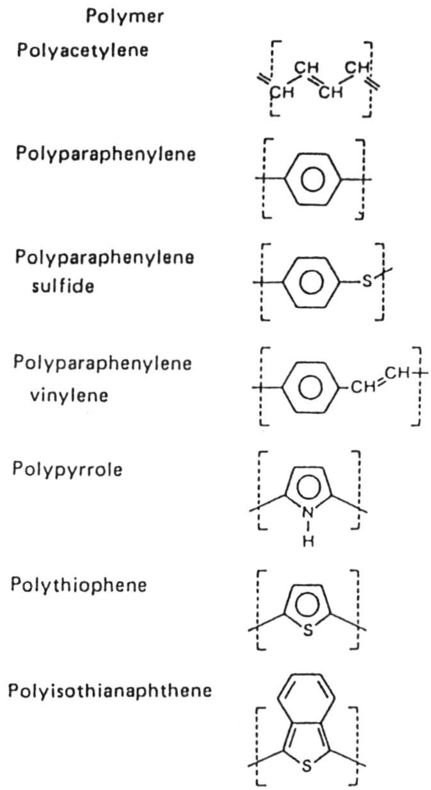

Polyacetylene Variants

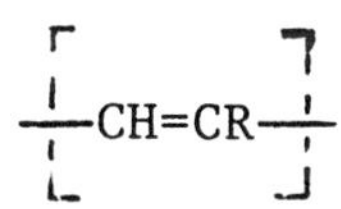

R = phenyl

R = $CH-CH_2-OH$

R = $CH_2NHCH_2\emptyset$

Fig. 18. Structures of various conducting polymers. (Taken in part from Ferraro and Williams, 1987.)

role as has been illustrated for $(CH)_x$. For example, in the polymerization reaction for some of these variants, different types of polymers are possible, depending on the type of catalyst used. Identification of these polymers has been made using infrared and ^{1}H-NMR spectra. Three types of PPA have been identified (Furlani *et al.*, 1986). Different types of polymers for PBPA have also been identified by infrared spectroscopy (Furlani *et al.*, 1987). Figure 19 illustrates infrared spectra for several PBPA polymer preparations made with various catalysts. Other uses of infrared and Raman spectroscopy have been made in the identification of the dopant counterions in iodine-doped polymers. Ferraro *et al.* (1984) identified the nature of the iodine moiety in iodine-doped PPA. The nature of the iodine moiety in POHP and PBPA has been identified by Ferraro *et al.* (1987a) and summarized in Table XIV. Similar techniques have been used by Oddi *et al.* (1985) in identifying the nature of the dopant counterions BF_4^-, SO_4^-, ClO_4^- in polypyrrole.

Lattice dynamical calculations for doped and undoped PPP have been performed and compared with results obtained from vibrational spectra (Zannoni and Zerbi, 1985). The vibrational spectra are consistent with a molecular structure with π electrons becoming delocalized with each ring, similar to findings for biphenyl. Vibrational assignments for PPP in its doped state were made for oligomers (bi to hexa), and the intensities of their infrared bands compared. The observed infrared frequencies were found to be the same in going from oligomers to the polymers, an observation supporting the existence of a common force field with independent $k = 0$ phonon dispersion curves. The infrared intensity ratios demonstrated no substantial change with increasing chain length, atomic charges and charge fluxes involved in the infrared-active motions (C. Castiglioni, M. Gussoni, J. T. Lopez Navarette, and G. Zerbi, unpublished observations). Doping of PPP, as determined from these vibrational data, was found to be consistent with the formation of conjugational defects of the bipolaronic type. The study of infrared intensities and

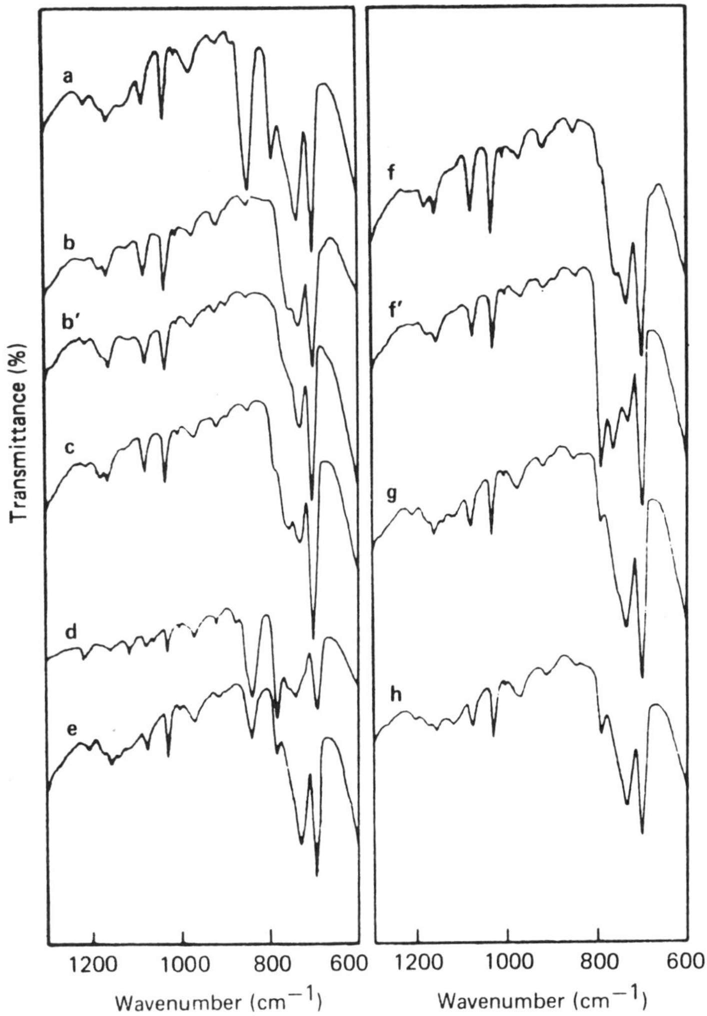

Fig. 19. Infrared spectra (Nujol mulls) of PBPA obtained with various catalysts: (a) $[Rh(cod)(bipy)]PF_6$; (b) $[Rh(cod)(bipy)]ClO_4$ (PBPA insoluble in CCl_4); (b′) $[Rh(cod)(bipy)]ClO_4$; (c) [Rh(cod) (bipy)] $[B(Ph)_4$; (d) $[Rh(nbd)(bipy)]PF_6$; (e) $[Rh(nbd)(dipyam)]PF_6$; (f) $[Rh(cod)(bpz)]PF_6$ (PBPA insoluble in CCl_4); (f′) [Rh(cod)(bpz)] PF_6; (g) $[Rh[cod)(dipyam)]PF_6$; (h) $[Pt(C{\equiv}CCH_2NHCN_2Ph)_2(PPh_3)_2$. [From Furlani *et al.* (1987), by permission of the publishers, Butterworth & Co. (Publishers) Ltd.

Table XIV Summary of the Dopant Moiety in Various Doped Organic Polymers

Compound	Dopant	Dopant moiety	Instrumentation used[a]	Reference
a 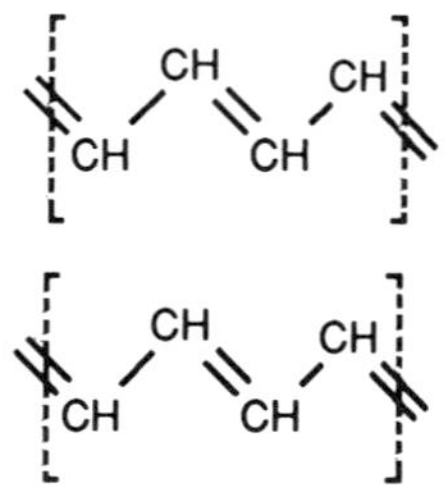	I_2 I_2	I_3^- I_5^-	Raman, Mössbauer Raman, Mössbauer	Hsu *et al.* (1978) Hsu *et al.* (1978) Lefrant *et al.* (1979) Harada *et al.* (1980)
b 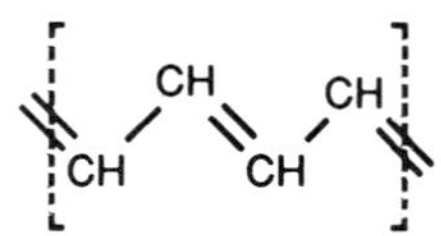	Br_2 $FeCl_3$	Br_3^-, Br_5^- $FeCl_4^-$	Raman, Mössbauer ir, Raman, EPR, Mössbauer	Faulques and Lefrant (1983) Pron *et al.* (1981, 1983) Chen *et al.* (1987)
c 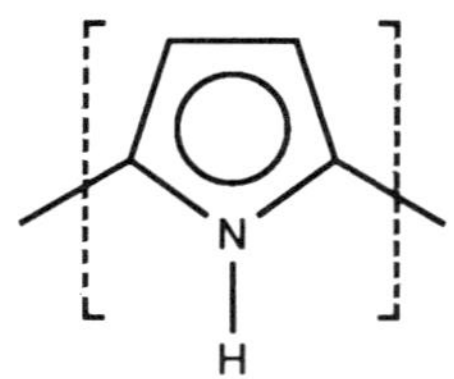 Polypyrrole	$(NBu_4)BF_4$ H_2SO_4 $HClO_4$	BF_4^- SO_4^{2-} ClO_4^-	UV, visible, ir	Oddi *et al.* (1985)

d 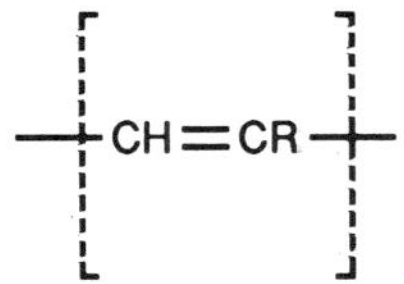 R = phenyl	I_2 (all doping levels)	I_5^-	Raman, FIR (FT-IR)	Ferraro *et al.* (1984)
e 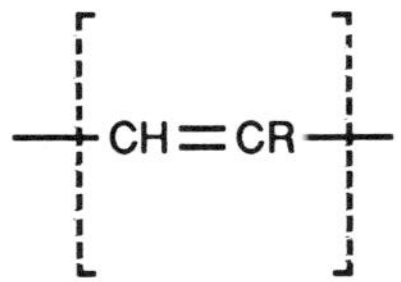 R = CH–CH_2–OH	I_2 (low level)	I_3^-	Raman, FIR (FT-IR)	Ferraro *et al.* (1987a)
	I_2 (high level)	I_5^-	Raman, FIR (FT-IR)	Ferraro *et al.* (1987a)
f 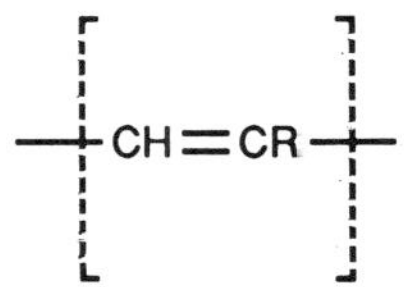 R = $CH_2NHCH_2Ø$	I_2	I_5^-	Raman, FIR (FT-IR)	Ferraro *et al.* (1987a)

[a]EPR, electron paramagnetic resonance spectroscopy; FIR, for-infrared; FT-IR, Fourier transform infrared method; ir, infrared; UV, ultraviolet.

frequencies were also applied to polyphenylenevinylene (Zerbi *et al.*, 1987).

For reviews on the subject of polymer electrical conductors, see Chien (1984), Ferraro and Williams (1987), and Pecile *et al.* (1985).

V. FT-IR MICROSPECTROSCOPY OF SYNTHETIC ELECTRICAL CONDUCTORS

Synthetic electrical conductors present a formidable challenge to proper characterization by use of infrared spectroscopy. These materials being metallic are opaque, and thus transmission methods are not possible. Furthermore, for some materials, the synthetic yield is extremely small, and only microcrystalline-sized samples are available. Most infrared sampling techniques are impossible because of these constraints. However, FT-IR microsampling techniques have been used [see Inabe *et al.* (1985) and Almeida *et al.* (1987), who applied the technique for M(mac) conductors, and Ferraro *et al.* (1987b), who applied the technique to organic charge-transfer superconductors]. The technique involves interfacing a microscope with a FT-IR spectrometer. The Marks group (see Inabe *et al.*, 1985; Almeida *et al.*, 1987) obtained single-crystal polarized reflectivity measurements of H_2(Pc)I in the 6000- to 1000-cm^{-1} range, operating in the reflectance mode. Crystal sizes were $1.0 \times 0.07 \times 0.07$ mm. Some results for H_2(Pc)I single crystals are illustrated in Fig. 20. Both $R_{\parallel}$ and $R_{\perp}$, polarized reflectance spectra are obtained. These results helped to elucidate information concerning the electronic structure of H_2(Pc)I. These results indicated that a metal ion is not required for high electrical conductivity in M(mac) conductors.

Ferraro *et al.* (1987b) applied similar microscopy techniques to the charge-transfer superconductors such as

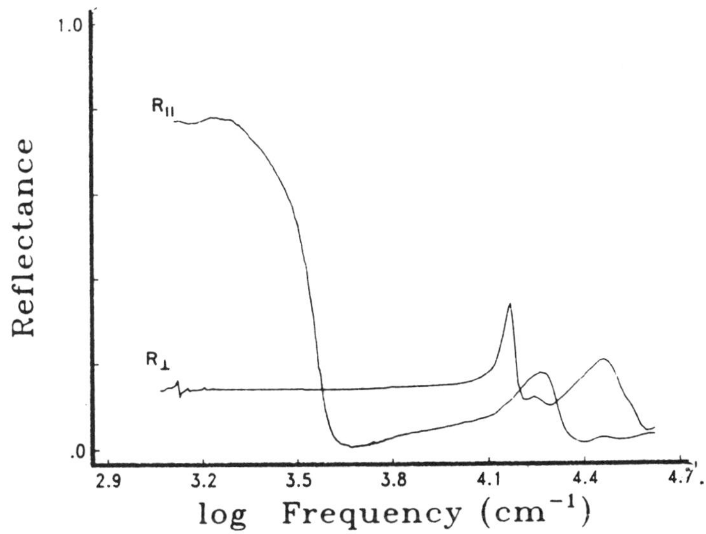

Fig. 20. Reflectance spectrum of an $H_2(Pc)I$ single crystal, and $R_{\|}$ and $R_{\perp}$ indicate reflectance polarized parallel to and perpendicular to the macrocycle stacking direction, respectively. [Reprinted with permission of Pergamon Press, Inc. and authors, *Solid Sate Commun.* **54**, 501 (1985), T. Inabe *et al.*]

β-$(ET)_2X$, where X = I_3^-, AuI_2^-, and IBr_2^-, and conductors where X = $AuBr_2^-$, $AuCl_2^-$, I_2Br^-, ICl_2^-, ReO_4^-, ClO_4^-, and CH_3—⬡—SO_3^-, and $(PT)_2X$ where X = I_3^-, IBr_2^-, I_2Br^-, and ICl_2^-. The technique was found useful for room temperature characterization of differences between the various crystal structures that produce superconductors (β structure) and those that produce semiconductors (α structure). Figure 21 shows the reflectance spectrum for α-$(ET)_2IBr_2$ and Fig. 22 shows the reflectance spectrum of β-$(ET)_2AuI_2$. The β-structure spectrum demonstrates vibrational features, at ~1200–1400 cm^{-1}, that are missing in the spectrum of the α structure.

The two examples presented in this section demonstrate how useful FT-IR microsampling techniques can be in characterizing synthetic electrical conductors.

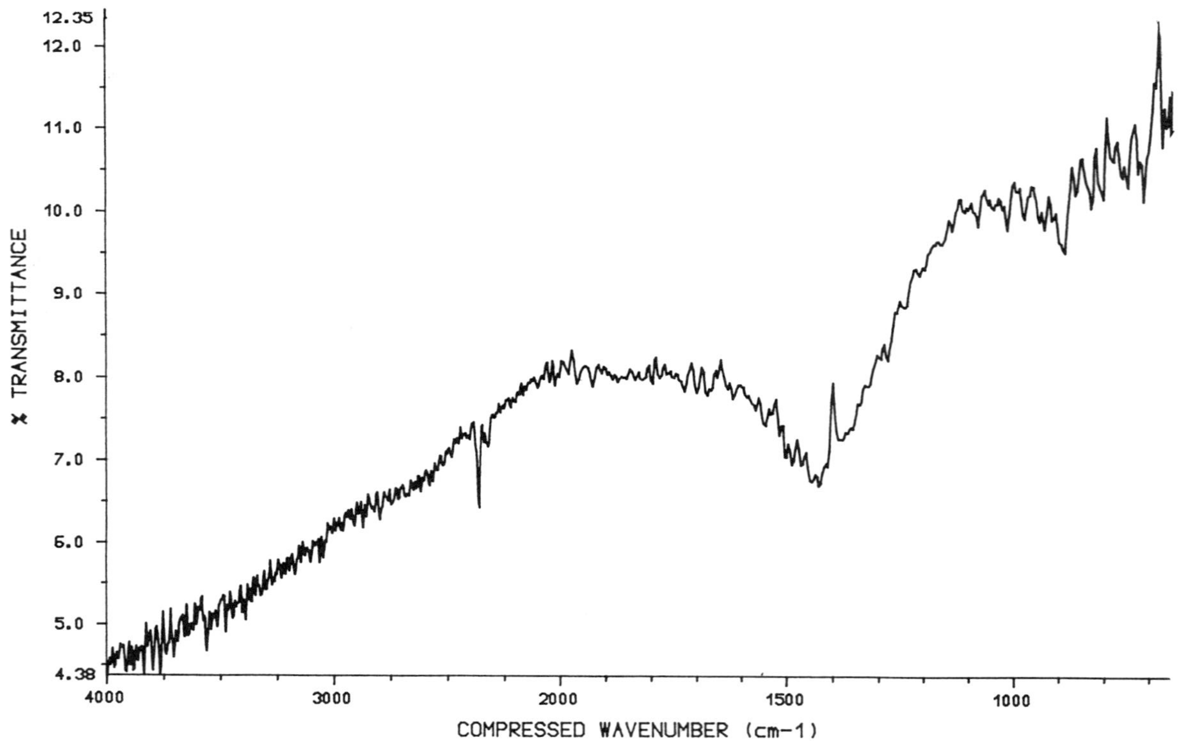

Fig. 21. Reflectance spectrum (298 K) for α-$(ET)_2IBr_2$. (From Ferraro *et al.*, 1987b.)

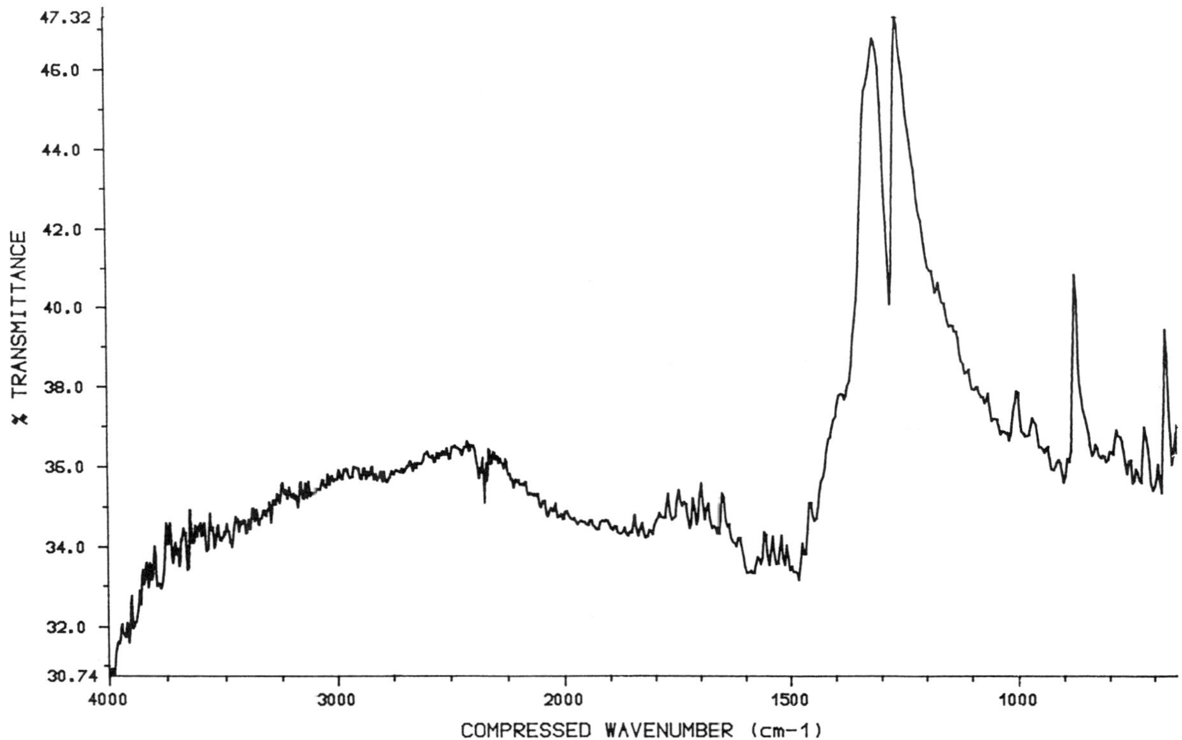

Fig. 22. Reflectance spectrum (298 K) for β-$(ET)_2AuI_2$. (From Ferraro *et al.*, 1987b.)

ACKNOWLEDGMENTS

This work was performed under the auspices of the Office of Basic Energy Sciences, Division of Materials Sciences, of the U.S. Department of Energy under the contract W-31-109-ENG-38 (Argonne National Laboratory). The authors would like to thank Mrs. Virginia Bowman and Mrs. Janet Bormet for their excellent editorial assistance with this manuscript.

REFERENCES

Allcock, H. R. (1985). *Chem. Eng. News* **63** (11), 22.

Allen, W. N., Decorpo, J. J., Saalfeld, F. E., Wyatt, J. R., and Weber, D. C. (1979). *Synth. Met.* **1,** 371.

Almeida, M., Kanatzidis, M. G., Tonge, L. M., Marks, J. T., Marcy, H. O., McCarthy, W. J., and Kannewurf, C. R. (1987). *Solid State Commun.* **63,** 457.

Armand, M. B., Chabagno, J. M., and Duclot, M. J. (1978). *Ext. Abstr.—Int. Conf. Solid Electrolytes, 2nd* Abstr. 6.5.

Armand, M. B., Chabagno, J. M., and Duclot, M. J. (1979). *In* "Fast Ion Transport in Solids" (P. Vashista, J. N. Mundy, and G. K. Shenoy, eds.), pp. 131–136. North-Holland, Amsterdam.

Armand, M. B., Duclot, M. J., and Rigand, P. (1980). *Int. Meet. Solid Electrolytes Solid State Galvanic Cells, 3rd Natl. Meet.* C116.

Bechtold, G., Genzel, L., and Roth, S. (1985). *Solid State Commun.* **53,** 1.

Blanchet, G. B., Fincher, C. R., Chung, J. C., and Heeger, A. J. (1983). *Phys. Rev. Lett.* **50,** 1938.

Bozio, R., and Pecile, C. (1980). *In* "The Physics and Chemistry of Low-Dimensional Solids" (L. Alcacer, ed.), pp. 165–186. Reidel, Dordrecht, The Netherlands.

Bozio, R., Zanon, I., Girlando, A., and Pecile, C. (1973). *J. Chem. Soc., Faraday Trans. 2* **69,** 239.

Bozio, R., Girlando, A., and Pecile, C. (1975). *J. Chem. Soc., Faraday Trans. 2* **71,** 1237.

Bozio, R., Girlando, A., and Pecile, C. (1977). *Chem. Phys. Lett.* **52,** 503 (1977).

Bozio, R., Zanon, I., Girlando, A., and Pecile, C. (1979). *J. Chem. Phys.* **71,** 2282.

Bozio, R., Meneghetti, M., and Pecile, C. (1982). *J. Chem. Phys.* **76,** 5785.

Bozio, R., Pecile, C., Scott, J. C., and Engler, E. M. (1985). *Mol. Cryst. Liq. Cryst.* **119,** 211.

Castiglioni, C., Gussoni, M., Miragoli, M., and Zerbi, J. (1985). *Mol. Cryst. Liq. Cryst.* **117,** 295.

Challener, W. A., and Richards, P. L. (1984). *Solid State Commun.* **51,** 765.

Chappell, J. S., Bloch, A. N., Bryden, W. A., Maxfield, M., Poehler, T. O., and Cowan, D. O. (1981). *J. Am. Chem. Soc.* **103,** 2442.

Chen, Z., Shen, Z., Chen, C., Zhao, Y., and Yang, M. (1987). *Makromol. Chem.* **188,** 345.

Chiang, C. K., Fincher, C. R., Park, Y. W., Heeger, A. J., Shirakawa, H., Louis, E. J., Gau, S. C., and MacDiarmid, A. G. (1977). *Phys. Rev. Lett.* **39,** 1098.

Chiang, C. K., Gau, S. C., Fincher, C. R., Park, Y. W., MacDiarmid, A. G., and Heeger, A. J. (1978a). *Appl. Phys. Lett.* **33,** 18.

Chiang, C. K., Park, Y. W., Heeger, A. J., Shirakawa, H., Louis, E. J., and MacDiarmid, A. G. (1978b). *J. Chem. Phys.* **69,** 5098.

Chiang, C. K., Druy, M. A., Gau, S. C., Heeger, A. J., Louis, E. J., MacDiarmid, A. G., Park, Y. W., and Shirakawa, H. (1978c). *J. Am. Chem. Soc.* **100,** 1013.

Chien, J. C. W. (1984). "Polacetylene—Chemistry and Physics and Material Science," pp. 1–634. Academic Press, New York.

Davidson, W. H. T. (1955). *J. Chem. Soc.* p. 3270.

Dupon, R., Papke, B. L., Ratner, M. A., Whitmore, D. W., and Shriver, D. F. (1982). *J. Am. Chem. Soc.* **104,** 6247.

Eckhardt, H., Shacklette, L. W., Szobota, J. S., and Baughman, R. H. (1984). *Mol. Cryst. Liq. Cryst.* **117,** 401.

Ehrenfreund, E., Vardeny, Z., Brafman, O., and Horovitz, B. (1985). *Mol. Cryst. Liq. Cryst.* **117,** 367.

Eldridge, J. E., and Bates, G. S. (1986). *Phys. Rev. B: Condens. Matter* **34,** 6992.

Etemad, S., Pron, A., Heeger, A. J., MacDiarmid, A. G., Mele, E. J., and Rice, M. J. (1981). *Phys. Rev. B: Condens. Matter* **23,** 5137.

Faulques, E., and Lefrant, S. (1983). *J. Phys., Colloq. (Orsay, Fr.), Suppl.* **44** (6), C3–337.

Faulques, E., Lefrant, S., Rachdi, F., and Bernier, P. (1984). *Synth. Met.* **9,** 53.

Fenton, D. E., Parker, S. M., and Wright, P. V. (1973). *Polymer* **14,** 589.

Ferraro, J. R. (1982). *Coord. Chem.* **43,** 205–232.

Ferraro, J. R., and Williams, J. M. (1987). "Introduction to Synthetic Electrical Conductors," pp. 1–337. Academic Press, Orlando, Florida.

Ferraro, J. R., Martin, K., Furlani, A., and Russo, M. V. (1984). *Appl. Spectrosc.* **38,** 267.

Ferraro, J. R., Furlani, A., and Russo, M. V. (1987a). *Appl. Spectrosc.* **41,** 830.

Ferraro, J. R., Wang, H. H., Ryan, J., and Williams, J. M. (1987b). *Appl. Spectrosc.* **41,** 1377.

Fincher, C. R., Peebles, D. L., Heeger, A. J., Druy, M. A., Matsumara, Y., MacDiarmid, A. G., Shirakawa, H., and Ikeda, S. (1978). *Solid State Commun.* **27,** 489.

Fincher, C. R., Ozaki, M., Heeger, A. J., and MacDiarmid, A. G. (1979). *Phys. Rev. B: Condens. Matter* **19,** 4140.

Fitchen, D. B. (1985). *Mol. Cryst. Liq. Cryst.* **117,** 411.

Furlani, A., Licoccia, S., Russo, M. V., Marsich, N., and Camus, A. (1986). *J. Polym. Sci., Polym. Chem. Ed.* **24,** 991.

Furlani, A., Paolesse, R., Russo, M. V., Camus, A., and Marsich, N. (1987). *Polymer* **28,** 1221.

Galtier, M., and Montaner, A. (1985). *Mol. Cryst. Liq. Cryst.* **117,** 307.

Gibson, H. W., Weagley, R. J., Prest, W. M., Mosher, R., and Kaplan, S. (1983). *J. Phys. (Les Arcs, Fr.), Colloq.* **44,** C3-123.

Girlando, A., and Pecile, C. (1973). *Spectrochim. Acta, Part A* **29A,** 1859.

Girlando, A., Bozio, R., and Pecile, C. (1974). *Chem. Phys. Lett.* **25,** 409.

Gussoni, M., Castiglioni, C., Miragoli, M. Lugli, G., and Zerbi, G. (1985). *Spectrochim. Acta, Part A* **41A,** 371.

Hanfland, M., Brillante, A, Girlando, A., and Syassen, K. (1988). *Phys. Rev. B: Condens. Matter* **38,** 752.

Harada, I., Tasumi, M., Shirakawa, H., and Ikeda, S. (1978). *Chem Lett.* p. 1411.

Harada, I., Furukawa, Y., and Tasumi, M. (1980). *J. Chem. Phys.* **73,** 4746.

Harada, I., Furukawa, Y., Arakawa, T., Takeuchi, H., and Shirakawa, H. (1985). *Mol. Cryst. Liq. Cryst.* **117,** 335.

Hardy, L. C., and Shriver, D. F. (1986). *J. Am. Chem. Soc.* **108,** 2887.

Hatano, M., Kambara, S., and Okamoto, S. (1961). *J. Polym. Sci.* **51,** 526.

Herbstein, F. H., Kapon, M., and Schwotzer, W. (1983). *Helv. Chim. Acta* **66,** 35.

Horovitz, B. (1981). *Mol. Cryst. Liq. Cryst.* **77,** 285.

Horovitz, B. (1982). *Solid State Commun.* **41,** 729.

Hsu, S. L., Signorelli, A. J., Pez, G. P., and Baughman, R. H. (1978). *J. Chem. Phys.* **69,** 106.

Inabe, T., Lyding, J. W., Moguel, M. K., and Marks, T. J. (1983a). *J. Phys. (Les Ulis, Fr.)* **C3,** 625.

Inabe, T., Lyding, J. W., and Marks, T. J. (1983b). *J. Chem. Soc., Chem. Commun.* p. 1084.

Inabe, T., Lomax, J. F., Lyding, J. W., Kannewurf, C. R., and Marks, T. J. (1984). *Synth. Met.* **9,** 303.

Inabe, T., Marks, T. J., Burton, R. L., Lyding, J. W., McCarthy, W. J., Kannewurf, C. R., Reisner, G. M., and Herbstein, F. H. (1985). *Solid State Commun.* **54,** 501.

Inagaki, F., Tasumi, M., and Miyazawa, T. (1975). *J. Raman Spectrosc.* **3,** 335.

Ito, T., Shirakawa, H., and Ikeda, S. (1974). *J. Polym. Sci., Polym. Chem. Ed.* **12,** 11.

Iwahuna, K., Kuzmany, H., Wudl, F., and Aharon-Shalom, E. (1982). *Mol. Cryst. Liq. Cryst.* **79,** 39.

Iwamoto, R., Saito, Y., Ishihana, H., and Tadokoro, H. (1968). *J. Polym. Sci. Part A-2* **6,** 1507.

Jacobsen, C. S. (1986). *J. Phys. C* **19,** 5643.

Jacobsen, C. S., Williams, J. M., and Wang, H. H. (1985). *Solid State Commun.* **54,** 937.

Jacobsen, C. S., Tanner, D. B., Williams, J. M., Geiser, U., and Wang, H. H. (1987). *Phys. Rev. B: Condens. Matter* **35,** 9605.

James, D. B., Stein, R. S., and Macknight, W. J. (1979). *Bull. Am. Phys. Soc.* **24,** 479.

Joyner, R. D., and Kenney, M. E. (1960). *Inorg. Chem.* **82,** 5790.

Jurgenson, C. W., and Drickamer, H. G. (1986). *Chem. Phys. Lett.* **125,** 554.

Kanicki, J., Vander Donckt, E., and Boue, S. (1981). *J. Chem. Soc., Faraday Trans. 2* **77,** 2157.

Kaplunov, M. G., Yagubskii, E. B., Rosenberg, L. P., and Borodko, Y. G. (1985). *Phys. Status Solidi* **89,** 509.

Koch, B., Geserich, H. P., Ruppel, W., Schweitzer, D., Dietz, K. H., and Keller, H. J. (1985). *Mol. Cryst. Liq. Cryst.* **119,** 343.

Koenig, J. L., and Angood, A. C. (1970). *J. Polym. Sci.* **8,** 1787.

Kozlov, M. E., Pokhodnia, K. I., and Yurchenko, A. A. (1987). *Spectrochim. Acta, Part A* **43A,** 323.

Krauzman, M., Poulet, H., and Pick, R. M. (1986). *Phys. Rev. B: Condens. Matter* **33,** 99.

Kuroda, H., Yakushi, K., Tajima, H., and Saito, G. (1985). *Mol. Cryst. Liq. Cryst.* **125,** 135.

Kuroda, Y., and Kobo, M. (1957). *J. Polym. Sci.* **26,** 323.

Kuroda, Y., and Kobo, M. (1959). *J. Polym. Sci.* **36,** 453.

Kuzmany, H. (1980). *Phys. Status Solidi* **97,** 521.

Kuzmany, H. (1981). *Chem. Scr.* **17,** 155.

Kuzmany, H., and Knoll, P. (1985). *Mol. Cryst. Liq. Cryst.* **117,** 385.

Lefrant, S., Lichtmann, L. S., Temkin, H., Fitcher, D. G., Miller, P. C., Whitewell, G. E., and Burlitch, J. M. (1979). *Solid State Commun.* **29,** 191.

Lefrant, S., Faulques, E. Lauchlan, L., Kletter, M. J., and Etemad, S. (1982). *Mol. Cryst. Liq. Cryst.* **83,** 1149.

Lefrant, S., Faulques, E., Briuio, G. P., and Mulazzi, E. (1985). *Solid State Commun.* **53,** 583.

Lichtmann, L. S. (1981). Ph.D. thesis. Cornell University, Ithaca, New York.

Lichtmann, L. S., and Fitcher, D. B. (1979). *Synth. Met.* **1,** 139.
Lichtmann, L. S., and Fitcher, D. B. (1981). *Bull. Am. Phys. Soc.* **26,** 485.
Lichtmann, L. S., Sarhangi, A., and Fitcher, D. B. (1980). *Solid State Commun.* **36,** 869.
Lichtmann, L. S., Sarhangi, A., and Fitcher, D. B. (1981). *Chem. Scr.* **17,** 149.
Lin, K. J., and Parsons, J. L. (1969). *Macromolecules* **2,** 529.
Lin, L., Marks, T. J., Kannewurf, C. R., Lyding, J. W., McClure, M. S., Ratajack, M. T., and Whang, T. (1980). *J. Chem Soc., Chem. Commun.* p. 954.
McClure, M. S., Liu, L., Whang, T., Ratajack, M. T., Kannewurf, C. R., and McQuillan, B., Street, G. S., and Clarke, T. C. (1982). *J. Electron. Mater.* **11,** 471.
Marks, T. J. (1976). *J. Coat, Technol.* **48,** 53.
Marks, T. J. (1978). *Ann. N.Y. Acad. Sci.* **313,** 594.
Marks, T. J. (1980). *Bull. Am. Phys. Soc.* **25,** 315.
Marks, T. J. (1985). *Science* **227,** 881.
Marks, T. J., and Kalina, D. W. (1982). *In* "Extended Linear Chain Compounds" (J. S. Miller, ed.), Vol. 1, pp. 197–331. Plenum, New York.
Matsuara, H., and Miyazawa, T. (1968). *Bull. Chem. Soc. Jpn.* **41,** 1798.
Matsuara, H., and Miyazawa, T. (1969). *J. Polym. Sci.* **1,** 1735.
Matsuzaki, S., Kuwata, R., and Toyoda, K. (1980). *Solid State Commun.* **33,** 403.
Maxfield, J., and Shepherd, I. W. (1975). *Polymer* **16,** 505.
Mele, E. J., and Rice, M. J. (1980). *Phys. Rev. Lett.* **45,** 926.
Meneghetti, M., Bozio, R., and Pecile, C. (1986). *J. de Phys.* **47,** 1377.
Meyer, G., and Wöhrle, D. (1974). *Makromol. Chem.* **175,** 714.
Meyer, G., Hartmann, M., and Wöhrle, D. (1975). *Makromol. Chem.* **176,** 1919.
Montaner, A., Galtier, M., Benoit, C., and Aldissi, M. (1981). *Solid State Commun.* **39,** 99.
Mulazzi, E., Brivio, G. P., and Tiziani, R. (1985a). *Mol. Cryst. Liq. Cryst.* **117,** 343.
Mulazzi, E., Brivio, G. P., Lefrant, S., and Faulques, E. (1985b). *Mol. Cryst. Liq. Cryst.* **117,** 351.
Mulliken, R. S. (1955). *J. Chem. Phys.* **23,** 1997.
Naarmann, H. (1987a). *Chem. Eng. News* **65,** No. 25, 20.
Naarmann, H. (1987b). *New Sci.* **114,** No. 1564, 34.
Ng. H. K., Timusk, T., Jérome, D., and Bechgaard, K. (1984). *Phys. Rev. B: Condens. Matter* **32,** 8041.
Oddi, L., Capelletti, R., Fieschi, R., Fontana, M. P., Ruani, G., Bocchi, V., and Gardini, G. P. (1985). *Mol. Cryst. Liq. Cryst.* **118,** 179.

Papke, B. L., Ratner, M. A., and Shriver, D. F. (1981). *J. Phys. Chem. Solids* **42,** 493.

Papke, B. L., Ratner, M. A., and Shriver, D. F. (1982a). *J. Electrochem. Soc.* **129,** 1434.

Papke, B. L., Ratner, M. A., and Shriver, D. F. (1982b). *J. Electrochem. Soc.* **129,** 1694.

Park, Y. W., Derenstein, A., Chiang, C. K., Heeger, A. J., and MacDiarmid, A. G. (1979). *Solid State Commun.* **29,** 747.

Pecile, C., Zerbi, G., Bozio, R., and Girlando, A. (eds.) (1985). *Mol. Cryst. Liq. Cryst., Parts A and B: Conduct. Polym. I and II* **118,** 1-450 and 1-486.

Petersen, J. L., Schramm, C. J., Stojakovic, D. R., Hoffman, B. M., and Marks, T. J. (1977). *J. Am. Chem. Soc.* **99,** 286.

Pron, A., Kulszewicz, I., Billund, D., and Srzyluski, J. (1981). *J. Chem. Soc., Chem. Commun.* p. 783.

Pron, A., Bernier, P., Billund, D., and Lefrant, S. (1983). *Solid State Commun.* **46,** 587.

Rabolt, J. F., Clarke, T. C., and Street, G. B. (1979). *J. Chem. Phys.* **71,** 4614.

Rabolt, J. F., Clarke, T. C., and Street, G. B. (1982). *J. Chem. Phys.* **76,** 5781.

Rice, M. J. (1976). *Phys. Rev. Lett.* **37,** 36.

Rice, M. J., Pietronero, L., and Brüesch, P. (1977). *Solid State Commun.* **21,** 757.

Riseman, S. M., Yaniger, S. I., Eyring, E. M., Macinnes, D., MacDiarmid, A, G., and Heeger, A. J. (1981). *Appl. Spectrosc.* **35,** 557.

Salaneck, W. R., Thomas, H. R., Bigelow, R. W., Duke, C. B., Plummer, E. W., Heeger, A. J., and MacDiarmid, A. G. (1980). *J. Chem. Phys.* **71,** 2044.

Scaringe, R. P., Schramm, C. J., Stoijakovic, D. R., Hoffman, B. M., Ibers, J. A., and Marks, T. J., (1980). *J. Am. Chem. Soc.* **102,** 6702.

Schoch, K. F., Kundalkar, B. R., and Marks, T. J. (1979). *J. Am. Chem. Soc.* **101,** 7071.

Schügerl, F. B., and Kuzmany, H. (1981). *J. Chem. Phys.* **74,** 953.

Shirakawa, H., and Ikeda, S. (1971). *Polym. J.* **2,** 231.

Shirakawa, H., and Kobayashi, T. (1983). *J. Phys. (Les Arcs, Fr.), Colloq.* **44,** C3-3.

Shirakawa, H., Ito, T., and Ikedo, S. (1973). *Polym. J.* **4,** 460.

Shirakawa, H., Louis, E. J., MacDiarmid, A. G., Chiang, C. K. and Heeger, A. J. (1977). *J. Chem. Soc., Chem. Commun.* p. 578.

Shirakawa, H., Ito, T., and Ikeda, S. (1978). *Makromol. Chem.* **179,** 1565.

Shriver, D. F., and Farrington, G. C. (1985). *Chem. Eng. News* **63** (20), 42.

Shriver, D. F., Papke, B. L., Ratner, M. A., Dupon, R., Wong, T., and Brodwin, M. (1981). *Solid State Ionics* **5,** 83.

Spindler, R., and Shriver, D. F. (1986). *Macromolecules* **19,** 347.

Stojakovic, D. R. (1977). Ph.D. thesis. Northwestern University, Evanston, Illinois.

Sugai, S., and Saito, G. (1986). *Solid State Commun.* **58,** 759.

Sugano, T., Yamada, K., Saito, G., and Kinoshita, M. (1985). *Solid State Commun.* **55,** 137.

Tadokoro, H., Chatani, Y., Yoshihara, T., Tahara, S., and Murshashi, S. (1964). *Makromol. Chem.* **73,** 109.

Tajima, H., Yakushi, K., Kuroda, H., Saito, G., and Inokuchi, H. (1984). *Solid State Commun.* **49,** 769.

Tajima, H., Yakushi, K., Kuroda, H., and Saito, G. (1985). *Solid State Commun.* **56,** 251.

Tajima, H., Kanbara, M., Yakushi, K., Kuroda, H., and Saito, G. (1986). *Solid State Commun.* **57,** 911.

Takahashi, Y., and Tadokoro, H. (1973). *Macromolecules* **6,** 672.

Takenaka, T. (1971). *Spectrochim. Acta, Part A* **27A,** 1735.

Tanaka, M., Watanabe, A., and Tanaka, J. (1980). *Bull. Chem. Soc. Jpn.* **C3,** 645.

Tanaka, M., Shimizu, M., Saito, Y., and Tanaka, J. (1986). *Chem. Phys. Lett.* **125,** 594.

Teeters, D., and Frech, R. (1986). *Solid State Ionics* **18/19,** 271.

Teitelbaum, R. C., Ruby, S. L., and Marks, T. J. (1979). *J. Am. Chem. Soc.* **101,** 7568.

Teitelbaum, R. C., Ruby, S. L., and Marks, T. J. (1980). *J. Am. Chem. Soc.* **102,** 3322.

Tokura, Y., Okamoto, H., Mitani, T., Saito, G., and Koda, T. (1986). *Tech. Rep. of Institute for Solid State Physics Ser. A* **1635.**

Torrance, J. B. (1979). *In* "Molecular Metals" (W. E. Hatfield, ed.), pp. 7–14. Plenum, New York.

Ugawa, A., Yakushi, K., Kuroda, H., Kawamoto, A., and Tanaka, J. (1986). *Chem. Lett.* p. 1975.

Van Duyne, R. P., Cape, T. W., Suchanski, M. R., and Siedle, A. R. (1986). *J. Phys. Chem.* **90,** 739.

Vardeny, Z., Brafman, O., and Hrenfreund, E. (1985). *Mol. Cryst. Liq. Cryst.* **117,** 355.

Vincent, C. A. (1987). *Prog. Solid State Chem.* **17,** 145.

Williams, J. M., Wang, H. H., Emge, T. J., Geiser, U., Beno, M. A., Leung, P. C. W., Carlson, K. D., Thorn, R. J., Schultz, A. J., and Whangbo, M.-H. (1987). *In* "Progress in Inorganic Chemistry" (S. L. Lippard, ed.). **35,** 51.

Wozniak, W. T., DePasquali, G., Klein, M. U., Sweany, R. L., and Brown, T. L. (1975). *Chem. Phys. Lett.* **33,** 33.

Wright, P. V. (1976). *J. Br. Polym.* **7,** 319.

Yoshihara, H., Tadokoro, H., and Murahashi, S. (1964). *J. Chem. Phys.* **41,** 1964.
Zannoni, G., and Zerbi, G. (1982a). *Chem. Phys. Lett.* **87,** 50.
Zannoni, G., and Zerbi, G. (1982b). *Chem. Phys. Lett.* **87,** 55.
Zannoni, G., and Zerbi, G. (1983a). *Chem. Phys. Lett.* **97,** 147.
Zannoni, G., and Zerbi, G. (1983b). *Solid State Commun.* **48,** 871.
Zannoni, G., and Zerbi, G. (1983c). *Solid State Commun.* **47,** 213.
Zannoni, G., and Zerbi, G. (1983d). *J. Phys. (Les Arcs, Fr.), Colloq.* **44,** C3-503.
Zannoni, G., and Zerbi, G. (1983e). *J. Mol. Struct.* **100,** 485.
Zannoni, G., and Zerbi, G. (1983f). *J. Mol. Struct.* **100,** 505.
Zannoni, G., and Zerbi, G. (1984). *Solid State Commun.* **50,** 55.
Zannoni, G., and Zerbi, G. (1985). *J. Chem. Phys.* **82,** 31.
Zerbi, G. (1987). *In* "Dynamics of Molecular Crystals" (J. Lascombe, ed.). Elsevier, Amsterdam.
Zerbi, G., and Dellepiane, G. (1987). *Gazz. Chim. Ital.* **117,** 591.
Zerbi, G., and Zannoni, G. (1983). *J. Phys. (Les Arcs, France), Colloq.* **44,** C3-273.
Zerbi, G., Zannoni, G., Gussoni, M., and Castiglioni, C. (1985). *Mol. Cryst. Liq. Cryst.* **117,** 287.
Zerbi, G., Castiglioni, C., Sala, S., and Gussoni, M. (1987). *Synth. Met.* **17,** 293.

3

FT-IR Microsampling Techniques

K. Krishnan
S. L. Hill
BioRad Digilab Division
Cambridge, Massachusetts

I. INTRODUCTION

With the advent of the FT-IR spectrometers and the concomitant advantages of speed and sensitivity, infrared absorption spectroscopy has become one of the preeminent tools for molecular characterization and identification. Using conventional macroscopic sampling techniques, it is easily possible to obtain high-quality infrared spectra from sample amounts as small as a few micrograms. With the rapid advances in manufacturing technology during the past few years, the need has arisen for the characterization of samples in subnanogram quantities by infrared spectroscopy. Thus, during the late 1970s and early 1980s, a number of papers were published in journals dealing with FT-IR microsampling techniques (Anderson and Wilson, 1975; Cournoyer *et al.*, 1977; Lacy, 1979, 1983). These techniques involve mounting the sample under study onto a very small pinhole (50 to 250 μm in diameter) located in the focus of a transmission beam condenser by micromanipulation techniques and recording the transmission spectrum of the sample. Unfortunately, micromanipulation techniques using a beam condenser were time-consuming and generally required a skillful operator. These disadvantages were largely overcome by the FT-IR microscope sampling technique, introduced in 1983. Since that time, the technique has found wide acceptance and has become the routine method for the characterization and identification of samples available in picogram quantities. In this chapter, the infrared microscope sampling technique and its range of applications are reviewed. In keeping with the spirit of the book, no attempt will be made to discuss in detail the theory and the design principles of the microscope. These topics have been addressed in the published literature, and the interested reader should refer to the relevant articles cited in the body of this text.

Even though the first FT-IR microscope sampling accessory was introduced in 1983, the concept of the infrared microscope incorporating all-reflecting optics dates back over four decades. Barer *et al.* (1949) reported spectra for a number of microsamples using an all-reflecting microscope developed by Burch (1947) on a dispersive infrared spectrometer. This microscope used a Cassegrain condenser to focus the infrared beam onto the sample. A similar Cassegrain element was used as the microscope objective. This group was also the first to use a polarizer in the optical path of the microscope and record the dichroic spectra of single crystals. Blout *et al.* (1949) discussed in detail the relevant theory and the performance characteristics of such a microscope, and reported the microtransmission spectra of a number of biological materials and fibers. Wood (1950) and Blout and Bird (1951) also recorded the microtransmission spectra of a number of biological compounds and the dichroic spectra of single crystals. In all of these studies, the microscope was located before the monochromator, and the infrared beam coming out of the microscope was redirected toward the normal detector of the infrared spectrometer. Coates *et al.* (1953) described an infrared microscope sampling accessory where an external detector was incorporated on the microscope itself. In this design, a Cassegrain condenser focused the infrared beam on the sample; the radiation transmitted by the sample was collected by a Cassegrain objective and focused on a small target thermocouple by another Cassegrain condenser. Incorporating a microscope of similar design, Nanometrics, Inc. (Sunnyvale, CA) introduced a microprocessor-based infrared spectrometer in the late 1970s. This nondispersive instrument used a number of interference filters for recording the infrared spectrum.

II. EXPERIMENTAL CONSIDERATIONS

As mentioned in the Introduction, the first commercial FT-IR microscope sampling accessory was introduced in 1983 by BioRad Digilab Division (Cambridge, MA). Since that time, a number of other commercial infrared microscope sampling accessories have become available. These have been introduced by commercial FT-IR manufacturers (Analect, Bruker, Mattson, JEOL, and Jasco) and by an FT-IR accessory manufacturer (Spectra-Tech Inc.). Except for minor variations, all of these accessories are similar in design and are capable of operation in the transmission or reflection modes. Because of the authors' familiarity, the succeeding descriptions in this article will be based on the infrared microscope accessory produced by the BioRad Digilab Division. Unless otherwise stated, the spectra shown in this article were also recorded while using this accessory on the Digilab line of FT-IR spectrometers. All of the reported spectra were recorded at 8-cm^{-1} resolution, for measurement times ranging from 1 to 5 min.

Figure 1 shows the schematic optical diagram of the Bio-Rad Digilab Division microscope sampling accessory for the transmission mode of operation. As shown in Fig. 2, the microscope sits atop the FT-IR sample compartment; the infrared beam from the FT-IR instrument is focused onto a sample placed on a standard microscope x–y stage. The infrared beam that passes through the sample is collected by a 36× Cassegrain objective. The objective produces an image of the sam-

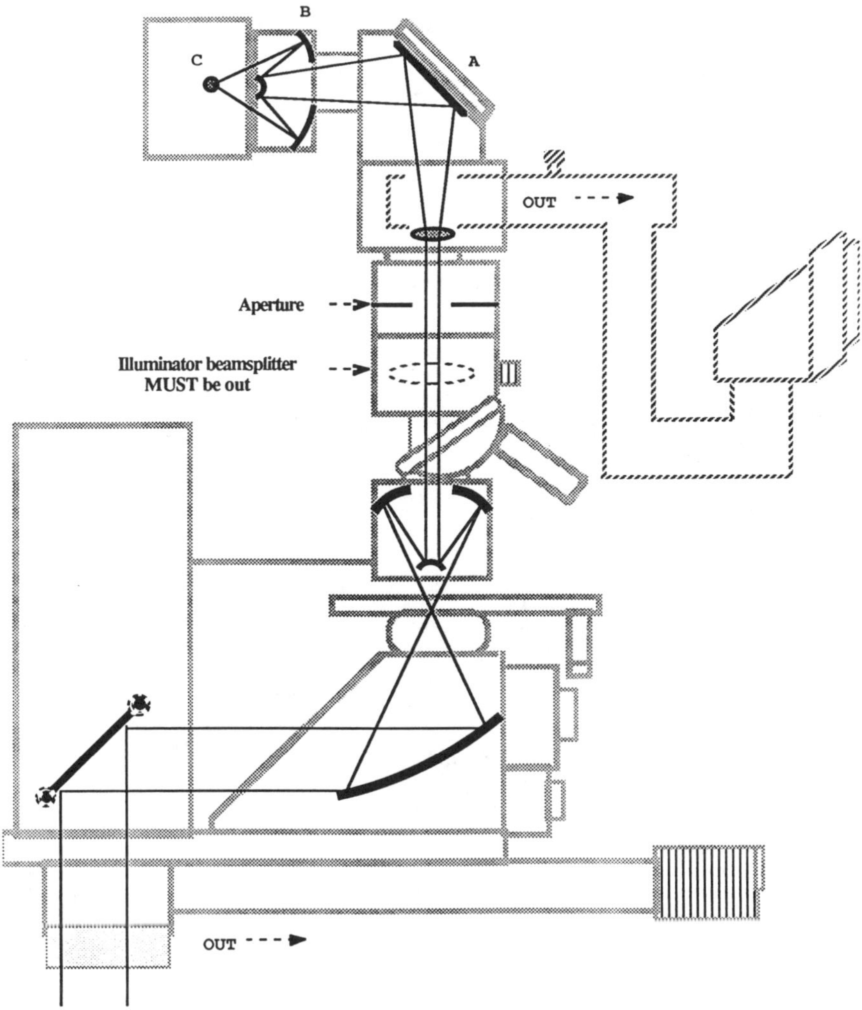

Fig. 1. The optical system of the UMA-300A infrared microscope in the transmission mode of operation.

ple within the barrel of the microscope, and a variable aperture (circular iris or circular diaphragm) is placed in this image plane. The radiation passing through the aperture is focused on a small area (250 × 250 μm) liquid nitrogen-cooled MCT (mercury cadmium telluride) detector by another 36×

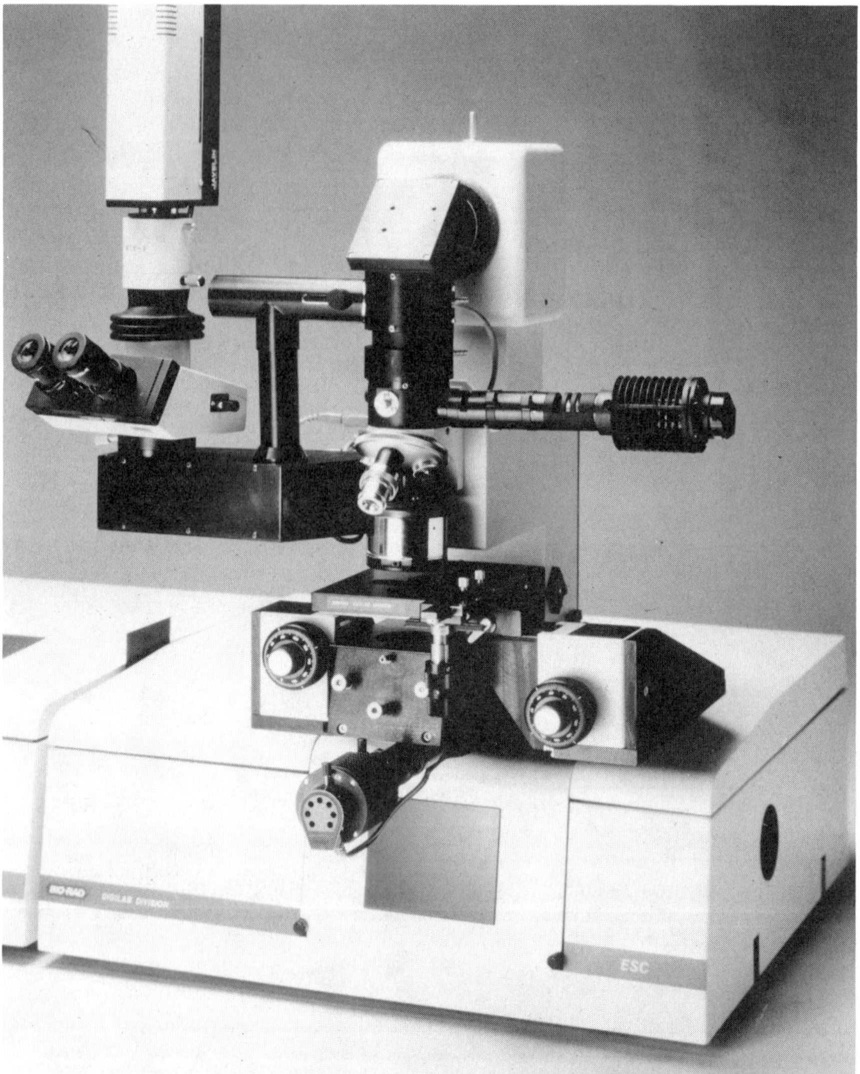

Fig. 2. The infrared microscope sampling accessory.

Cassegrain condenser. The microscope includes a switchable 4× or 10× glass objective and a collinear visible illumination and observation system (normally incorporating a 10× eyepiece) that can be used for the visual inspection of the sample. For recording the spectrum of a microsample such as an inclusion on a polymer film, the sample is first placed on the microscope stage, and the desired area of the sample located by using the low-power glass objective. The 36× Cassegrain objective is then switched into place and provides a total visual magnification of 360×. With the desired region of the sample, e.g., the inclusion, centered in the viewing field, the variable aperture is closed down so that only the inclusion is visible in the field of view. The isolation of the desired area on the sample image is optically equivalent to masking the sample with a pinhole. The result will be the spectrum of the desired part of the sample.

In some cases, particularly in the case of very small, isolated samples, stray light may be collected by the microscope. The issue of stray light has been discussed by Coates *et al.* (1953). Under special circumstances, it may be possible to reduce stray light by employing a second aperture in the optics that focus the infrared radiation onto the sample (Messerschmidt, 1988); however, this arrangement reduces the sensitivity of the accessory. The transmission of the microscope accessory is relatively low, typically in the 5 to 10% range. To compensate for this poor throughput, a narrow-range (4000–700 cm^{-1}) MCT detector is usually employed. For commonly used infrared detectors, the detector noise is proportional to the square root of the area of the detector element. To reduce noise and enhance the sensitivity of the microscope sampling accessory, a small-area (250 × 250 μm) MCT detector element is normally used. Since the microscope accessory just described has a one-to-one correspondence between the areas of the sample and the focused image at the detector, the maximum area of the sample to be examined is limited by the size of the detector element. Thus, one can improve the sensitivity of the microscope sampling accessory for the measurement of very small samples by employing detector elements of even smaller dimensions.

By manually switching one of the mirrors in the optical train, the microscope can be converted from the transmission to the reflection mode (Fig. 3). In this case, the infrared beam passes through a hole on the side of the Cassegrain objective and is focused on the sample by a small mirror in the objective. The reflected radiation is collected in the same manner as it is in the transmission mode. The reflectance spectra from smooth, dielectric samples will usually exhibit distorted bandshapes due to the dispersion of the refractive index of the sample. The analysis of these spectra will be greatly facilitated by the use of the Kramers-Kronig (K-K) transformation.

For the measurement of dichroic spectra using the infrared microsampling accessory, a rotatable, wire-grid polarizer can be inserted in the optical path of the microscope, just in front of the detector (Krishnan, 1988). For specialized experiments that involve heating the sample, a miniature heater can be placed on the sample stage. For experiments involving polymer stretching, a miniature polymer stretcher can be used. When one wants to map a large sample on a microscopic scale, the manual x–y stage of the microscope can be replaced by a motorized, computer-controlled stage. Such computerized stages and software routines for graphical display of the results have been described by Beduhn and White (1986) and by Harthcock and Atkin (1988).

Figure 4 illustrates the sensitivity levels currently achievable by the transmission FT-IR microscope sampling method. The sample used in this case was a polystyrene film 20 μm thick. The transmission spectrum of the polystyrene was recorded (at 8-cm^{-1} resolution, 5-min measurement time) with the rectangular variable aperture set to sample a 5 × 5 μm area of the film. Under these conditions, some energy losses due to diffraction effects would be expected, particularly in the long wavelength range. Figure 4 (bottom) illustrates this effect. This transmission spectrum was produced by using a background spectrum recorded with the aperture set to a sam-

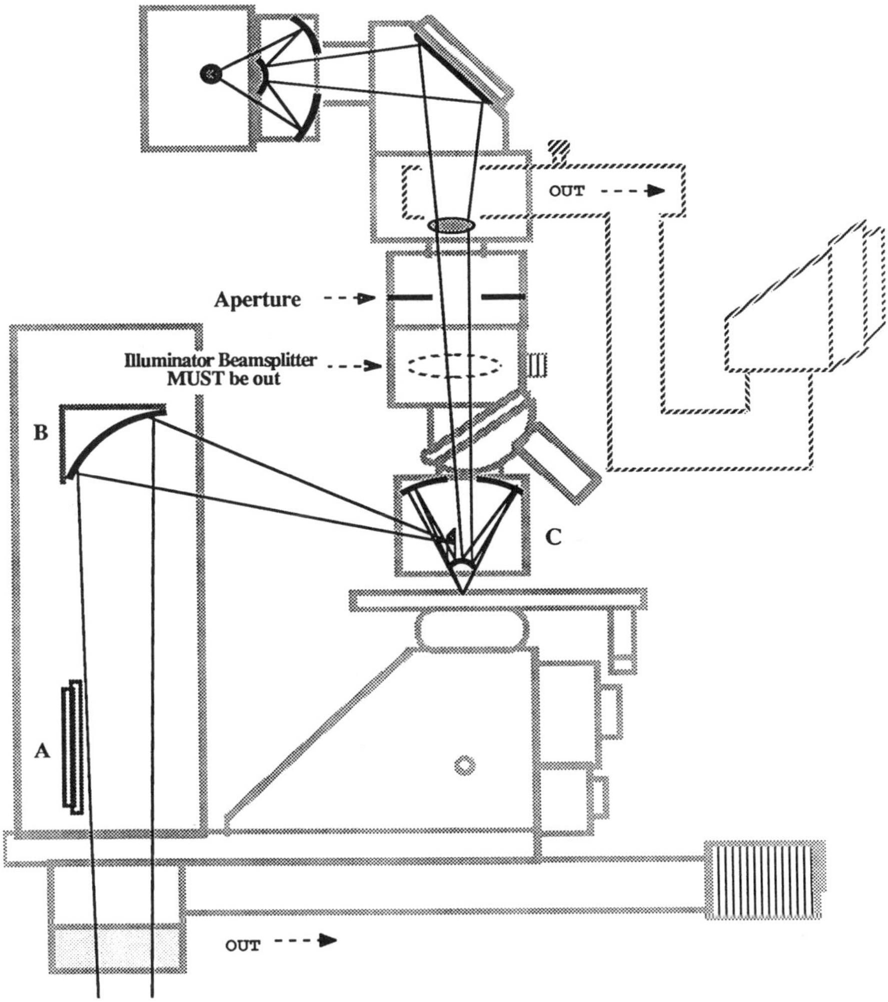

Fig. 3. The optical system of the UMA-300A infrared microscope in the reflection mode of operation.

pling area of 50 × 50 μm. This mismatch between the sample and the background dimensions leads to an unbalancing of the diffraction losses and a sloping baseline for the spectrum. However, when the spectrum of the sample and the background are both recorded at the same aperture settings, the

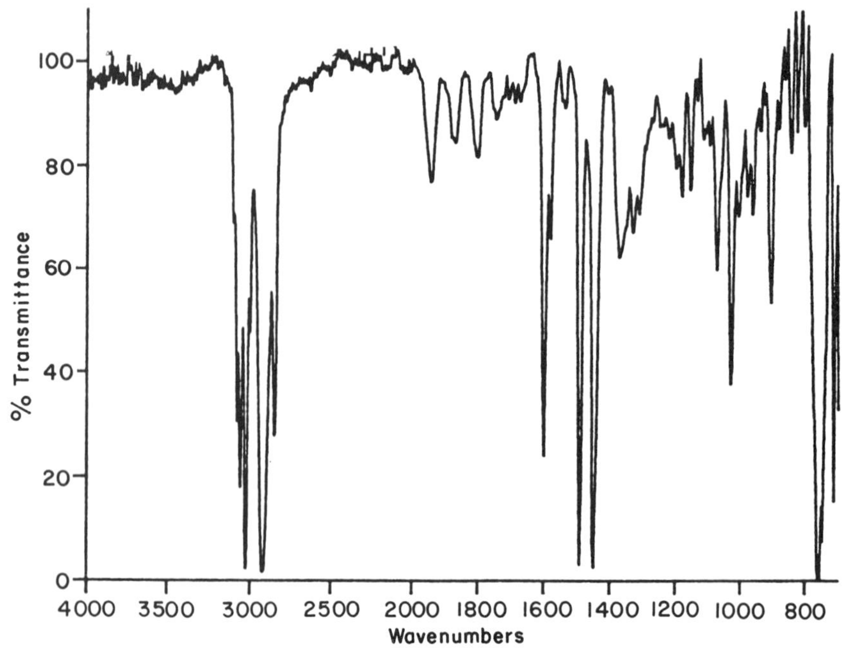

100
80
60
40
20
0
% Transmittance
4000 3500 3000 2500 2000 1800 1600 1400 1200 1000 800
Wavenumbers

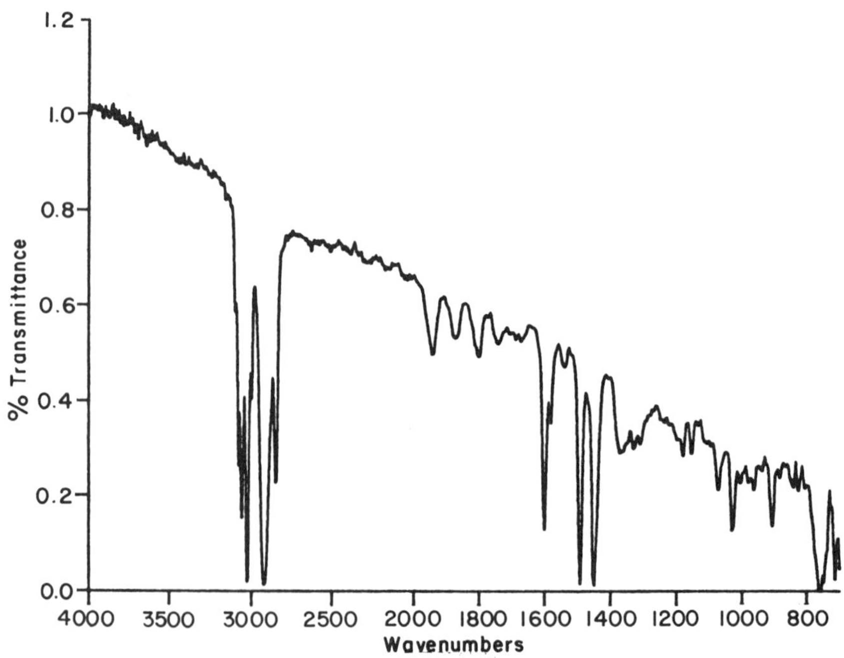

1.2
1.0
0.8
0.6
0.4
0.2
0.0
% Transmittance
4000 3500 3000 2500 2000 1800 1600 1400 1200 1000 800
Wavenumbers

5UPOLY2

diffraction losses generally balance out, and the resulting spectrum (Fig. 4, top) exhibits a flat baseline. Thus, the spectra show that it is possible to produce high-quality infrared spectra even from such small sampling areas.

III. APPLICATIONS

A. Forensic Applications

One of the principal applications of the infrared microscope sampling technique involves the analysis and identification of microscopic particles. As such, the technique lends itself easily to the study of forensic materials such as controlled drug particles, automotive paint chips, inks, fragments of polymers, textile fibers, papers, inorganics, adhesives, and explosives. A transmission spectrum for such a material may be obtained by carefully isolating the particle from its environment using standard techniques for small particle manipulation (Teetsov, 1977; Humecki, 1988) and placing it on a suitable infrared transmitting plate positioned on the microscope stage.

Using such techniques, Krishnan *et al.* (1985) have reported the infrared spectrum of a single particulate of cocaine. The spectrum for a $40 \times 30 \times 10$ μm crystal is shown in Fig. 5. One can identify this spectrum as that of cocaine by standard spectral search methods.

Krishnan (1986) has also used the infrared microscope sampling technique to record the transmission spectra of a number of paint particles. Figure 6 shows an easily identifiable spectrum of an acrylic enamel particle with dimensions $40 \times 40 \times 12$ μm. Prior to the advent of the infrared microscope

---

Fig. 4. Microtransmission spectra of a polystyrene film for a sampling area of 5×5 μm. (Top) Aperture size for the background is 5×5 μm. Note the baseline flatness when the same aperture size is used for both the sample and the background. (Bottom) Aperture size for the background is 50×50 μm.

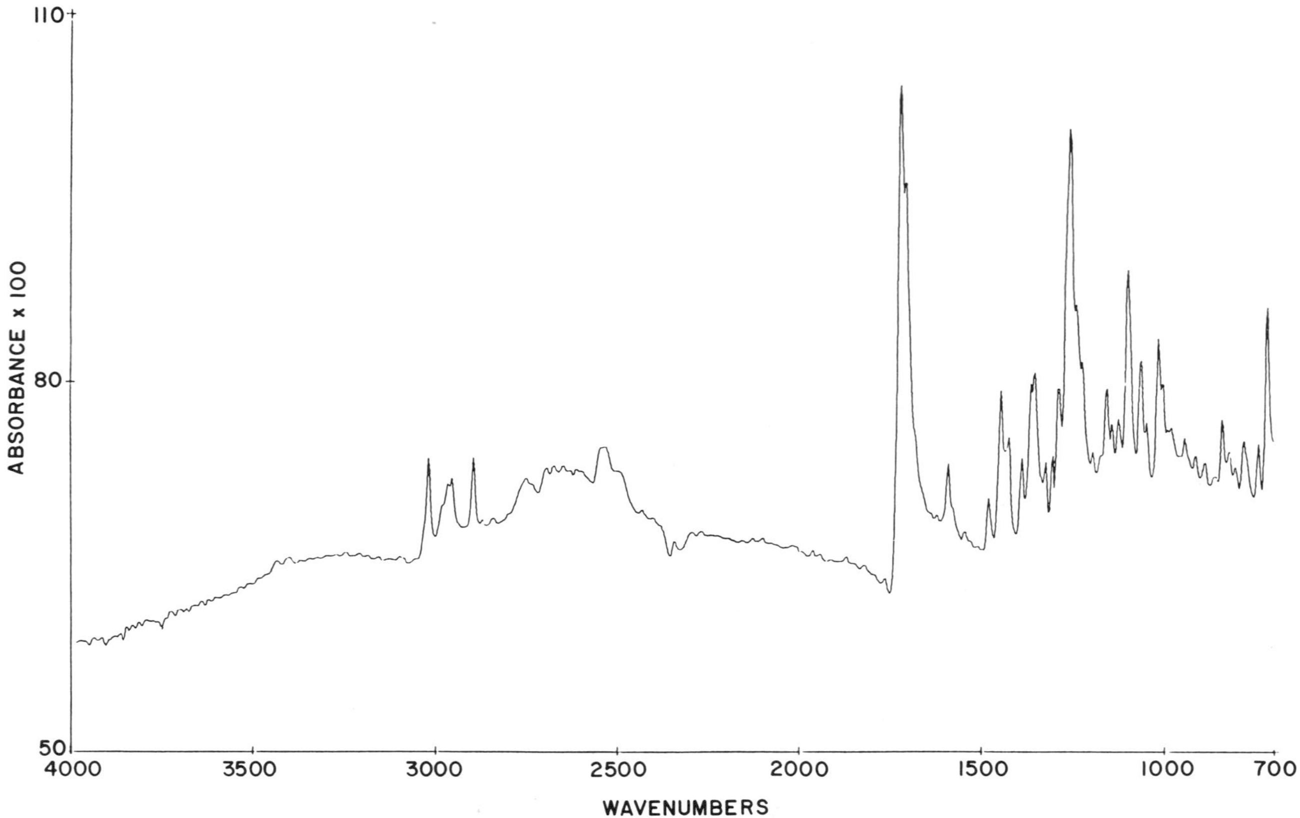

Fig. 5. Microtransmission spectrum of a single 40 × 30 × 10-μm crystal of cocaine. [From Krishnan *et al.* (1985). Reprinted with permission of the publishers, SPIE.]

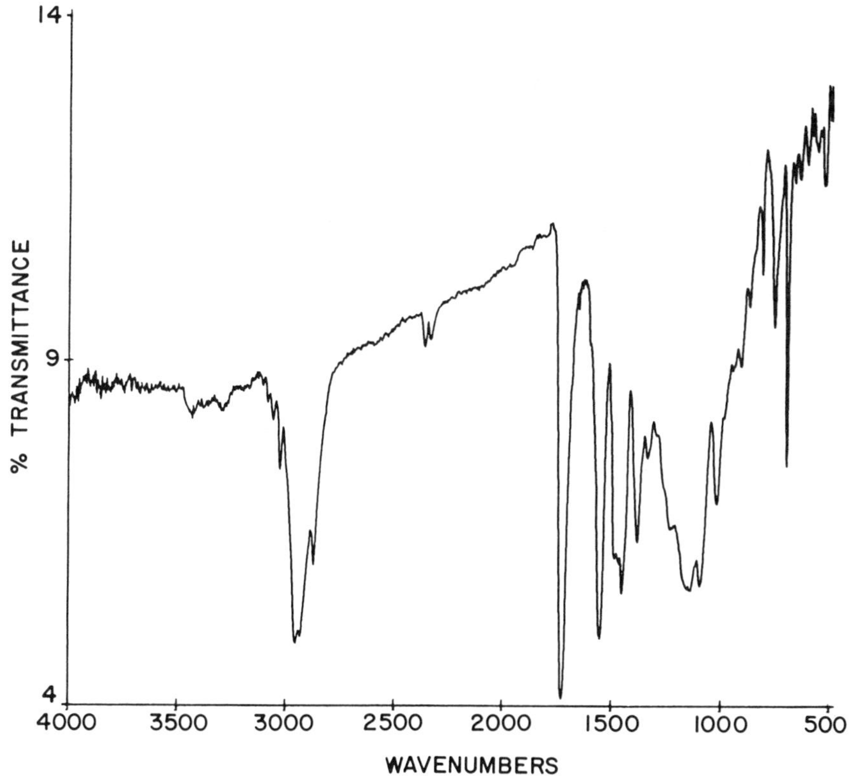

Fig. 6. Microtransmission spectrum of a 40 × 40 × 12-μm acrylic enamel paint chip. [From Krishnan (1986). Reprinted with permission of the publishers, SPIE.]

sampling technique, the standard method for recording the spectra of automotive paint chips involved the use of a diamond anvil cell (Rodgers *et al.*, 1976; Krishnan and Ferraro, 1982). Even with the infrared microscope sampling technique, it may be necessary to use the diamond cell in cases where the sample thickness is too large. Krishnan (1984, 1986, 1987) has described the use of a small diamond anvil cell (High Pressure Diamond Optics, Tucson, AZ) with the transmission infrared microscope sampling technique; he has demonstrated its application to the study of polymer fibers.

Ink spots, such as those from genuine and counterfeit currency notes, may also be easily characterized with the infrared microscope sampling technique. Sommer *et al.* (1988) have recorded the transmission spectra of the inks of various colors used on a U.S. one dollar bill. In this case, the inks were physically removed from the dollar bill before their transmission spectra were recorded. However, in some forensic applications, it may be necessary to record the spectrum of the ink *in situ*. Figures 7 and 9 illustrate such an application. Figure 7 shows the transmission spectrum of normal writing paper through a black ink spot, through a blank region of the

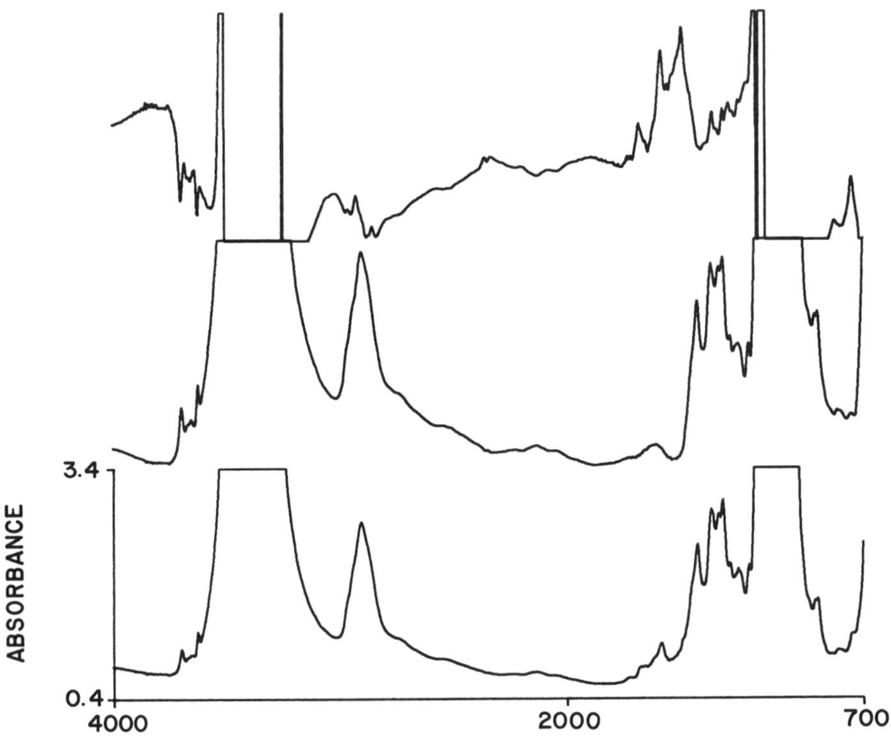

Fig. 7. Microtransmission spectra of a black ink spot on writing paper. From bottom to top, spectra of black ink spot on paper, blank paper, and the difference.

paper, and the difference. The paper sample is so thick that complete absorption occurs for the major cellulose and water absorption bands. However, in the 3000- to 1220-cm^{-1} region, one can clearly see the absorption bands due to the ink in the difference spectrum.

The microreflectance technique can be used for the *in situ* analysis of microscopic particles, particularly on metal substrates. In the latter case, one actually records the reflection–absorption (RAS) spectrum. Figure 8 shows the RAS spectrum of a particle on a metal bullet. This spectrum could easily be identified as a nylon-type material. The reflectance spectrum could also be used for recording the *in situ* RAS spectrum of inks on the tips of ball point pens. Figure 9 shows the transmission difference spectrum of the black ink on paper and the RAS spectrum of the same ink on the tip of a ball point pen. One can see the excellent correspondence between the two spectra.

B. General Characterization of Particulate Matter

Since the infrared microscope sampling technique is a powerful tool for the study of small particles, the method has been used in the study of individual crystallites, coals, organic charge-transfer superconductors, and environmental contaminants.

One of the advantages of the infrared microscope sampling technique is the ability to record the spectra of powdered samples with no sample preparation, such as grinding with KBr or Nujol. The powdered sample may be sprinkled onto a suitable infrared-transmitting salt plate, and the spectrum of a suitable small particle selected under visual observation may be recorded. Figure 10 shows the transmission spectrum of a single particulate of perlyene, $C_{20}H_{12}$. For comparison, the diffuse reflectance of the same sample finely dispersed in KBr

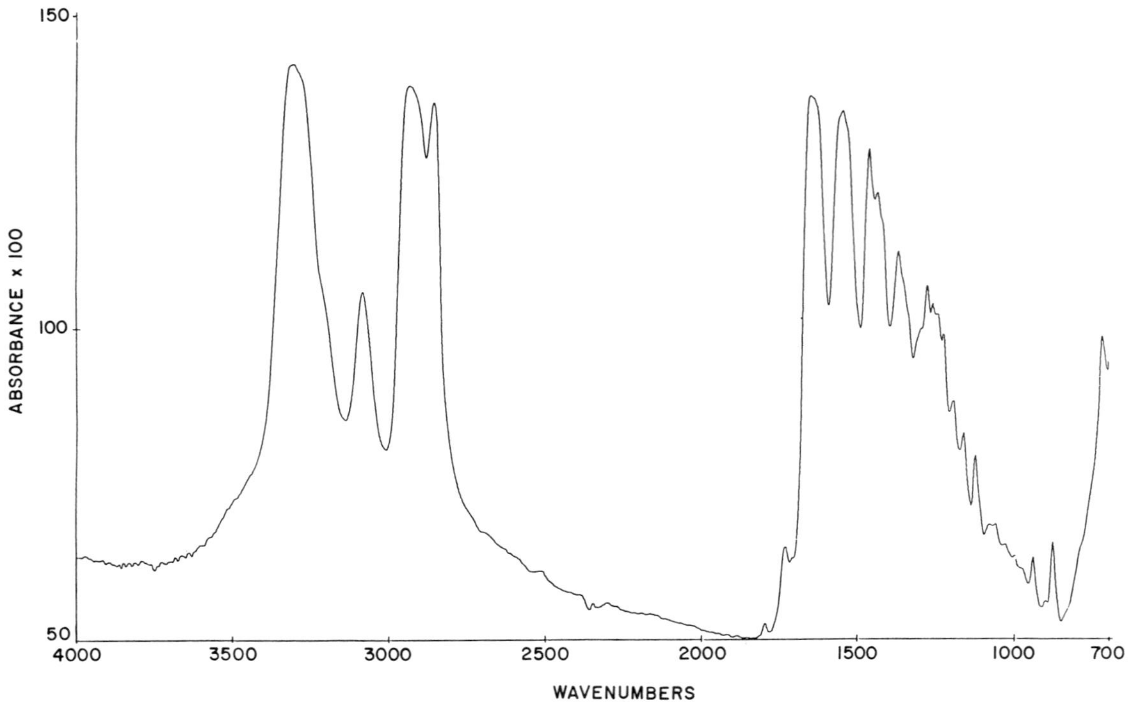

Fig. 8. Microreflectance spectrum of a nylon particle on the surface of a bullet.

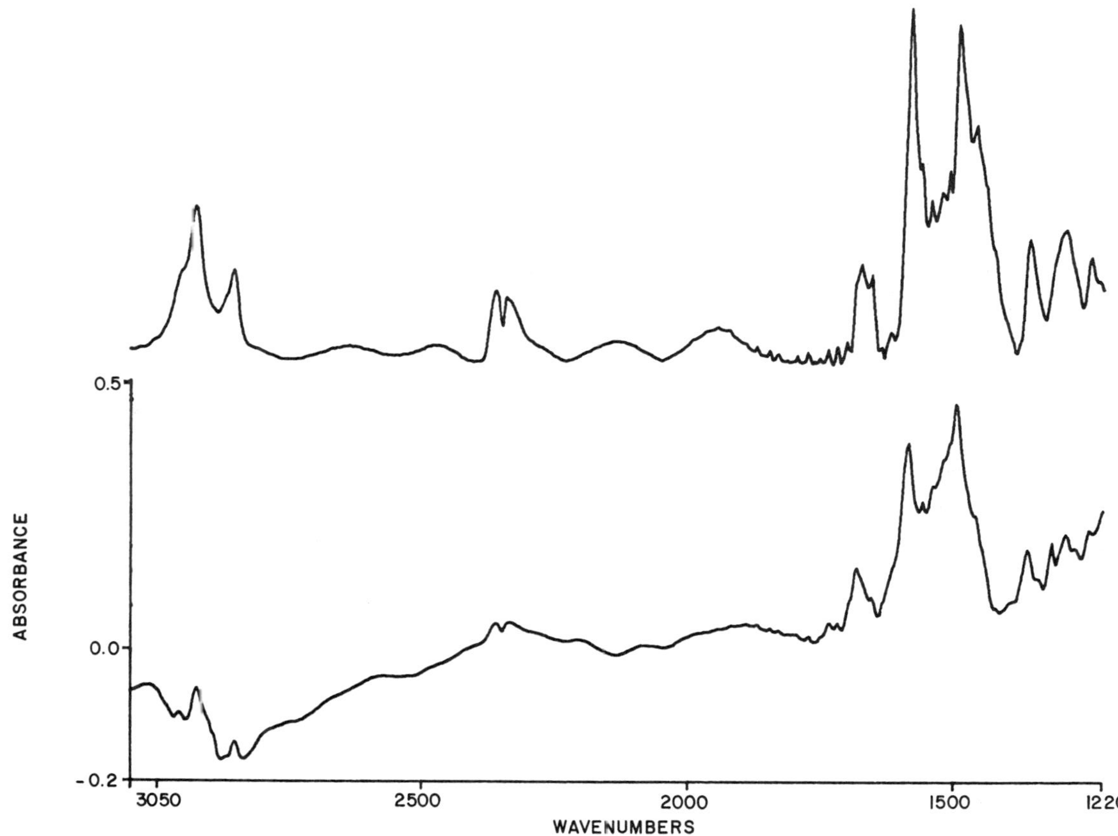

Fig. 9. (Bottom) Microtransmission spectrum of black ink on paper (top spectrum of Fig. 7). (Top) The reflection spectrum of the same ink on the tip of a ballpoint pen.

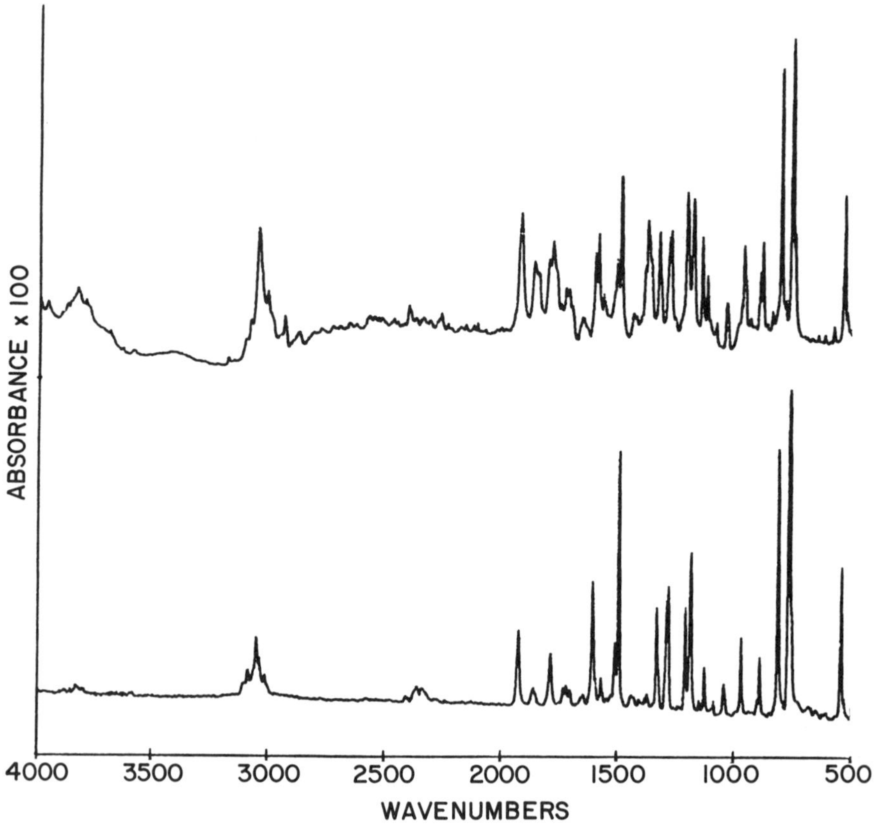

Fig. 10. Comparison of the microtransmission (bottom) and diffuse reflectance (top) spectra of perylene ($C_{20}H_{12}$).

powder is also shown. One can see from the figure that, while the two spectra appear to be very similar in the number and positions of the absorption bands, the relative intensities in the two spectra, particularly below 2000 cm^{-1}, are quite different. These differences may be due to the morphological changes induced by the grinding process.

Krishnan *et al.* (1985) have reported on the microtransmission spectra of coal particles. Figure 11 shows the spectra

Fig. 11. Microtransmission spectra for three adjacent 80 × 40-μm regions of a 250 × 40 × 10-μm coal particle. [From Krishnan *et al.* (1985). Reprinted with permission of the publishers, SPIE.]

recorded from a single coal particle. The dimensions of the particle were 250 × 40 × 10 μm. The spectra were recorded over three adjacent 80 × 40 μm sections. The lower spectrum in the figure shows high mineral concentration, evidenced by the absorption bands near 1000 and 3700 cm^{-1}. The middle spectrum seems to be free of these mineral bands, and there is evidence of a carbonyl stretching band at approximately 1740

cm^{-1}. The carbonyl band becomes very strong in the top spectrum. This pattern indicates that the coal particle is from the vicinity of an oxidation ring, and the spectra in the figure indicate the gradual transition from mineral-rich to oxidized coal.

The microreflectance technique may be used effectively in cases where the particle under study is too thick or opaque. Figure 12 (top) shows the microreflectance spectrum of a polymer particle with dimensions 200 × 200 × 200 μm. One can see that the appearance of the spectrum is distorted because of the dispersion of the refractive index. As mentioned previously, such spectra may be corrected by performing a K-K transformation.

When electromagnetic radiation is reflected from a smooth dielectric surface, Fresnel's relations apply to the reflection process, and the reflection coefficient, **r**, at any one frequency is defined as

$$\mathbf{r} = [(\mathbf{n} - 1)/(\mathbf{n} + 1)]^2$$

where **n** is the complex refractive index, $\mathbf{n} = n - ik$. Here n is the refractive index, and k is the extinction coefficient. The conventional infrared spectrum is the dispersion of the extinction coefficient, that is, a plot of k as a function of the frequency. The recorded reflectance spectrum, **R**, is the reflection coefficient as a function of frequency. The K-K dispersion relations shown below allow the calculation of the optical constants—n and k—from the reflectance spectrum:

$$n = (1 - \mathbf{R}) / (1 - 2\mathbf{R}\cos\phi + \mathbf{R}), \quad \text{and}$$

$$k = (-2\mathbf{R}\sin\phi) / (1 - 2\mathbf{R}\cos\phi + \mathbf{R})$$

Here, Ø is the phase shift introduced by the reflection process. Thus, by using the K-K transformation, the reflectance spectra can be corrected to yield the k spectrum, which will resemble "normal" infrared spectra.

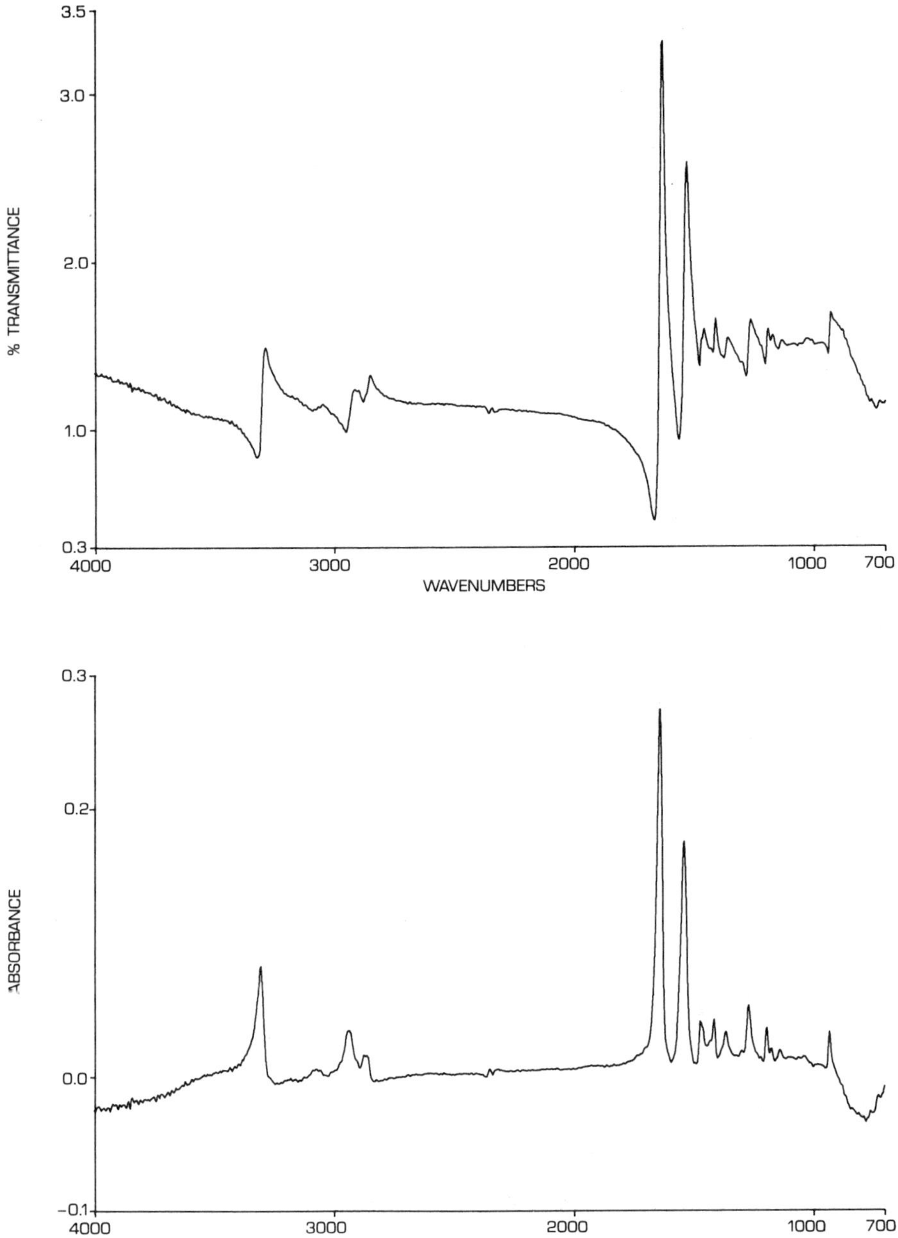

Fig. 12. Microreflectance spectrum (top) and Kramers-Kronig corrected absorbance spectrum (bottom) of a 200 × 200 × 200-μm extruded nylon pellet.

Figure 12 (bottom) shows the K-K corrected spectrum of the polymer particle, which can now be easily identified as nylon. Ferraro *et al.* (1987) have used the microreflectance/K-K transformation technique for the study of the dichroic spectra of a number of organic charge-transfer superconductors. Figure 13 shows the spectrum of a typical organic charge-transfer superconductor.

C. Identification of Contaminants

The ability of the infrared microscope sampling technique to deal with small particles makes it a powerful tool for the study of environmental and industrial contaminants. Contaminants on solid objects can be studied by removing them from their substrates and measuring transmission, or they can be studied on the substrate (*in situ*) by reflectance. Contaminant particles in air or liquids can be trapped on suitable filter media and studied by either method.

Using these techniques, Lang *et al.* (1988) have characterized a number of commonly occurring particles in dust by transmission spectroscopy. Among the identified components of dust are different forms of cellulose fibers, polymer debris, and ephithelial cells. Ramsey and Hausdorff (1981), Scott and Ramsey (1982), and Popek and Ramsey (1983) have used the infrared microscope sampling technique in conjunction with a nondispersive infrared spectrometer for the characterization and identification of contaminants found on printed circuit (PC) boards. They have removed contaminants from the PC boards and have recorded their transmission spectra. They have found these contaminants to be solder flux, proteins, and particles of nylon, polyethylene, polycarbonate, and other polymers. Using an FT-IR spectrometer, Lang and Katon (1986) have also studied the contaminants on PC boards. Humecki (1988) and Smyrl *et al.* (1988) have used the

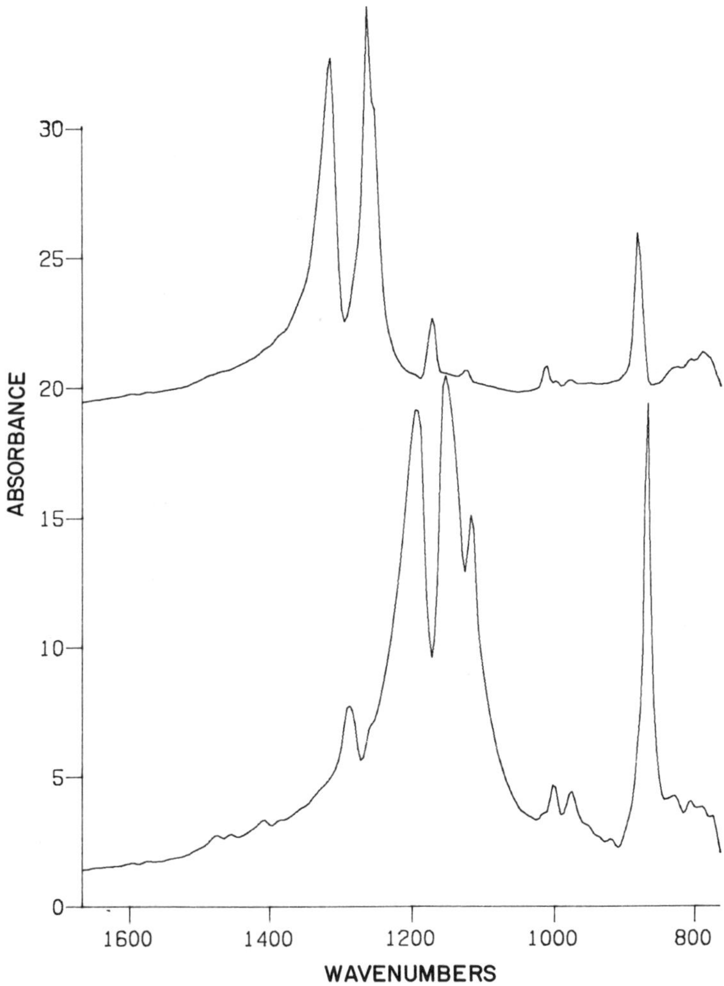

Fig. 13. Comparison of parallel (bottom) and perpendicular (top) polarized microreflectance spectra for a single crystal of κ-$Et_2Cu(NCS)_2$, an organic charge-transfer superconductor. The original reflectance spectra (not shown) have been corrected by a Kramers-Kronig transformation (Ferraro *et al.*, 1989).

microtransmission technique for the identification of contaminants encountered in industrial processes.

Ishitani (1985) and Ishida *et al.* (1987) have shown that the microreflectance technique can be used for the identification of contaminants on PC boards *in situ*. Figure 14 (Ishitani, 1985) shows a photomicrograph of the contaminant and its reflectance spectrum. The authors have identified the spectrum as a polyamide. The microreflectance technique has also been used by Harthcock and Atkin (1988) for studying the "fish-eyes" on painted polymer panels. The fish-eye in this case was on the painted side of an extruded polyurethane panel. By comparing the reflectance spectra from the fish-eye, a well-painted region, and the unpainted region, they were able to conclude that the fish-eye occurred over an area where the mold release agent had not been washed properly. Along similar lines, Krishnan (1987) has reported the microreflectance spectra from the "crater" and the "good" areas of a painted metal aluminum panel. These spectra are shown in Fig. 15. The microreflectance technique has been used by Smyrl *et al.* (1988) to characterize oily streaks on uranium metal and by

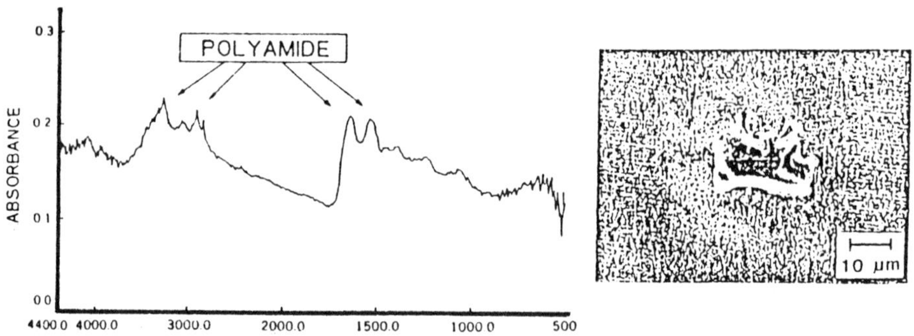

Fig. 14. Photomicrograph of (right) and microreflectance spectrum for (left) a 10 × 20-μm contaminant on a PC board. The particulate was identified as a polyamide compound [From Ishitani (1985). Reprinted with permission of the publishers, SPIE.]

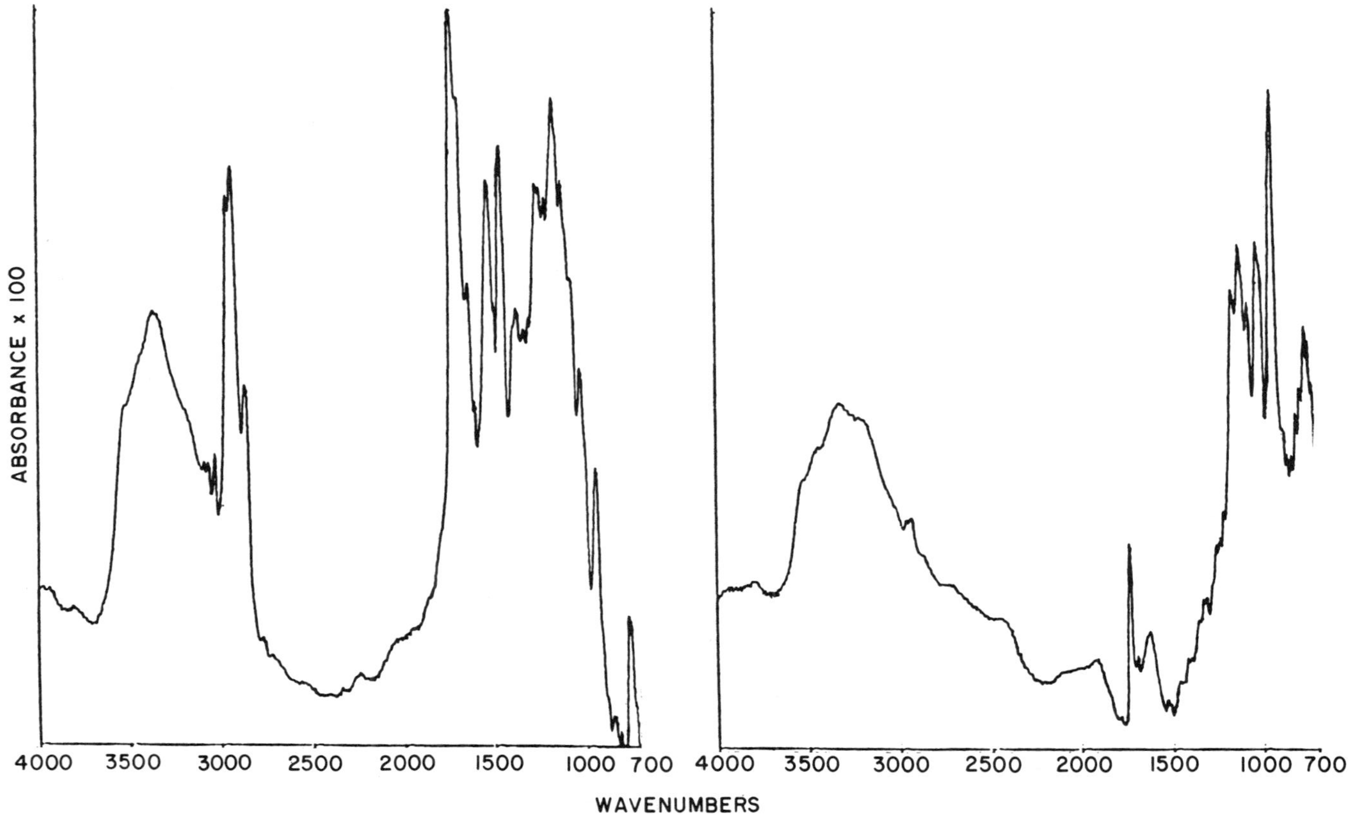

Fig. 15. Microreflectance spectra from a good area (left) and a crater (right) of a painted metal panel. [From Krishnan (1987). Reprinted with permission of the publishers, Plenum Press, New York.]

Wooten and Hughes (1988) in the study of oil residues on ball bearings and similar rotor parts.

D. Polymer Characterization

The infrared microscope sampling technique may be used for a number of polymer characterization techniques. These include the study of industrial contaminants in polymers, the identification of individual layers in multilayer polymer laminates, the characterization of single-filament polymer fibers, and the measurement of the dichroic spectra as a function of polymer stretching. Some of these applications are described in this section.

1. Imperfections in Polymers

Industrially produced polymers often contain imperfections such as foreign particles, haze, and dimples. These imperfections are often microscopic in nature. When the polymer sample is thin, these imperfections may easily be studied by the microtransmission method. Krishnan (1987) has illustrated this application. Figure 16 shows the spectra of a hazy spot of dimensions 150 × 150 μm, a clear spot of the same dimensions, and the difference from a polyethylene sample that is used as the sheath in electrical transmission cables. One can easily identify a number of absorption bands in the difference spectrum and thus ultimately identify the cause of the hazy spots. Similar analyses of polymer contaminants has also been reported by Schiering (1988).

Another example of the characterization of imperfections is shown in Figs. 17 and 18. The sample under study in this case was a polished plate (2 mm thick), of a glass–epoxy

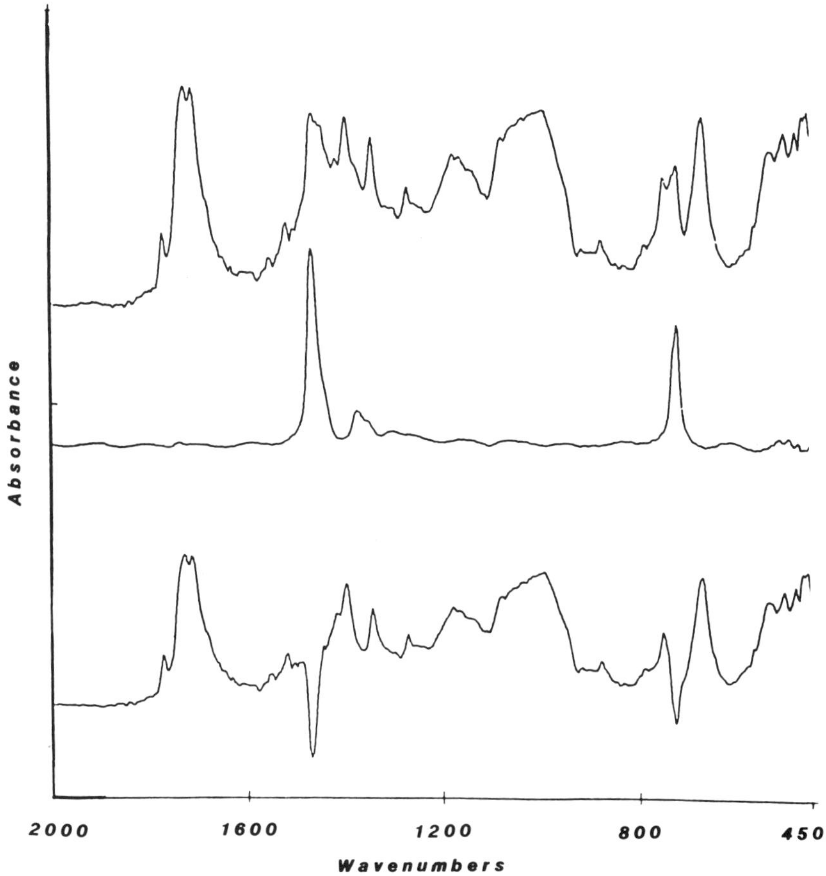

Fig. 16. Microtransmission spectra of polyethylene insulation from electrical wire. From top to bottom, spectra of a hazy area, a clean area, and the difference. [From Krishnan (1987). Reprinted with permission of the publishers, Plenum Press, New York.]

composite of the type used in airplane and automobile frames. Under visual microscopic examination, this otherwise colorless sample exhibited some purplish veins. Since the sample thickness is large, and glass is opaque to much of the mid-infrared spectrum, microreflectance spectra were recorded

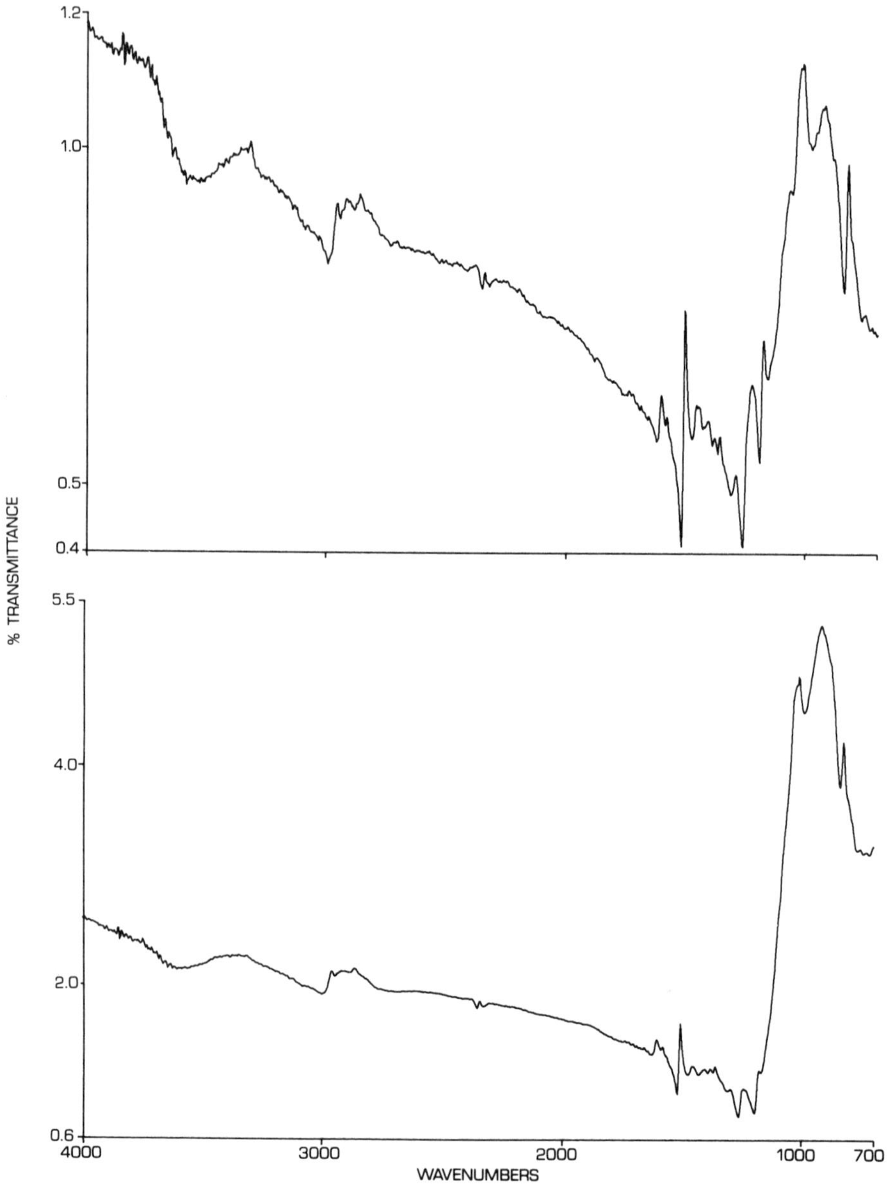

Fig. 17. Microreflectance spectra of a "veined" region (top) and a normal region (bottom) of an epoxy-silica composite plate.

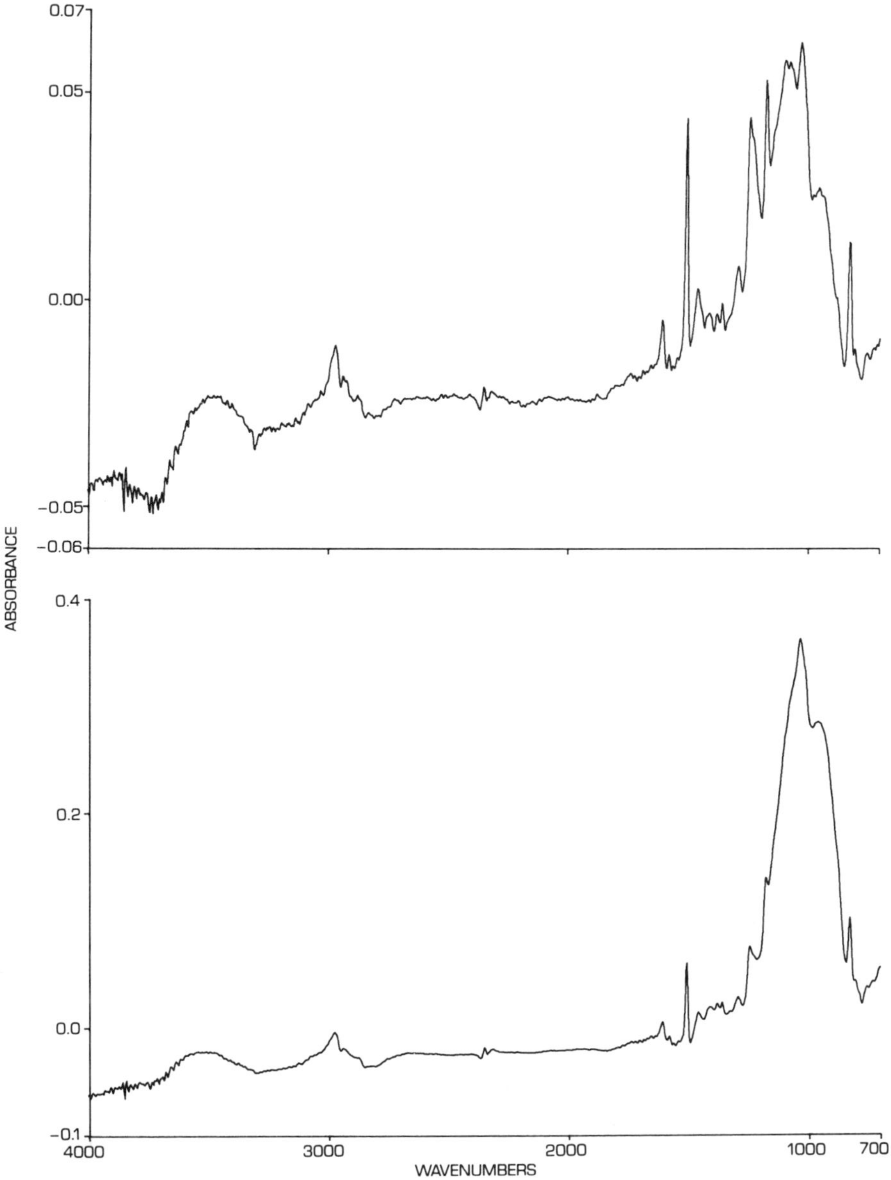

Fig. 18. Kramers-Kronig transforms of the microreflectance spectra of Fig. 17: "veined" region (top) and normal region (bottom). The strong organic absorptions in the "veined" region are due to excess epoxy in the composite.

from an area 160 × 80 μm in the region of the veins and from a clear area (Fig. 17) of the sample. The reflectance spectra were then corrected by K-K transformation to yield the "normal-looking" spectra. One can easily see from Fig. 18 that the area of the veins contains larger amounts of epoxy than does the normal area. Obviously, the composite is very inhomogeneous, and the veins appear to be due to the segregation of epoxy in the composite.

2. Multilayer Polymers

Most polymer packaging materials consist of multiple polymer layers (laminates). Cross sections 5–15 μm thick can be produced with a microtome. The aperturing ability of the infrared microscope sampling accessory enables investigators to record high-quality transmission spectra of the individual polymer layers in these cross sections. Harthcock *et al.* (1986), Krishnan (1986, 1987), and Reffner *et al.* (1987) have characterized a number of polymer laminates using the microtransmission technique. Figure 19, reported by Krishnan, shows the transmission spectra of the four-layer laminate: polyethylene terephthalate(PET)–polyethylene (PE)–aluminum–polyethylene. One can see from the figure that identifiable spectra could be recorded from the PET and PE layers that are less than 15μm thick.

Another example of a multilayer polymer characterization is shown in Fig. 20. This analysis involved a variety of

Fig. 19. Microtransmission spectra for the three polymer layers represented by the schematic of a four-layer, 100-μm-thick laminate film. The spectra correctly identify, from top to bottom, polyethylene terephthalate (14 μm thick), polyethylene (15 μm thick), and polyethylene (61 μm thick). [From Krishnan (1986). Reprinted with permission of the publishers, SPIE.]

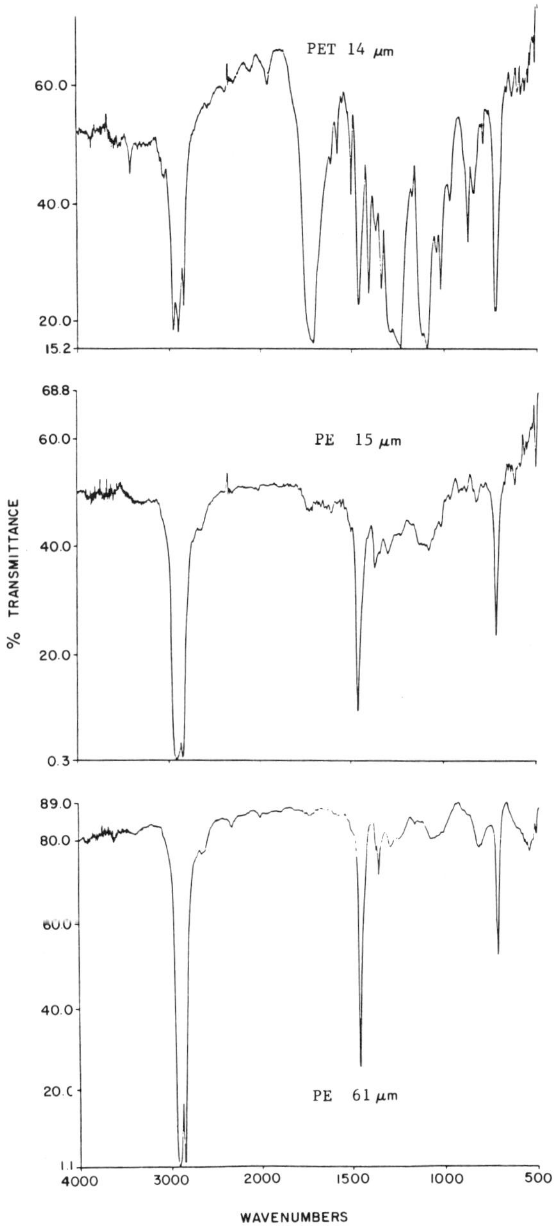
PET 14 μm
PE 15 μm
PE 61 μm
% TRANSMITTANCE
WAVENUMBERS
4000
3000
2000
1500
1000
500

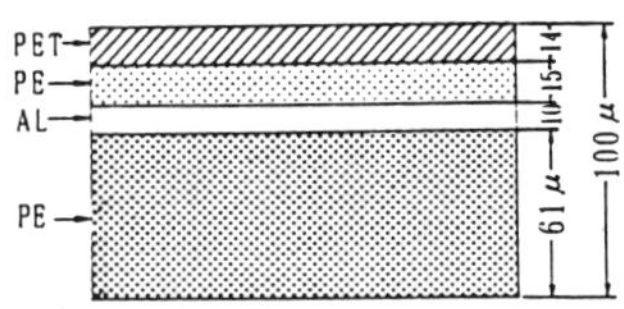
PET
PE
AL
PE
14
15
10
61 μ
100 μ

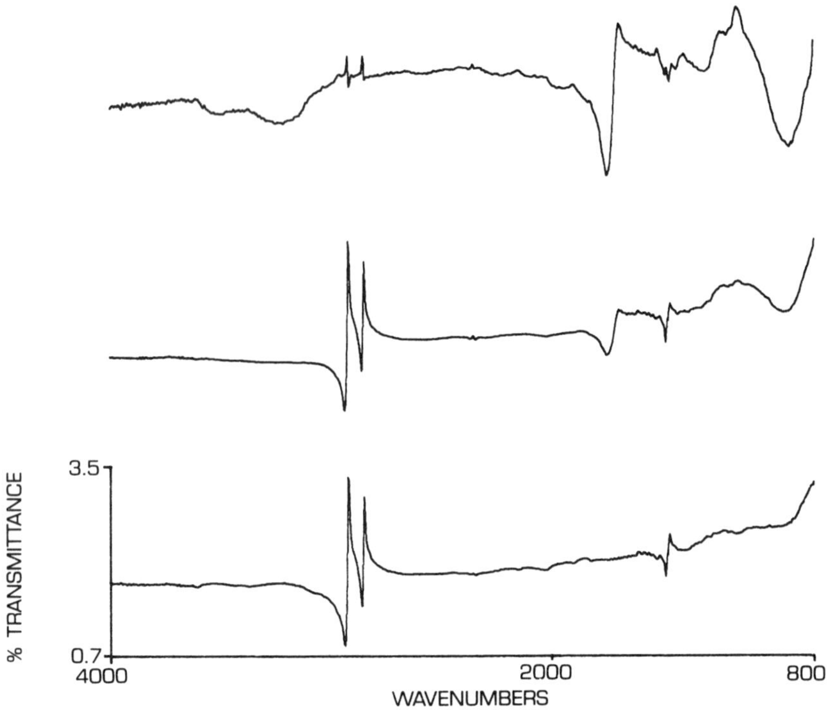

Fig. 20. Microreflectance spectra of ungrafted (bottom) and grafted (middle) polyethylene beads 4 mm in diameter, along with the difference spectrum (top).

grafted and ungrafted polyethylene spheres with a nominal diameter of approximately 4mm. This type of sample is used in biochemical experiments involving proteins and enzymes. Microreflectance spectra were taken for several samples, and the K-K transformation was performed upon each. The resulting spectra of the ungrafted and the grafted samples and the difference between the two is shown in Fig. 21. The microreflectance spectrum of the graft (the difference spectrum of Fig. 21) and the *in situ* photoacoustic spectrum are compared in Fig. 22. One can see excellent correspondence between the two spectra.

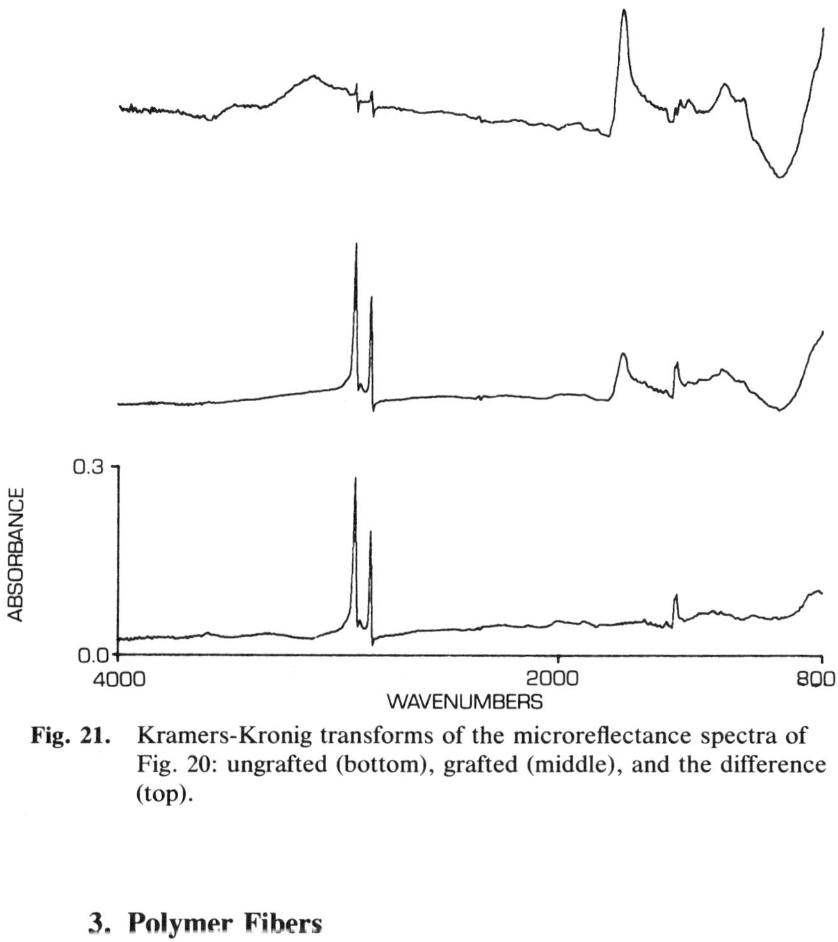

Fig. 21. Kramers-Kronig transforms of the microreflectance spectra of Fig. 20: ungrafted (bottom), grafted (middle), and the difference (top).

3. Polymer Fibers

Transmission spectra of single filaments of fibers may easily be recorded using the infrared microscope sampling method. Blout *et al.* (1949) were the first to record the infrared transmission spectrum of a single fiber, that of polyvinyl alcohol, using a dispersive infrared spectrometer. Coates *et al.* (1953) have recorded the spectra of acrylic-type fibers of less than 20 μm in diameter, also using a dispersive infrared spectrometer. Krishnan (1984) reported the first FT-IR transmission spectrum of a single fiber (a filament of PET 15 μm in diameter).

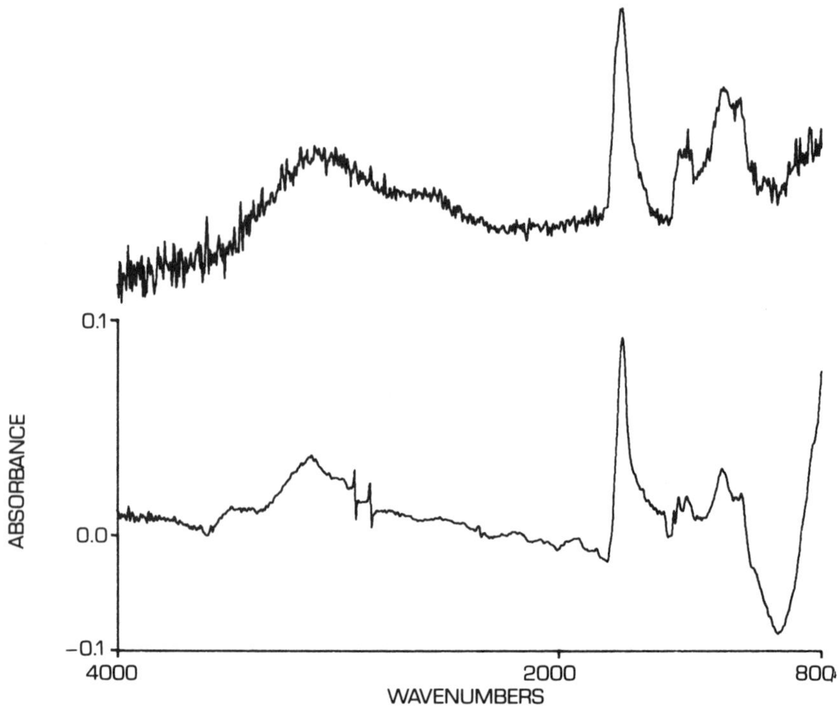

Fig. 22. Comparison of the photoacoustic (top) and Kramers-Kronig transformed microreflectance difference (bottom) spectra for the grafted polymer surface on polyethylene beads 4 mm in diameter.

Subsequently, a number of single-fiber FT-IR spectra have been reported in the literature (Krishnan *et al.*, 1985, 1986, 1987; Lang *et al.*, 1988; Reffner *et al.* 1987; Smyrl *et al.*, 1988). Figure 23 illustrates the sensitivity of the FT-IR microscope sampling technique for the study of single fibers. The figure shows the spectrum of a single Dralon fiber recorded with measurement times of 0.25 sec (one scan) and 70 sec (256 scans). One can see from the figure that an identifiable spectrum could be recorded in a single scan!

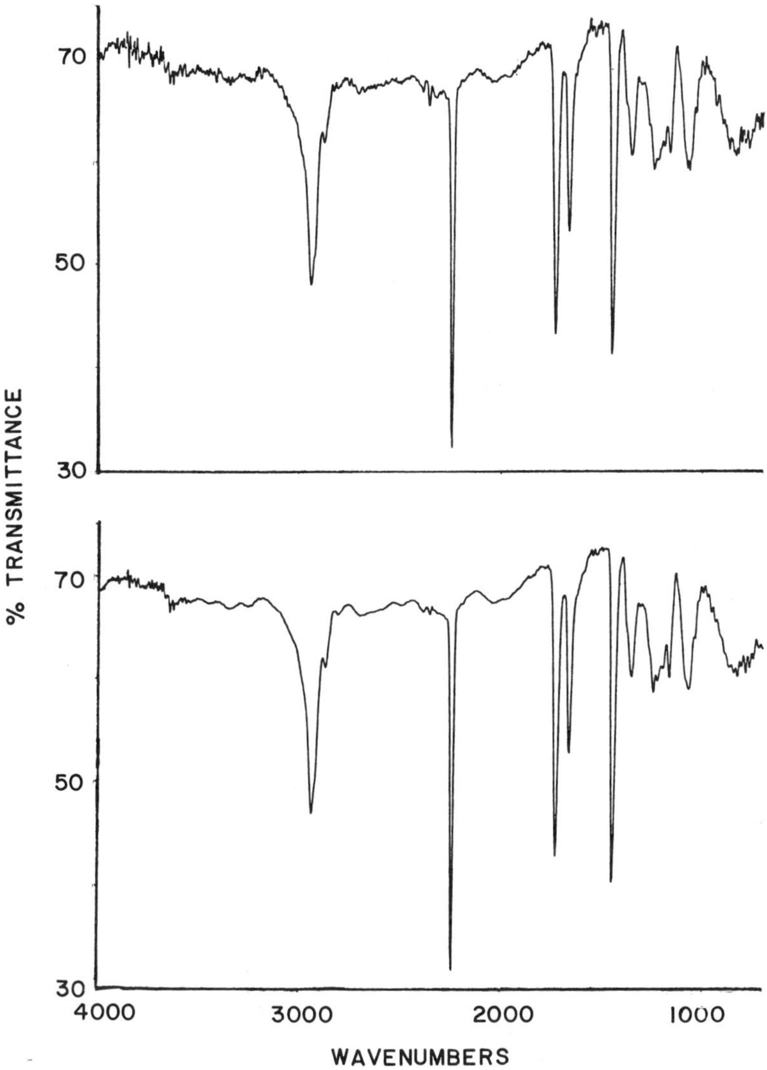

Fig. 23. Microtransmission spectra for a Dralon fiber 15 μm in diameter. The spectra were measured for 1 scan (top) and 256 scans (bottom).

4. Dichroic Studies

Infrared polarization (dichroic) measurements can provide valuable information on molecular orientations. The experimental considerations involved in these measurements have been outlined elsewhere in this chapter. Barer *et al.* (1949) and Coates *et al.* (1953) have reported the dichroic spectra of a number of crystalline materials recorded using the transmission infrared microscope sampling technique. Krishnan (1984) has recorded the dichroic transmission spectra of a PET fiber 15 μm in diameter using FT-IR spectroscopy.

The infrared microscope sampling technique can also be used for the recording of the dichroic spectra of ultrasmall samples such as single polymer fibers as a function of applied stress. Brasch and Lustiger (1988) have investigated the phenomenon of "necking," which commonly occurs after stretching a polymer. The neck defines the boundary between amorphous and crystalline regions of the polymer. Hill *et al.* (1987) have also studied the dichroic spectra of a few polymer samples as a function of stretching. Figures 24 and 25 show the dichroic spectra of a polypropylene film as a function of stretching. One can see the dichroism, and hence the molecular orientation, of the sample changes with stretching. Figure 25 shows the dichroic ratio of two representative bands in the polypropylene spectrum, 1220 cm^{-1} and 997 cm^{-1}, as a function of stretching. These figures show that a parallel band gradually changes in relation to a perpendicular band as a function of stretching.

Another example of structural changes with applied stress is shown in Fig. 26. The sample used in this case was polybutylene terephthalate (PBT). It is known that the PBT molecule changes from the alpha to the beta form upon elongation (Holland-Moritz and Stach, 1988). In the present study, when the PBT film was stretched while on the microscope stage, a "neck" could be seen on the stretched sample. A number of transmission spectra were recorded using a sampling area of 80 × 20 μm over the necking region. These

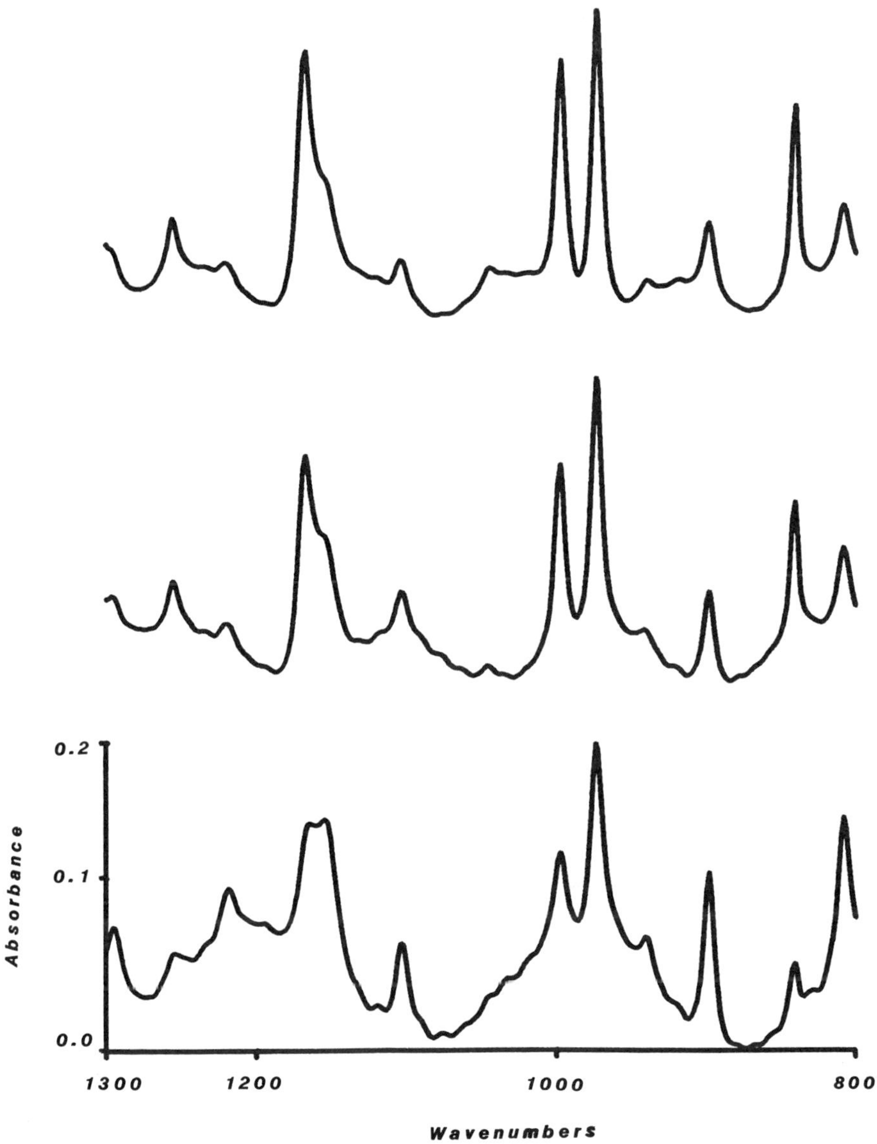

Fig. 24. Polarized microtransmission spectra of a polypropylene film under stress in a polymer stretching device. The electric vector is oriented perpendicular to the stretching direction. The spectra correspond to low (bottom), medium (middle), and high (top) levels of draw.

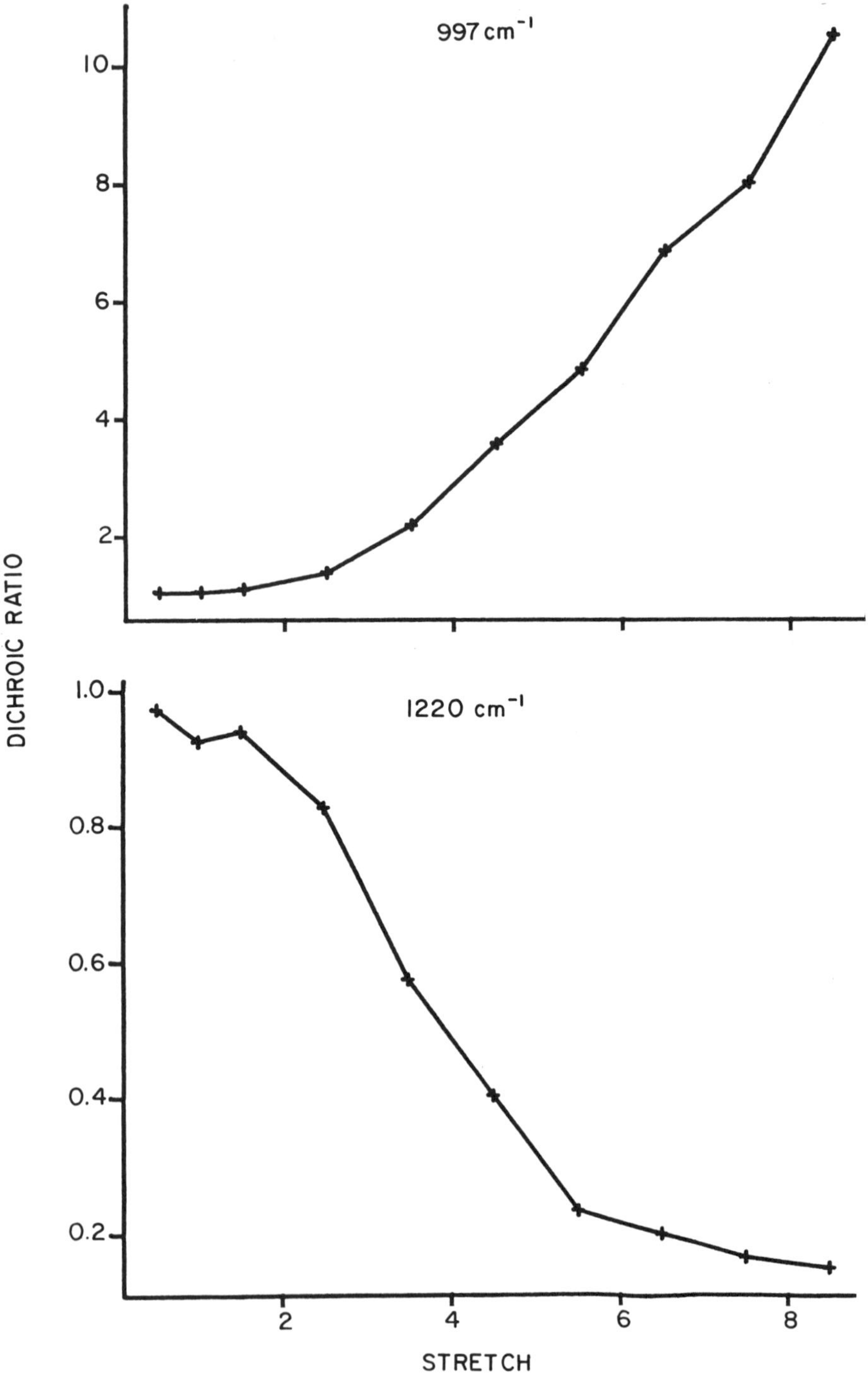

997 cm⁻¹
1220 cm⁻¹
DICHROIC RATIO
STRETCH
10
8
6
4
2
1.0
0.8
0.6
0.4
0.2

spectra are shown in Fig. 26. It can be seen that the transition from the alpha to the beta form occurs suddenly at the "neck"; that is, the transition between the two forms takes place within a 40-μm distance around the "neck." As far as the authors are aware, this is the first report on such a phenomenon.

Another example of the study of the spectral changes as a function of draw is found in our study of a single Kevlar fiber filament 15 μm in diameter. When stretched, the fiber elongates and then breaks. The spectra shown in the figure were recorded before the fiber broke. Figure 27 shows a schematic of the breakdown region just before the break occurred, along with transmission spectra for the deformed region, the transition zone, and the normal, undistorted fiber. The figure illustrates the ease with which such seemingly difficult studies may be carried out.

E. Biological Applications

Biological materials are complex and microscopic, but they can easily be studied with the infrared microscope sampling technique. Indeed, one of the early applications of the infrared microscope sampling technique involved the investigation of biological samples (Barer *et al.* 1949; Blout *et al.*, 1949; Wood, 1950; Blout and Bird, 1951; Coates *et al.*, 1953). These investigations dealt with frog muscles and rat tissues. All of these studies were carried out using dispersive infrared spectrometers. A few applications of the infrared microscope sampling technique using FT-IR spectrometers have been reported in the literature. Herres (1985) has recorded the transmission spectrum of a cluster of spores from the fungus *Lycoperdon perlatum*. Levinson *et al.* (1985) have identified

---

Fig. 25. Comparison of the dichroic ratios of the polypropylene absorption bands at 1220 cm^{-1} and 997 cm^{-1}. The dichroic ratio of the 1220-cm^{-1} (perpendicular) band decreases as a function of stretch, whereas that of the 997-cm^{-1} (parallel) band increases.

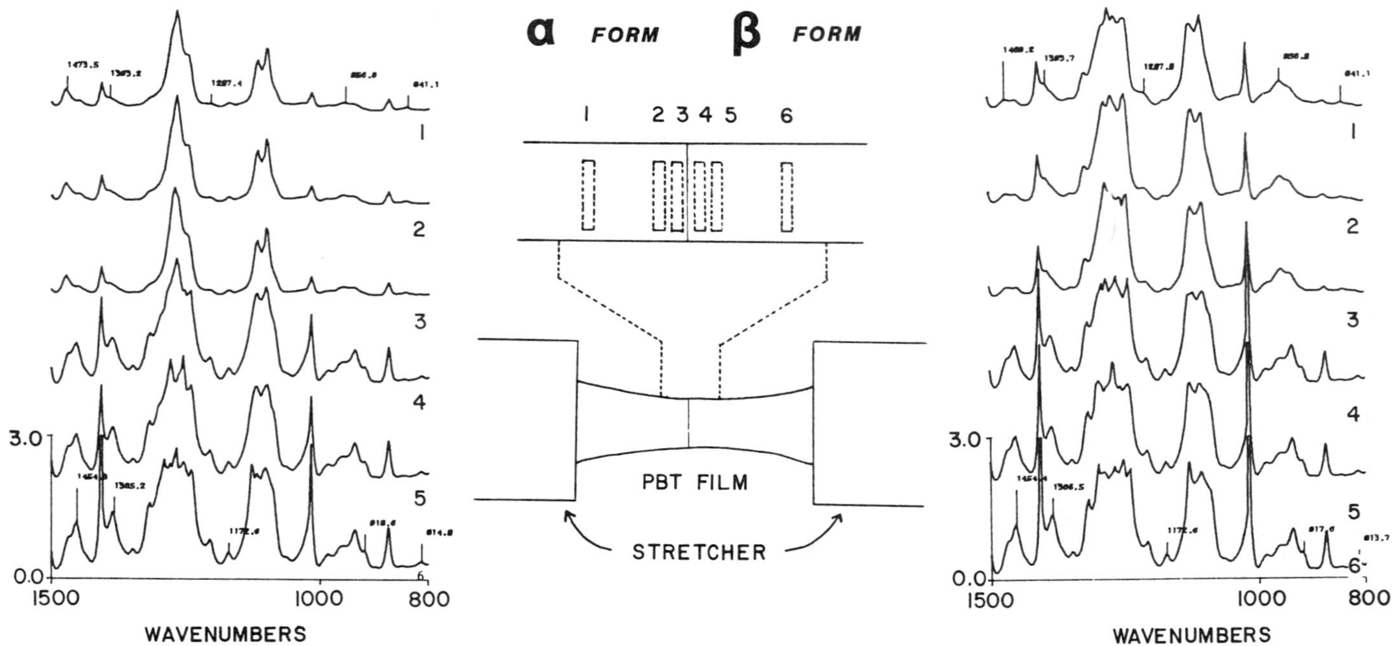

Fig. 26. Comparison of the parallel (right) and perpendicular (left) polarized microtransmission spectra for polybutylene terephthalate (PBT). A schematic diagram of the stretched polymer and the sampling regions is shown in the middle. The figure shows the sharp transition from the alpha (bottom) to the beta (top) forms.

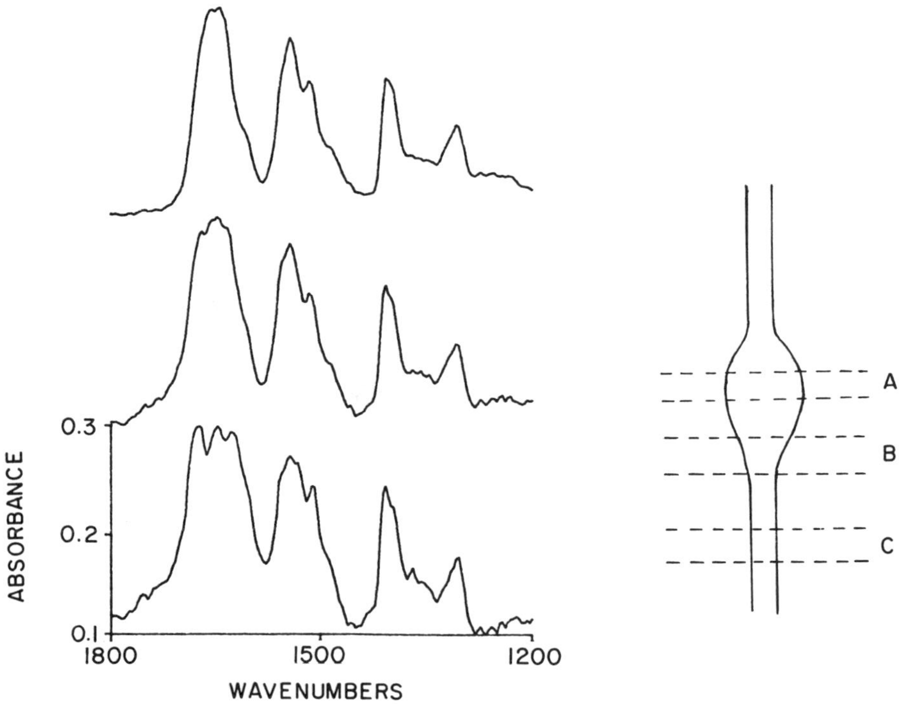

Fig. 27. Polarized microtransmission spectra for unstretched (bottom; C), transition (middle; B), and deformed (top; A) regions of a single Kevlar fiber. The electric vector is oriented perpendicular to the stretching direction.

calcium oxalate in human tissue sections. Dluhy and Mendelsohn (1988) have studied rat femur sections and the problem of calcination.

Hill *et al.* (1987) have reported the infrared spectra of normal and Alzheimer disease-affected brain tissue. Under microscopic examination, some plaque-like deposits could be seen on the neurofibular walls of the diseased tissue. Figure 28 shows the average transmission spectra of the normal and the diseased cells along the neurofibular walls from a number of cells. Figure 29 shows the difference spectrum between the diseased and normal cells. It has been postulated that aluminum silicate deposits along the walls of the Alzheimer-afflicted brain cells; and in the difference spectrum some

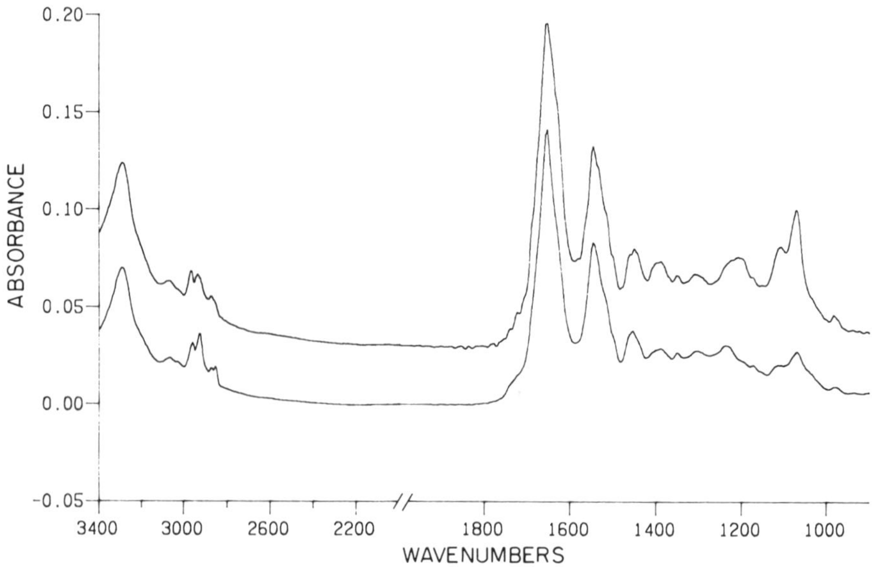

Fig. 28. Microtransmission spectra for healthy (bottom) and Alzheimer-diseased (top) brain tissue sections. The spectra are the average of 64 arbitrary regions selected with a motorized stage. A β-pinene adhesive was used to support the tissue section, and its spectral features have been subtracted away.

absorption bands could indeed be seen in regions of the silicate absorption. However, caution must be exercised in forming firm conclusions based upon the spectra of one set of samples.

Another example of the application of the microtransmittance technique to the study of biological samples is shown in Figs. 30–32. Figure 30 depicts the cross section of an arterial wall. A cross section (5 μm thick) of the arterial wall was mounted on a BaF_2 plate, and the transmission spectrum of the epithelial deposit on the inner wall of the artery was recorded. The original and the baseline corrected spectra of this epithelial deposit is shown in Fig. 31. Figure 32 shows a series of spectra recorded over sections 100 × 100 μm in area that reveal the transition from the smooth muscle to the epithelial wall that is laden with fatty deposits.

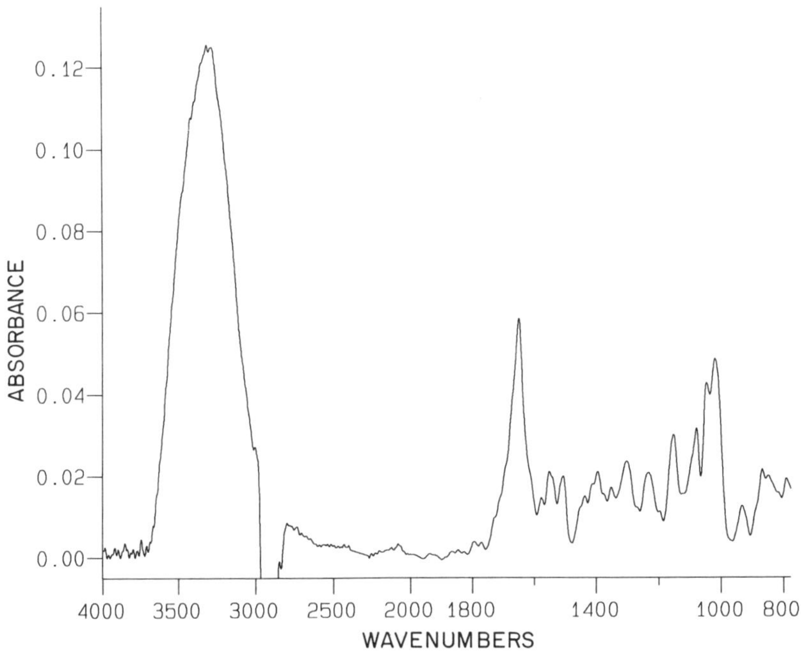

Fig. 29. Microtransmission difference spectrum of an Alzheimer-diseased neuron tissue (80 × 20 μm) has absorption bands that correspond with those of aluminum silicate. The spectra of the adjacent brain tissue and the adhesive were separately subtracted.

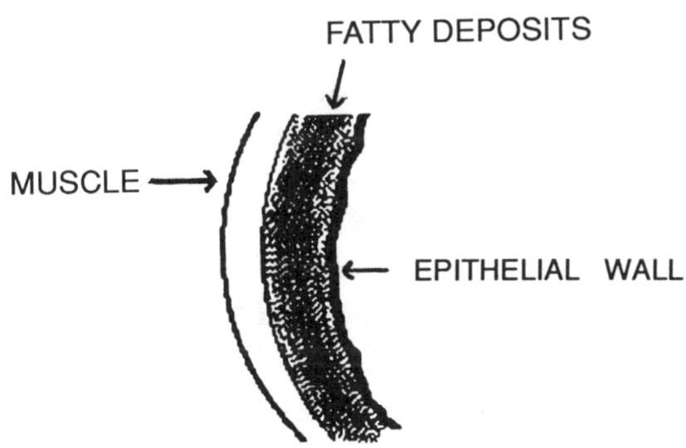

Fig. 30. A schematic of a microtome cross section of an arterial wall.

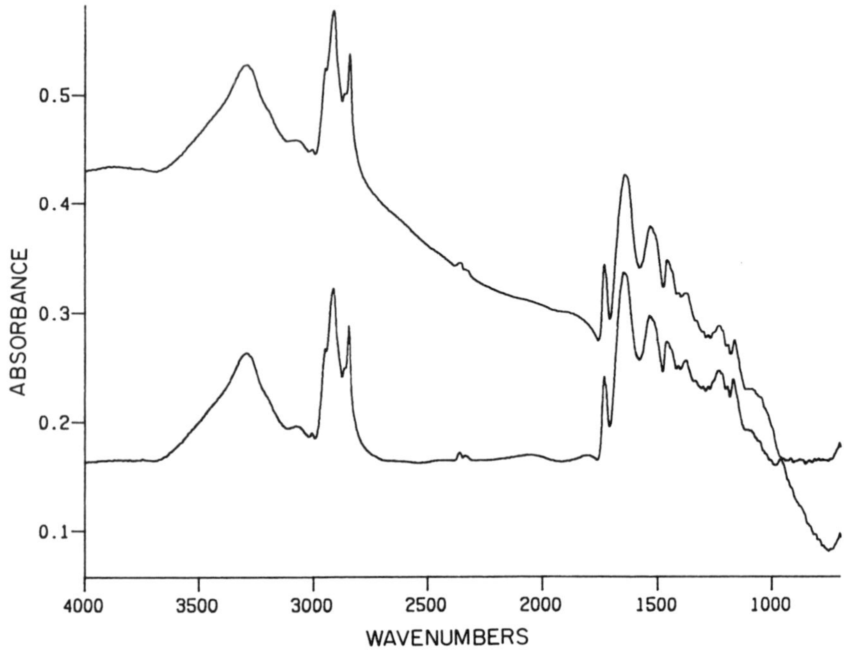

Fig. 31. The original microtransmission spectrum of the epithelial deposits on the arterial wall (top) and after baseline correction (bottom).

The microreflectance technique can be used in cases where the biological sample could not be cut easily into thin sections. Figures 33 and 34 illustrate the reflectance technique in dental studies. Figure 33 shows the microreflectance spectrum of a polymer used as a moldable denture base. From the K-K corrected spectrum, the filler could be identified as polymethylmethacrylate from the spectral search report shown in Fig. 34. Some other examples of the application of the microreflectance technique to biological samples are shown in the following figures. Figure 35 shows the reflectance and the K-K corrected spectrum of a urine stone. Figure 36 shows similar data for a kidney stone. Figure 37 is the search report for the K-K corrected spectrum of the kidney stone, identifying the material in the stone to be calcium carbonate.

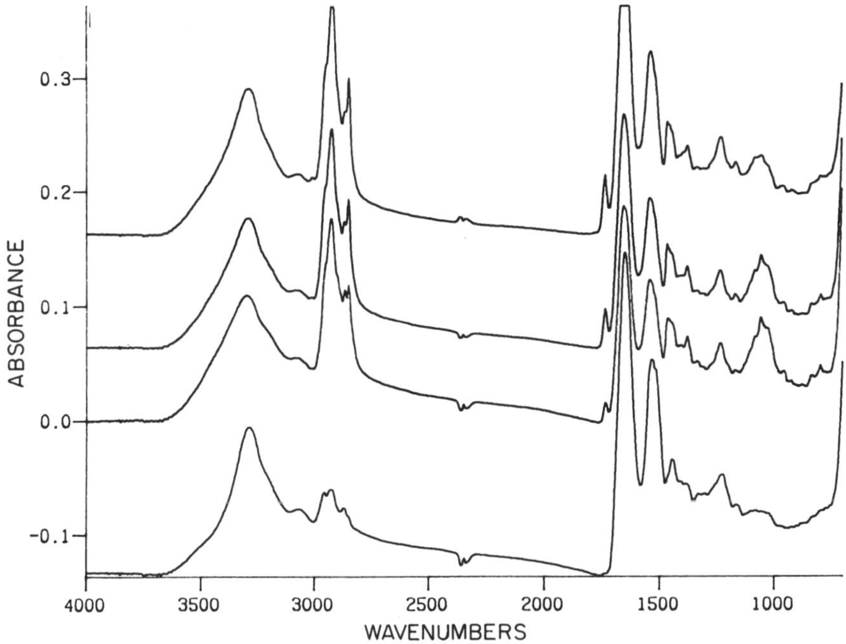

Fig. 32. Comparison of microtransmission spectra from the inner wall (top) to the smooth outer muscle (bottom) of the arterial wall. The absorption near 1700 cm^{-1} indicates the distribution of fatty acids on the arterial wall.

F. Applications in the Near-Infrared Region

The infrared microscope sampling accessory, when coupled to a near-infrared FT-IR instrument, can be used effectively in the near-infrared region (10,000–3000 cm^{-1}). The standard MCT detector in most cases can be used for the near-infrared measurements; if higher sensitivity is needed, the MCT detector can be replaced by an InSb detector. The range of applications in the near-infrared region include the characterization of thick polymer and biological samples and particles from their overtone and combination spectra, and the study of low-energy electronic transitions. The latter spectra could be from

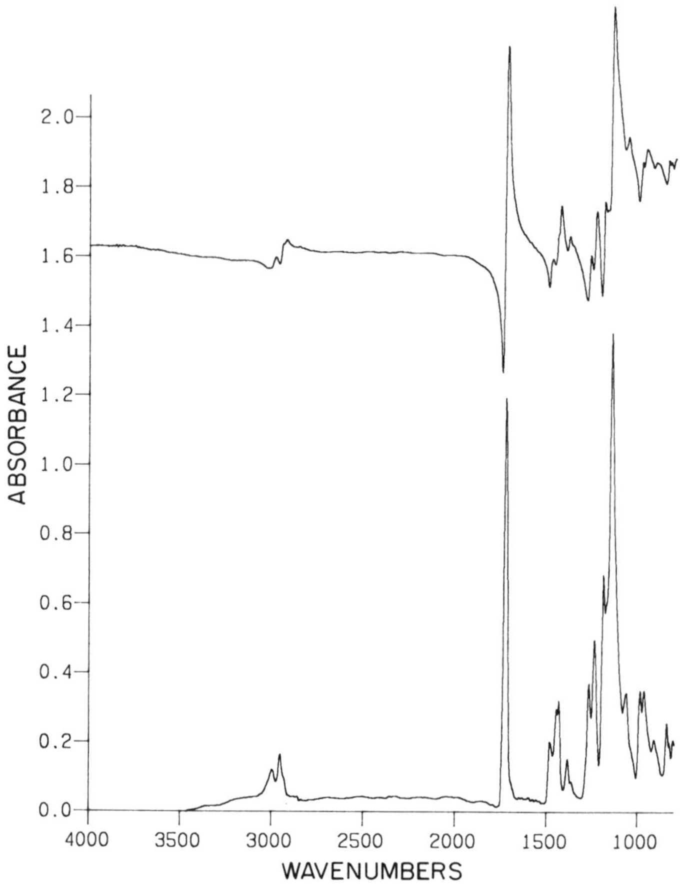

Fig. 33. Microreflectance spectrum of a molded denture base (top) and its Kramers-Kronig transform (bottom).

ionic and rare earth crystals or from semiconducting materials. Hitchman *et al.* (1988) have studied the dichroic electronic spectra of a single crystal (150 × 50 μm) of K_2NiO_2 in the near-infrared. The sample, which is highly hygroscopic, was kept sealed in a fused silica capillary tube under a nitrogen atmosphere. A wire-grid polarizer and an InSb detector were used in recording the dichroic spectra. These spectra are shown in Fig. 38. Hitchman *et al.* were able to assign all these

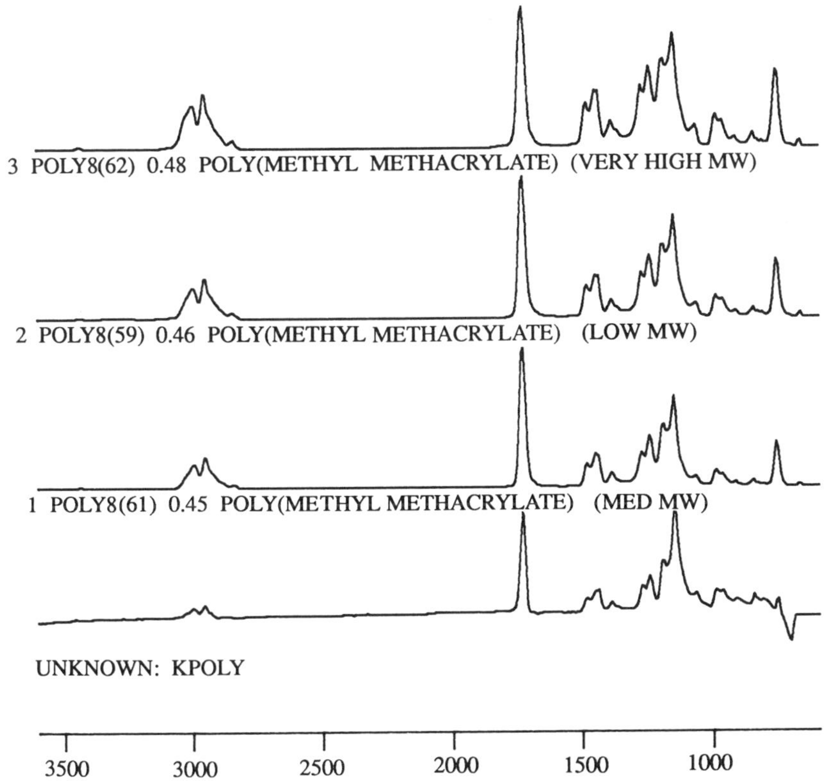

Fig. 34. Library search identification of a molded denture plate confirms the composition as polymethylmethacrylate. The data base for this identification was an independent 8-cm^{-1} digitized polymer spectral library.

bands to various electronic transitions, and they concluded that there is significant configuration interaction between the metal $3d_z^2$ and the 4s orbitals in the linear NiO_2^{2-} ion.

There have also been unpublished reports on the microscopic mapping of energy level transitions in Group III-V semiconductors. This study is similar to the mapping study of the EL2 level in GaAs reported by Dobrilla and Blakemore (1985).

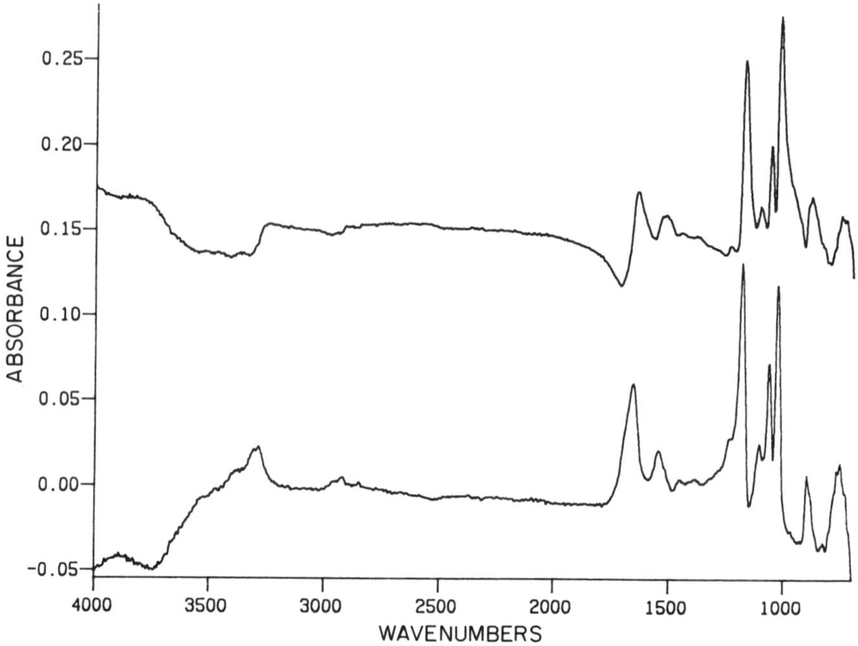

Fig. 35. Microreflectance spectrum of a urine stone (top) and its Kramers-Kronig transform (bottom).

G. Semiconductor Measurements

The infrared microscope sampling technique can be used in a number of ways during semiconductor device manufacturing. One of the major applications, that of identifying contaminants on finished PC boards, has been described already. In this section we will describe the applications of the technique to semiconductor material characterization.

1. Epitaxial Thickness Measurements

Most semiconductor devices are fabricated using epitaxial methods. These epitaxial layers could be built on silicon,

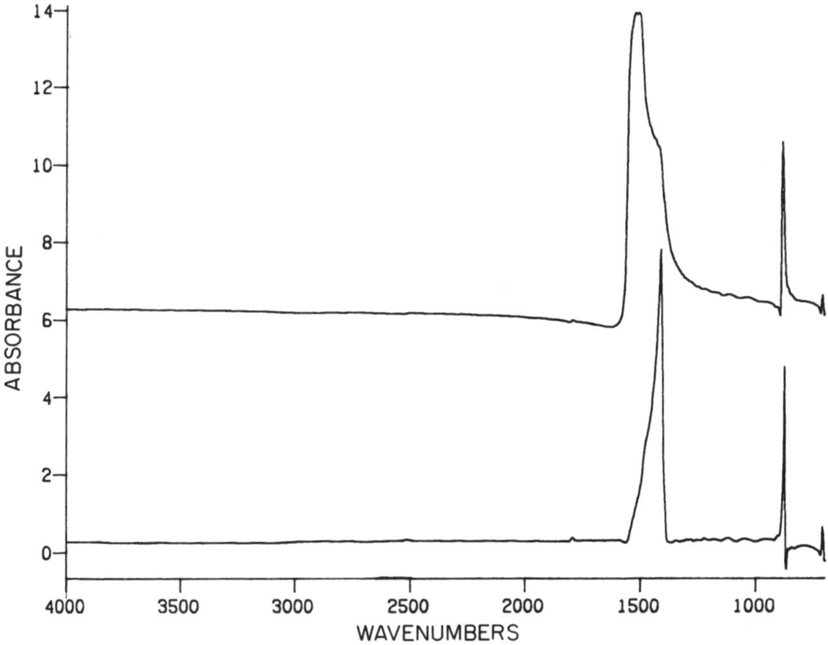

Fig. 36. Microreflectance spectrum of a kidney stone (bottom) and its Kramers-Kronig transform (top).

Group III-V, or Group II-VI materials. Reflectance FT-IR spectroscopy is a rapid, nondestructive method for the determinations of these epitaxial thicknesses (Krishnan *et al.*, this volume). On a macroscopic level, these measurements are carried out over sampling areas ranging in diameter from 3 to 12 mm. With the microsampling technique, one can make these measurements over sampling areas as small as 10×10 μm. Thus, the infrared microscope sampling technique is ideally suited for the determination of the epitaxial layers over buried layers and on microstructures such as MCT array detectors and Group III-V superlattice structures. Krishnan and Kuehl (1984), and Krishnan and Mundhe (1983), and Krishnan (1988) have described these microscopic epitaxial measurements in detail.

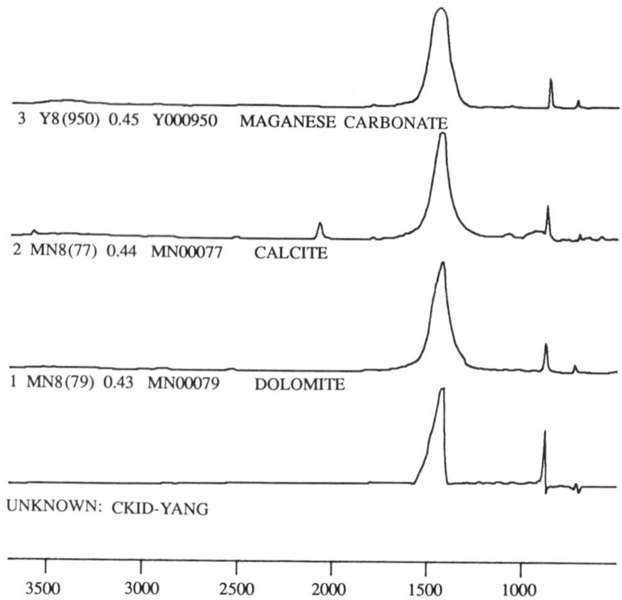

Fig. 37. Library search identification of a kidney stone suggests that it is composed of calcium carbonate. The data base used for this identification was the Sadtler ™ 8-cm^{-1} digitized mineral spectral library.

Figure 39 shows the determination of epitaxial thickness over a buried layer 20 × 20 μm in area. The figure shows the interferogram recorded (at 8-cm^{-1} resolution, 3-sec measurement time) over the buried layer, a reference area immediately adjacent to the buried layer, and the difference. One can clearly see the two side lobes, or secondary interferograms, characteristic of the epitaxial thickness in the difference interferogram. The spacing between the side lobes yields a value of 1.25 μm for the epitaxial thickness. The infrared microscope sampling technique can also be used to probe the inhomogeneity of the epitaxial layer. Figure 40 (Krishnan, 1988) shows the variation (as shown by the position of the side lobes) of

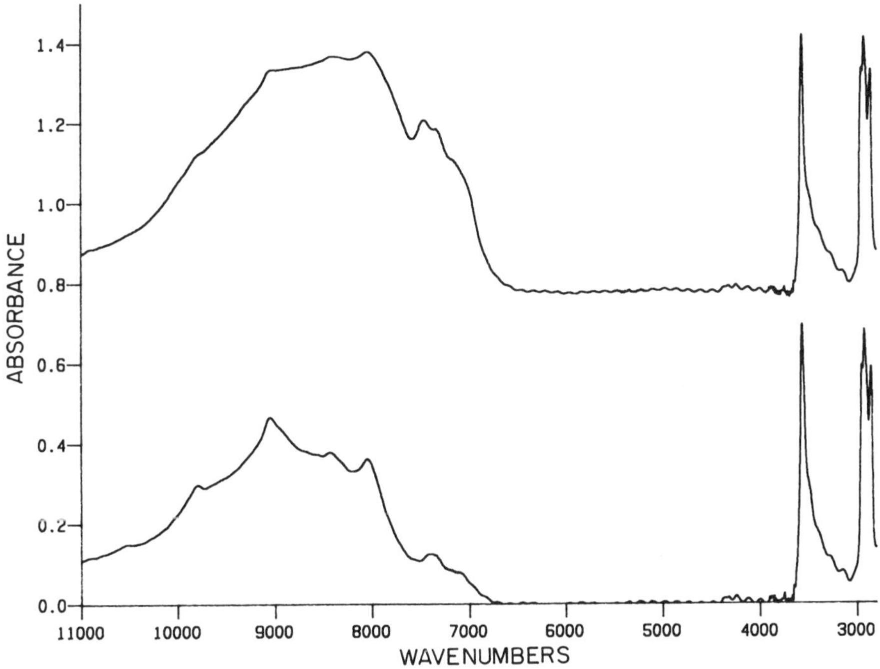

Fig. 38. Near-infrared dichroic spectra for a single crystal of K_2NiO_2. The electric vector is oriented parallel (top) and perpendicular (bottom) to the single crystal.

the epitaxial thickness across the diameter of a silicon wafer 100 mm in diameter. Figure 41 illustrates the measurement of an MCT epitaxial layer on a cadmium telluride substrate. The sample used in this case was one of the elements, 20 × 25 μm in linear dimensions, in an MCT array detector. The figure shows the reflected interferogram from the MCT element, a reference CdTe element, and the difference. Once again, the side lobes characteristic of the MCT epitaxial thickness could easily be seen. From the locations of the side bursts, the thickness of the epitaxial layer was estimated to be 18.5 μm.

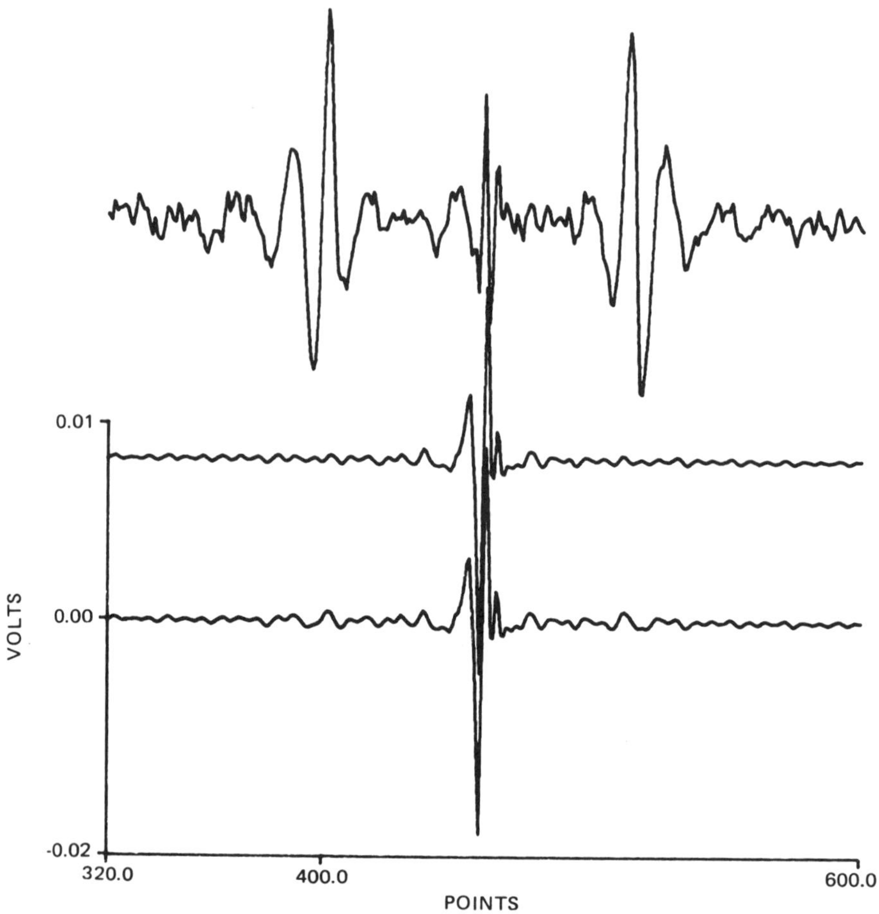

Fig. 39. Reflected interferograms for a silicon epitaxial film across a buried layer: (bottom) interferogram from the buried layer; (middle) reference interferogram from an area adjacent to the buried layer; (top) the difference interferogram clearly showing the side bursts due to the buried layer thickness. [From Krishnan, (1988). Reprinted with permission of the publisher, Marcel Dekker, Inc., New York.]

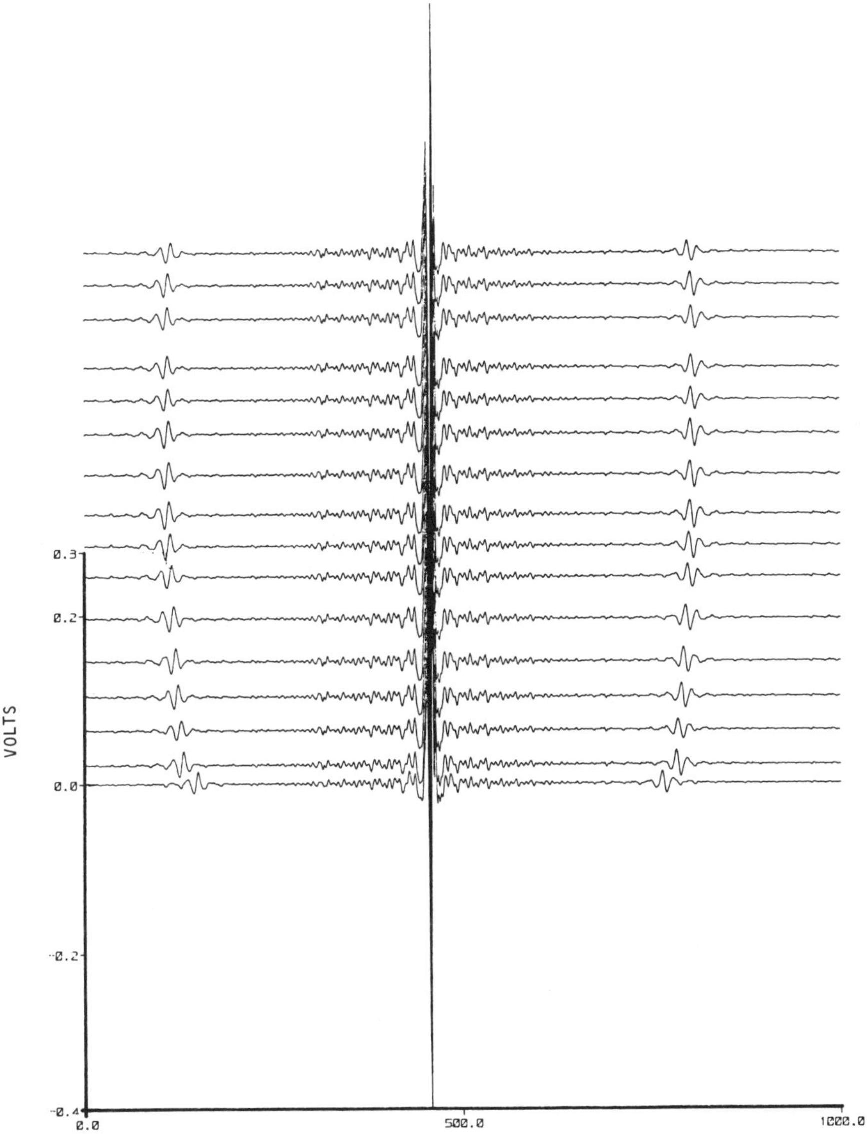

Fig. 40. Variation of the epitaxial thickness on a 4-in. silicon wafer. The nominal thickness of the epi layer is 30 μm. The single-scan interferograms shown were recorded at 5-mm intervals over one diameter of the wafer with the sampling area set to 100 × 100 μm [From Krishnan (1988). Reprinted with permission of the publishers, Marcel Dekker, Inc., New York.]

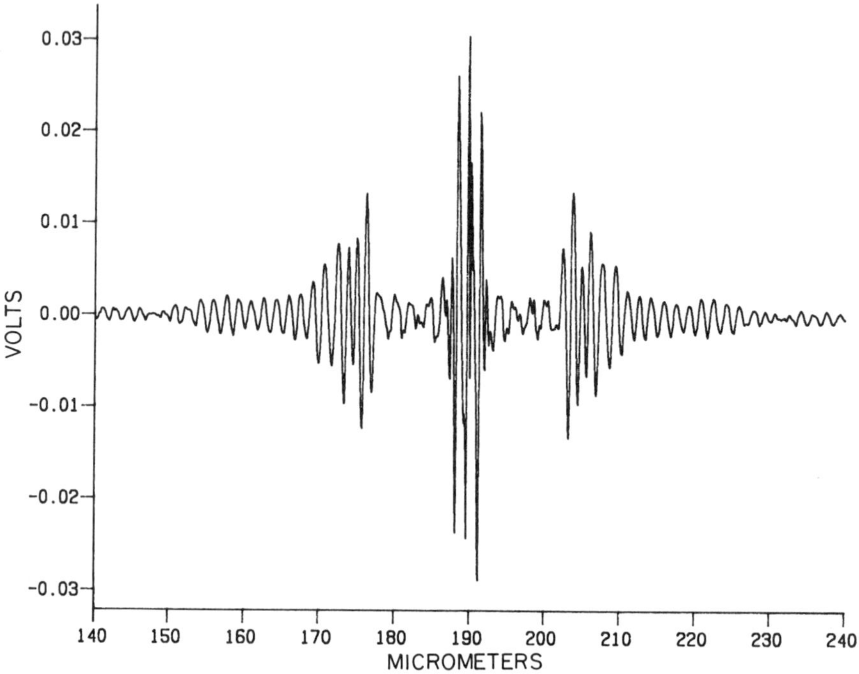

Fig. 41. Difference interferogram for an MCT epitaxial layer on a CdTe substrate.

2. Determination of the Concentrations of Interstitial Oxygen and Substitutional Carbon in Silicon

Transmission FT-IR spectroscopy is used as rapid, precise, and nondestructive technique for the measurements of the oxygen and carbon concentrations in silicon (Krishnan *et al.*, this volume, and references therein). Krishnan and Kuehl (1984), Krishnan and Mundhe (1983), Kim and Smetana (1986), and Yao and Witt (1987) have described the application of the FT-IR infrared microscope sampling technique to the study of these concentrations on a microscopic scale. Using a sampling area of 100 $\times$ 100 μm, Kim and Smetana (1986) have studied the effect of a large axial magnetic field applied during the Czochralski growth of the silicon crystal (AMCZ silicon). They found that the application of the magnetic field had the effect of reducing the interstitial oxygen concentrations and increasing the carbon concentrations. They were able to measure the oxygen segregation coefficient from the accumulated data. Yao and Witt (1987) have shown that the oxygen concentration could be studied by the microtransmission techniques over sampling areas as small as 40 × 40 μm. Figure 42 shows the infrared absorption band at 1107 cm^{-1} (due to interstitial oxygen in silicon) recorded as a function of distance along a section 2 mm thick cut parallel to the axis of an AMCZ silicon crystal similar to those used by Kim and Smetana (1986). One can see the significant variations of the oxygen concentration in this sample. Figure 43 graphically shows the variation of the concentration of oxygen in a quadrant of a silicon crystal 100 mm in diameter. The measurements were performed with sampling areas of 100 × 100 μm over the whole wafer using an automated *x–y* stage on the infrared microscope sampling accessory, while the data reduction was carried out on a personal computer.

3. Other Semiconductor Applications

In addition to the applications described in the preceding section, the transmission infrared microscope sampling technique can be used for the determination of the phosphorus and boron concentrations in phosphosilicate (PSG), borosilicate (BSG), and borophosphosilicate (BPSG) passivation layers on a microscopic level (Zearing and Coates, 1981; Krishnan and Kuehl, 1984; Krishnan, 1988). In finished silicon devices employing a buried-layer structure, there may be areas where the passivation layer is over lightly doped substrate areas. In such regions silicon will transmit the infrared radiation and facilitate the transmission measurements. The microsampling

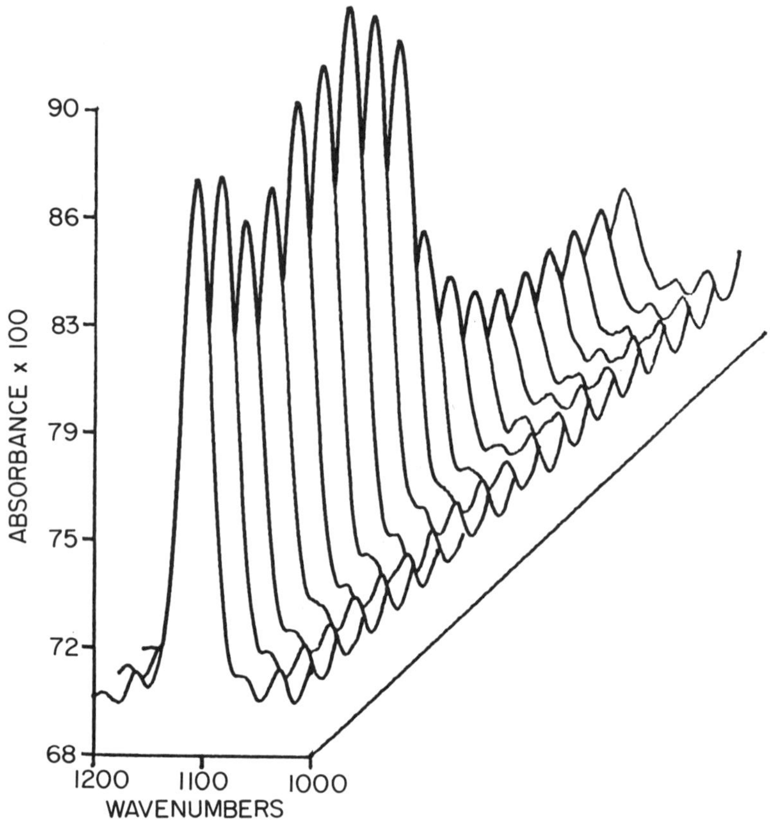

Fig. 42. Difference spectrum showing the variation of the oxygen concentration along the growth axis of an AMCZ silicon sample. [From Krishnan (1988). Reprinted with permission of the publishers, Marcel Dekker Inc., New York.]

technique allows the possibility of measuring the P and B concentrations on such finished devices. The transmission infrared microscope sampling technique can also be used for the mapping of atomic hydrogen concentration in silicon nitride passivation layers, the carbon impurity concentration in GaAs, and, in the near-infrared, the mapping of the EL2 level in GaAs.

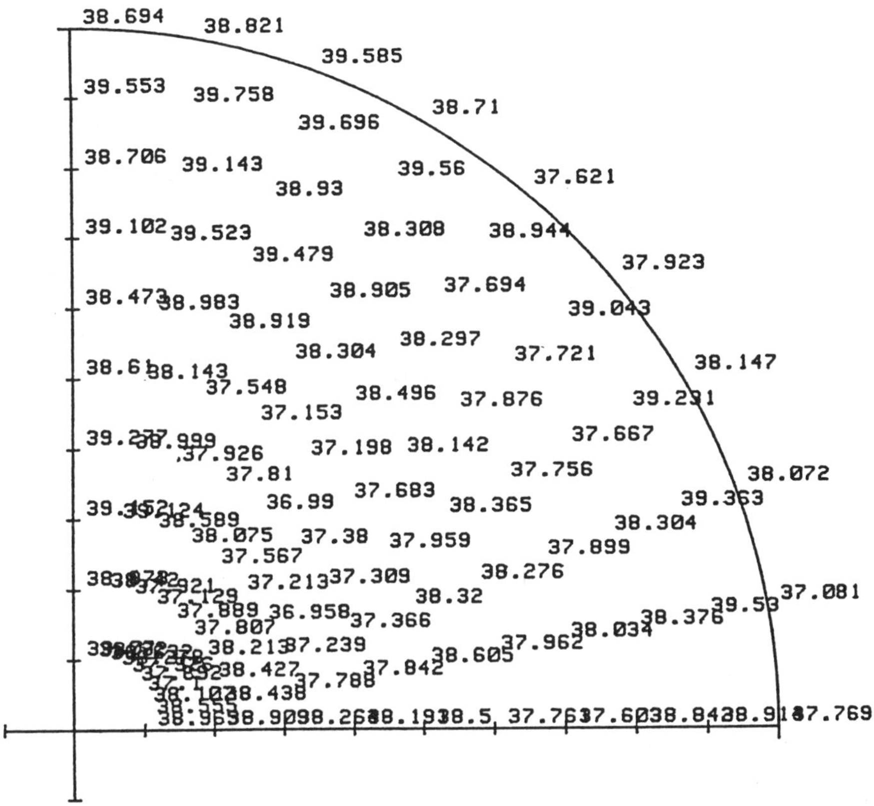

Fig. 43. Distribution of interstitial oxygen in a CZ silicon wafer as a function of *x–y* position. A motorized stage was used to automate the analysis.

H. Nonroutine Measurements

The infrared microscope sampling technique has been used for the development of specialized applications, which are likely to become more numerous in the future. Some of these current applications will be outlined in this section.

1. DSC-FT-IR Studies

Mirabella (1986) has used a differential scanning calorimeter (DSC) in conjunction with the infrared transmission microscope sampling accessory to study the melting of polyethylene. He used a DSC sample cup with NaCl windows and studied the melting process by collecting the transmission spectrum of the sample in 1-min intervals during the melting process. Mirabella (1988) has also studied the degradation of a small amount of poly(vinyl alcohol) over a 3-hr period using the technique.

2. Low-Temperature GC-FT-IR Spectroscopy

It has been shown by Reedy *et al.* (1985a,b) that the sensitivity of the gas chromatograph FT-IR (GC-FT-IR) technique can be increased considerably by employing low-temperature, matrix isolation techniques in conjunction with conventional microreflectance techniques. Brown and Wilkins (1988) have described a low-temperature GC-FT-IR technique that may allow the recording of good-quality spectra of gas chromatographic eluents in the low-nanogram levels. Griffiths and coworkers (1986, 1987) have used the infrared microscope sampling accessory in conjunction with a liquid nitrogen-cooled sampling stage for trapping the neat gas chromatographic eluents to demonstrate the high sensitivity capabilities of the technique. Because of the fairly high temperature (77 K) of the trapping stage, this technique can only be used for nonvolatile compounds. However, for nonvolatile, high-molecular-weight compounds, the infrared microscope sampling technique in conjunction with a low-temperature sample stage is capable of very low detection limits. Figure 44 shows the transmission spectrum of an 1-ng sample of diisobutylphthalate eluting from a gas chromatograph, trapped onto a spot 100 μm in diameter on the liquid nitrogen-cooled sampling stage. The spectrum of the sample was recorded using the in-

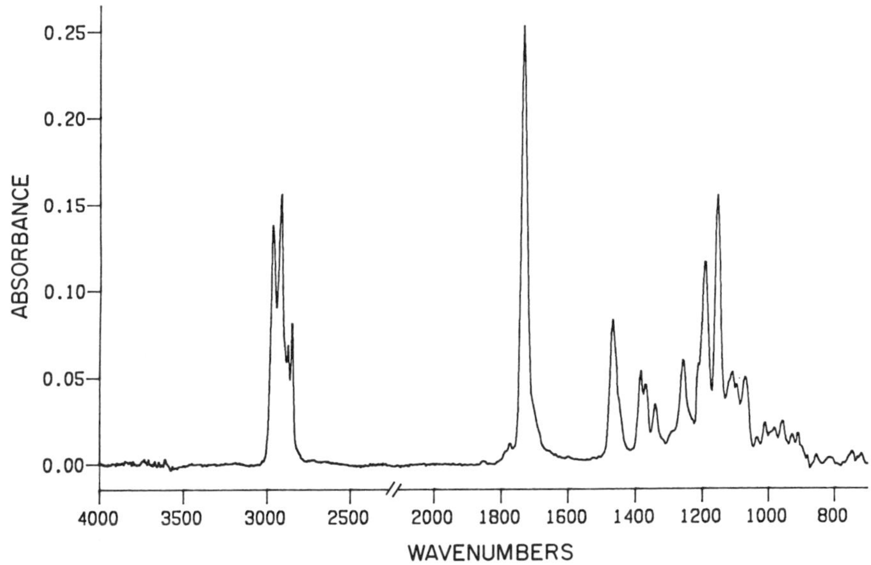

Fig. 44. Real-time GC-FT-IR absorbance spectrum of a 1-ng fraction of di-isobutyl phthalate using the Tracer™ (BioRad/Digilab Division, Cambridge, MA) low-temperature GC-FT-IR accessory. The measurement time for the spectrum is 2 sec.

frared microscope sampling optics on a commercial low-temperature GC-FT-IR accessory. One can see that the high quality of this spectrum bodes well for the future of this GC-FT-IR technique.

IV. CONCLUSION

The current status of the infrared microscope sampling technique and its applications have been outlined in this chapter. Developments are currently underway to use the technique for solving more challenging problems. Compton *et al.* (1988) have described the possibility of using the infrared microscope sampling technique in conjunction with mid-infrared transmitting fibers 100 μm in diameter for the *in situ* attenuated total internal reflectance spectral study of remote samples. Such a technique, when mature, will allow for the possibility of studying in place the curing of the epoxy formulations in automobile and airplane frames. The low-temperature GC-FT-IR technique outlined in Section III,H,2 promises the possibility of high-sensitivity detection of gas chromatographic eluents, particularly those with high molecular weights. Typically, these eluents have been the most difficult to study by conventional light-pipe GC-FT-IR techniques.

The infrared microscope technique also may enable investigators to study the inhomogeneity of support catalysts. Wolf and Sant (1988) have reported initial results using the technique for the study of the reaction of carbon monoxide with a number of alumina support catalysts. They have been able to record the spectra over metal-rich and metal-deficient regions of the catalyst sample. Doubtless other applications such as dichroic studies of polymers as a function of draw and of temperature on microscopic levels, the study of the different phases in polymer blends as a function of temperature, will also find wide use. These are but a few of the many potentially important applications of the infrared microscope sampling technique.

REFERENCES

Anderson, D. H., and Wilson, T. E. (1975). *Anal. Chem.* **47,** 2482.

Barer, R., Cole, R. H., and Thompson, H. W. (1949). *Nature (London)* **163,** 198.

Beduhn, D. L., and White, R. L. (1986). *Appl. Spectrosc.* **40,** 628.

Blout, E. R., and Bird, G. R. (1951). *J. Opt. Soc. Am.* **41,** 547.

Blout, E. R., Bird, G. R., and Grey, D. S. (1949). *J. Opt. Soc. Am.* **39,** 1052.

Brasch, J. W., and Lustiger, A. (1988). *In* "Infrared Microspectroscopy: Theory and Applications" (R. G. Messerschmidt and M. A. Harthcock, eds.), pp. 103–114. Dekker, New York.

Brown, R. S., and Wilkins, C. L. (1988). *Anal. Chem.* **60,** 1483.

Burch, C. R. (1947). *Proc. Phys. Soc.* **59,** 47.

Coates, V. J., Offner, A., and Siegler, E. H., Jr. (1953). *J. Opt. Soc. Am.* **43,** 984.

Compton, D. A. C., Hill, S. L., Wright, N. A., Druy, M. A., Piche, J., Stevenson, W. A., and Vidrine, W. A. (1988). *Appl. Spectrosc.* **42,** 972.

Cournoyer, R., Shearer, J. C., and Anderson, D. H. (1977). *Anal. Chem.* **49,** 2275.

Dluhy, R., and Mendelsohn, R. (1988). *Anal. Chem.* **60,** 269A.

Dobrilla, P., and Blakemore, J. S. (1985). *J. Appl. Phys.* **56,** 208.

Ferraro, J. R., Wang, H. H., Ryan, J., and Williams, J. M. (1987). *Appl. Spectrosc.* **41,** 1377.

Ferraro, J. R., Wang, H. H., Geiser, U., Kini, A. M., Beno, M. A., Williams, J. M., Hill, S., Whangbo, M.-H., and Evain, M. (1989). *Solid State Commun.,* in press.

Griffiths, P. R., Pentoney, S. L., Giorgetti, A., and Shafer, K. H. (1986). *Anal. Chem.* **58,** 1349A.

Griffiths, P. R., Pentoney, S. L., Jr., and Fraser, D. J. J. (1987). Paper 806, presented at the 1987 Pittsburgh Conference, Atlantic City, New Jersey.

Harthcock, M. A., and Atkin, S. C. (1988). *In* "Infrared Microspectroscopy: Theory and Applications" (R. G. Messerschmidt and M. A. Harthcock, eds.), pp. 21–39. Dekker, New York.

Harthcock, M. A., Lentz, L. A., Davis, B. L., and Krishnan, K. (1986). *Appl. Spectrosc.* **40,** 210.

Herres, W. (1985). *Chimia* **39,** 64.

Hill, S. L., and Krishnan, K. (1988). Paper 373 presented at the Pittsburgh Conference and Exposition on Analytical Chemistry and Applied Spectroscopy, New Orleans, Feb. 23–26.

Hill, S. L., Krishnan, K., Pulsinelli, P., Sutcliffe, J., and Moosey, J. (1987). Paper 356 presented at the Pittsburgh Conference and Exposition on Analytical Chemistry and Applied Spectroscopy, Atlantic City, Mar. 9–13.

Hitchman, M. A., Stratemeir, H., and Hoppe, R. (1988). Manuscript in preparation.

Holland-Moritz, K., and Stach, W. (1988). Personal communication.

Humecki, H. (1988). *In* "Infrared Microspectroscopy: Theory and Applications" (R. G. Messerschmidt and M. A. Harthcock, eds.), pp. 51–72. Dekker, New York.

Ishida, H., Ozaki, Y., Kamoto, R., Ishitani, A., Iriyama, K., Takagi, I., Tsukie, E., Shibata, K., Ishihara, F., and Kameda, H. (1987). *Microbeam Anal.* **22,** 189.

Ishitani, A. (1985). *Proc. Soc. Photo-Opt. Instrum. Eng.* **553,** 25.

Ishitani, A. (1986). *Oyo Butsuri* **55,** 473.

Katon, J. E., Pacey, G. E., and O'Keefe, J. F. (1986). *Anal. Chem.* **58,** 465.

Kim, K. M., and Smetana, P. (1986). *J. Electrochem. Soc.* **133,** 71.

Krishnan, K. (1984). *Polym. Prepr., Am. Chem. Soc., Div. Polym. Chem.* **25,** 182.

Krishnan, K. (1986). *Proc. Soc. Photo-Opt. Instrum. Eng.* **665,** 252.

Krishnan, K. (1987). *In* "Fourier Transform Infrared Characterization of Polymers" (H. Ishida, ed.), pp. 97–111. Plenum, New York.

Krishnan, K. (1988). *In* "Infrared Microspectroscopy: Theory and Applications" (R. G. Messerschmidt and M. A. Harthcock, eds.), pp. 139–151. Dekker, New York.

Krishnan, K., and Ferraro, J. R. (1982). *In* "Fourier Transform Infrared Spectroscopy" (J. R. Ferraro and L. J. Basile, eds.), Vol. 3, pp. 149–209. Academic Press, New York.

Krishnan, K., and Kuehl, D. T. (1984). *ASTM Spec. Tech. Publ.* **850,** 325.

Krishnan, K., and Mundhe, R. (1983). *J. Soc. Photo-Opt. Instrum. Eng.* **42.**

Krishnan, K., Hill, S. L., and Gelfand, L. S. (1985). *Proc. Soc. Photo-Opt. Instrum. Eng.* **553,** 338.

Lacy, M. E. (1979). *Proc. Contam. Control Conf., 5th.*

Lacy, M. E. (1983). *Proc. Inst. Environ. Sci.,* Anaheim, California, Oct. 16–19.

Lang, P. L., and Katon, J. E. (1986). *Microbeam Anal.* **21,** 47.

Lang, P. L., Katon, J. E., Bonanno, A. S., and Pacey, G. E. (1988). *In* "Infrared Microspectroscopy: Theory and Applications" (R. G. Messerschmidt and M. A. Harthcock, eds.), pp. 41–50. Dekker, New York.

Levinson, D. A., Crocker, P. R., and Allen, S. D. (1985). *Eur. Spectrosc. News* **62,** 18.

Messerschmidt, R. G. (1988). *In* "Infrared Microspectroscopy: Theory and Applications" (R. G. Messerschmidt and M. A. Harthcock, eds.), pp. 1–19. Dekker, New York.

Mirabella, F. M., Jr. (1986). *Appl. Spectrosc.* **40,** 417.

Mirabella, F. M., Jr. (1988). *In* "Infrared Microspectroscopy: Theory and Applications" (R. G. Messerschmidt and M. A. Harthcock, eds.), pp. 85–92. Dekker, New York.

Popek, K. M., and Ramsey, J. N. (1983). *Proc. Soc. Photo-Opt. Instrum. Eng. Tech. Symp. East.*

Ramsey, J. N., and Hausdorff, H. H. (1981). *Microbeam Anal.* **16,** 91.

Reedy, G. T., Bourne, S., and Cunningham, P. T. (1985a). *Anal. Chem.* **57,** 1086.

Reedy, G. T., Ettinger, D. G., and Schneider, J. F. (1985b). *Anal. Chem.* **57,** 1602.

Reffner, J. A., Coates, J. P., and Messerschmidt, R. G. (1987). *Am. Lab.* **19,** 86.

Rodgers, P. G., Cameron, R., Cartwright, N. S., Clark, W. H., Deak, J. S., and Norman, E. W. W. (1976). *Can. Forens. Sci. J.* **9,** 1.

Schiering, D. W. (1988). *In* "Infrared Microspectroscopy: Theory and Applications" (R. G. Messerschmidt and M. A. Harthcock, eds.), pp. 229–243. Dekker, New York.

Scott, R. M., and Ramsey, J. N. (1982). *Microbeam Anal.* **17,** 91.

Smyrl, N. R., Howell, R. L., Hembree, D. M., Jr., and Oswald, J. C. (1988). *In* "Infrared Microspectroscopy: Theory and Applications" (R. G. Messerschmidt and M. A., Harthcock, eds.), pp. 211–228. Dekker, New York.

Sommer, A. J., Lang, P. L., Miller, B. S., and Katon, J. E. (1988). *In* "Infrared Microspectroscopy: Theory and Applications" (R. G. Messerschmidt and M. A. Harthcock, eds.), pp. 245–258. Dekker, New York.

Teetsov, A. S. (1977). *Microscope* **25.**

Wolf, E. E., and Sant, R. (1988). Personal communication.

Wood, D. L. (1950). *Rev. Sci. Instr.* **21,** 764.

Wooten, D. L. and Hughes, D. W. (1988). *In* "Infrared Microspectroscopy: Theory and Applications" (R. G. Messerschmidt and M. A. Harthcock, eds.), pp. 259–272. Dekker, New York.

Yao, K. H., and Witt, A. F. (1987). *J. Cryst. Growth* **80,** 453.

Zearing, D. J., and Coates, V. J. (1981). *Proc. Soc. Photo-Opt. Instrum. Eng.* **276,** 249.

4

Possibilities and Limitations of FT-Raman Spectroscopy

B. Schrader
Institut für Physikalische und Theoretische Chemie
Universität Essen
Federal Republic of Germany

I. INTRODUCTION

Scientific infrared spectroscopy is now about 100 years old (Möller and Rothschild, 1971). In spite of the pioneer work by Coblentz, who published the first collection of infrared spectra and discussed characteristic frequencies in 1905, infrared spectroscopy was at that time not accepted by the chemists. They thought it too difficult to build the necessary spectrometers and to record spectra. However, when Raman spectroscopy was discovered in 1928 (Raman and Krishnan, 1928), it was relatively easy to record spectra using the mercury arcs, spectrographs, and photoplates that are standard equipment in many laboratories. Therefore, scientists expected its immediate application in the analytical laboratory and also its combination with the complementary infrared spectroscopy (Schaefer and Matossi, 1930). Kohlrausch's work (1943) showed clearly the prospect for studies of molecular structures with Raman spectroscopy. Nevertheless, Raman spectroscopy did not find acceptance in the analytical laboratory. The sample amount was too large, and the fluorescence of impurities too often masked the Raman spectrum completely.

After 1945, infrared spectroscopy turned out to be a useful tool as a result of the development of servo amplifiers, which made infrared spectrometers commercially available. After the discovery of the laser light sources in 1960 (Maiman, 1960), a "renaissance" of Raman spectroscopy occurred, since the red radiation of the ruby laser at 694 nm allowed recording of Raman spectra of colored samples (Schrader and Stockburger, 1965), even photoelectrically (Schrader and Stockburger, 1966). Furthermore, very small amounts of sample were sufficient (Schrader and Meier, 1966). Nevertheless, fluorescence continued to be a problem and thus prevented wide acceptance of Raman spectroscopy by chemists.

The present decade brought interferometers and the Fourier transform (FT) technique to chemists. These developments made a wealth of new applications of infrared spectroscopy accessible. This is the topic of this book. Unexpectedly, these developments produced new and exciting prospects for Raman spectroscopy.

Hirschfeld and Chase (1986) demonstrated that Raman spectra, excited with lasers in the near-infrared region, may be recorded with the FT-IR instruments originally designed for absorption spectroscopy in the near-infrared region. They demonstrated that the fluorescence, usually associated with Raman spectroscopy, obviously disappeared. Raman spectra of fluorescing dyes and even of explosives were obtained, demonstrating that the thermal load by the illumination with the laser was not too large. They proved that the prejudice of many scientists against this technique (Hirschfeld, 1976, 1979; Hirschfeld and Schildkraut, 1974) was apparently not justified. Immediately after this publication, several groups started work with this technique and instrument manufacturers began designing FT-Raman spectrometers.

The anticipated advantages of near-infrared FT-Raman spectroscopy are the absence of fluorescence combined with large throughput, high resolution, and good wavenumber accuracy. It is too early to confirm well-founded expectations that FT-Raman will be a successful cooperating partner and complement to the infrared technique with further special merits of its own. This chapter deals with anticipated applications. I will show that careful design of the instrument and sample facilities combined with suitable sample preparation is necessary. Since the new instruments and especially the necessary accessories are not yet commonly available, some statements are made without proof. I will concentrate on the "classical" (linear) Raman effect and on its applications in chemistry.

II. PRINCIPLES

A. Raman Spectroscopy

Chemical bonds are elastic and atoms have masses, therefore molecules show vibrations. Their frequencies depend on the molecular geometry, the bond orders, and the atomic masses. The collection of all frequencies, the vibrational spectrum, is a "fingerprint," a unique means of identification of molecules. When a molecule has n atoms there are $3n - 6$ independent vibrations, for linear molecules $3n - 5$. These so-called normal vibrations can be observed by different techniques: inelastic neutron scattering, fluorescence, but especially by infrared and Raman spectroscopy. A vibration is "infrared active" when it modulates the molecular dipole moment. It is "Raman active" when the molecular polarizability is modulated. To explain Raman spectroscopy, I present two models.

Figure 1 shows a molecule under the influence of the electric field of an electromagnetic wave. Since electrons are shifted by this field against the nuclei (this molecular property is called polarizability), the wave induces a dipole moment modulated with the frequency of the wave. Therefore, the molecule emits a wave with this frequency, the so-called Rayleigh radiation. When the molecule performs a vibration during this process that changes its polarizability, the Ray-

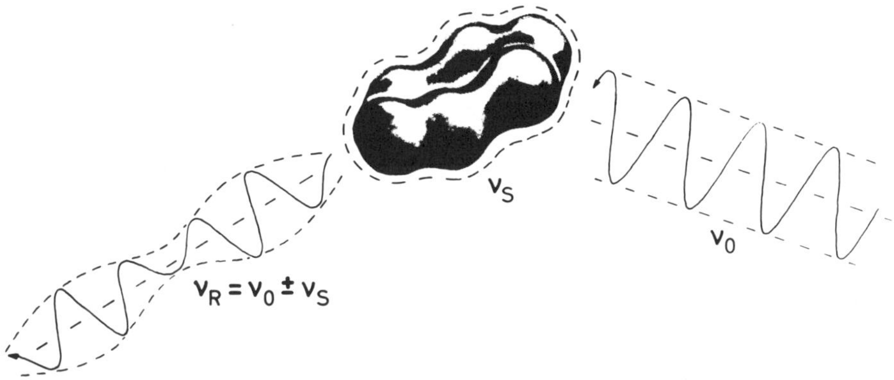

Fig. 1. The Raman effect as modulation of a wave. An exciting wave with frequency ν_0, vibrational frequency modulating the molecular polarizability ν_S, modulated wave $\nu_R = \nu_0 \pm \nu_S$.

leigh radiation is modulated with the frequency of this vibration. This radiation analyzed by a spectrometer shows a central peak (the Rayleigh peak) and two sidebands for each vibration, shifted to larger and smaller frequency values by the frequency of the molecular vibration: the Raman lines. It is useful to know that the polarizability has the dimension of a volume and that its value is nearly equal to the molecular volume. The intensity of the Raman lines is proportional to the square of the change of the molecular polarizability by the vibration.

The second model is demonstrated by Fig. 2. Here the interaction of a light quantum of energy $h\nu_0$ with the molecule is shown. There may be an elastic impact that yields a light quantum with the same energy, but there are also two kinds of inelastic collisions. One excites a molecular vibration and leaves a quantum of lower energy; the other is the result of the impact of a vibrating molecule with the light quantum. It yields a quantum with higher energy and nonvibrating molecule. Since the relative probability of an impact with a vibrating and a nonvibrating molecule is described by Boltzmann's law, the intensity of the scattered radiation with lower energy

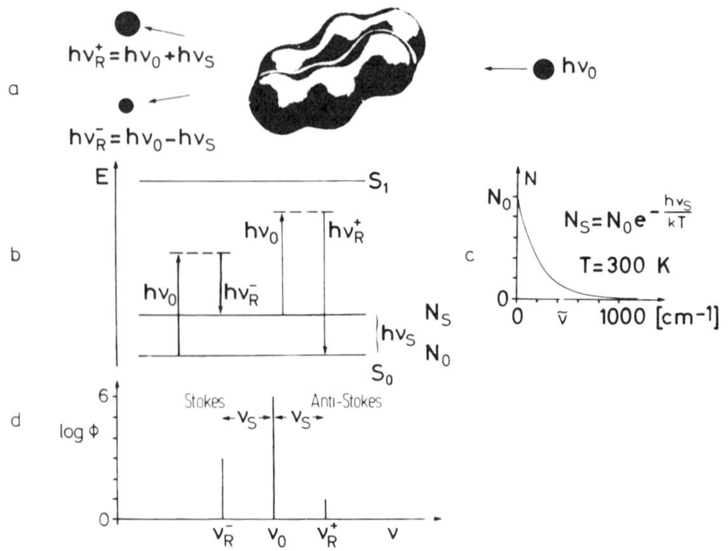

Fig. 2. The Raman effect as a consequence of an impact of a light quantum $h\nu_0$ with a molecule: (a) principle; (b) term scheme; (c) Boltzmann's law; (d) intensity of observed spectral lines Φ.

of the light quanta is larger than that with the higher energy. The relative intensity of both lines can be measured to determine the temperature of the molecules during Raman scattering. Usually only the stronger (red-shifted) spectrum, the so-called Stokes spectrum is recorded; the blue-shifted spectrum is called the anti-Stokes Raman spectrum.

Quantum mechanics supplies an exact theory of the Raman effect, which is comprehensively described in some books (Brandmüller and Moser, 1962; Califano, 1976; Long, 1977).

B. Chemical Applications of Raman Spectroscopy

Classical (linear and nonresonant) Raman spectroscopy is independent of the frequency of the exciting line. Kohlrausch (1943) recorded practically the same spectra by excitation with the mercury line at 436 nm on his photoplates that we record with our spectrometers using the Ar-ion laser at 488 and 515 nm, the ruby laser at 694, or the NdYAG (yttrium-aluminum-garnet, doped with neodymium) laser at 1064 nm. When spectra of the same compound look different, this may be due to the spectral sensitivity of the spectrometer or to the (pre-) resonance Raman effect, which amplifies some lines in contrast to others. Therefore, most collections of Raman spectra may be employed generally for work with Raman spectra, independent of the wavelength used for excitation. Useful collections are published by different authors and companies (Dollish *et al.*, 1974; Kohlrausch, 1943; Landolt-Börnstein, 1955; Sadtler Research Laboratory; Schrader and Meier, 1974–1976; Schrader, 1989).

There is no special NdYAG-laser Raman spectroscopy—the special aspects are only the circumstances of excitation and recording. All the knowledge accumulated hitherto with respect to the chemical applications of Raman spectroscopy and the combination with the infrared technique can be applied here. Some books and articles give general surveys of the chemical application of Raman (with infrared) spectroscopy (Colthup *et al.*, 1975; Dollish *et al.*, 1974; Freeman, 1974; Grasselli *et al.*, 1981; Schrader, 1973, 1980, 1983; Weidlein *et al.*, 1982). Publications dealing with sample preparation (Schrader *et al.*, 1971) and with the limit of detection (Schrader *et al.*, 1981) are applicable as well.

1. Identification of Compounds

Figure 3 shows the infrared and the Raman spectra of crystalline cystine (Schrader and Meier, 1974–1976). Since the molecule has 26 atoms, its vibrational spectrum has $3 \cdot 26 - 6 = 72$ normal vibrations. Both spectra show many bands at practically the same frequency but with quite different intensities, thereby demonstrating that they give complementary images of the vibrational spectrum. In the infrared spectrum, vibrations of polar groups are strong, like that of the ammonium or

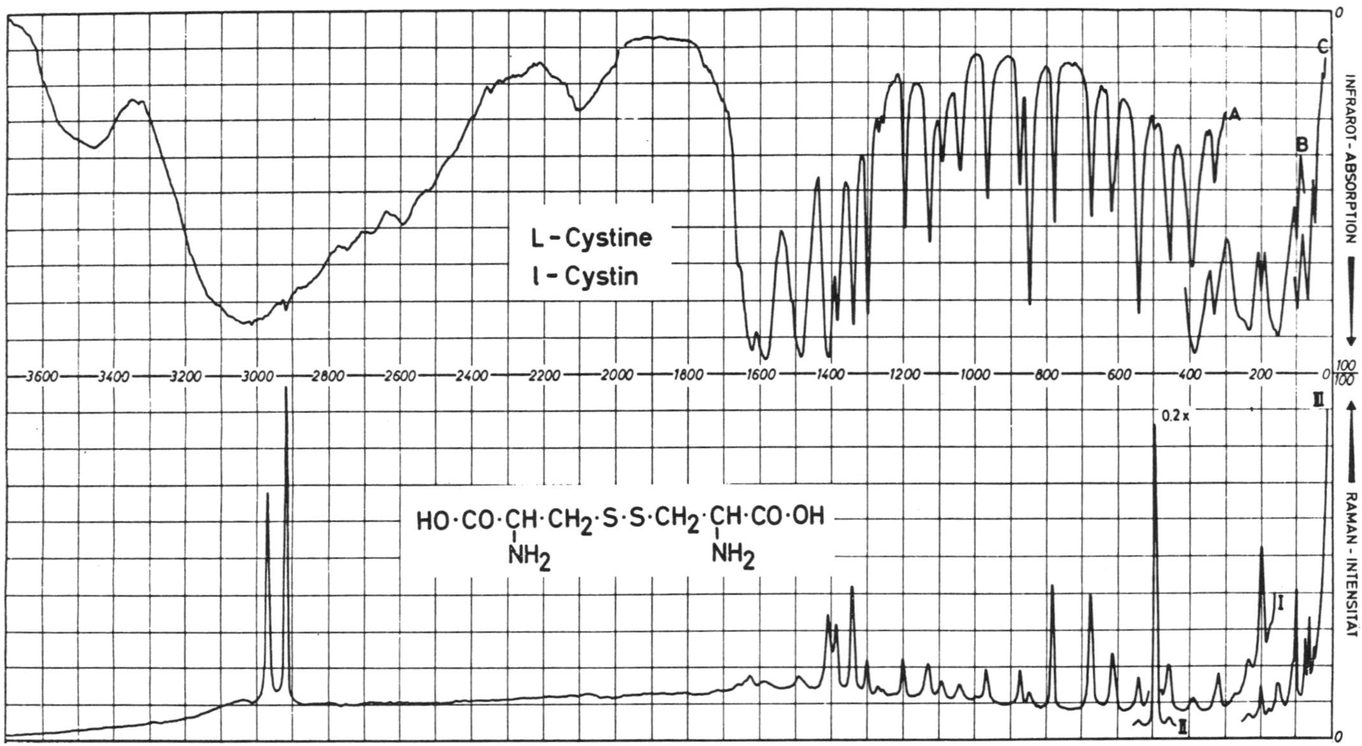

Fig. 3. Infrared (top) and Raman (bottom) spectra of crystalline cystine (Schrader, 1989).

carboxylic group of the zwitterionic state of this amino acid near 3000 and 1600 cm^{-1}. In the Raman spectrum, strong bands belong to vibrations of nonpolar groups like the C—H bonds at 2900 and the S—S bond at 500 cm^{-1}. Both spectra exhibit sharp bands below 200 cm^{-1}, a pattern indicating that the sample is crystalline. These bands disappear when the sample is dissolved or melts.

We see that infrared and Raman spectra are complementary. Nonsymmetrical molecules show their normal vibrations, with different intensities, in the infrared and the Raman spectra. In principle, either type of vibrational spectroscopy can be used as a fingerprint identification of molecules. But there are important differences. Solutions in water are investigated more easily by Raman spectroscopy, since water gives only a weak Raman spectrum (contrary to its strong infrared spectrum). There also are no difficulties with dissolution of the material of the cuvettes. Furthermore, Raman spectra can be recorded from samples in closed containers, e.g., ampoules containing sensitive, dangerous, or expensive samples (Schrader, 1973), and in remote containers or pipes by using fiber optics. The identification of gemstones (Nassau, 1981), inclusions in minerals, dust particles (Rosasco, 1980), or even single particles of aerosols (Schrader, 1986) are also worth mentioning. Collections of Raman spectra (Dollish *et al.*, 1974; Kohlrausch, 1943; Landolt-Börnstein, 1955; Sadtler Research Laboratory; Schrader, 1979) may be used as reference material for visual or computerized comparisons.

2. Quantitative and Multicomponent Analyses

Since an emission process in an optically thin medium is employed, calibration curves for quantitative Raman analyses of one or more components may be practically linear over several orders of magnitude. However, intermolecular interactions and nonlinear responses of the detectors make careful calibration necessary. Scattering cross sections of gases are published (Schrötter and Klöckner, 1979). In order to lower the limit of detection, general rules should be observed (Schrader *et al.*, 1981).

3. Determination of Structures by Evaluation of Characteristic Vibrations

Characteristic vibrations have been known for a long time (Coblentz, 1905). They gain more specificity when, in addition to the frequencies, the intensities in the infrared and the Raman spectra are also considered. Table I (Schrader, 1973)

Table I Characteristic Frequencies and Raman and IR Intensities of Groups in Organic Compounds[a]

Vibration[b]	Frequency (cm^{-1})	Intensity[c] Raman	Infrared
ν(O—H)	3650–3000	1	3
ν(N—H)	3500–3300	2	2
ν(≡C—H)	3300	1	3
ν(=C—H)	3100–3000	3	2
ν(—C—H)	3000–2800	3	3
ν(—S—H)	2600–2550	3	1
ν(C≡N)	2255–2220	2–3	0–3
ν(C≡C)	2250–2100	4	0–1
ν(C=O)	1820–1680	1–3	4
ν(C=C)	1900–1500	2–4	0–1
ν(C=N)	1680–1610	3	2
ν(N=N) (aliph. derivat.)	1580–1550	2	0
ν(N=N) (arom. derivat.)	1440–1410	2	0
ν_a([C—]NO_2)	1590–1530	2	3
ν_s([C—]NO_2)	1380–1340	4	2
ν_a([C—]SO_2])	1350–1310	0–1	3
ν_s([C—]SO_2(—C])	1160–1120	3	3
ν([C—)SO(—C])	1070–1020	2	3
ν(C=S)	1250–1000	3	1
$\delta(CH_2)$, $\delta_a(CH_3)$	1470–1400	2	2
$\delta_s(CH_3)$	1380	1–2 3[d]	2–3

Table I *(continued)*

Vibration[b]	Frequency (cm^{-1})	Intensity[c] Raman	Infrared
ν(CC) (aromatics)	1600, 1580	2–3	2–3
	1500, 1450	1–2	2–3
	1000	3[e]	0–1
ν(CC) (alicycles, aliphatic chains)	1300– 600	2–3	1–2
ν_a(C—O—C)	1150–1060	1	3
ν_s(C—O—C)	970– 800	2–3	0–1
ν_a(Si—O—Si)	1110–1000	0–1	4
ν_s(Si—O—Si)	550– 450	4	0–1
ν(O—O)	900– 845	3	0–1
ν(S—S)	550– 430	3	0–1
ν(Se—Se)	330– 290	3	0–1
ν(C[arom.]—S)	1100–1080	3	2–3
ν(C[aliph.]—S)	790– 630	3	2–3
ν(C—Cl)	800– 550	3	3
ν(C—Br)	700– 500	3	3
ν(C—I)	660– 480	3	3
δ_s(CC) (aliph. chains C_n)			
$n = 3 \ldots 12$	400– 250	2–3	0–1
$n > 12$	$2495/n$		
lattice vibrations in molecular crystals and polymers	200– 10	0–4	0–3

[a]From Schrader (1980).
[b]ν, stretching; δ, bending; s and a for symmetric and antisymmetric, respectively.
[c]0, very weak or inactive; 1, weak; 2, middle; 3, strong; 4, very strong.
[d]On C=C bond.
[e]for mono- and 1,3,5 derivatives.

gives typical frequency ranges of characteristic vibrations together with their typical relative intensities in the Raman and infrared spectra. Symmetric vibrations and those of bonds between equal atoms (C≡C, C=C, N=N, S—S, O—O) usually show strong Raman and weak infrared bands (Schrader, 1973). Conjugated dienes exhibit their conformation by evaluation of their in-phase and out-of-phase stretching vibrations in the infrared and Raman spectra (Schrader and Ansmann,

1975). Also conjugated enones reveal their s-cis or s-trans conformation (Oelichmann *et al.*, 1982). The substitution pattern of benzene derivatives is shown clearly by characteristic Raman bands (Schrader and Meier, 1972). The substitution pattern of aliphatic and alicyclic halogen compounds may be concluded from a joint evaluation of the Raman and infrared bands (Schrader and Meier, 1975). The structure and substitution pattern of steroids may be derived from the evaluation of the Raman bands (Schrader and Steigner, 1973). Characteristic Raman bands of polymers are also valuable for structural studies (Siesler and Holland-Moritz, 1980). General tabular collections of characteristic Raman bands are found in some books (Dollish *et al.*, 1974; Freeman, 1974; Landolt-Börnstein, 1955; Weidlein *et al.*, 1981, 1982). Automatic evaluation of the characteristic vibrations in the infrared and Raman spectra to determine structure is possible (Schrader and Meier, 1972).

4. Symmetry of Molecules, Aggregates, and Crystals

Vibrations of molecules in the liquid and gaseous state show bands in the Raman spectrum, characterized by their frequency, their intensity, and their "depolarization factor." Bands are "polarized" or "depolarized," depending on whether or not the vibrations are symmetrical with respect to all symmetry operations of the molecule (totally symmetric vibrations). When molecules have a center of symmetry, a "rule of mutual exclusion" is valid. In other words, Raman-active vibrations of such molecules cannot be infrared active, and vice versa. For crystals, liquid crystals, or crystalline domains of polymeric material, the symmetry of the unit cells determines which vibrations are Raman or infrared active, and which polarization properties they have. Orthorhombic crystals, when observed with appropriate polarizers, show three different infrared and six different Raman spectra. These spectra give a very detailed picture of the crystal, its components, the molecules, and their interactions. Further information may be found in several publications (Califano, 1976; Colthup *et al.,* 1975; Schrader, 1978, 1980; Siesler and Holland-Moritz, 1980).

5. Model Calculations

Using the methods discussed in preceding sections, some information can be gained about molecular structures. In principle, the vibrational spectra contain information about all aspects of the molecular structure. This information can be evaluated comprehensively by calculations of frequencies and infrared and Raman intensities of models of the molecule under investigation and by refining the parameters through comparisons of the calculated and the observed spectra. The methods are discussed in books by Califano (1976) and Schrader *et al.* (1980). Several publications deal with molecules (Ansmann and Schrader, 1976; Klaeboe *et al.,* 1985; Oelichmann *et al.,* 1981; Schrader, 1988; Schrader and Pacansky, 1983; Schrader *et al.,* 1980; Spiekermann *et al.,* 1980) and crystals (Bleckmann *et al.,* 1971; Bougeard *et al.,* 1973; Schrader and Bergmann, 1967; Takahashi *et al.,* 1967). Large molecules (with 60 atoms and more) can be investigated when models with simplified sets of parameters are employed for the evaluation of the frequencies and the intensities (Schrader, 1988; Schrader *et al.,* 1984).

III. INSTRUMENTATION

A. SPECTROMETERS

FT-Raman spectrometers usually employ the cw (continous wave) NdYAG laser for the excitation of the Raman spectra with radiation at a wavelength of 1064 nm. The necessary power is less than 1 watt; but sometimes more than 5 watts may be delivered by the source. It is advisable to introduce a glass plate at an angle equal to the Brewster angle in the resonator. The emerging radiation will then be linearly polarized. By introducing a so-called $\lambda/2$ plate in the emerging beam, the direction of the polarization may be changed by 90°. Furthermore, an interference filter (primary filter) or a Pellin-Broca-prism (Hallmark *et al.*, 1987; McClain, 1973; Zimba *et al.*, 1987) may be used to eliminate plasma and false laser lines. A sample compartment contains the necessary facilities for samples under different conditions. They are described in the following chapter.

The radiation emerging from the sample contains the Raman spectrum and, in addition, radiation of the wavelength of the exciting line with an intensity several orders of magnitude stronger than the Raman lines. The FT spectrometers cannot deal with the Raman spectrum under this condition. They transform the statistical noise of the exciting radiation into high-intensity noise in the whole spectrum, hiding the Raman lines completely. This effect may be called "multiplex disadvantage."

Therefore, a "secondary filter" has to be introduced, to eliminate the exciting radiation completely. This problem is an old one in Raman spectroscopy of crystal powders, dating back to the time when they were investigated with ordinary spectrographs and photoplates. Different constructions are described in the literature and are now being considered for application in the FT-Raman instruments. Double polychromators with reciprocal dispersion are effective, but they are also expensive and difficult to align (Hirschfeld and Chase, 1986; Kohlrausch, 1943). Other possible solutions have been discussed: color filters (Hirschfeld and Chase, 1986), interference long-wave-pass edge filters (Chase, 1986, 1987), interference filters in reflection (Hirschfeld and Chase, 1986; Meier *et al.*, 1972), Bragg reflection from monodisperse polystyrene spheres (Flaugh *et al.*, 1984), Scheibe aggregates (J-aggregates) of dyes in solution (Buchanan and Honigs, 1986), and systems that discriminate Raman and fluorescence processes by their different time constants (Hirschfeld and Chase, 1986). Especially convenient would be processes like the low-pressure gas absorption of iodine vapor, which is very convenient for the stabilized green argon laser line. A most effective secondary filter was described in 1929 by Rasetti, who completely eliminated the mercury resonance line at 253.6 nm in Raman spectra of gases with a drop of mercury put into the spectrograph (Rasetti, 1930). Most instruments working with the NdYAG-laser radiation use interference filters. Last but not least, I should mention that the intensity of the Raman lines can be enhanced and that of the exciting radiation reduced effectively by optimizing the sample facilities. This technique is discussed in the following section.

Nearly all FT-IR instruments on the market have already been used for recording Raman spectra: Bomem (Hallmark *et al.*, 1987), Bruker (Schrader and Simon, 1987), Digilab, Nicolet (Hirschfeld and Chase, 1986), Perkin-Elmer (Hendra and Mould, 1988; Parker *et al.*, 1988). Some companies develop special Raman spectrometers or Raman modules as complements to conventional infrared spectrometers; e.g., Fig. 4 shows the layout for the Bruker FT-Raman module. The interferometer and especially the detector have to be optimized for the near-infrared region. Ideal detectors for this range have not been found as yet. Cooled PbS, Ge, and InGaAs detectors are used. They suffer either from their high noise-equivalent power (NEP) limited within the range of the Raman spectrum (about 1–2 μm), strongly varying spectral sensitivity, or from a sensitivity to the muons from cosmic radiation.

An advantage of the FT-Raman spectrometers is the convenience of computer manipulation of the spectra. Most of the programs developed for infrared spectroscopy may be used. In addition, some special procedures have to be introduced for plotting in relative wavenumber scales, correction of the spectral sensitivity, the recording of depolarization ratios, and the determination of the sample temperature from the Stokes/anti-Stokes intensity ratio (Schrader, 1980, 1983).

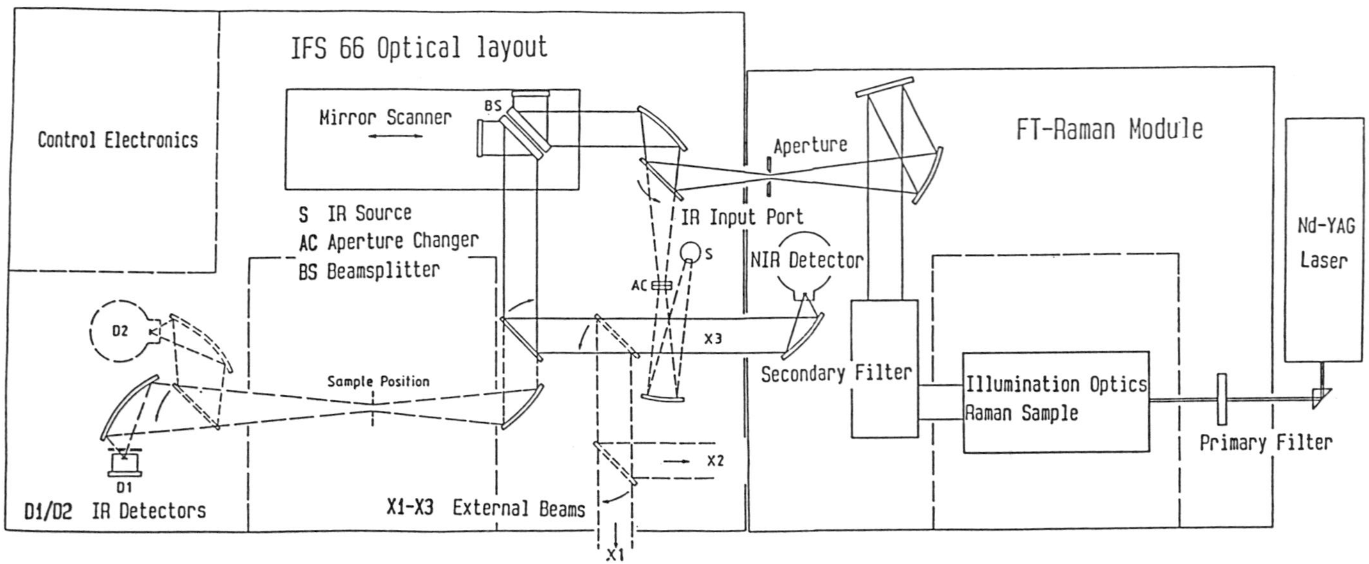

Fig. 4. Layout of the BRUKER FT-Raman module (Schrader and Simon, 1987).

B. Suitable Sample Techniques

The convenient sample facilities that have been developed allow simple preparation of small samples in all states and yield high intensities of the Raman lines with low intensities of the exciting radiation. Of special interest are the facilities for liquids and for crystal powders. They are discussed in this section.

1. Sampling Techniques for Liquid Samples

The optimal sample arrangement for liquids places a small sample at the focus of the laser beam and transfers a maximal part of its Raman radiation by optics with a large angle of acceptance into the interferometer (Schrader, 1980). The necessary volume of the sample corresponds to that of the focal region. The depolarization factor is most conveniently observed with a right-angle arrangement of the optical axes of the exciting versus the Raman radiation. The most efficient liquid cell for the right-angle arrangement is a micro cell (Schrader, 1980; Schrader and Meier, 1966).

The usual rectangular or cylindrical Raman cuvettes do not have the maximal possible efficiency. They suffer from an aperture error, which prevents an exact image of a Jacquinot aperture at the focal region of the samples, from reflection losses, and from the fact that the observed solid angle of the Raman radiation is reduced by refraction at the surface of the cell. Therefore, a spherical cell was designed and is shown in Fig. 5 (Schrader, 1984). It has the following advantages: (1) It has (an "aplanatic" system) no aperture error. (2) It has small reflection losses, independent of angle and polarization, that can be further reduced by coating. (3) The solid angle of the observed Raman radiation is equal to the solid angle of the imaging system. (4) The necessary amount of liquid sample is of the order of 1 microliter. (5) The cell can be made from

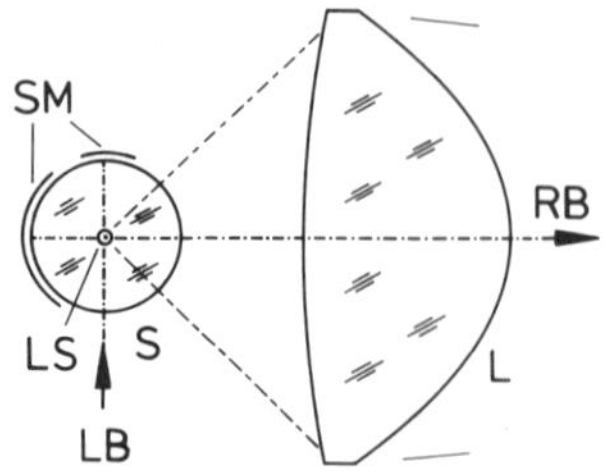

Fig. 5. Spherical cell (S) for liquids (Schrader, 1984). LB, laser beam; LS, liquid sample; SM, surface mirrors; L, high aperture lens; RB, Raman beam.

sapphire, a good thermal conductor. Therefore, it can be cooled effectively. (6) Surface mirrors nearly double the exciting radiation density and increase the observed flux of the Raman radiation. They stay clean and are by themselves ideally adjusted to the center of the sphere. (7) The adjustment of the cell in the spectrometer is simple (by observing the beams, reflected by the surface mirrors). Furic and Durig (1988) also have described the advantages of the spherical Raman cell.

2. Sampling Techniques for Crystal Powders

Most known chemical compounds are solids; therefore, the techniques for the investigation of Raman spectra of solids under optimal conditions are of major importance. This statement is especially true because the usual straightforward sample arrangement that observes the backscattered radiation from an illuminated surface sends a very large amount of scattered laser radiation to the spectrometer together with a very low intensity of the Raman radiation. Therefore, a useful Raman spectrometer should be able to record spectra of poorly scattering samples, like substances on catalysts. A short discussion of the theory of optimal powder arrangements for Raman spectroscopy is therefore justified.

Kubelka and Munk (1931) developed the theory of the optical properties of white pigments. In an analogous way, Schrader and Bergmann (1967) have developed the theory of the Raman scattering of crystal powders. The properties of a flat layer of a crystal powder of the thickness *d* is described by a scattering module *r*, an absorption module *a*, and a Raman scattering module *s*. The solution of the differential equations describing the elementary processes lead to equations describing the balance of the following radiation fluxes dependent on the thickness of the layer. I_0 is the exciting radiation, hitting the surface of the layer ($I_{rel} = 1.0$); J_P is the "unshifted" scattered radiation from the illuminated surface; J_R, the Raman radiation from this surface. I_P is the exciting radiation emerging from the rear; and I_R, the Raman radiation. Figure 6 shows the changes in these light fluxes with changing thickness of the layer. The results are calculated with the parameters set to values similar to that of typical Raman samples. The Raman intensity is enhanced by a large factor, in order to show it on the same diagram.

One can clearly see that more than 90% of I_1, the flux of the exciting radiation, is diffusely reflected from the front surface. In other words, a large part of the exciting radiation is usually lost. Only a small part is transformed into Raman radiation, which is emitted as J_R from the front surface. The

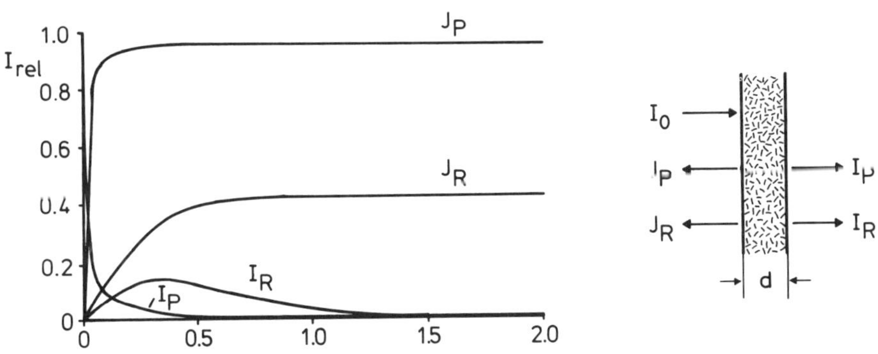

Fig. 6. Radiation balance of a layer of crystal powder of thickness *d* (Schrader and Bergmann, 1967). I_0, exciting radiation ($I_{rel} = 1$); J_P, exciting radiation, diffusely reflected from the surface; I_P, exciting radiation, emerging from the rear; J_R, Raman radiation from the front; I_R, Raman radiation from the rear.

exciting radiation emerging from the back of the layer (I_P) has a very low intensity. The Raman radiation emerging from the back (I_R) has an intensity maximum at the "optimal thickness." For the radiation emerging from the back, the ratio of Raman to exciting radiation flux is larger by a factor of about 100 than the ratio for the radiations emerging from the front of the sample.

Since the flux of the exciting radiation entering the interferometer has to be as small as possible for FT-Raman spectroscopy, it is advisable to observe the radiation from the back of a sample that is illuminated at the front. Its thickness should be at least the optimal thickness.

Figure 7 shows micro cells for crystal powders; these cells are designed for an optimal figure of merit and employ the radiation from the back of a layer that is illuminated at the front (Schrader, 1984). The powder is contained in a polished stainless steel cylinder, which may include a tube for a cooling fluid. The illuminated side is covered with a half sphere made from sapphire or glass; the surface of the half sphere is coated with a metallic mirror, except for a small unmirrored disk at the axis. The laser beam is transmitted through this disk onto the surface of the powder. The diffusely reflected exciting ra-

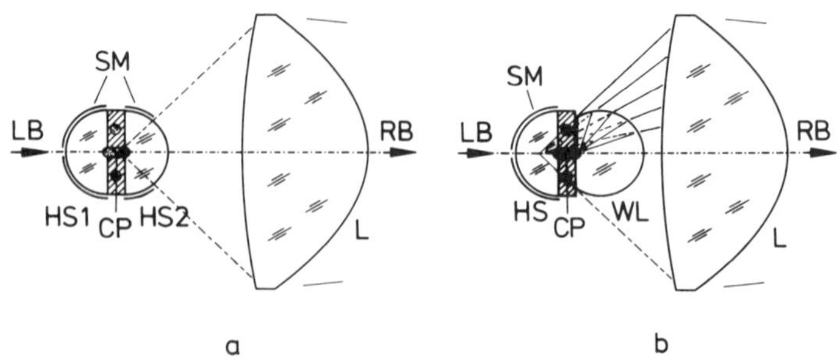

Fig. 7. Spherical cells for crystal powders (Schrader, 1984). LB, laser beam; CP, crystal powder in highly reflecting metal container, with pipe for cooling fluid; SM, surface mirror; L, high aperture lens; RB, Raman beam; HS1, HS2, HS, half spheres; WL, Weierstrass lens.

diation is reflected back to the powder by the surface mirror. As a result of this multiple reflection arrangement, a high percentage of the exciting radiation is, together with the Raman radiation from the front surface, employed for the illumination of the sample. The radiation emerging from the back is, as shown in Fig. 7a, partly imaged into the spectrometer. The rest of the radiation is reflected back to the sample. Alternatively, as shown in Fig. 7b, the radiation may be completely sent to the spectrometer by means of a Weierstrass lens (Weierstrass, 1856). This system makes use of the aplanatic points of a sphere at $r \cdot n$ and r/n, where r is the radius of the sphere and n its refractive index.

C. Fiber Optics Sampling

The modern techniques of communication have made possible the transport of radiation in the near-infrared over kilometers with fibers of high transmittance. In the wavelength range of the Raman spectrum, excited by a NdYAG laser (1000 to 1800 nm), the fiber optic transmittance has its maximum. Therefore, fiber optics may be employed for "remote Raman spectroscopy" (McCreery *et al.*, 1983; Plaza *et al.*, 1986; Schwab and McCreery, 1984). Samples at different places in a factory may be analyzed on-line without transportation and with immediate feedback of the results, e.g., for production or pollution control. The laser radiation is transported to the sample with one fiber. If the distance is very long, an interference filter should be placed near the sample head to remove the Raman scattering of the fiber itself. Alternatively, diode-pumped lasers (Begley *et al.*, 1987) can be employed at each sample head. The Raman radiation is transported to the spectrometer by a bundle of fibers with a cross section large enough to illuminate the interferometer properly. Figure 8 shows typical arrangements of two-way bundles, with one fiber for the laser radiation and several fibers for the transport

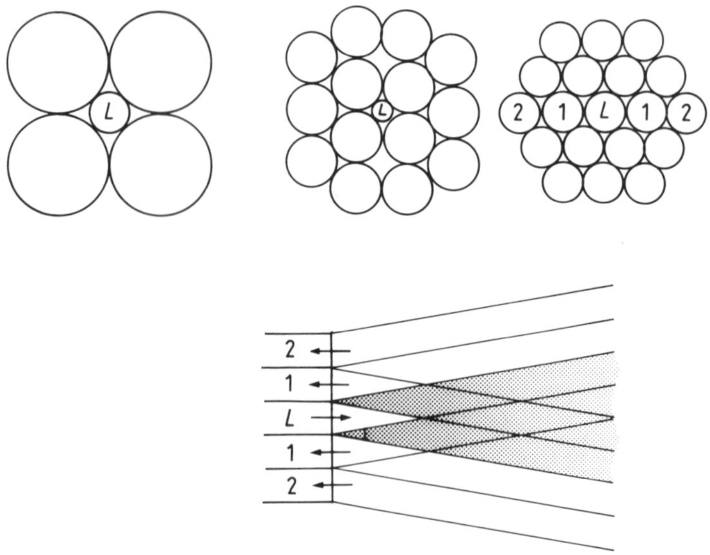

Fig. 8. Two-way fiber optics for FT-Raman spectroscopy. L, fiber for the laser beam; 1, 2, fibers for Raman radiation.

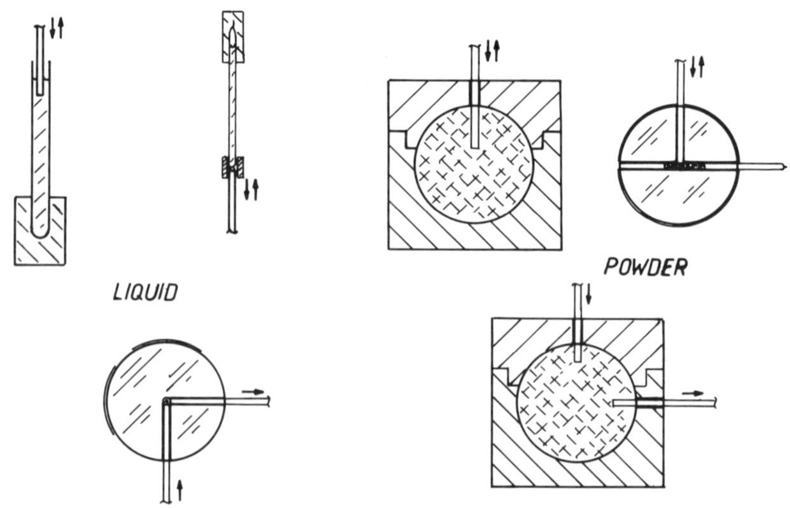

Fig. 9. Sample arrangements for Raman spectroscopy for liquids (left) and powders (right) with fiber optics; with two-way fibers (above), and with one-way fibers (below).

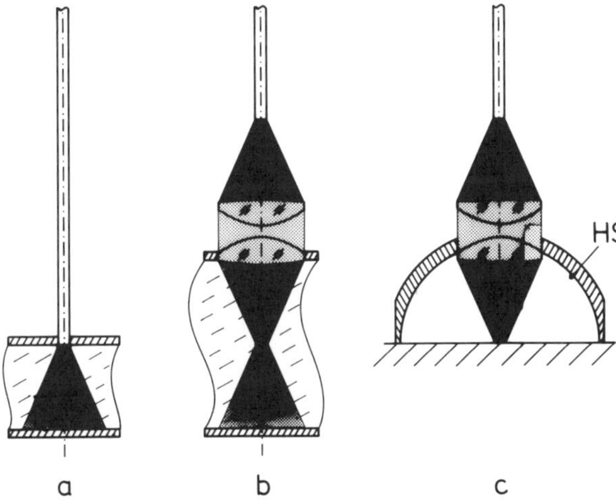

Fig. 10. Sample arrangements for process and production control with fiber optics. (a,b) Control of liquid systems; (c) control of surfaces. HS, half spherical mirror.

of the Raman radiation to the spectrometer. Figures 9 and 10 show sample heads for laboratory samples and for production control.

IV. APPLICATIONS AND LIMITATIONS

A. Conventional and Novel Applications

In the following section, the Raman spectra that are reproduced were recorded with early prototypes of a Bruker Raman module (Schrader and Simon, 1987) like that shown in Fig. 4. Figure 11 shows the Raman spectrum nitrobenzene recorded with 10 scans in 30 sec, and Fig. 12 shows the Raman spectrum of polytetrafluoroethylene.

The performance of the secondary filter determines the low-frequency limit of the Raman spectra. Usually this limit is at 350 cm^{-1}. It is, possible, however, to construct filters

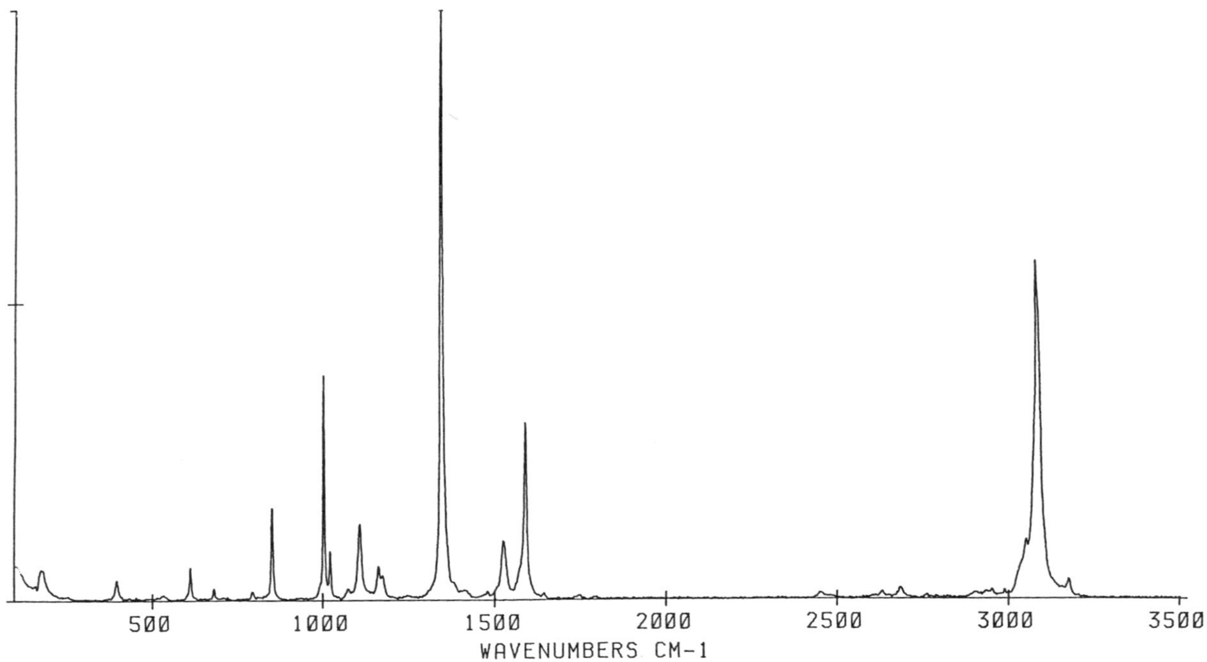

Fig. 11. Raman spectrum of liquid nitrobenzene; recording time, 30 sec.

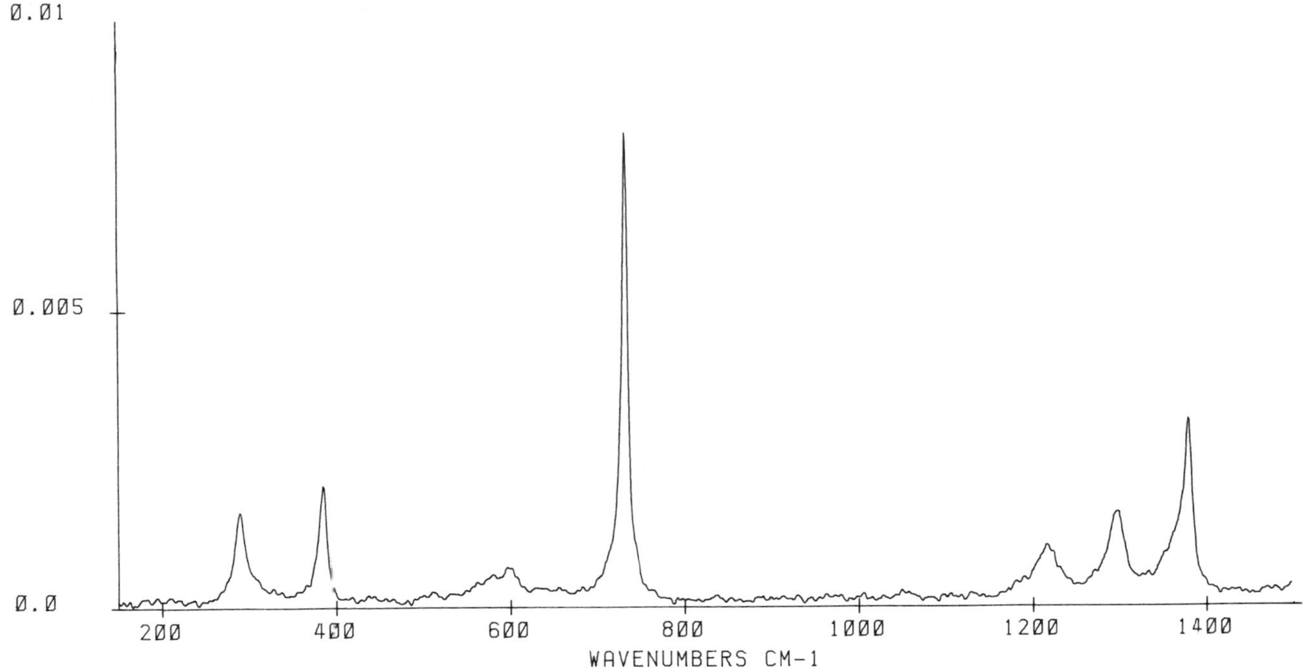

Fig. 12. Raman spectrum of polytetrafluoroethylene (Teflon).

that allow the observation of lines at frequencies down to 100 cm^{-1}, as shown in Fig. 13. Also the anti-Stokes Raman spectrum may be recorded together with the Stokes spectrum; see Fig. 14. This practice has the advantage that the temperature of the sample can be determined automatically from the intensity ratio in both spectra by employing Boltzmann's law. This knowledge is of great importance, since large laser powers may be necessary to record spectra with a sufficient signal-to-noise ratio. When the laser power is controlled by watching the sample temperature, the danger of pyrolysis, evaporation, boiling, or melting may be largely reduced.

One of the main advantages of near-infrared FT-Raman spectroscopy is that Raman spectra of fluorescing samples may be recorded. Figure 15a shows the Raman spectrum of rhodamine 6G, a well-known, strongly fluorescing, laser dye, and the spectrum of the solvent, ethanol (Fig. 15b). Fluorescence is only excluded when the samples do not have absorption bands of the π electron system at the wavelength of the laser source, 1064 nm. When it occurs, as for some cyanine dyes (Polland *et al.*, 1983), fluorescence may well be observed (Fig. 16).

Several important applications of near-infrared FT-Raman spectroscopy have already been described. Because FT spectra are highly reproducible, Raman spectra of mixtures can be analyzed by subtraction of the spectra of the components (Chase, 1986). Spectra of explosives have been recorded safely, in spite of the danger of overheating with the laser radiation (Hirschfeld and Chase, 1986). Spectra of colored inorganic and organic compounds, of proteins, of nucleic acids, and of many kinds of polymers have been recorded (Hendra and Mould, 1988; Hallmark *et al.*, 1987). It is especially interesting that it is now possible to record spectra of paints in wet and dry state, of catalyst systems under working conditions, and even of "surface-enhanced" spectra of molecules on metal electrodes (Hendra and Mould, 1988). The analysis of derivatized pore glass by near-infrared FT-Raman

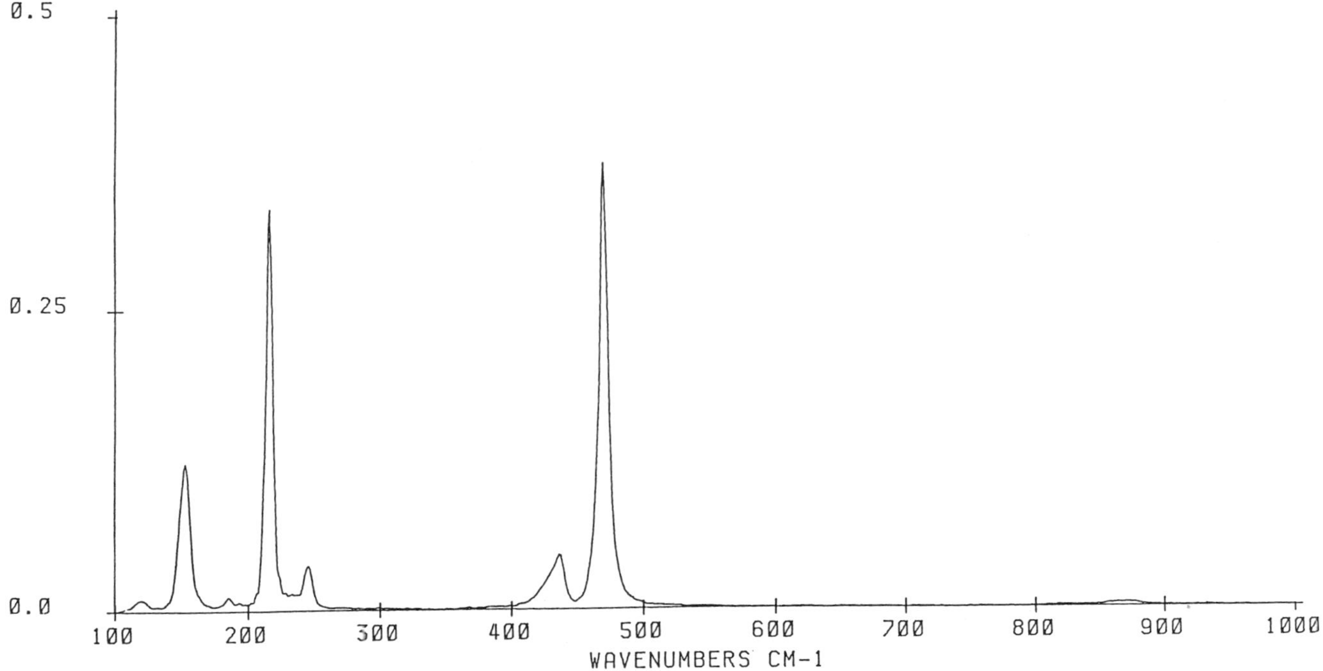

Fig. 13. Raman spectrum of crystalline sulfur.

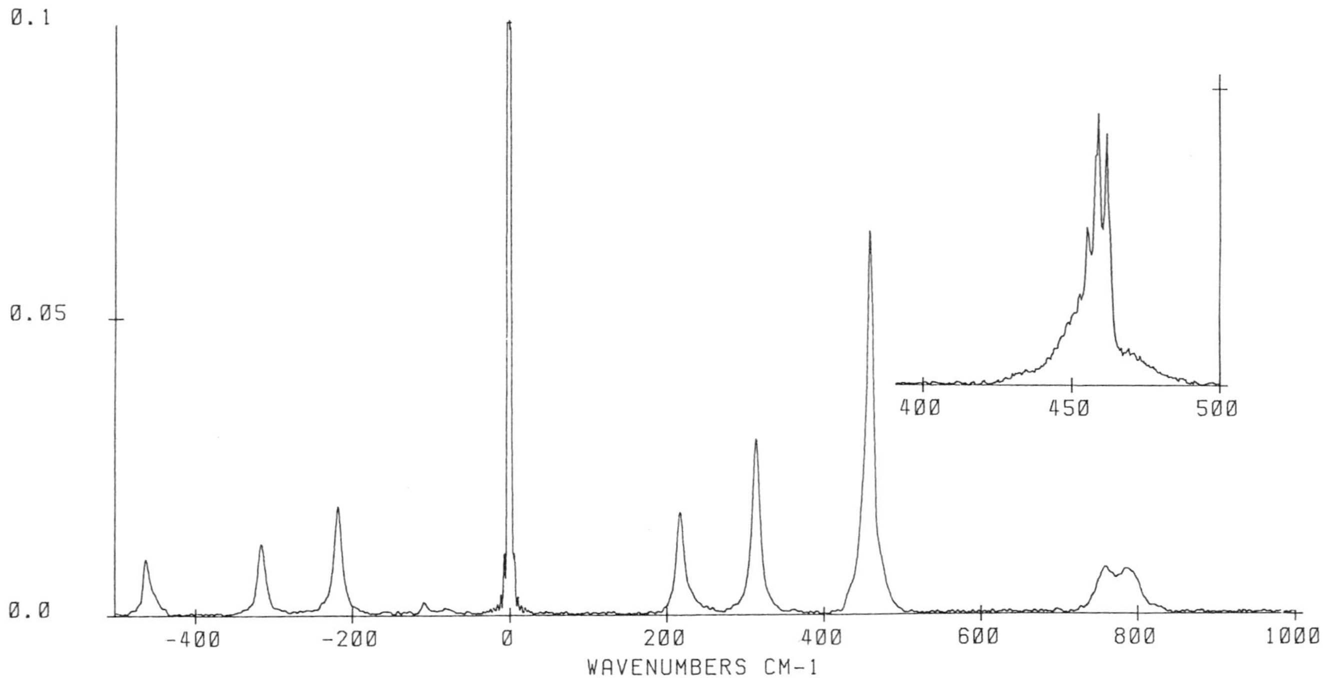

Fig. 14. Stokes and anti-Stokes Raman spectra of carbon tetrachloride, recorded with 100 scans. Inset: Stokes band at 456 cm^{-1}, recorded with a resolution of 1 cm^{-1} and only 35 scans.

spectroscopy has been described recently (Archibald and Honigs, 1987).

Some possible new applications of known Raman techniques to near-infrared FT-Raman spectroscopy have not yet been demonstrated. Gas analyses may be possible, even the determination of temperatures of gases (Belz *et al.*, 1987). Of some interest is the analysis of inclusions in geological samples (Wopenka and Pasteris, 1986) or the investigation of single particles of aerosols (Schrader, 1986). Adsorption effects at liquid–solid interfaces in suspensions can be studied (Witke, 1982). High-resolution FT-Raman spectroscopy with ultrashort laser pulses may yield information about intramolecular energy transfer processes (Graener and Laubereau, 1985).

Diode-pumped NdYAG lasers are excellent candidates for the future role of light source for near-infrared FT-Raman spectroscopy, because their size, weight, efficiency, stability, flexibility, reliability, and lifetime (Begley *et al.*, 1987) are much better than the lasers pumped with "classical" light sources.

B. Limitations

It has already been demonstrated that near-infrared FT-Raman spectroscopy can deal with most old and many new problems typical of Raman spectroscopy. Nevertheless, the ν^4 factor and the high NEP-value of the detector are the reason that, compared with classical Raman spectroscopy, larger power densities of the exciting radiation are necessary for the recording of spectra with the same signal-to-noise ratio. Even extremely small particles of absorbing material (e.g., metal or carbon) in the sample may initiate thermal decomposition. This process produces macromolecules from organic samples that have graphite-like structure with polycyclic aromatics as intermediates. Some samples will show fluorescence, even

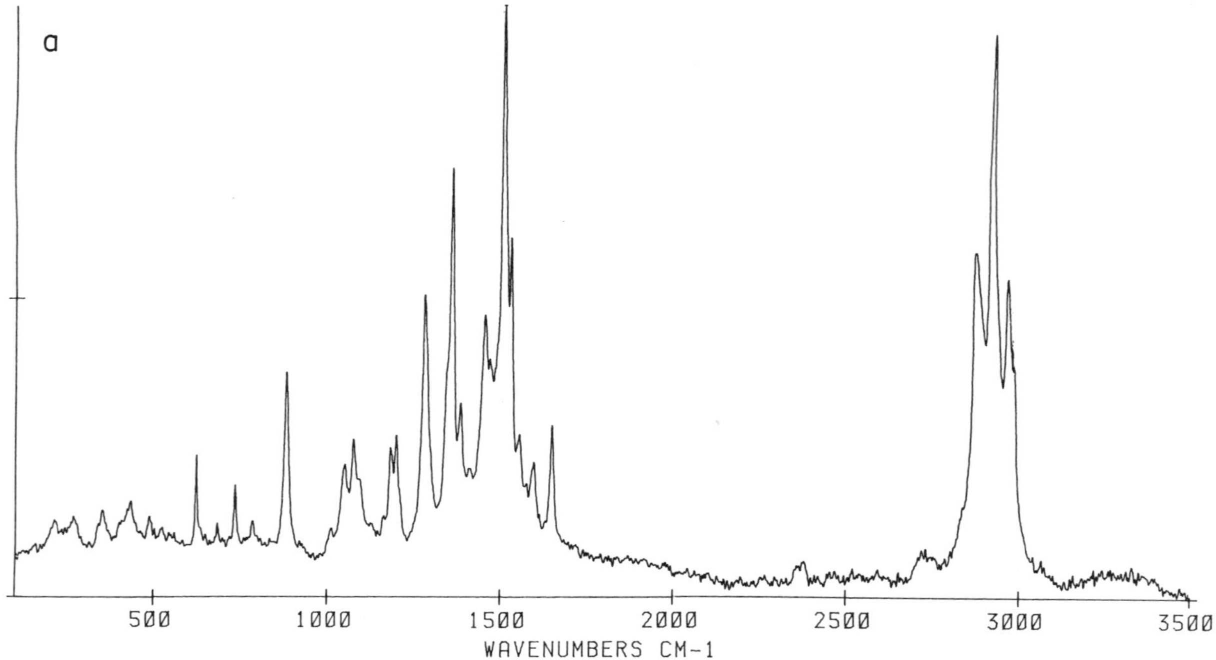

Fig. 15. Raman spectra of (a) the laser dye rhodamine 6G in ethanol and (b) ethanol.

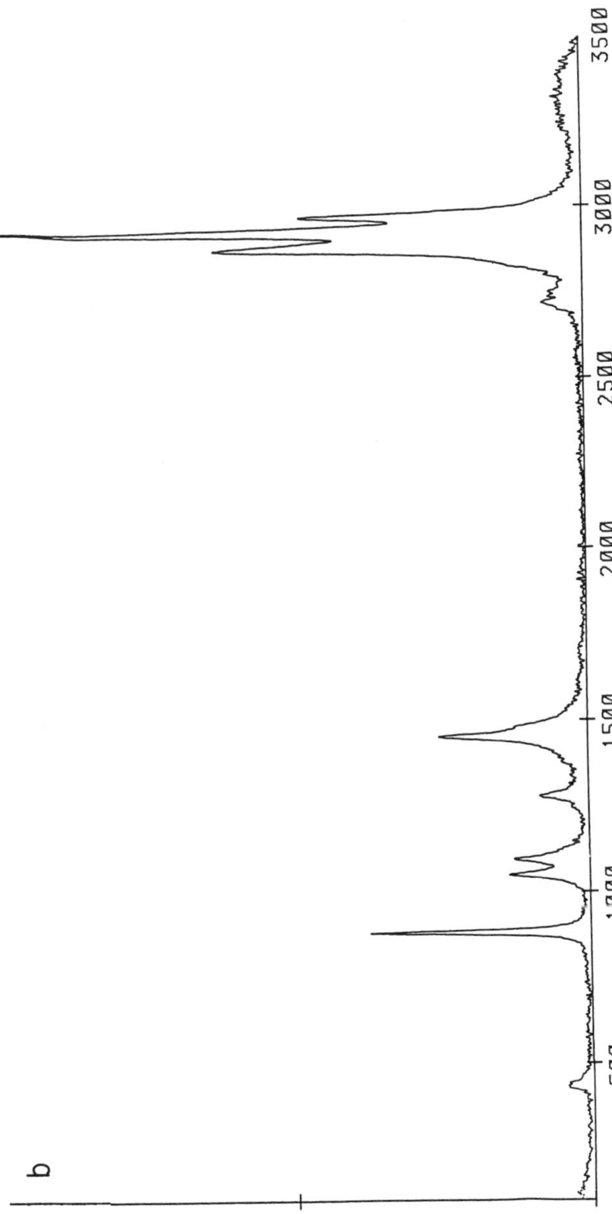
b
500
1000
1500
2000
2500
3000
3500
WAVENUMBERS CM-1

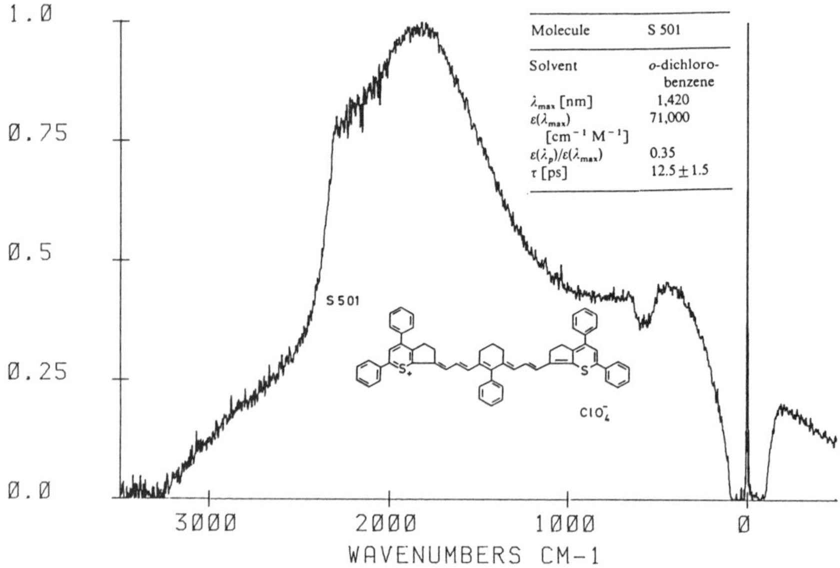

Fig. 16. Fluorescence spectrum of cyanine dye S 501 (Polland *et al.*, 1983), recorded with the BRUKER experimental Raman module; excitation at 1064 nm.

when the NdYAG-laser line at 1064 nm is employed. Fluorescence has a quantum yield of 6 to 10 orders of magnitude larger than the Raman effect. Therefore, the danger of fluorescence is, in fact, small, but not completely excluded. Another source of background spectra that may mask the Raman spectrum is the thermal emission of the near-infrared combination vibrations due to Planck's and Kirchhoff's law or just the blackbody thermal emission.

V. PROSPECTS

It has been demonstrated recently that near-infrared Raman spectra may well be recorded by employing conventional grating spectrometers (Fujiwara *et al.*, 1986; Porterfield and Campion, 1988). It has been shown that the throughput advantage of interferometers is small because large slit-widths can be used. It was stated further that grating spectrometers are vastly superior to FT instruments for low-frequency vibrations (Porterfield and Campion, 1988). Figure 14 shows, however, that properly designed FT instruments may well show Raman lines at frequencies of 150 cm^{-1}. Because of the multiplex advantage, such near-infrared FT-Raman spectrometers will be superior to a grating instrument with respect to the signal-to-noise ratio, allowing smaller recording times or higher resolution. Of special importance are the reproducible frequency scale, allowing precise spectral subtractions, and the fact that large spectral ranges (Stokes and anti-Stokes spectrum) need the same recording time as small ranges. Equally important are the compatibility of infrared and Raman spectroscopy, which employ the same instrument (one computer), the same programs, and similar formats, and the capability of recording infrared and Raman spectra easily on the same diagram. An alternative suggestion claims similar advantages for a Hadamard transform near-infrared Raman spectrometer (Tilotta *et al.*, 1987; Waters, 1987, Treado and Morris, 1989).

REFERENCES

Ansmann, A., and Schrader, B. (1976). *J. Raman Spectrosc.* **5,** 281.

Archibald, D. D., and Honigs, D. E. (1987). *Microbeam Anal.* **22,** 123.

Begley, D. L., Krebs, D. J., and Ross, M. (1987). *Proc. Soc. Photo-Opt. Instrum. Eng.* **712,** 42.

Belz, H. H., Gutberlet, H., Schallert, B., and Schrader, B. (1987). *Appl. Spectrosc.* **41,** 1009.

Bleckmann, P., Schrader, B., Meier, W., and Takahashi, H. (1971). *Ber. Bunsenges. Phys. Chem.* **75,** 1279.

Bougeard, D., Bleckmann, P., and Schrader, B. (1973). *Ber. Bunsenges. Phys. Chem.* **77,** 1059.

Brandmüller, J., and Moser, H. (1962). "Einführung in die Ramanspektroskopie." Steinkopff, Darmstadt, FRG.

Buchanan, B. R., and Honigs, D. E. (1986). *Appl. Spectrosc.* **40,** 564.
Califano, S. (1976). "Vibrational States." Wiley, New York.
Chase, D. B. (1986). *J. Am. Chem. Soc.* **108,** 7485.
Chase, D. B. (1987). *Anal. Chem.* **59,** 881A.
Coblentz, W. W. (1905). "Investigations of Infra-red Spectra." Carnegie Inst., Washington, D.C.
Colthup, N. B., Daly, L. H., and Wiberley, S. E. (1975). "Introduction to Infrared and Raman Spectroscopy," 2nd Ed. Academic Press, New York.
Dollish, F. R., Fateley, W. G., and Bentley, F. F. (1974). "Characteristic Raman Frequencies of Organic Compounds." Wiley (Interscience), New York.
Flaugh, P. L., O'Donnel, S. E., and Asher, S. A. (1984). *Appl. Spectrosc.* **38,** 847.
Freeman, S. K. (1974). "Applications of Lasser Raman Spectroscopy." Wiley (Interscience), New York.
Fujiwara, M., Hamaguchi, H., and Tasumi, M. (1986). *Appl. Spectrosc.* **40,** 137.
Furic, K., and Durig, J. R. (1988). *Appl. Spectrosc.* **42,** 175.
Graener, H., and Laubereau, A. (1985). *Springer Proc. Phys.* **4,** 11.
Grasselli, J. G., Snavely, M. K., and Bulkin, B. J. (1981). "Chemical Applications of Raman Spectroscopy." Wiley, New York.
Hallmark, V. M., Zimba, C. G., Swalen, J. D., and Rabolt, J. F. (1987). *Spectroscopy (Eugene, Oreg.)* **2,** 40.
Hendra, P., and Mould, H. (1988). *Int. Lab.,* **34.**
Hirschfeld, T. (1976). *Appl. Spectrosc.* **30,** 68.
Hirschfeld, T. (1979). *In* "Fourier Transform Infrared Spectroscopy" (J. R. Ferraro and L. J. Basile, eds.), Vol. 2, 193. Academic Press, New York.
Hirschfeld, T., and Chase, B. (1986). *Appl. Spectrosc.* **40,** 133.
Hirschfeld, T., and Schildkraut, E. R. (1974). *In* "Laser Raman Gas Diagnostics" (M. Lapp, ed.), pp. 379–388. Academic Press, New York.
Klaeboe, P., Schrader, B., Bougeard, D., Paetzold, P., and von Plotho, C. (1985). *Spectrochim. Acta* **41,** 53.
Kohlrausch, K. W. F. (1943). "Ramanspektren." Akademische Verlagsgesellschaft, Leipzig, FRG. [Reprinted in 1972 by Heyden, London.]
Kubelka, P., and Munk, F. (1931). *Z. Tech. Phys.* **12,** 593.
Landolt-Börnstein, N. (1955). "Zahlenwerte und Funktionen," 6. Aufl. Bd. I/2 und I/3, Molekeln I und II. Springer, Berlin 1951. Bd I/4 Kristalle. Springer, Berlin 1955.
Long, D. A. (1977). "Raman Spectroscopy." McGraw-Hill, London.
McClain, W. M. (1973). *Appl. Opt.* **12,** 153 and 1405.

McCreery, R. L., Fleischmann, M., and Hendra, P. (1983). *Anal. Chem.* **55,** 146.

Maiman, T. H. (1960). *Nature (London)* **187,** 493.

Meier, W., Schrader, B., and Pisárcik, M. (1972). *Messtechnik (Halle)* **5,** 119.

Möller, K. D., and Rothschild, W. G. (1971). "Far-Infrared Spectroscopy." Wiley (Interscience), New York.

Nassau, K. (1981). *J. Gemmol.* **17,** 306.

Oelichmann, H.-J., Bougeard, D., and Schrader, B. (1981). *J. Mol. Struct.* **77,** 149 and 179.

Oelichmann, H.-J., Bougeard, D., and Schrader, B. (1982). *Angew. Chem., Int. Ed. Engl.* **21,** 639.

Parker, S. F., Williams, K. P. J., Hendra, P. J., and Turner, A. J. (1988). *Appl. Spectrosc.* **42,** 796.

Plaza, P., Dao, N. Q., Jouan, M., Fevrier, H., and Saisse, H. (1986). *Appl. Opt.* **25,** 3448

Polland, H. J., Elsaesser, T., Seilmeier, A., Kaiser, W., Kussler, M., Marx, N. J., Sens, B., Drexhage, K. H. (1983). *Appl. Phys. B* **32,** 53.

Porterfield, D. R., and Campion, A. (1988). *J. Am. Chem. Soc.* **110,** 408.

Raman, C. V., and Krishnan, K. S. (1928). *Nature (London)* **121,** 501.

Rasetti, F. (1929). *Phys. Rev.* **34,** 367.

Rasetti, F. (1930). *Z. Phys.* **66,** 646.

Rosasco, G. J. (1980). *In* "Advances in Infrared Raman Spectroscopy" (R. J. H. Clark and R. E. Hestor, eds.), Vol. 7, 223. Heydon and Sons, London.

Sadtler Research Laboratory (1955). "Sadtler Standard Spectra." Sadtler Research Lab., Philadelphia.

Schaefer, C., and Matossi, F. (1930). "Der Ramaneffekt", *Fortschr. Chem., Phys. Phys. Chem.* **20,** 1–52.

Schrader, B. (1984). German patent appl. 3424108.6–52 (September 29, 1984); also U.S. patent 4714346 (December 22, 1987) and U.K. patent GB 2162961 B (January 6, 1988).

Schrader, B. (1973). *Angew. Chem., Int. Ed. Engl.* **12,** 882.

Schrader, B. (1978). *In* "Molecular Spectroscopy: A Specialist Periodical Report" (R. F. Barrow, D. A. Long, and D. J. Millen, eds.), Vol. 5, pp. 235–269. Chem. Soc., London.

Schrader, B. (1980). *Ullmanns Encykl. Tech. Chem. C* **5,** 303.

Schrader, B. (1983). *Fresenius' Z. Anal. Chem.* **314,** 363.

Schrader, B. (1986). *In* "Physical and Chemical Characterization of Individual Airborne Particles" (K. R. Spurny, ed.). Horwood/Wiley, Chichester, England.

Schrader, B. (1988). *Croat. Chem. Acta* **61,** 589.

Schrader, B., (1989). "Raman Infrared Atlas of Organic Compounds." (Second Edition). VCH Verlagsgesellschaft, Weinheim.

Schrader, B., and Ansmann, A. (1975). *Angew. Chem., Int. Ed. Engl.* **14,** 364.
Schrader, B., and Bergmann, G. (1967). *Z. Anal. Chem.* **225,** 230.
Schrader, B., and Meier, W. (1966). *Z. Naturforsch., A* **21a,** 480.
Schrader, B., and Meier, W. (1972). *Z. Anal. Chem.* **260,** 248.
Schrader, B., and Meier, W. (1974–1976). "Raman/IR—Atlas of Organic Compounds," 2 vols. Verlag Chemie, Weinheim, FRG.
Schrader, B., and Meier, W. (1975). *Z. Anal. Chem.* **275,** 177.
Schrader, B., and Pacansky, J. (1983). *J. Chem. Phys.* **78,** 1033.
Schrader, B., and Simon, A. (1987). *Bruker Rep.* **2,** Appl. Note 29.
Schrader, B., and Steigner, E. (1973). *In* "Raman Spectroscopy of Steroids" (E. Heftmann, ed.), pp. 231–243. Academic Press, New York.
Schrader, B., and Stockburger, M. (1965). *Naturwissenschaften* **52,** 298.
Schrader, B., and Stockburger, M. (1966). *Z. Anal. Chem.* **216,** 117.
Schrader, B., Meier, W., Steigner, E., and Zöhrer, F. (1971). *Z. Anal. Chem.* **254,** 257.
Schrader, B., Bougeard, D., and Niggemann, W. (1980). *In* "Computational Methods in Chemistry" (J. Bargon, ed.). Plenum, New York.
Schrader, B., Niggemann, W., Belz, H. H., and Schallert, B. (1981). *Z. Chem.* **21,** 249.
Schrader, B., Spiekermann, M., Hecht, L., and Bougeard, D. (1984). *J. Mol. Struct.* **113,** 49.
Schrötter, H. W., and Klökner, H. W. (1979). *Top. Curr. Phys.* **7,** 123.
Schwab, S. D., and McCreery, R. L. (1984). *Anal. Chem.* **56,** 2199.
Siesler, H. W., and Holland-Moritz, K. (1980). "Infrared and Raman Spectroscopy of Polymers." Dekker, New York.
Spiekermann, M., Bougeard, D., Oelichmann, H.-J., and Schrader, B. (1980). *Theor. Chim. Acta* **54,** 301.
Takahashi, H., Schrader, B., Meier, W., and Gottlieb, K. (1967). *J. Chem. Phys.* **47,** 3842.
Tilotta, D. C., Hammaker, R. M., and Fateley, W. G. (1987). *Spectrochim. Acta, Part A* **43A,** 1493.
Treado, P. F., and Morris, M. D. (1989). *Appl. Spectrosc.* **43,** 190.
Waters, D. N. (1987). *Appl. Spectrosc.* **41,** 708.
Weidlein, J., Müller, U., and Dehnicke, K. (1981). "Schwingungsfrequenzen I." Thieme, Stuttgart, FRG.
Weidlein, J., Müller, U., and Dehnicke, K. (1982). "Schwingungsspektroskopie." Thieme, Stuttgart, FRG.
Weierstrass, K. (1856). "Mathematische Werke," Vol. 3, p. 175. [Reprinted in 1967 by Johnson Reprint Corp., New York; see also Born, M. (1933). "Optik," p. 58. Springer-Verlag, Berlin.].
Witke, K. (1982). *Appl. Spectrosc.* **36,** 471.
Wopenka, B., and Pasteris, J. D. (1986). *Appl. Spectrosc.* **40,** 144.
Zimba, G. G., Hallmark, V. M., Swalen, J. D., and Rabolt, J. R. (1987). *Appl. Spectrosc.* **41,** 721.

5

Vibrational Circular Dichroism: Comparison of Techniques and Practical Considerations

Timothy A. Keiderling
Department of Chemistry
University of Illinois at Chicago
Chicago, Illinois

I. INTRODUCTION

Natural optical activity is one of the most structurally sensitive molecular properties available for spectroscopic probing, and, as such, its measurement has been extensively exploited in stereochemical analyses. Most spectroscopic techniques (for example, infrared and Raman spectroscopies) sense detailed aspects of molecular geometry only via perturbation of energy levels and selection rules. However, molecular optical activity originates as a direct consequence of the geometrical arrangement and interaction of the atoms in a molecule. In other words, the observable optical activity can be considered to have a first-order dependence on the molecular geometry. In recent years, the standard technique for assessing optical activity has been circular dichroism (CD) spectroscopy, or measurement of the differential absorption of left and right circularly polarized light ($\Delta A = A_L - A_R$). Circular dichroism is related to the previously used optical rotatory dispersion (ORD) measurement via a Kramers-Kronig transform. It is now preferred over ORD as a result of the development of high-sensitivity CD instrumentation and the direct relationship of CD to the properties of the specific states being excited in the transition rather than to dispersion properties that are formally dependent on integration over transitions to all excited states.

In principle, the CD of different transitions, either electronic or vibrational in nature, can be coupled with a theoretical model relating the rotational strength, R^k, of a transition to geometrical properties of the molecule to give structural information about the system being studied. Experimentally,

$$R^k = 0.23 \times 10^{-38}\left[\int \frac{\Delta\epsilon^k}{\nu^k}\, d\nu\right] \tag{1}$$

in units of $(\text{esu}\cdot\text{cm})^2$ where $\Delta\epsilon^k = \epsilon_L^k - \epsilon_R^k$ in molar extinction units and ν^k is the frequency of a transition to a state $|k\rangle$ from the ground state, $|0\rangle$. Theoretically,

$$R^k = \text{Im}(\underset{\sim}{\mu}^k \cdot \underset{\sim}{m}^k) \tag{2}$$

where $\underset{\sim}{\mu}^k$ is the electric and $\underset{\sim}{m}^k$ the magnetic dipole moment of the transition, and Im indicates taking the imaginary component of the complex product.

Before the 1970s, all CD experiments were limited to the visible and ultraviolet regions of the spectrum and thus were used to study transitions to excited electronic states of the molecule. As such, particularly for organic and biopolymer molecules that are the targets of most chiroptical studies, the number and type of transitions that could be studied with CD was very limited. In fact, it is usually necessary to confine such studies to molecules containing a chromophore, such as a carbonyl. Such chromophores are usually intrinsically symmetrical, so the CD observed is a result of its perturbation by the rest of the molecule. The prime advantage of optical activity for structural studies thus becomes a higher order effect.

There are molecules that contain intrinsically chiral chromophores, but these are few in number.

Whereas most organic molecules have only one or two accessible chromophores in the ultraviolet (UV) region, by contrast, all optically active molecules have extensive infrared absorption and Raman scattering spectra. Chiroptical variants of these vibrational spectroscopic methods can be used to derive molecular structural information from all of these states in a manner analogous to that used in electronic CD. Furthermore, these vibrational transitions are well resolved, relatively easily assigned (at least in the functional group region), and provide local, or groups of such, probes of the chirality. Finally, the vibrational optical activity measured is a ground state property, so it should yield structural information derived from the molecular state of primary interest to chemists. The wealth of structurally sensitive data available from this phenomenon as measured in the infrared—now commonly called vibrational circular dichroism (VCD)—implies that more detail regarding solution-phase molecular geometries can be potentially obtained.

This prospect of application to stereochemical problems has led a number of research groups worldwide to develop the technology to measure VCD as well as the analogous light-scattering phenomenon, Raman circular intensity differential scattering or Raman optical activity (ROA). Much of the effort since the first true solution-phase VCD reported by Holzwarth *et al.* (1974) has centered on technique development. Vibrational circular dichroism is difficult to measure both because it is a weak phenomenon and because it occurs in the infrared region, where light polarization and detection are technologically somewhat more difficult than in the ultraviolet and visible regions. Rotational strengths (R^k) for vibrational transitions are expected to be three to four orders of magnitude smaller than for electronic transitions because of the dependence of the magnetic moment on nuclear rather than electron mass. However, the measurability of VCD, i.e., the signal-to-noise ratio (S/N), is determined by $\Delta A/A$ or, effectively, by the ratio of the VCD to the absorbance. For isolated single bands, it can be shown that $\Delta A/A = 4R^k/D^k$; or that S/N will be proportional to the ratio of the rotational strength, $R^k = Im(\underset{\sim}{\mu}^k \cdot \underset{\sim}{m}^k)$, to the dipole strength, $D^k = |\underset{\sim}{\mu}^k|^2$. Since vibrational transitions have D^k values two to three orders of magnitude lower than those found for typical electronic transitions, it is not surprising that VCD has been shown to be measurable in the infrared region using instruments whose sensitivity approaches that commercially available in the ultraviolet and visible spectral regions.

A substantial amount of work in the VCD field has been reported over the past 15 years, and a number of extensive reviews of VCD have already been published. These include theory (Stephens and Lowe, 1985; Polavarapu, 1984a), experimental technique (Polavarapu, 1985; Nafie and Diem, 1979a; Nafie and Vidrine, 1979, 1982; Stephens and Clark, 1979), and more general applications (Freedman and Nafie, 1987; Nafie, 1981, 1984; Keiderling, 1981a). Several reviews of ROA are also available (Nafie and Zimba 1987; Barron and Vrbancich, 1985; Nafie, 1984; Barron, 1976, 1979, 1984, 1987, 1989; Barron and Buckingham, 1975). Therefore, this chapter does not attempt to be a complete review of the field but offers instead a comparative discussion of techniques used in VCD measurements in particular and a presentation of a selection of example data from The University of Illinois at Chicago (UIC) laboratory. The examples chosen emphasize biopolymer applications, an aspect of VCD that is relatively new, has a practical, empirical nature, and has only been briefly reviewed before (Keiderling, 1986).

A. Significance for Biochemical Studies

It is a basic tenet of chemistry that structure and function are intimately related. Under this precept, numerous studies have been undertaken to elucidate the structure of chemical systems in order to understand how and why they exhibit their observed activity. These studies have been instrumental in the development of models of chemical processes. In addition, many of these processes are accompanied by structural changes in the molecules involved. Monitoring of such changes has been vital to the detection of such processes and to an understanding of their dynamic and equilibrium properties.

Structural tools available to the chemist are many, with those of a spectroscopic nature being most adaptable to the study of solution-phase systems. The information potentially accessible with VCD is, in some cases, not now available from other spectroscopic techniques. Hence VCD studies offer a promise of new insight into molecular structure. Conventional circular dichroism and vibrational (infrared and Raman) spectroscopies have been shown to be very useful for determining qualitative aspects of molecular structures. However, each technique has limitations that can be somewhat overcome by combining the two approaches into VCD studies. This strength of VCD is particularly true for biochemical studies of the secondary structures of proteins ad nucleic acids.

Conventional CD is a study of transitions to excited electronic states only a few of which, such as the π-π^* and n-π^* of aromatics and carbonyls, are accessible for UV-CD in proteins and nucleic acids. Spectra involving these transitions usually involve overlapping, broad, featureless bands that consequently have interfering CD signals. Although CD has been used extensively to determine the α-helical content of proteins and polypeptides (Johnson, 1985; Yang *et al.*, 1986), this interfering nature has compromised the results, for example, in cases where aromatic side chains provide a significant contribution to the spectrum. Complications in determining the nature of transitions to π-excited states have similarly limited CD studies of nucleic acids, particularly theoretical analyses. However, the great strength of CD for biological studies has been and continues to be the use of its structural sensitivity as an *empirical probe* of conformation.

In the vibrational region of the spectrum, infrared and Raman transitions are typically much better resolved and are more straightforward to assign and interpret than are electronic transitions. The nature and polarization of various characteristic vibrations has been firmly established in many cases. Extensive infrared and Raman studies have demonstrated that certain spectral features are characteristic in frequency or intensity for expected structural aspects of either proteins or nucleic acids. However, in ordinary infrared and Raman spectroscopies, the spectral differences between related structures are quite small; so, even though a great amount of spectral detail is available, sensitivity to conformational change is limited.

Vibrational circular dichroism studies combine the advantages of each of these techniques by providing the detail of infrared (or Raman) data with the added conformational sensitivity and two-signed nature of CD. Thus, with VCD, the number and variety of transitions that can be used as structurally sensitive probes of the ground electronic state is significantly increased. These discrete spectral features sample several molecular structural elements whose spectra are strongly influenced by both their local conformation *and* by the relative juxtaposition of these elements to chemically similar groups. Furthermore, such transitions are typically resolved enough to enable different transitions corresponding to different subunits of a large molecule to be studied independently (Yasui and Keiderling, 1986a). Thus, the VCD studies undertaken on biopolymer systems should complement and extend the valuable insight into biochemical structure that has been provided by conventional CD and vibrational spectroscopies.

Clearly, there are other techniques for study of protein and nucleic acid structure in solution that also give conformationally sensitive information. Vibrational circular dichroism studies are envisioned not as replacing these but as offering *additional information* to aid in the overall problem of conformational analysis. Perhaps the most widely exploited spectroscopic technique today for solution structural analysis is nuclear magnetic resonance (NMR). At high resolution, resonances of many individual nuclei can now be assigned and followed through chemical variation, even in polymeric systems. However, this amount of detail is not what is required in some cases. Rather, what is needed for questions of secondary structure is a combined or average physical property of repeated structural units in a chain; however, the property must still reflect each subunit's displacement from and orientation with respect to other like structural units. For such questions, VCD is an ideal technique, because several such modes that can be probed with VCD arise from relatively localized atomic motions in the molecule. Work from the UIC group has implied that VCD data exhibits a short-range sensitivity to conformation that complements longer range effects sensed with electronic CD (Yasui *et al.,* 1986a; Yasui and Keiderling, 1986a,b). The possibility of fine-tuning this information through coupling of the two techniques thus becomes apparent and makes our studies much more significant.

The biopolymer-oriented research at UIC emphasizes *empirical correlation* of VCD spectra with secondary structure. This approach has historically been the most profitable route employed in both parent fields, electronic CD and vibrational (i.e., infrared and Raman) spectroscopies. Until a reliable, stable data set and an understanding of environmental effects are available, detailed theoretical analysis will be premature. Development of spectral patterns consistent (or not) with theoretical expectations will eventually be useful. However, application to real problems of biopolymer conformation will come faster via an empirical approach. Unlike electronic CD, we can correlate data for several different, resolved spectral features in a single molecule. And unlike conventional infrared or Raman spectroscopies, each of these features will have a sharp and possibly a different dependence on stereochemistry.

The cost of all this increase in stereochemical sensitivity and spectral diversity is an increased difficulty in measurement over that found with the more conventional techniques.

Thus, the application of VCD to empirical conformational analysis is only limited by the measurability of spectra for available molecular systems and by attainable instrumental S/N. Random noise is not the only problem to overcome here. For both VCD and ROA, there is a significant artifact problem with false spectral bands, which often correspond well to absorption (or scattering) bands of the molecule. The S/N problems have been addressed over the years through development of higher quality instrumentation (Nafie *et al.*, 1976; Su *et al.*, 1980; Devlin and Stephens, 1987; Diem *et al.*, 1988) and through implementation of FT-IR VCD capability as first proposed and demonstrated by Nafie and co-workers (Nafie and Diem, 1979b; Nafie and Vidrine, 1979, 1982; Nafie *et al.*, 1979, 1981; Lipp *et al.*, 1982; Lipp and Nafie, 1984). The artifact problem of FT-IR VCD noted earlier by Lipp and Nafie (1984) and by Polavarapu *et al.* (1984) has recently been effectively solved (Malon and Keiderling, 1988; Malon *et al.*, 1988a), thus allowing spectra to be obtained on a single enantiomer that demonstrates the applicability of FT-IR VCD to biomolecular studies. We can now measure VCD with a high S/N ratio over the entire mid-infrared region in reasonable time period for model systems as well as for biological molecules in solution. In some cases, particularly for single isolated bands and for the near-infrared modes (e.g., C—H, N—H, and O—H stretches), the older, dispersive style of instrument proves to be more useful than the FT-IR method. In this chapter I will discuss both approaches to experimental VCD measurement and attempt to identify the strengths and weaknesses of each technique by contrasting them.

B. Brief Historical Summary

Before detailing the technological aspects of FT-IR VCD, it may be useful for those readers who are new to the field to have a terse outline of the major milestones in the field. More detail can be found in the previously noted reviews. The first true molecular ROA (Barron and Buckingham, 1973; Barron *et al.,* 1973a,b) and VCD (Holzwarth *et al.,* 1974; Hsu and Holzwarth, 1973) spectra were reported about 15 years ago. At that time, both techniques appeared to be experimentally extremely difficult and yielded weak, noisy spectra. Subsequent developments in instrumentation have made routine VCD spectroscopy possible over large spectral regions. Some of the major advances in experimental VCD techniques include development of improved photoelastic modulators (Cheng *et al.,* 1975, 1976; Nafie *et al.,* 1976), brighter light sources (including the carbon rod) (Su *et al.,* 1980), and the development of FT-IR VCD (as noted earlier). In the ROA field, major advances were made via a dual-beam-input optics design to minimize artifacts (Hug, 1981) and, most important, multichannel detection capability to reduce experiment duration (Hug and Surbeck, 1979, 1982; Brocki *et al.,* 1980; Boucher *et al.,* 1977). Very recently, 180° backscattering ROA has been remeasured and found to decrease artifacts and significantly enhance S/N over the conventional 90° scattering experiments (Hecht, *et al.* 1989; Hug, 1982). Successful magnetic VCD of achiral molecules has also been reported (Keiderling, 1981b; Devine and Keiderling, 1983, 1984, 1985, 1986, 1987), as have magnetic ROA (Barron, 1975, 1977; Barron *et al.,* 1982; Barron and Vrbancich, 1982, 1983, 1985).

Theory, at least in VCD, at first led experiment; now it is attempting to keep pace. Several calculational approaches exist for VCD; these schemes vary from simple coupled-oscillator models (Holzwarth and Chabay, 1972) and an empirical (fixed partial) charge model (Schellman, 1973) to various "corrected" models that attempt to account for bond currents (Abbate *et al.,* 1981; Moskovits and Gohin, 1982), molecular orbital motion (Nafie and Walnut, 1977a,b; Nafie and Polavarapu, 1981a,b), polarizability (Barnett *et al.,* 1980a,b), and other contributions to the VCD. These methods have recently been comparatively reviewed along with parameterized models based on infrared intensity theories (Stephens and Lowe,

1985). Various applications of these models for interpretation of VCD spectra have been put forward by several groups. The fundamental theoretical problem of VCD arises because, at the Born-Oppenheimer (BO) level of approximation, it can be shown that the electronic contribution to the VCD magnetic dipole matrix elements disappears. This result leads to a need for approximate models that avoid use of non-BO level calculations. Methods to correct for these non-BO effects have been proposed (Nafie and Freedman, 1983, 1987; Stephens, 1985). The latter has developed into a calculational scheme (Stephens, 1987; Amos *et al.,* 1987; Jalkanen *et al.,* 1987b, 1988a) and has been demonstrated to be fairly successful for small-molecule systems (Lowe *et al.,* 1986a,b; Kawiecki *et al.,* 1988; Jalkanen *et al.,* 1987a, 1988b).

The theory for ROA is somewhat less advanced, because, after the fundamental formalism was set down, few calculationally tractable approaches to ROA have appeared (Prasad and Nafie, 1979; Barron and Clark, 1982; Nafie and Freedman, 1981; Freedman and Nafie, 1983a,b; Escribano *et al.,* 1987). Even fewer applications of the available models to actual spectra exist. The most promising results have been dependent on transferred intensity parameters or on models that are difficult to fully specify. Recently, Bose, *et al.* (1989) have reported *ab initio* calculation of the ROA and achieved agreement in sign for several bands in the (+)-R-methylthiirane spectrum. Finally, for both ROA and VCD, empirical correlation of spectra with structure has been done successfully for a number of molecules in a manner paralleling the early development of electronic CD as a structural chemical tool (see reviews noted earlier).

II. EXPERIMENTAL DESIGN

As is the case for infrared absorption experiments, two main techniques are employed for VCD experiments—dispersive and Fourier transform (FT-IR) measurement. In this section I outline the basic experimental designs employed in the two approaches, significant technical innovations, and relative advantages and disadvantages.

A. Dispersive VCD Technique

Most dispersive infrared CD spectrometers have been constructed in a similar fashion. Because of the high sensitivity required, dedicated instruments rather than modified commercial instruments have been constructed for dispersive VCD measurement. To point out the special design considerations of most use, the spectrometer used at the University of Illinois at Chicago (Keiderling, 1981a) is discussed here as an example. Variances will be noted as appropriate.

The UIC optical system is presented diagrammatically in Fig. 1. The source optics, monochromator, focusing optics, sample, and detector are encased in flushable housings on a stable table top. Reflection optics are used in all except the detector focusing position in the UIC instrument to avoid the infrared transmission and chromatic problems of lenses. Other instruments in their present incarnations have adopted this approach or modest variants of it (Chabay and Holzwarth, 1975; Diem *et al.,* 1978, 1988; Marcott, 1979; Havel, 1981; Devlin and Stephens, 1987). In addition, the number of mirrors can be minimized to reduce reflective loss. Although such losses are marginal per reflection, Diem *et al.* (1988) have indicated that using a minimum number of mirrors and enhancing their reflectivity can aid S/N improvement.

Several types of light sources have been used in VCD. Tungsten–halogen (W-I) lamps provide high power, have high color temperature, are inexpensive, and are useful in the near-infrared region, $\lambda < 4$ μm (Osborne *et al.,* 1973; Nafie *et al.,* 1976). But a 300-W xenon arc with an integral sapphire window and rear parabolic reflector has been shown to give better S/N in the 3-μm region (Su, 1982; Su and Keiderling, 1980;

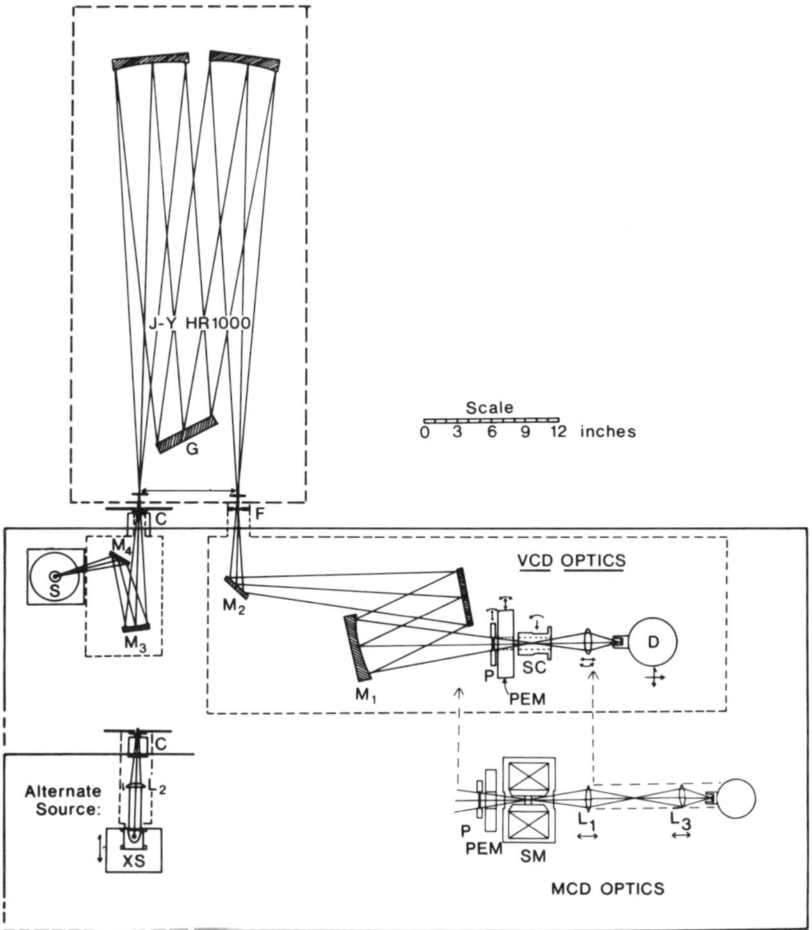

Fig. 1. The optical assembly of the UIC dispersive VCD spectrometer. The major components are S, carbon rod light source; XS, alternate xenon-arc source; C, mechanical chopper; G, grating (120 × 140 mm^2); F, low-pass filter; P, polarizer; PEM, modulator; SC, sample cell; SM, super conducting magnet; D, detector; M_1–M_4, aluminized mirrors; L_1–L_3, lenses.

Stephens and Clark, 1979). Although this Xe source is useful to the sapphire cutoff at $\lambda \sim 6$ μm, its limited spectral range, moderately short lifetime, and high replacement cost have led us to abandon its use. In practical terms, the line structure of the Xe spectrum causes some difficulty in normalization for VCD determination (see later) and for absorption measurements in those spectral regions where interference exists.

At UIC, a carbon rod light source has been constructed that is now used for all of our VCD experiments (Su *et al.*, 1980). It provides a hot blackbody capable of attaining significantly higher power (3–4 kW) and higher color temperature (2400°K) than traditional infrared sources, such as the glowbar used by Chabay and Holzwarth (1975); but it must be isolated from the atmosphere. The lamp's spectral range is thus primarily limited by the demountable window on its housing. We typically use NaCl or KBr windows, which afford total coverage to the ultimate wavelength limitations imposed by our modulators (ZnSe). A solid Cu back reflector more recently added to our cooled lamp housing affords a 10–20% increase in output power.

This source is superior to all others we have used in the mid-infrared at $\lambda < 5$ μm and gives VCD with satisfactory S/N in the shorter wavelength regions (Su, 1982). Its lack of spectral structure makes it more convenient than the Xe source for routine VCD. Maintenance costs of the carbon rod source are low, only involving regular (approximately weekly) replacement of very inexpensive rods. However, the source is physically quite large, and the rods draw a significant amount of power to attain the optimal color temperature, most of which is dissipated in the water-cooled housing. Specifically, we use rods that are 8mm in diameter and 100 mm long and have the center 50 mm machined down to a semicircular cross section. Devlin and Stephens (1987) have used a somewhat different design. Our source is typically powered with ~300 A at ~8 V from a constant voltage DC power supply but is capable of operation at 400 A and at 10 V. The higher currents shorten rod life and, as might be expected from considerations of the blackbody formula, do not give a significant increase in light output in the 5- to 12-μm region. Such rods can be run with an AC power supply with equivalent performance (Boyd *et al.*, 1974; Smith *et al.*, 1978), but some minimal power line ripple can be expected.

The monochromator used in the UIC instrument has a 1-m focal length, a high angular aperture (f/6.8), and digital control (Jobin-Yvon, HR-1000). Such a design requires the use of fairly large mirrors and similar constraints on the other optics, but it maximizes the collection of light and, hence, the S/N. The long focal length of the monochromator combined with maximum groove density gratings allow the use of relatively wide slits. This arrangement, in turn, permits high power transmission through the instrument from the large, blackbody source used while still maintaining adequate resolution for VCD purposes. In the mid-infrared, slits as wide as 8 mm can be used which match the focused diameter of the carbon rod. The slit widths result in spectral resolutions ranging from 6 to 12 cm^{-1}, depending on wavelength and grating used. VCD spectra are most easily obtained with gratings blazed for each spectral region and used in first order. A light chopper is placed in front of the entrance slit to modulate the light intensity as required for the infrared detectors. Multilayer-coated, long-wave-pass, germanium substrate filters (OCLI) are mounted at the exit slit to block higher order diffraction.

Most dispersive VCD spectrometers have similar optical designs. Somewhat of an exception is a variant that has been recently described by Diem *et al.* (1988) and that uses a much shorter focal length monochromator (0.32 m), a hot Nernst glower source (Artcor, Inc., Model 103-5), and appropriately small slits. Such a down-scaling in focal length can significantly reduce cost and flushing requirements. Furthermore, the power requirements of a Nernst glower are much more modest than those for the carbon rod. Their design is capable of high S/N mid-infrared VCD and should be straightforwardly adaptable to other spectral regions. It is expected that dispersive instruments constructed in the future would follow this approach from a cost-effectiveness basis. The shorter optical system is perhaps somewhat less flexible in terms of potential spectral resolution; but, in realizable terms, higher resolution VCD spectra are better obtained with the FT-IR approach (see next section).

As practiced with all modern instrumentation, CD is a modulation experiment whereby the state of the polarization of the light beam being measured is varied between left and right circular, and the differential signal that is in-phase with that modulation is detected. In VCD, linearly polarized light is made elliptically polarized through phase retardation in a photoelastic modulator. Most instruments employ grid polarizers (Cambridge Physical Sciences, PTR Optics) that are deposited on BaF_2, ZnSe, or other transmitting substrates. In our experiments, AgCl-based polarizers (Perkin-Elmer) have had a limited lifetime, as evidenced by substantially reduced polarization efficiency after 1–2 years in the lab. Grid polarizers provide high angular and linear aperture as well as an adequate extinction ratio for infrared wavelengths beyond ~3 μm. Some early experiments (Nafie *et al.*, 1976) used $LiIO_3$ crystal polarizers (Interactive Radiation) in the near-infrared, 2- to 4.5-μm, region. Although these polarizers are capable of attaining quite high polarization extinction ratios, their angular aperture is very restrictive and would drastically reduce the light throughput and hence S/N in the UIC, and other, high optical aperture instruments. In addition, they are much more expensive and hygroscopic than grid polarizers.

Stress-optic or photoelastic modulators (PEM) that are available and appropriate for the infrared are constructed with CaF_2 or ZnSe optical elements (Hinds International) these modulators allow $>\lambda/4$ retardation to the transmission cutoff (~600 cm^{-1} for ZnSe). Such PEMs operate under sinusoidal modulation that, compared with square-wave modulation, has only a small effect (reduction in magnitude) on the signal magnitude developed. The convenience and other practical aspects of these modulators far outweigh this disadvantage. Now, PEMs have become standard in all commercial CD instrumentation. At the longer wavelengths, (>10 μm), more stress must be developed in the optical crystal to attain circularly polarized light. We have found that this level of stress leads to substantial heating of the ZnSe crystal, thereby introducing some instability in the modulation if not monitored and controlled with air cooling. Anti-reflection (AR) coatings on the ZnSe modulators can be profitably used to enhance transmission.

The stress developed in a PEM can be adjusted to result in a $\lambda/4$ retardation at the center of the crystal aperture. This stress falls off over the face of the crystal like a cosine, thus providing a large aperture area with adequately constant retardation that, in addition, is relatively insensitive to variation of angle of incidence. As the spectrum is scanned, this stress can be varied to maintain a constant level of circular polarization in the beam (Diem *et al.*, 1988), as is done on commercial UV-CD instruments. Instead, we pick an appropriate stress value and calibrate over the wavelength range being sampled. Since the wavelength range scanned with the dispersive instrument is typically only a few hundred cm^{-1}, the calibration value changes relatively little over the spectral segment measured. Use of a modulator made of CdTe has been proposed to enable further penetration toward the far-infrared (Hoffman *et al.*, 1987), but no VCD data are yet available from this system.

In the UIC dispersive instrument, our optimal optical arrangement for midinfrared VCD includes a BaF_2 substrate grid polarizer 42 mm in diameter (Cambridge) and a ZnSe modulator (JCK-III) with a large cross section and a single-layer AR coating optimized for 10 μm. This setup yields the maximum attainable aperture and light throughput. While desirable in general, this optical arrangement is particularly important for magnetic VCD measurements where the modulator must be several inches from the sample position in order to accommodate the magnet bore length (see section II.C.). In the near-infrared and often for $\lambda = 6$ μm measurements (e.g., C=O stretch), we use a CaF_2 modulator of smaller aperture. Its higher light throughput due to lower reflection loss (when it can be located near the instrument focus) is sometimes an advantage. Also it has proved to be more stable than one of our ZnSe modulators for studies involving comparative quantitative measurements such as are needed for kinetic analyses (Chickos *et al.*, 1986).

Sampling devices for VCD are essentially the same as for any transmission infrared experiment, with the additional precaution that care must be taken to minimize the stress imparted to the windows. Stress in the windows typically leads to a birefringence-induced signal that is evident as a baseline offset or, at times, as a distortion of the VCD bandshape. We have used sealed liquid cells, variable-path cells, demountable-window cells with Teflon or lead spacers, and gas cells with windows fixed onto a glass or metal cylinder with RTV (Dow-Corning), wax, or epoxy. Experiments on mulls, pellets, and films have been undertaken (Diem *et al.*, 1979; Su, 1982; Sen and Keiderling, 1984a). In general, data obtained with these approaches are not simply related to solution- or gas-phase data. In our opinion, such solid-state VCD is dominated by nonisotropic intermolecular interaction effects and is not susceptible to simple interpretation from a molecular point of view. Successful comparative empirical studies of controlled systems have been carried out to monitor variations in related molecules (Narayanan *et al.*, 1985, 1986). In addition, there have been two reports of matrix isolation VCD where some improvement in bandwidth is seen (Schlosser *et al.*, 1982; Henderson and Polavarapu, 1986).

We have also constructed double-walled cell holders to contain demountable (spacer) cells for variable temperature operation (Su, 1982; Narayanan, 1987). It is important to use a second pair of windows on such a housing to provide an insulation between the cell and the ambient-temperature sample compartment. Flow of cooled or heated fluid from a controlled temperature bath (Neslab) through the cell holder then can provide sufficient control of the sample temperature.

In a CD experiment, the sample acts as a transducer of polarization modulation into intensity modulation. The efficiency of this transduction at any given wavelength is related to the CD of the sample. To detect the modulated infrared intensity, high-sensitivity, cooled solid-state detectors with moderately fast response (due to the modulation frequency) are used. To focus the beam onto the detector, we use f/1.2–1.5, CaF_2 or ZnSe focusing lenses, the choice of lens depending on the spectral region of interest. At $\lambda < 5$ μm, InSb cooled to 77 K is the preferred detector material. It can be obtained as a photovoltaic device; the size and shape of its sensitive area can be selected to match the optical beam cross section that is appropriate to the moderate focus provided by the focusing lens. In the mid-infrared (~5–11 μm), $HgCd_x$-Te_{1-x} (MCT) detectors cooled with liquid nitrogen provide high sensitivity and reasonably large, photoconductively active areas. The achievable size of an MCT detector is limited by its resistance, which is inversely related to the surface area. Thus, large detectors tend to draw a large bias current in the photoconductive mode. The usable spectral region can be affected by changing doping ratios, but we have done best in terms of S/N with so-called narrow band detectors having a maximum in D* at ~11 μm (Infrared Associates). Larger areas can usefully be obtained by creating arrays of smaller detectors whose combined shape and size matches the slit image. The signals obtained from each with its own preamplifier are then summed in an auxillary amplifier. In restricted parts of the MCT spectral region, the S/N also can be improved by placing a cooled band or high-pass optical filter and a restrictive aperture directly in front of the detector (mounted on the cold shield inside the dewar). Even CaF_2 is a moderately effective cold filter for measurement in the 6- to 8-μm region.

For the near-infrared we use a 2 × 10 mm InSb detector (Spectronics), and for the mid-infrared we use two 3 × 3 mm MCT elements mounted vertically (Infrared Associates). The Diem *et al.* (1988) shorter focal length instrument is optimized with a 0.5 × 5 mm MCT detector area, which can have significantly less noise than does our design. For longer wavelengths, we have a Si:As detector (Santa Barbara) with a

2 × 8 mm aperture mounted in a liquid helium dewar. This detector is similar to one reported by Devlin and Stephens (1987), but that one is smaller and is mounted on a closed cycle refrigerator, an arrangement that yields a more convenient mode of operation. The biasing needed for the UIC Si:As detector is about 100 V, a demand that places a high restriction on power supply stability. We have found it useful to use batteries to provide a stable source of bias current.

The signal developed at the detector is processed by the electronics assembly schematically outlined in Fig. 2. The low-frequency component (I^{LF}) due to the chopped light beam (fixed at 150 Hz in the UIC instrument) and the high-frequency signal (I^{HF}) due to the CD (30–60 kHz, depending on the modulator) are amplified and separated (with an L-C bandpass filter tuned to ω_M) for the phase-sensitive detection with lock-in amplifiers. Use of a chopper frequency that is not a multiple of the line frequency is important. Others have used asynchronous choppers at lower frequencies (Diem *et al.*, 1988; Devlin and Stephens, 1987). To avoid overloading the CD lock-in, the chopping frequency is made as low as possible within the constraint required for low 1/*f* noise.

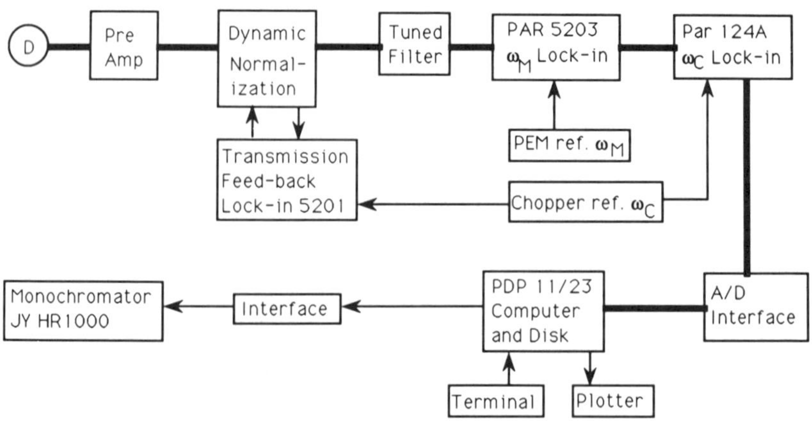

Fig. 2. The electronics assembly for the UIC dispersive VCD spectrometer. See the text for further detail.

The UIC instrument uses a normalization amplifier that holds constant the output signal from the first 150-Hz lock-in at a constant amplitude by varying the total signal gain. The raw signal at 150 Hz is a measure of the instrumental transmission. The variation in gain applied by the normalization amplifier to the input signal makes the signal output into the high-frequency channel proportional to I_{HF}/I_{LF}. This, in turn, is proportional to ΔA (for low values of ΔA) by

$$I_{HF}/I_{LF} = 2\, J_1(\delta_0) \tanh[1.15\, \Delta A] \sin(\omega_M t) \qquad (3)$$

where δ_0 is the phase retardation of the modulator, J_1 is a first-order Bessel function, and ω_M is the modulator frequency (Nafie *et al.*, 1976).

We have found it advantageous to use two lock-in amplifiers in series in the CD channel to gain sensitivity as well as to discriminate against stray signals at ω_M. This technique is possible because the CD signal itself is also modulated by the chopper. Because of the two very different frequencies used in the UIC experiment, the ω_M lock-in can be operated with a $<$1-msec time constant at its output and, thus, will pass an AC signal at the chopping frequency that is proportional to the CD. Ground loops and other stray signals are discriminated against by this approach, and the necessary gain needed to create a useful output VCD signal for digitization is spread over two lock-ins, a scheme that reduces the problem of overload.

The resultant VCD spectrum can be displayed on a strip-chart recorder and is digitized (A/D) and transmitted to a dedicated computer (DEC PDP-11/23) for storage. The computer also controls the data collection by scanning the monochromator and monitoring several points in the instrument; it can also be used to display or plot the resultant VCD. Averaging of several scans in a repetitive scan mode improves S/N and provides a consistency check. We typically collect data while scanning in both directions for efficiency, and we use a software routine to correct for spectrometer backlash and wavelength calibration. The final form of the data is obtained by measuring a CD baseline with a racemic sample (if available) and thereby correcting the spectra for residual CD artifacts. Baselines in the current setup are effectively flat over moderate wavelength ranges; and, for relatively small sample absorbances, it is not difficult to correct the sample scans when measuring moderate to large VCD signals.

In their new design, Diem *et al.* (1988) have used a more digital approach. Both signals are amplified with lock-ins, but dynamic normalization is not used. Rather, normalization is done after lock-in amplification and A/D conversion, but in each case the gain is varied as the spectrum is scanned to maintain high digital precision in each output. They also use two lock-ins in series in the VCD channel, with the final stage of amplification being digitally controllable. These signals are then digitized and subsequently ratioed by the computer. The various gain factors used are taken into account in producing the final spectrum. Controlling scan rate as well as lock-in parameters by computer is central to such a design since it is necessary to stop the scans to give the electronics time to settle after each gain change. This approach does lead to optimal S/N in the digital ratioing process.

In the UIC dispersive instrument, a simple lock-in with some flexibility in time constant (PAR 5101) can be used for the 150-Hz sample transmission measurement—because of the high signal levels obtained. Low time constants ($\sim$10 msec) are needed for proper operation of the feedback elements in the normalization circuit. We currently use a two-stage design for the feedback with a fast audio-attenuator (Motorola) followed by a slower photodiode-LED assembly both acting as a variable gain portion of a feedback amplifier. Attainable dynamic range is on the order of 10^3. In their alternate design, Diem *et al.* (1988) use a unique digital lock-in (Ludl Electronic Products) that is very fast and capable of computer-controlled gain changes.

For the VCD channel, high sensitivity and high overload

capability are needed, with the intermediate lock-in being capable of operation with a low time constant ($\leq$1 msec) to pass the 150-Hz modulated VCD. We currently use a PAR 5203 followed by a PAR 124A. The final lock-in stage needs to have stable, low noise output characteristics since the DC level is the measure of VCD. We have used a heterodyne-based system for this purpose with equal success (PAR 186A). The initial lock-in needs to have low time constant and good overload capabilities. Due to the two stages of amplification, high sensitivity is not needed for either one.

Infrared detectors have a slight polarization sensitivity that, at the levels needed to detect VCD, gives rise to artifact signals at ω_M. These artifacts could be due to imperfect polarization of the beam; to residual static birefringence in the modulator, sample cell, or optics; to retardation introduced upon reflection; or to detection of multiply reflected light beams. In general, the artifacts can be minimized by careful alignment and by the use of moderate focusing such as is compatible with large-area detectors. This effect led to the use at UIC of a focusing lens rather than the off-axis ellipsoidal mirror that was employed in the first extensive VCD experiments reported (Nafie *et al.*, 1976). For reliable measurement of the VCD for a sample with very small ΔA values, baselines of racemic samples measured under exactly the same conditions (cell, pathlength, concentration, temperature) should be compared to the data obtained from the optically active sample. For molecules with large- to medium-size ΔA's, this baseline is less important in dispersive VCD experiments provided the extent of the instrumental artifacts is known and relatively transmissive samples ($A < 1$) are used. Sometimes, non-optically active samples with absorption spectra similar to that of the optically active sample (in a given spectral region) can be used to obtain a medium-sensitivity baseline. Correlation of the single-beam transmission of the instrument plus sample with the VCD being measured in real time can be very useful in the identification of artifact signals.

B. FT-IR-Based VCD Measurement

VCD can also be measured with an FT-IR-based instrument, an approach first proposed (Nafie and Diem, 1979b) and demonstrated by Nafie and co-workers (Nafie *et al.*, 1979, 1981; Nafie and Vidrine, 1979, 1982; Lipp *et al.*, 1982). The concept generally used is optically similar to the dispersive design described earlier. The wide optical band-pass output of a fast-scanning interferometer is intensity modulated over a range of Fourier frequencies that is appropriate to the band-pass and the mirror speed. For VCD experiments, this output is passed through a polarizer and modulator that polarization-modulate the beam at a frequency (ω_M) that is much higher than the Fourier frequencies being generated. It then passes through the sample and is focused onto the detector. A nonzero VCD in the sample will generate an intensity-modulated signal at a frequency of ω_M at the detector. This signal will have components (sidebands) at Fourier frequencies that are appropriate to those light frequencies that have dichroic absorption.

Passing the signal thus developed and preamplified through a low-pass filter, as would be typically found in standard FT-IR signal processing networks, yields a standard interferogram of the single-beam transmission of the sample (plus instrument). On the other hand, passing the signal through a lock-in amplifier with a sufficiently low output time constant effectively creates a VCD detection channel. The high-frequency band-pass filtering and rectification of the ω_M component of the signal by the lock-in results in a signal from this VCD channel that is an interferogram corresponding to the polarization-modulated signal alone, as developed at the detector. Ideally, all of the polarization-modulated signal should result from differential absorption or CD of the sample. Since ΔA (at the low values typical of VCD) is proportional (Eq. 3) to the ratio of the polarization-modulated signal (I_{HF}) to the sample transmission signal (I_{LF}), digitally ratioing the two single-beam spectra obtained by an FFT of these two interferograms yields the raw VCD spectrum of the sample.

The theory of this measurement for FT-IR has been extensively reviewed by Nafie and Vidrine (1982) and by Polavarapu (1985) in this series, hence repetition here would be superfluous. In what follows, I discuss the optical and electronic parameters used in the UIC instruments and compare their design and performance with other FT-IR-based and dispersive instruments, as appropriate. We have actually assembled two separate FT-IR VCD set-ups at UIC, and their comparison has led us to propose a solution of the absorption-correlated artifact problem for FT-IR VCD. This, in turn, opens up the possibility of studying biopolymeric systems for which enantiomers are rarely available for conventional baseline correction. Both UIC instruments are based on standard Digilab FTS-40/60 optical benches. Other successful, routinely used instruments have been based on the Nicolet 7199 and 6000 benches; Digilab FTS-14 and Perkin-Elmer 1800 benches have also been used for VCD. It is our opinion that any modest FT-IR optical bench could prove adaptable for VCD given a modicum of flexibility in sampling and detectors. Optics are not the major problem for FT-IR VCD, but software is.

One of the setups at UIC, denoted here as the "mirror setup," has a cutout in the standard sample compartment so that a plate containing an optical filter, grid polarizer, stress-optic modulator, and sample cell holder can be inserted. This arrangement places the modulator near the focus of the normal f/3 optics in the FTS-40 sample compartment. Following the sample is the standard ellipsoidal mirror (hence, the denotation) and a small-area MCT detector (1 mm in diameter, $D^* = 6.9 \times 10^{10}$, Infrared Assoc.). We have typically used a BaF_2 substrate polarizer (Cambridge, GP-228) and a 70-kHz ZnSe modulator (Hinds, JCK-II) with this setup, but other polarizers and modulators have been used with roughly equal success. In particular, the high modulation frequency we have chosen is not necessary for good S/N FT-IR VCD, as we have demonstrated by comparative runs with alternate modulator/polarizer assemblies.

The second setup, schematically illustrated in Fig. 3, will be denoted as the "lens setup." Here the polarization optics are contained in a separate sample compartment, which is accessed via the external beam and has been modified by Digilab, to our design, based on their standard GC accessory. The main difference between the lens setup and the mirror setup is that the optical train in the external sample compartment of the lens setup utilizes an f/5 focusing mirror before the sample and a BaF_2 lens (1 inch in diameter, f/1) to focus the infrared beam onto the detector. Because of this milder focusing optic, we have chosen to use a larger area MCT detector for the lens setup (4×4 mm, $D^* = 1.4 \times 10^{10}$, Infrared Assoc.). In the lens setup, we typically use a Ge substrate grid polarizer (GP-226) and 38-kHz ZnSe modulator. In both instruments, long-

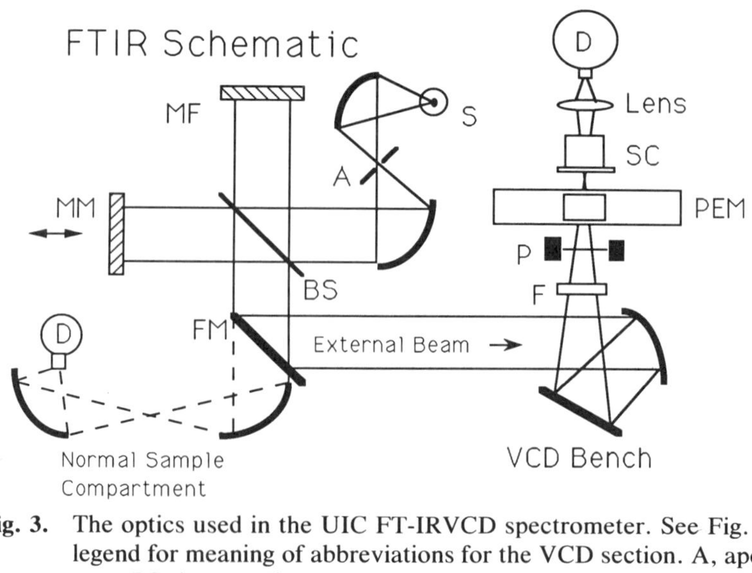

Fig. 3. The optics used in the UIC FT-IRVCD spectrometer. See Fig. 1 legend for meaning of abbreviations for the VCD section. A, aperture; BS, beam splitter; MM, moving mirror; MF, fixed mirror; FM, flip mirror to external sample compartment.

wave-pass, Ge-substrate optical filters (OCLI) were chosen for the spectral range of interest and mounted in front of the polarizer. The filter and linear polarizer do combine to limit the aperture of the sample compartment beam. At the source focus, an aperture appropriate for 2-cm^{-1} resolution was typically used for our experiments. The larger "open" aperture of the Digilab design gives more light throughput, but we have noted a slight degradation in baseline when using it. For aqueous samples, we often need to use the "open" position because of solvent absorption.

Recently we have constructed a second VCD sample compartment (Croatto, 1989) that is much larger than the original and is accessed by means of a flip mirror installed in the first compartment (lens setup). This new bench has even milder focusing (3-inch diameter gold coated spherical mirror with a 45-inch f.l., effectively f/16) and a 2-inch diameter, 2 inch f.l. ZnSe imaging lens onto the detector (2.5 mm square, $D^* \sim 2 \times 10^{10}$). The larger beam diameter that results necessitates use of a large aperture polarizer and modulator (JCK-III, Hinds) but permits adequate space for use of a magnet in the sample area, or of longer path cells, such as gas cells or a temperature variation apparatus.

Other instruments have used basically the same approach as we have employed in the mirror design. Nafie and co-workers (Lipp *et al.*, 1982) have used a separate sample compartment with ellipsoidal mirror focusing, and Polavarapu (1985) has used a standard sample compartment with the milder reflective focusing characteristic of the Nicolet 6000. Optical band-pass limitation proves to be useful, since the frequencies >1800 cm^{-1} carry little spectral information for most organic molecules. Restriction with a filter permits use of higher preamplifier and lock-in gains without overload of the lock-in or saturation of the detector–preamplifier combination. FT-IR VCD of C—H stretch and similar higher frequency modes have rarely been reported (Nafie and Vidrine, 1982; Nafie *et al.*, 1979). Since the Fourier frequencies increase with infrared frequency, the DC band-pass of the lock-in, which is controlled by the output time constant, sharply attenuates the higher frequency Fourier sidebands on the polarization modulation (ω_M) signal that correspond to the higher optical frequencies of the near-infrared. It should be noted here, for clarity, that, like any normal fast-scanning FT-IR interferometer, at a given mirror speed the interferogram of the VCD will require fast detector and subsequent amplifier response to measure higher optical frequencies. Slower speeds can be used, but, at least in our Digilab instruments, slow mirror velocities lead to stability problems. We, in fact, typically employ the fastest mirror speed available ("20 kHz" or ~0.6 cm/sec) to speed data collection and thus allow more extensive signal averaging. This choice does significantly reduce high-frequency sensitivity. Improved response and S/N in the C=O stretch region can be obtained with 0.3 cm/sec ("10khz").

The analog signal processing we have used follows that used in other laboratories. The detector output is amplified with a standard Digilab MCT preamp, with two modifications: bandwidth has been increased to >100 kHz and a variable gain adjustment has been added. The former characteristic is required because of the PEM frequencies used, and the latter is useful for optimizing the gain conditions to fill the analog to digital (A/D) convertor. In practice, a generally useful preamp gain setting is determined and set, and the software-controllable gain (via DC amplifiers in the analog signal path of the FTS-60) along with the lock-in amplifier gain are later used to adjust for sample variations. The detector signal (in a differential format) is then sent, under software control, to either the FT-IR analog amplifiers and A/D convertor for measurement of the sample single-beam transmission interferogram or to the VCD electronics.

In the VCD path (Fig. 4), the ω_M signal is processed through a 20-kHz high-pass filter prior to detection with a lock-in amplifier (PAR 124A). We typically operate with a $Q = 20$ band-pass input, 1- to 5-mV sensitivity, and minimum output time constant (~400 μsec). High sensitivity is not re-

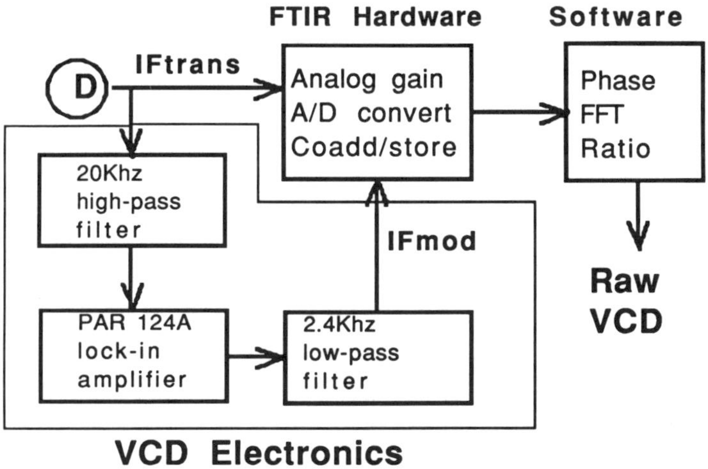

Fig. 4. The electronic modifications used for VCD measurement on the UIC FT-IRVCD spectrometer. The components are discussed in the text.

quired for this lock-in, but good out-of-band overload characteristics, low output noise, and, specifically, low output time constant are needed. Here judicious choice of preamp gain can maximize the overload rejection and optimize the S/N. A 10-V output level is also useful in conjunction with the standard Digilab FT-IR A/D convertor. The requirement of low output time constant eliminates many heterodyne-type lock-in systems from consideration for FT-IR VCD. We have successfully used simpler (i.e., cheaper) lock-ins but have had occasional problems with overloading. The high-pass filter used before the lock-in is needed to attenuate the transmission interferogram, which would overload the first stage of the lock-in preamplifier. A different style of lock-in with a filter as a first stage could be used to avoid this problem. The lock-in gain is empirically chosen to approximately fill the A/D convertor yet avoid overload. If this proves impossible, less lock-in gain is used and DC gain on the spectrometer is used to make up the difference. In practice, our gain settings are limited by A/D overload in the FT-IR due to noise spikes in actual sample scans.

The lock-in output signal is passed into the FT-IR electronics through a low-pass filter. We then use an instrumentation amplifier to reconstitute the lock-in output as a differential signal for compatibility with the Digilab electronics. However, single-ended processing also works satisfactorily, resulting in just half the signal amplitude. The last low-pass filter element is optional, but we have found it to improve the S/N of the resultant VCD when tuned appropriately for the combination of spectral region and mirror speed used. In our experiments, a 2.5-kHz, 24-dB roll-off low-pass filter setting has been used with a 0.6-cm/sec mirror velocity for the 2000- to 850-cm^{-1} spectral range. It should be clear that this extensive amount of filtering has a significant effect on the phasing of the interferogram of the ω_M signal. Other laboratories have not adopted electronic filtering to such an extent and may require a simpler phase correction as a result. Now, we use it only as an option.

In our experiments, data are collected every other zero crossing of the HeNe reference signal for an undersampling ratio (UDR) of 2, with the gain-ranging disenabled (on the lens setup). A pure modulation interferogram of a differential signal should have no center burst, so gain-ranging would be counter-productive. Our usual procedure is to gather four sets of 4096 symmetrical (two-sided) interferograms for the VCD (modulation interferogram) of each enantiomer at 4-cm^{-1} resolution and four sets of 64 scans each for the transmission interferogram. It is, in fact, not necessary to collect so many scans to determine the qualitative features of the VCD. We usually do a test collect of 1024 scans, which takes about 20 min, gives us a very good idea of the spectrum, and helps us choose the optimal scan parameters. For typical data sets, the more extensive collections yield better S/N; and the self-consistency of the blocks is very useful for elimination of inconsistencies and time-dependent aspects of the experiment which might arise for a particular sample. These data sets are then averaged after phase correction, FFT, and normalization calculations. (The latter use the standard Digilab transmission computation routine to determine the ratio of the single-beam signal at the modulator frequency to the single-beam transmission of the instrument.) Our typical processing uses triangular apodization and a zero filling factor of 4. Much of Nafie and co-workers' data (Lipp *et al.*, 1982; Lipp and Nafie, 1984) appears to have been collected at every zero crossing (UDR = 1) and processed with Happ-Genzel apodization without zero filling. Polavarapu has proposed using a UDR = 4 sampling to enhance transmission of the VCD modulation side bands through the lock-in and to improve phase correction (Polavarapu, 1984b). We have not been able to test these contentions because of the limitations of the Digilab software. Our initial FFT calculations are done with the standard software and phase-correction routines. However, as will be discussed in the next section, these are sometimes inadequate, particularly for the data from the lens setup, in which case a transferred phase-correction routine is used for the FFT computation.

An example of the kind of real interferograms that can be seen with this procedure is shown in Fig. 5, (IF_{Mod}). This trace represents 2100 data points from the center of a symmetrically collected, 4-cm^{-1} resolution (8192 points) interferogram in the VCD or polarization-modulated mode. It is clear that significant intensity exists in IF_{Mod}, even for data points far from the very small residual center burst. By way of comparison, the IF_{Trans} trace shown in Fig. 5 is the single-beam transmission interferogram of the same sample; it falls off quickly for large path differences. After phasing and transformation of these interferograms, the two spectra in Fig. 6 result, FFT_{Mod} being the modulated spectrum and FFT_{Trans} the single-beam transmission. The effect of our optical cutoff filter (at ~2000 cm^{-1}) and the MCT long-wavelength cutoff (at ~900 cm^{-1}, the BaF_2 sample cell also cuts off) are obvious in this 0- to 4000-cm^{-1} representation of FFT_{Trans}. The large peak in FFT_{Mod} corresponds to a significant absorption band in the FFT_{Trans} spectrum at ~1650 cm^{-1}, which is the amide I band

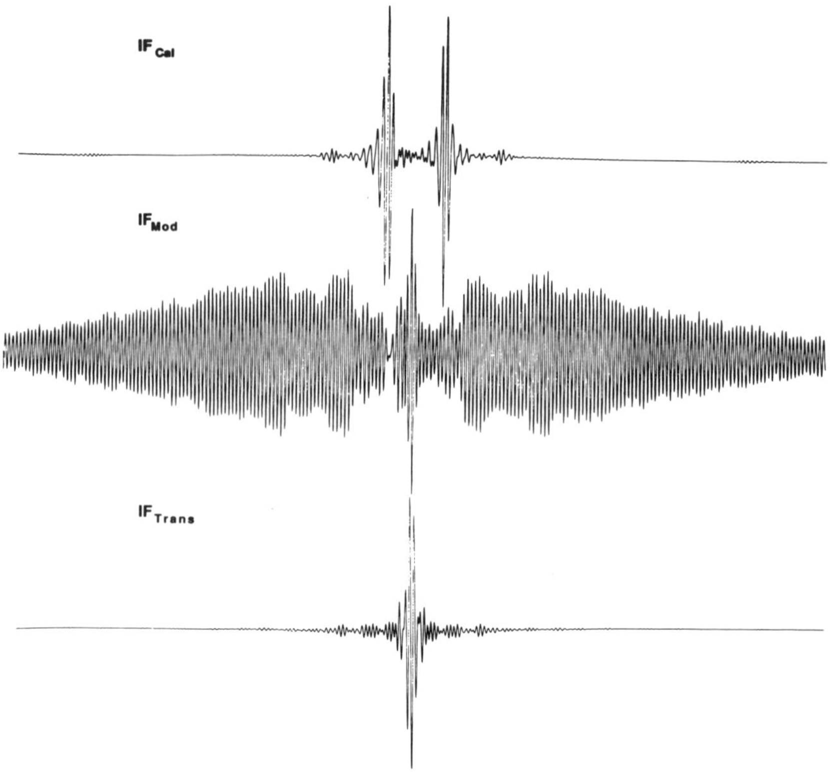

Fig. 5. Partial interferograms obtained on the lens setup. Displayed for each are the center 2100 points of a 4-cm^{-1} resolution (8192 point) symmetric interferogram. IF_{Cal} is a polarization modulated interferogram of the birefringent plate-polarizer calibration apparatus showing a nearly zero center burst (64 scans, 500 mV lock-in sensitivity). IF_{Mod} is the modulated interferogram (4096 scans, 1 mV) of gramicidin-S under nearly "ideal" polarization offset conditions (optimal alignment). Note the slow decay of the signal with increasing path difference relative to the rate of decay in either of the other interferograms. IF_{Trans} is the single-beam transmission interferogram of the same sample (64 scans).

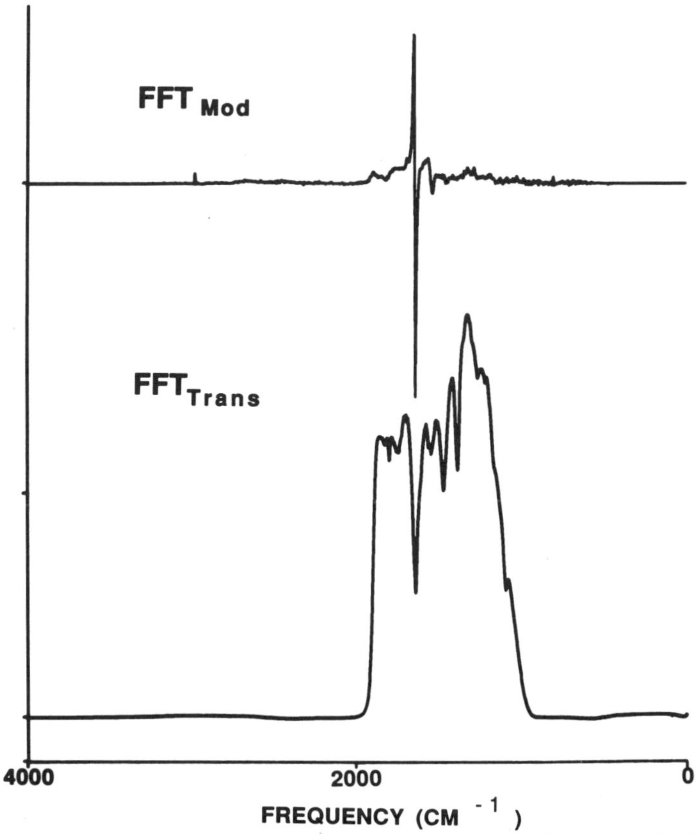

Fig. 6. Fourier transformation of the lower two interferograms in Fig. 5 displayed from only 0 to 4000 cm^{-1}. FFT$_{Mod}$ shows a large VCD feature at ~1650 cm^{-1}, and FFT$_{Trans}$ illustrates the sample absorption bands occurring in the window created by the low-pass filter (~1900 cm^{-1}) and the MCT cutoff (~850 cm^{-1}).

(C=O stretch) of gramicidin-S, a cyclic oligopeptide. Ratioing FFT$_{Mod}$ to FFT$_{Trans}$ gives the resultant VCD shown in Fig. 7 for the 1900- to 1000-cm^{-1} region of relevance, considering the optical cutoff constraints. It should be noted that this is a single block of 4096 scans uncalibrated and uncorrected

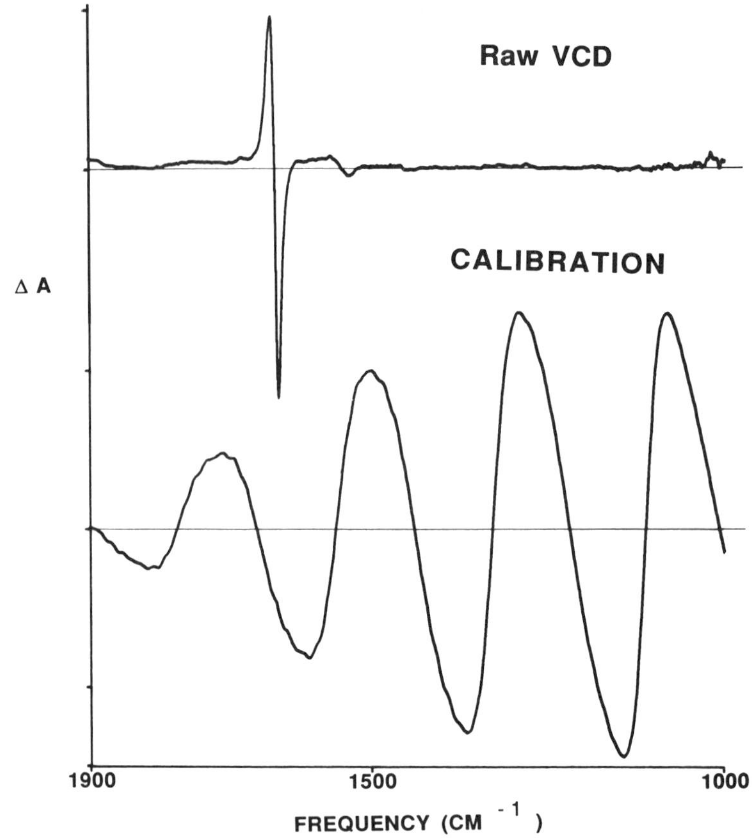

Fig. 7. Ratioed FT-IR VCD spectra of gramicidin-S in DMSO-d_6 (I_{HF}/I_{LF}) corresponding to the two traces in Fig. 6 (Raw VCD) and to the calibration setup (CALIBRATION). The raw VCD spectrum is *not* corrected for baseline. Subtraction of a solvent baseline results in only a negligible difference, corresponding to some flattening above 1700 cm^{-1}. More typical alignments show some slope to the baseline with a few broad, rolling features (Malon and Keiderling, 1988). These differences are manifested as a more pronounced center burst in the interferogram.

for baseline effects, which, in this case, clearly must be minimal.

As suggested in the previous section, it is necessary to calibrate VCD instruments to account for the wavelength dependence of the modulator and, even more important, for the various gain factors in the amplification used. In FT-IR VCD those gain factors have additional wavelength dependence as a result of the attenuation of the Fourier frequencies by the lock-in. There are no molecular standards for VCD like those employed in the electronic (UV-CD) case, so an optical apparatus was created to generate a controllable CD-like signal of known relationship to the gain factors (Osborne *et al.*, 1973)

FT-IR VCD calibration spectra are run using a birefringent plate (CdS) and grid polarizer with the same general setup as first used for dispersive VCD spectra (Nafie *et al.*, 1976). Because the calibration signal from this device is necessarily two-signed and because the interferogram has no center burst, the standard phase-correction routine fails to properly compute the FFT of this signal. The IF_{Cal} trace in Fig. 5 illustrates a typical calibration interferogram on our instrument. The zero path difference (ZPD) occurs midway between the two large side lobes that dominate the interferogram. The fast decay of IF_{Cal} is consistent with the regular, nearly sine-like, oscillating pattern of the calibration spectrum (Fig. 7).

McCoy and deHaseth (1988) have recently reinvestigated the phase-correction problem of VCD, which was evident as early as the first attempt by Nafie *et al.* (1979) to produce a VCD spectrum. They have proposed a phase-correction method that restricts the arctan function in the phase determination to a range of $-\pi/2$ to $\pi/2$ and thus allows the resultant spectrum to have negative components. We do not have this computational capability; thus, we simply follow an early suggestion of Nafie and Vidrine (1982). A phase array for transfer of phase correction is obtained from a similarly obtained scan of an assembly consisting of a stressed ZnSe plate and polarizer that, through an electronic path identical to that used for VCD, generates a pseudo-VCD signal having only one sign.

Using this transferred phase array and resetting the zero path difference point of the collected calibration interferogram to its point of reflection symmetry, the calibration signal can be successfully transformed. After division by the corresponding single-beam transmission of the calibration device, the expected shape of the calibration signal results (Fig. 7, lower trace). The calibration factor used to correct VCD spectra must account for the ratio of gains used to collect the data. The values used are typically obtained from the magnitude of this curve at the midpoint of the zero crossings (Nafie *et al.*, 1976), but they can be more exactly determined by choosing alternate polarization axes. The factor for correcting the observed VCD to proper ΔA units is not a constant over the wavelength range we access; so, typically, the average value for the specific region of most interest in the spectrum being obtained is used. Alternatively, the correction can be made by multiplication with a function that represents the calibration value dependence on wavelength. This function is more complex than the Bessel function dependence (Nafie *et al.*, 1976) that is seen in dispersive VCD, a characteristic which is due to the effect of the lock-in and filters on the intensities at the Fourier frequencies corresponding to the various wavelengths.

To be useful, such transfer of phase must be made from an interferogram collected via the same optical *and* electronic path as would be used for the VCD measurement. In addition, it is vital that both the phase standard and the calibration interferogram to be corrected have the same ZPD point. In interferometers with a white light source, this correction is not a problem, because the ZPD can be determined independently of the data set being collected. Unfortunately, in modern, simpler interferometers that use fringe counting alone to keep track of the mirror position, determination of the ZPD can be a problem. For calibration scans, as described above, the signals are large enough that a single scan gives a reproducible maximum peak. The interferogram symmetry is such that it is usually straightforward to shift the computer-selected ZPD point to the actual point of symmetry. However, a single VCD scan of a real sample on our instrument (lens setup) has no detectable center burst that can be used to initialize the scan, so the computer collects data somewhat "in the dark". But, since the co-addition is controlled by the fringe counting, a reasonable interferogram results (after ~1000 scans). However, even after signal averaging, the point of symmetry is often less than obvious in sample scans (see Fig. 5) because of severe phase shifting and the convolution of artifacts with the sample's VCD spectrum. Thus, some care is needed to assure oneself that the resultant data set is sensibly treated. For IF_{Cal}, the sharply peaked side lobes permit manual determination of the ZPD from symmetry considerations without significant problem.

With our mirror setup, however, a significant, stable center burst is seen on all scans, as has been reported by other laboratories. This center burst is due to the large, polarization-induced offset of the spectrum present in the sharply focused (ellipsoidal mirror-based) optical system used there. We have shown that this offset is also correlated to the detection of large absorption artifacts (Malon and Keiderling, 1988). These are present in the VCD of racemic samples and, unfortunately, do not reproducibly subtract out from the spectrum of an optically active sample, even if a racemic scan is available (Malon *et al.*, 1988a).

In short, we have, with the lens setup, traded off some stability and ease of phase calculation for a dramatic reduction in artifacts and the new possibility of baseline-correcting FT-IR VCD with just a solvent scan. On instruments with a stable ZPD point and carefully controlled transferred phase-correction capability, this trade-off should be advantageous. This advance in technique makes biochemical FT-IR VCD a realistic possibility and thus greatly expands the potential applications open to the VCD field. Some of our work in this particular area will be noted in the final section of this chapter.

C. Magnetic VCD Considerations

All of the preceding discussion has concerned measurement of VCD spectra for naturally optically active systems. Results of such vibrational measurements directly relate to molecular stereochemistry. In the realm of electronic excitations, a corollary technique of magnetic CD (MCD) has developed over the last 25 years as a useful means of determining excited electronic state symmetries and thus, more indirectly, molecular structure (Stephens, 1974, 1976). Such MCD experiments are essentially measurements of the first- and second-order Zeeman effects, using differential absorption intensity rather than frequency shifts as a means of detection. This method results in a trade-off of accuracy for sensitivity. As a result, splittings on the order of 1 cm^{-1} can be detected in electronic transitions having bandwidths on the order of 10^3 cm^{-1}.

Several years ago at UIC, it was demonstrated that a similar phenomenon could be seen in the vibrational (infrared) region of the spectrum, which we have denoted magnetic VCD (MVCD) (Keiderling, 1981b; Devine and Keiderling, 1983, 1984). To accomplish MVCD measurement, a magnet with its field oriented parallel to the direction of propagation of the light must be inserted into the sample position of a VCD spectrometer (Fig. 1). In practice, we have used two, different superconducting solenoid magnets with their bores mounted horizontally, both of which allow us to place standard-design, but in-house constructed (nonmagnetic), demountable sample cells at the center of the magnetic field. Both of our systems have a homogeneity of 1% over 1 cm^3 at the center of the bore, where (in our case) the field is maximum. This level of homogeneity is sufficient for MVCD measurements at the resolution and amplitude stability attainable in available VCD instruments.

Our early measurements used a Varian magnet of a compact design that routinely could be used at 4 T (40kG). This magnet had a restrictive room-temperature bore, which we modified by removal of its 77 K shield and addition of superinsulation, so that a cell made with windows 25 mm in diameter could be inserted (Devine, 1984). Smaller windows are difficult to purchase in a variety of materials; and, furthermore, they seriously degrade the optical throughput because of the limited aperture. Subsequently, we have employed an Oxford Instruments, modified SM-1 magnet capable of 8 T at 4.2 K and 10 T at 2 K with a room-temperature bore 45 mm in diameter (Croatto and Keiderling, 1988b). The higher field of this magnet permits high S/N measurement, because MVCD measurements are linear in applied magnetic field strength. Furthermore, its higher linear aperture coupled with a modest bore length results in only a minimal restriction of the f/3.5 light beam formed by the sample focusing optics in our dispersive instrument. This magnet has an ~30-hr liquid helium hold time, so long MVCD runs without interruption with extensive signal averaging are possible under computer control.

Absorption artifacts with MVCD spectra are not so serious a problem as they are in "normal" VCD, because baselines can be obtained on the same sample (without even moving it, if desired) by running an identical scan at zero field. We have found it advantageous (because of the factor of 2 increase in S/N) to run baseline scans at negative field, which can be readily obtained by repowering the magnet with the leads reversed. This technique does imply an increased cost for helium, since a zero field baseline could be run after the helium had boiled away; but it is equivalent to measurement of the baseline with the opposite enantiomer, as is sometimes done in "normal" VCD.

Such superconducting magnets always have a significant stray field that can affect other elements in the spectrometer and electronics. This problem is well known in MCD, where photomultiplier tubes must be heavily shielded and operated at a significant distance from the magnet. In our case, there are two separate considerations. Infrared detectors appear to be less sensitive to field strength than are photomultipliers. We have found that our InSb detector (being photo voltaic) exhibits an ω_M signal that scales with applied field and results in a baseline offset that is more than an order of magnitude larger than the MVCD signals. This field sensitivity demands magnetic shielding and an extension of the optical train to position the detector 30–50 cm from the magnet. We find that soft steel cylinders placed around the snout of our side-looking detector offer a reasonable level of shielding. Our MCT (photoconductive) detectors have little or no field sensitivity and can be operated without shielding at ~30 cm from the magnet. Such a physical separation, even if not needed for shielding, is additionally useful for allowing operator access to the sample position. We have mounted the first lens immediately adjacent to the magnet by use of a kinematic mount so that its removal for purposes of sample changing is quick and and its replacement is reliable.

In addition, as a second concern, the standard Hinds photoelastic modulator system is relatively field sensitive and will, in fact, shut off if placed directly adjacent to the magnet (i.e., in its ideal location). This sensitivity arises from the use of ferrite core torroids in the resonator circuit and can be adequately overcome by mounting the oscillator circuit in a separate housing that is coupled to the modulator head by a shielded cable of ~30 cm in length. At this distance and with this flexibility, an optimal location for the circuit can be determined empirically, and the optical element can be right next to the magnet bore. We have not found it necessary to magnetically shield the oscillator circuit further, but such an enhancement is straightforward.

Data collection, calibration, and sensitivity constraints for MVCD are the same as discussed earlier for "natural" VCD, with one addition. The sign of the field must be determined either from the magnet design or by calibration with the MCD of some known compound. There are several transition metal complexes with near-infrared MCD appropriate for this need, and we now have several test compounds with moderately large MVCD to further check the signs. These are generally standardized against "normal" VCD signs as calibrated with (+)-3-methylcyclohexanone or some other easily obtained compound, since the sign of the applied field for a given magnet, once determined, is easily controlled.

The above comments apply to our published MVCD data with the UIC dispersive instrument, but they also carry over in spirit to the FT-IR VCD instruments. This was the impetus for construction of the long f.l. bench described in the previous section. In addition, one must take care that the stray field does not affect the air bearing or the computer in the FT-IR. In our case the CRT colors proved to be quite sensitive to the field strength. Use of large aperture polarization optics enables us to keep the modulator ~30 cm away from the magnet, and the detector a variable distance away. Flushing such a large system can be a problem, but we have found that use of air dryers provides sufficient reduction of H_2O peaks.

III. CURRENT EXPERIMENTAL CAPABILITIES

A. Dispersive VCD

Dispersive VCD can be measured at optical frequencies down to or below 700 cm^{-1}, with detection limits approaching $\Delta A/A \sim 5 \times 10^{-6}$ in some regions. As an example of the range of spectra accessible with the UIC dispersive instrument, the VCD of (1R,2R) d_2-cyclobutane in the gas phase is presented in Fig. 8 over the C—H and C—D stretches and down through the mid-infrared to below 900 cm^{-1} (Annamalai *et al.*, 1984, 1985). This measurement involves various changes between two modulators, two detectors, three gratings, two lenses, and three filters. A good example of accessible S/N in the mid-infrared region for a highly soluble, small chiral species is shown in Fig. 9 for α-pinene in CS_2, which has become somewhat of a VCD standard for instrumental comparison because numerous versions of this spectrum are available (Diem *et al.*, 1988; Lipp *et al.*, 1982; Lipp and Nafie, 1984; Polavarapu *et*

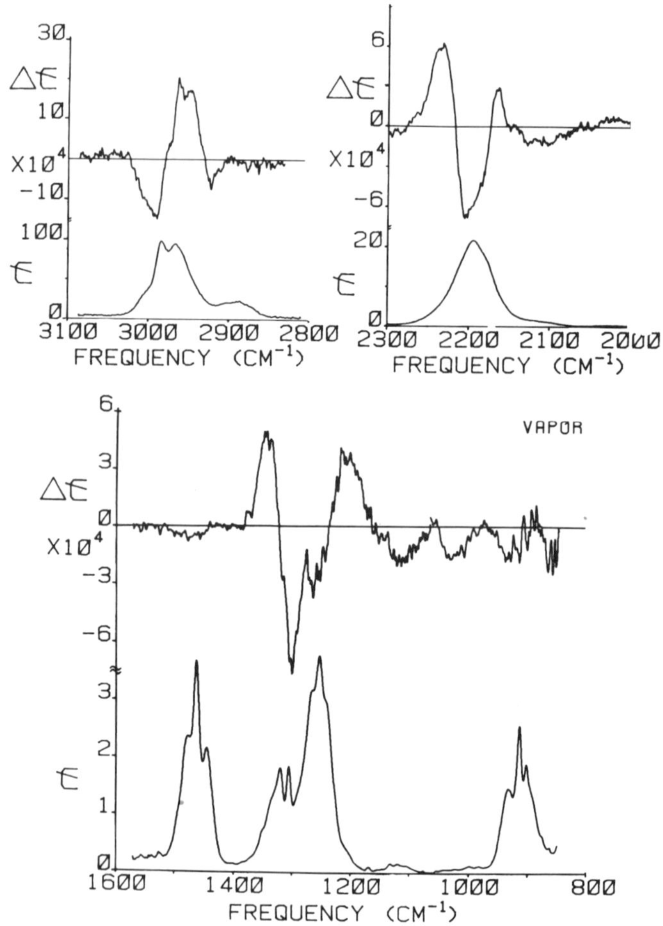

Fig. 8. VCD (top) and absorption (bottom) spectra of (1R,2R)-d_2-cyclobutane in the gas phase over the near- and mid-infrared. The near-infrared data were determined on a pure sample of ~2 mg, whereas the mid-infrared required substantially more and were measured for the (1S,2S) isomer at ~70% enantiomeric excess and replotted as (1R,2R).

al., 1984; Polavarapu, 1985; Malon and Keiderling, 1988). The spectrum in Fig. 9 is the result of averaging two, 3-sec time-constant scans. In our instrument, baseline artifacts are typically not severe in the longer wavelength region (8–11μm) but have been somewhat more of a problem around 6 μm. Recently, use of a new MCT detector (Infrared Assoc.) consisting of an array of four 3 × 3 mm elements, of which we select two to match the slit image, has improved this situation.

We have also been able to obtain VCD data below 900 cm^{-1} with a Si:As detector, an example of which is shown in Fig. 10 for 3-methylcyclohexanone. In our instrument with

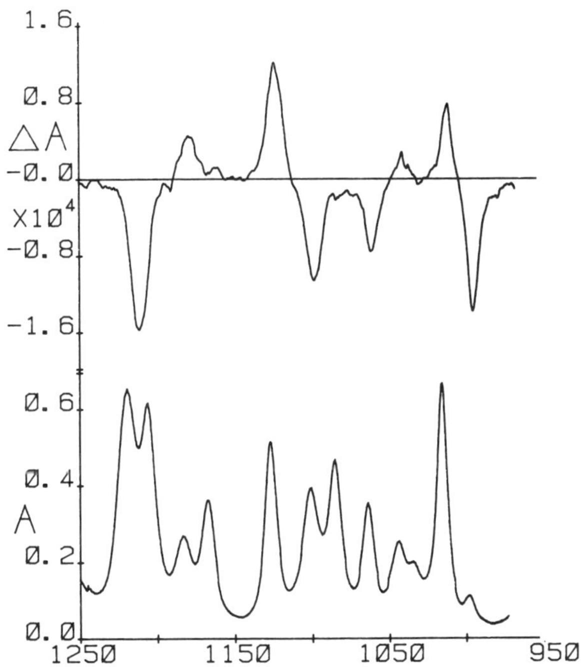

Fig. 9. VCD (top) and absorption (bottom) spectra over the accessible region (limited by solvent and MCT cutoffs) of the mid-infrared for (+) α-pinene (time constant = 3 sec, 2 scans averaged, baseline corrected with racemate).

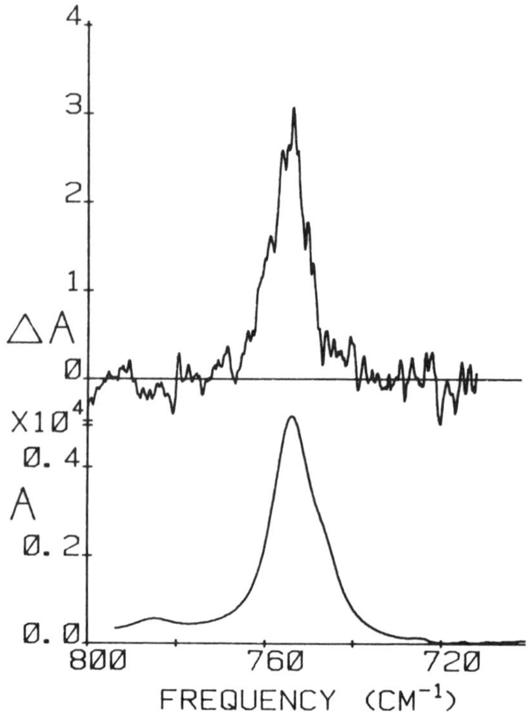

Fig. 10. VCD (top) and absorption (bottom) spectra for (+)3-methylcyclohexanone in CS_2 for the region beyond the MCT cutoff, as measured with a Si:As detector held at ~12 K. Data are from two scans averaged, obtained with a time constant of 3 sec, racemate baseline.

this detector, significant baseline artifacts develop in the 700-cm^{-1} region, a problem that makes its use less general than might have been hoped. The limited liquid helium hold-time of our in-house-constructed dewar is another limitation on its regular usage. Devlin and Stephens (1987) have reported use of this style of detector cooled with a closed-cycle refrigerator system [CTI Cryogenics]. They have attained equivalent results for 3-methylcyclohexanone, but their equipment has the decided advantages of a well-controlled constant temperature without the need for use of liquid helium.

A major limitation of the dispersive instrument is its resolution capability. We have not been able to collect VCD routinely at resolutions *substantially* better than 8 cm^{-1} in the mid-infrared. In our instrument, S/N at 4 cm^{-1} in the 1300-cm^{-1} region, for example, degrades to such an extent that signal averaging cannot be practically used to attain S/N levels approaching those obtained at low resolution. In a dispersive monochromator, the light level falls off as the square of the slit width which, in turn, is linear in the wavelength resolution. As a result of the dynamic normalization methods we use, the low light levels implicit in higher resolution scans imply that significant gain must be applied to the signal. It is our opinion that noise spikes present under these conditions instantaneously overload the lock-in circuitry, thereby causing a degradation of the resulting VCD that would be unexpected from light-level considerations alone. While this same problem is, in principle, true of the digital normalization used in the instrument of Diem *et al.* (1988), for their situation software modification can be made to limit gain changes to values at which the noise level is considered to be acceptable.

Another view of the capability of dispersive VCD is given in Fig. 11, which shows the MVCD of In(III)-tetraphenylporphyrin chloride in CS_2 over that portion of the mid-infrared region accessible using CS_2 as a solvent. The high quality of S/N available compared with that in our previously published porphyrin MVCD results (Croatto and Keiderling, 1988a) can be attributed to two things: the sample has very large MVCD signals and the magnet used was our improved 8-T Oxford Instruments design. With such large signals, higher resolution spectra (e.g., 4 cm^{-1}) could be reasonably obtained on the UIC dispersive spectrometer. We are currently modifying our FT-IR VCD instrument for MVCD capability in order to attain higher resolution data in that manner. MVCD in the near-infrared, e.g., C—H stretching, region tends to be

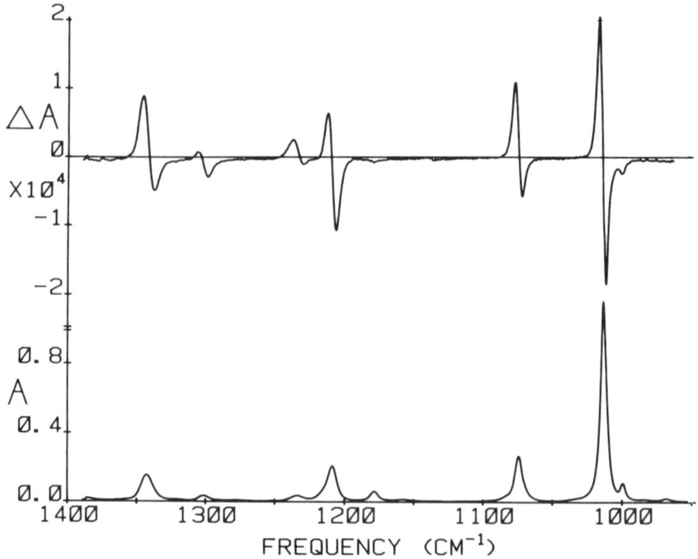

Fig. 11. MVCD (top) and absorption (bottom) spectra of In (III)-tetraphenylporphyrin chloride in CS_2 over the accessible mid-infrared. Two scans averaged, time constant 3 sec, field strength normalized to 40 kG.

significantly weaker even for porphyrin systems and presents a more severe experimental challenge.

B. FT-IR VCD

By contrast, Fig. 12 demonstrates our current capability for measurement of VCD over the entire mid-infrared region in a single experiment with the FT-IR instrument (lens set-up) for a $CDCl_3$ solution of a spiro dilactam compound (Blaha *et al.*, 1982; Malon *et al.*, 1988b). In Fig. 13, the VCD spectra of (+)-3-methylcyclohexanone in CS_2 is shown over the solvent-accessible range at nominal 8-,4-, and 2-cm^{-1} resolution to

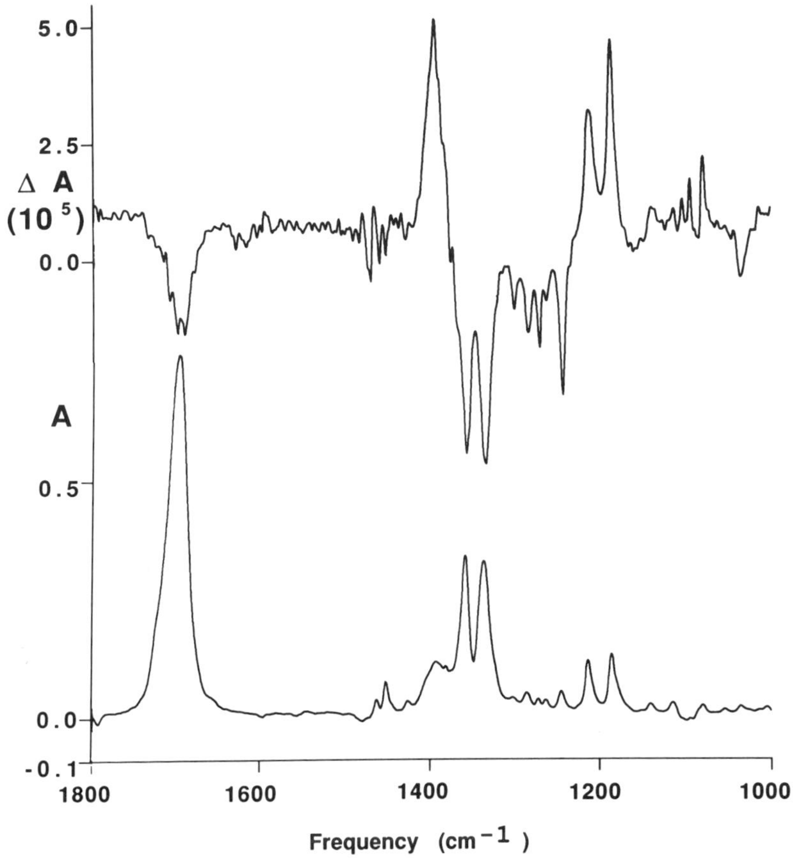

Fig. 12. VCD (top) and absorption (bottom) spectra over the entire FT-IRVCD mid-infrared region of a chiral spirodilactam dissolved in $CDCl_3$. Four blocks of 4096 symmetric scans at 4-cm^{-1} nominal resolution were collected for sample and baseline VCD. Baseline correction was obtained from an identical data collection on a racemic spirodilactam identical in structure except for one methyl group (Blaha *et al.*, 1982).

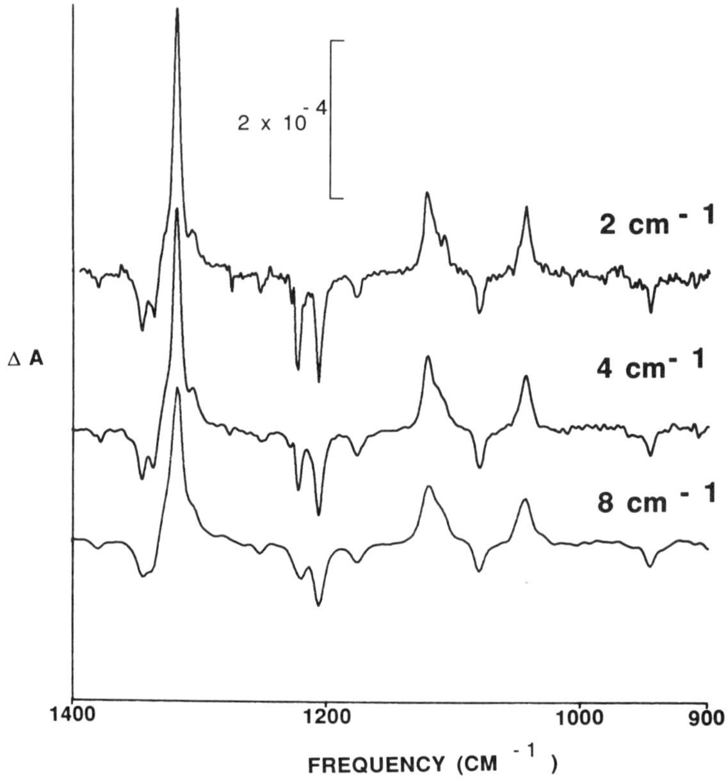

Fig. 13. FT-IRVCD spectra obtained for (+)3-methylcyclohexanone in CS_2 at 8-, 4-, and 2-cm^{-1} nominal resolutions. Baseline correction is by use of racemate. Four blocks of 4096 scans averaged for the polarization-modulated spectra at 4 and 8 cm^{-1}, but only two blocks at 2 cm^{-1}. The vertical bracket indicates a VCD magnitude of $\Delta A = 2 \times 10^{-4}$ for all three scans.

demonstrate the feasibility of acquiring higher resolution data. The changes with resolution in the FT-IR VCD for 3-methylcyclohexanone previously reported by Lipp and Nafie (1984) are in substantial agreement with those shown in Fig. 13. Using the lens set-up, a baseline correction has been obtained by subtraction of just a solvent scan Malon and Keiderling, 1988) that is equivalent to that obtained in the experiments shown in

Fig. 13 which were corrected by subtraction of the racemate spectrum. While higher resolution data, e.g., 2 cm^{-1}, is straightforwardly obtained using FT-IR VCD, in this case it yields no significant new information. Increasing resolution in the FT-IR spectrum does require added time, but for constant S/N, the time increase is roughly linear. Since Digilab uses fixed aperture sizes, to the point at which the aperture must be decreased (i.e., resolution better than 2 cm^{-1}) the detector sees a constant overall light level in our experiments. (For the data shown in Fig. 12, the 2-cm^{-1} VCD was determined with half as many scans, a factor that accounts for the change in noise level.) Also it should be noted that the triangular apodization we have used degrades resolution; but use of boxcar apodization with the Digilab software seems to seriously degrade S/N and cannot be viewed as a practical alternative with our instrument.

As is implicit in the figures, but not indicated there quantitatively, both the sample and baseline FT-IR spectra for these examples typically lie offset above the zero line (Malon and Keiderling, 1988). This offset and the consequent overall "positive" character of the spectrum correlate with an interferogram of this spectrum that has a well-defined center burst after co-addition of a sufficient number of scans. As such, the standard software phase-correction routine works well here. It should be noted that the sign of the spectrum is arbitrary since the phase-correction routine forces the spectrum to be net positive. However, under typical operating conditions, this sign is stable and is determined by the sense of the offset signal, which, in turn, is determined by the gross polarization sensitivity of the instrument and is only incrementally affected by the sample. In the lens set up, change of the sample cell and, to some degree, the solvent can alter this situation. The solvent effect seems to arise from changes in the spectral band-pass that is being sampled as a result of the fact that strong solvent absorption bands block out certain frequency regions. On the other hand, with our mirror set-up, and with most other FT-IR VCD instruments, there is always a strong center burst in the IF_{Mod} and large offset from zero in the resultant VCD, even if the sample and cell are removed. In the mirror setup, cell and solvent effects only have a minor influence on the baseline offset, which means that the VCD sign is typically well behaved and thus known from an initial determination.

However, it is possible under some conditions to adjust the optics so that the baseline is very close to and/or crosses zero. The latter situation leads to difficult problems in terms of the signs of the resultant VCD. The phase-correction routines will flip the baseline (and hence the VCD) sign as it crosses zero. This flip will give a baseline spectrum that is effectively "rectified" and VCD bands with a varying relative sign in a single spectrum. We have resorted to a transferred phase correction to solve this problem by using a calculation that closely parallels that used for the calibration routine discussed earlier for the quarter-wave-plate/polarizer assembly. With a transferred phase array, there can be sizable changes in the relative magnitudes for VCD bands in different spectral regions as compared with those obtained with the standard phase correction. This difference is most evident close to the optical frequency where the baseline undergoes a sign flip (i.e., the frequency at which the phase array values change by π). Figure 14 shows the FT-IR VCD of (+)-1,2-d_2-trans-1,2-dicyanocyclopropane with standard (above) and transferred (below) phase corrections on our lens set-up. The VCD sign flip is evident in transitions below ~1100 cm^{-1}. Relative magnitudes also differ substantially between the two calculated spectra, with the 1074-cm^{-1} band being sharply attenuated in the sign-flipped (Standard Phase) spectrum.

Contrary to reports from other laboratories (Lipp and Nafie, 1984), we have not observed sigmoidal effects on bandshapes from such a calculational procedure. We have, however, found fairly severe distortions in the transferred phase-corrected spectrum if the wrong ZPD position is used. Determination of the ZPD is a nontrivial problem in a system without a white-light source such as the Digilab FTS-60,

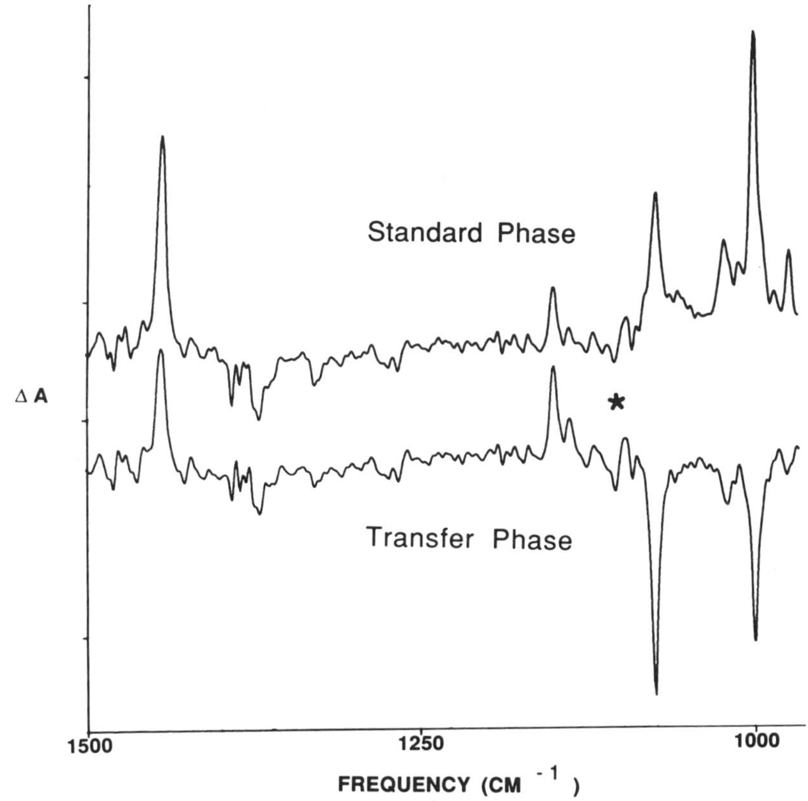

Fig. 14. FT-IRVCD of (+)-1,2-d_2-trans-1,2-dicyanocyclopropane in $CDCl_3$ with standard (top) and transfer (bottom) phase-correction illustrating the sign-flip problem. The asterisk indicates the approximate position of the sign flip. Note the relative magnitude change of the 1074-cm^{-1} band in the two calculated spectra. The slope in baseline is not evident here because the racemate baseline spectrum has been substracted.

which normally uses the center-burst of an initial scan to reference the range of mirror motion to be used for the co-added scans. Even when the ZPD is manually determined, there can be problems due to the strong phase corrections needed to account for the various levels of electronic filtering used in our

VCD setup (see Fig. 5, for example). In most cases, we have found that the symmetry of the interferogram can be deduced by looking at fringes away from the ZPD position, and their symmetry can be used to determine the ZPD. If this is not definitive, a limited number of FFTs can be made for those points that could reasonably be expected to correspond to the ZPD. The similarity of the result, or lack of it, to other blocks can be used to assure that a proper phase correction is being used.

While the phase correction routine of McCoy and deHaseth (1988) would allow the resultant VCD to have either sign, it is not yet clear if such a correction can be usefully applied to real VCD interferograms with indeterminate ZPD points. The best solution to this particular problem is development of a reliable independent determination of the ZPD as is done with separate white-light interferometer-containing instruments. Alternatively, a fixed scanning range could be used so that all the data sets would be identical. Then with truncation and/or zero-filling, an optimal size of the data set for computational purposes could be obtained.

C. Comparison of Techniques

At this stage in the development of the experimental aspects of VCD, we can make an informed comparison of the two general approaches to measurement of VCD as summarized in the preceding discussion. The early promise of the FT-IR technique has been realized; but, unlike the situation in absorption spectroscopy, FT-IR VCD has not become the dominant technique. If there were a competition between the two over the longer term, it seems that no winner could be predicted. Each experimental approach has its advantages and disadvantages, a situation that will maintain both in use for the foreseeable future. In a sense, the two methods are complementary, each meeting the needs of different specific experiments.

FT-IR VCD is likely to expand in usage for two main reasons. The attractiveness of collecting the whole mid-infrared spectra simultaneously, at moderate resolution, in an acceptable time frame will lead many future users to adapt this approach to VCD. However, the more important practical reason is the "near-commercial" character of such instrumentation. Given the experience of the originators of this technique, a few FT-IR companies are now able to supply instrumentation capable of VCD measurement or, more realistically, to provide a shopping list of needed components (i.e., those noted in this chapter) and the requisite optical and software enhancements of their respective FT-IR instruments to get the VCD assembly to function properly. This task is relatively straightforward, since "research-grade" FT-IR instruments often contain the necessary software as either standard or optional equipment and the optical changes can be minimal (particularly for sample compartment contained VCD operation).

The more important issue is which measurement situations (i.e., sample characteristics) are best suited to one or the other approach for VCD measurement. Typically one might use the time of experiment as the overiding consideration and be tempted to think that the FT-IR experiment has the overall advantage because of its simultaneous collection of the whole spectrum. Actually, we find that this is not a very useful criterion. Although individual FT-IR scans are quite fast, one needs to accumulate on the order of a few thousand scans to see the spectral features. However, mere detection of features is usually not sufficient, and many more scans are needed to reduce the noise level to allow reliable estimation of signal magnitude and bandshape. This, unfortunately, is a slowly converging process (S/N $\sim \sqrt{n}$ for n scans). Furthermore, it is usually necessary to obtain a baseline spectrum under the same averaging conditions for correction of the sample scan.

With indiscriminately mounted sampling and polarization optics, this step can prove to be vital and thus would demand use of a racemate to determine the baseline. With our improved optical system, often just a solvent scan will do, but control of the pathlength and the cell and solvent characteristics is still necessary.

The most useful VCD are obtained on molecular species in solution. In the infrared, solvent absorption bands pose significant interference problems that consequently allow one to obtain VCD data only in regions of spectral windows. Thus, other than in exceptional cases, a set of spectra should be obtained in different solvents to determine the "complete" VCD spectrum. However, in many cases, the "complete" VCD is not needed and data for one, or a few bands, will suffice to answer the question at hand. In these cases, measurement of VCD with a dispersive instrument can be much more efficient than with an FT-IR. Single bands can often be measured at modest resolution in a few minutes, and averaging over several scans for sample and baseline can be done on the timescale of an hour.

On the other hand, if higher resolution spectra are needed or if a "complete" VCD is required, the FT-IR VCD approach will be more efficient. The latter recommendation is, of course, dependent on whether one's setup has the capability of discriminating against artifacts and controlling the sign of the VCD measured. Furthermore, such a recommendation is limited to the mid-infrared region. At present, only the dispersive instruments work well enough for routine near-infrared work. Such experiments are often measured over 100–200 cm^{-1} within an isolated spectral region and thus do not take advantage of the simultaneous spectral acquisition capability of FT-IR. That situation, compounded with the FT-IR VCD sensitivity fall-off at high Fourier frequencies, strongly favors the dispersive approach.

The same may prove true for longer wavelength regions in that liquid helium-cooled detectors can be optimized for specific spectral band-passes and have degraded performance at the very high light levels characteristic of FT-IR measurements. To date, VCD experiments at wavelengths beyond ~11 μm have been done with both approaches, but the results, although apparently of higher S/N in the dispersive mode, are not really comparable because of the differences in detectors employed (Devlin and Stephens, 1987; Polavarapu, 1984b). A proposal has been made that far-infrared VCD can be measured using a Martin and Puplett (1969) polarizing interferometer (Dignam and Baker, 1981). Such an instrument has been built by BOMEM and used by Nafie *et al.* (1987) for preliminary experiments emphasizing linear dichroism measurement. However, at this writing, no VCD have been reported from it (other than calibration scans) in either the mid- or far-infrared regions. Alternative designs based on a step-scan polarizing interferometer have also been analyzed theoretically by Polavarapu (1988).

IV. EXAMPLE APPLICATION: VCD OF POLYPEPTIDES AND PROTEINS

Most VCD studies to date have been oriented toward development of the technique and the testing of theoretical models. Selection of molecules for such studies was aimed variously at their being available, having large signals, being exceptionally simple, or having a desired functionality. Exceptions to this trend are the extensive applications of Nafie and co-workers of the ring current model (Nafie and Freedman, 1986; Freedman *et al.*, 1985) to empirically analyze conformations of a number of compounds, including several transition metal complexes (Nafie *et al.*, 1983 Freedman *et al.*, 1985, 1987; Young *et al.*, 1986, 1988). This work has recently been reviewed by Freedman and Nafie (1987). Another empirical study of six-member ring systems was made by Laux *et al.* (1982); this study used coupling of CH_2 groups to explain observed C—H stretch VCD and the success of simple calculations for such molecules (Polavarapu and Nafie, 1980; Singh and Keiderling, 1981a,b; Marcott *et al.*, 1981; Freedman *et al.*, 1988). Several comparative VCD studies of cyclopropane- (Heintz and Keiderling, 1981; Yasui and Keiderling, 1987) and allene-based (Narayanan *et al.*, 1988) molecules have also resulted in identification of characteristic modes.

Another application that has developed is the elucidation of further detail regarding biopolymer conformation in solution. Most of our work in this area concerns poly- and oligopeptides and is beginning to yield new insight into their conformations. A brief survey of some of our applications of VCD to this area is presented in this final section of the chapter.

A. Polypeptides

After our early demonstration (Singh and Keiderling, 1981c) that α-helical polybenzyl-L-glutamate (PBG) gave rise to significant VCD in the amide I, II, and A regions, Lal and Nafie (1982) showed through a comparative study of a number of polypeptides that the bandshape and sign pattern of the amide I (primarily C═O stretch) and amide A (primarily N—H stretch) were characteristic of the sense of the helix (right- or left-handed twist) and not of the chirality of the α-carbon (L or D). Both the amide A and I bands had spectral shapes that were nonconservative couplets but had patterns of the opposite sign. For the amide I VCD, the dominant negative lobe lies higher in energy than the positive lobe for a right-handed α-helix. Subsequently, Sen and Keiderling (1984b) showed that the amide II (mixture of C—N stretch and H—N—C deformation) VCD had the same helical dependence but was single-signed and lower in energy than the absorption maximum. Furthermore, we demonstrated that while deuteration of the amide N—H yielded a small shift to lower energy for the 1650-cm^{-1} amide I absorption band, a significant VCD bandshape change to a three-peaked (− + −) pattern results. By contrast, the amide II shifts considerably from ~1550 to ~1450 cm^{-1} on *N*-deuteration, but its VCD remains negative.

While it is established that α-helices give VCD characteristic of the secondary structure, one of our major efforts following that work has been the determination of the VCD of other secondary structural types and the monitoring of transitions between them. Early electronic CD studies utilized poly-L-lysine (PLL) to characterize the spectrum of the polypeptide α-helix, β-sheet, and random coil (Greenfield and Fasman, 1969). Because of its polyionic nature, PLL readily undergoes transitions between these states with change in pH and/or in solvent. We have taken a similar approach in the investigation of the VCD of PLL using dispersive VCD (Yasui and Keiderling, 1986b), and these results have been confirmed by the independent study of Paterlini *et al.* (1986) using FT-IR VCD techniques.

Figure 15 shows the amide I VCD and absorption spectra of *N*-deuterated poly-L-lysine in D_2O at pD ~7.3 ("random-coil"), at pD ~11.2 after heating (anti-parallel β-sheet), and in 95% CD_3OD (right-handed α-helix). These spectra exemplify the basic structural types for polypeptides and their characteristic, deuterium-exchanged amide I (designated as amide I′) VCD. The α-helix has a three-peak (− + −) VCD pattern centered over a single absorption band as is typical for *N*-deuterated right-handed α-helices in contrast to the positive couplet pattern normally found in *N*-protonated α-helices (Lal and Nafie, 1982; Sen and Keiderling, 1984b). Early higher resolution FT-IR VCD of protonated PBG (Lipp and Nafie, 1985) also showed this third peak, but this bandshape has subsequently been shown to be susceptible to FT-IRVCD absorption artifacts, even with baseline correction, *if* measured with a mirror setup (Malon *et al.*, 1988a). While the β-sheet yields only negative VCD in this region, two separate bands occur at 1615 and 1690 cm^{-1}; these bands are characteristic of the amide I′ transition for the antiparallel β-sheet conformer. A very clear discrimination between these two highly structured conformers, α and β, is thus evident.

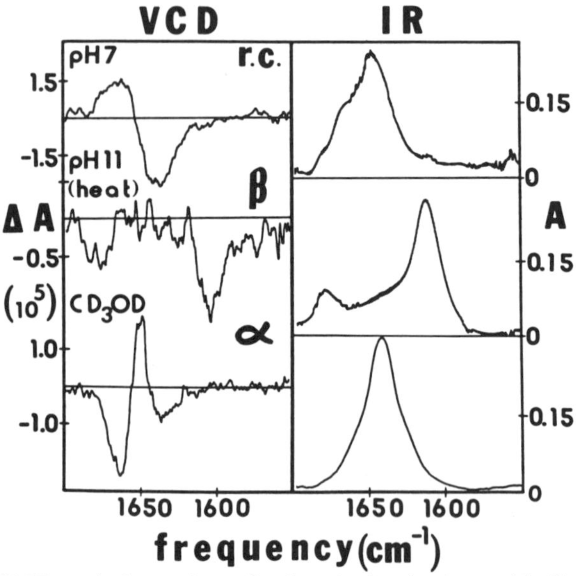

Fig. 15. VCD and absorption of poly-L-lysine in the amide I′ region in neutral D_2O (r.c.), basic D_2O/NaOD (β) and 95:5 CH_3OD:D_2O (α) exhibiting characteristic VCD for the three main polypeptide secondary structural types. α, helix; β, sheet; r.c., random coil.

In contrast to near-ultraviolet CD results, the VCD of random-coil PLL has a surprisingly large magnitude, based on its $\Delta A/A$ value, compared with that of the α-helix. It might be expected that a "random-coil" should have a very weak VCD if it is truly unordered. This expectation comes from our observations that blocked dipeptides and tripeptides in solution have vanishingly weak VCD in these bands and that the polypeptide VCD we have observed is primarily dependent on secondary structure, i.e., amide–amide interaction and not on the nature of the amino acids involved (local chirality). Paterlini *et al.* (1986) have interpreted this "random-coil" VCD as being indicative of formation of a short-range ordered, "extended helix" that has a left-handed sense but no long-range order (Tiffany and Krimm, 1968a, 1969; Balasubramanian, 1974). In fact, the "random-coil" PLL results are quite similar to the amide I VCD of left-handed α-helixes such as poly-benzyl-L-aspartate (Lal and Nafie, 1982; Sen and Keiderling, 1984b). Differentiation between these two possible conformations may be possible for systems in nonaqueous solution by parallel analysis of the amide A band. In that band, a broad, ill-defined band occurs for the "random coil" of poly-L-tyrosine in dimethyl sulfoxide (DMSO) (Yasui and Keiderling, 1986a), but a sharper, well-defined band is found in the left-handed helix of polybenzyl-L-aspartate (Lal and Nafie, 1982). Our more recent studies (Kobrinskaya *et al.*, 1988) indicate that poly-L-proline-II (PLP), a left-handed helical form, gives the same amide I sign pattern, but higher magnitude (see later) as do several "random-coil" polypeptides having charged (e.g., polyglutamic acid) or aromatic chains (e.g., polytyrosine). It is clear that these "random coils" must have a significant and characteristic order. In high-salt conditions, this signal can be diminished as the structure becomes truly disordered (Paterlini *et al.*, 1986), but it was not fully eliminated in our repeated experiments. Such results are evidence that VCD has a short-range sensitivity that is different from that seen with electronic CD. That truly unordered systems lack significant VCD is supported by our studies of blocked oligopeptides (see later), which have no VCD for very short species.

In the course of our PLL study, we found a new intermediate state at pD ~10.5 with a VCD consisting of a single negative band lying to the low-frequency side of the amide I′ ab sorption band. We took this as evidence of a new conformation, not previously noted with electronic CD, that probably was mixed in nature, being just locally coherent and lacking in long-range order. In addition, these PLL studies demonstrated the feasibility of attaining reasonable S/N when studying the VCD of aqueous (D_2O) solutions. In poly-L-glutamic acid, a similar pD change from neutral to acid solution results in a clear "coil"-to-helix transition (Dukor and Keiderling, 1989). Again a new intermediate state was found with monosignate, negative VCD at pD ~5.6, but this bandshape could be approximated by a linear combination of the "coil" and helix VCD. The bandshapes of the VCD of the α-helical and "coil" phases are identical to the PLL results (Fig. 15) and thus are characteristic of the secondary structure and apparently independent of the side chain. This result is precisely the characteristic needed to make VCD useful in the determination of protein secondary structure. Furthermore, it is evidence that the "random coil" must encompass a well-defined structure, at least at the local level, and probably that this is a conformer of a left-handed helical sense (although not particularly an α-helix).

An important question with regard to these results is their dependence on the length of the peptide chain. In conjunction with Prof. Roichi Katakai of Gunma University, Japan, we have studied a series of alternating sequence, α-helical oligopeptides, $(Met_2Leu)_n$,$n = 6$ to 11 (Yasui *et al.*, 1987a). Down to the 18-mer level ($n = 6$), there was no qualitative nor significant quantitative change in the VCD for any of the regions studied (amide I, II, or A) from the typical right-hand α-helical spectra.

In summary, our results indicate that, in sensing the secondary structure, VCD has a more local sensitivity than does electronic CD and thus may offer information complementary to that available from UV-CD. It also suggests that the variety of well-resolved chromophores accessible in VCD yield spectra having a distributed information content. In other words, analysis of different VCD (infrared) bands can offer different (independent) insight into problems of peptide stereochemistry.

The spectral separability of bands possible with VCD is especially useful for the study of polypeptides with aromatic side chains. The relatively resolved bands characteristic of the vibrational region of the molecular spectrum allows discrimination between the amide modes and those of most aromatic groups, while, in electronic CD, their interference prevents a conventional determination of the secondary structure. In Fig. 16 are shown the amide VCD spectra of poly-L-tyrosine

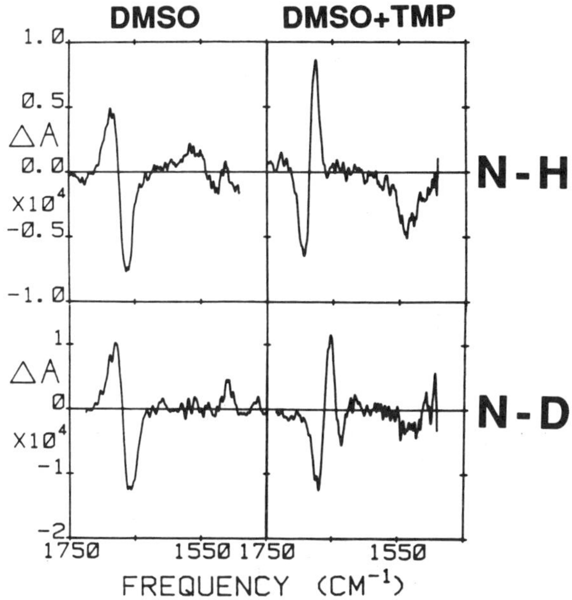

Fig. 16. VCD comparison of poly-L-tyrosine in DMSO ("random coil") and 1:1 DMSO:TMP (α-helix) for protonated (normal) and *N*-deuterated peptides in the amide I and II regions.

in DMSO and in a 50 : 50 DMSO–TMP (trimethyl phosphine) mixture over the amide I, II, and amide I′ (*N*-deuterated) regions (Yasui and Keiderling, 1986a). The coil–helix transition induced by the TMP is clearly one to a right-handed helix, a result that eliminates previous speculation, based on studies of frequency shifts alone, that a left-handed helix might be involved. Subsequent studies of poly-L-tryptophan (Yasui and Keiderling, 1988a) and of poly(L-Lys(Z)$_2$-L-napAla) with Prof. M. Sisido of Kyoto (Yasui *et al.*, 1987b), in nonaqueous solvents in both cases, yielded VCD spectra consistent with what is expected for a right-handed α-helix (protonated). The conformations of these polypeptides were originally determined by a complex CD–fluorescence method; but, by contrast, the VCD determination of secondary structure was done with one straightforward experiment. In these examples, the positive couplet seen in the amide I is relatively sharp and larger in magnitude than the negative single-signed amide II VCD (which is lower in wavenumber than the absorbance maximum). The amide II infrared absorption and VCD occasionally does indicate interference by the aromatic groups, but the flexibility of VCD to study multiple bands neatly sidesteps this potential problem.

We have also made a preliminary study of the conformation of poly-L-histidine as a function of pD. From pD ~1 to 4.4, a typical "random-coil" VCD is seen. However, at higher pDs, a transformation occurs and results in a VCD unlike any we have previously reported (Yasui and Keiderling, 1988a). The final state is probably β-like, but its spectrum is not like that in Fig. 15. Such a conclusion may imply that the PLL antiparallel β-sheet VCD is not characteristic of all β-structures or that VCD should be able to discriminate between various β-structural types. It also reemphasizes that the β-sheet VCD result has very few examples. The VCD seen for high-pH PLL has also been shown to be a component of the phosvitin VCD spectrum (heavily antiparallel β-sheet at low pH) (Yasui *et al.,* 1989b). But aside from these, no further examples of antiparallel or parallel β-sheets are available, so more work will be needed to clarify this issue.

The experiments discussed in the preceding paragraphs do demonstrate that the VCD spectra are yielding additional information beyond that available just with electronic CD or with infrared and Raman data. Thus, based on these polypeptide studies, our long-range goal of learning more detail about protein structure via VCD studies seems reasonable. On the other hand, more systematic surveys of various structural types remain necessary to use this sensitivity in a reliable manner. Such studies are underway in our laboratory.

For example, poly-L-proline (PLP) exists in two forms, designated as I and II. In the solid state, form I is a right-handed 10_3-helix of *cis*-amides and form II is a left-handed 3_1-helix of *trans* amides. Considerable work has been reported on the conformation of PLP in solution, particularly as determined with electronic CD. Controversy has arisen over the relationship of these spectra to those obtained from random-coil polypeptides (Tiffany and Krimm, 1968b, 1969; Balasubramanian, 1974). In trifluoroethanol (TFE), both forms have a negative VCD couplet (Kobrinskaya *et al.*, 1988) similar to that found for the "random-coil" phase of PLL and other such molecules. The PLP II amide I VCD is about twice the magnitude of that of PLP I and is shifted ~10 cm^{-1} to higher energy. Similar results in terms of magnitude and frequency are obtained for PLP II in D_2O (Fig. 17, dotted line). The

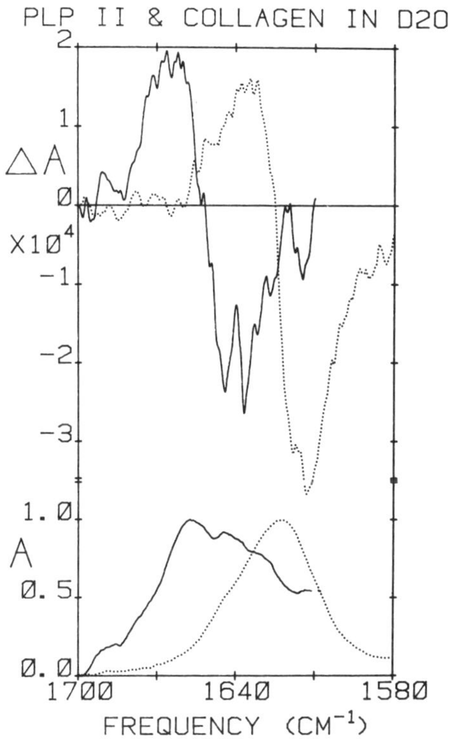

Fig. 17. VCD (top) and absorption (bottom) spectra of poly-L-proline (form II, dotted line) and collagen (solid line) in D_2O at room temperature in the amide I.

left-handed nature of the PLP II conformations, its *trans*-amide repeat unit, and its common amide I VCD bandshape with "random-coil" polypeptides support the existence of the proposed left-handed "extended-helix" conformation (Tiffany and Krimm, 1968a, 1969) for coherent segments of the "random coils" we have studied. A study of the length dependence of $(Pro)_n$ oligomers indicates that for $n \geq 3$ a pattern is developed similar to that of the PLP-II VCD (Dukor and Keiderling, 1989b). But the magnitudes are smaller than those in PLP-II.

Collagen is a natural protein related to PLP II since it has a triple, left-handed 3_1-helix and exhibits an electronic CD consistent with that of PLP II. Indeed the collagen VCD couplet also mimics that found for PLP II, with the frequency of the amide I absorbance maximum being shifted to higher energy and the absorption broadened as a result of the presence of both secondary and tertiary amides in collagen (Fig. 17, solid line). When the sample temperature is raised to ~45°C, the collagen VCD collapses to a magnitude less than one-third that of the negative feature in Fig. 17, which would be appropriate for formation of a more unordered form. PLP II is more stable in D_2O, requiring both high salt concentration and high temperature conditions to lose its characteristic VCD (Kobrinskaya *et al.*, 1988). These results, together with our phase transition studies on other ordered polypeptides, further support the original proposal of Tiffany and Krimm (1968a, 1969) that polyionic polypeptides have significant local order that appears to be helical in nature.

B. Oligopeptides

The dependence of our polypeptide results on polymer length and end effects is important to their ultimate utility in protein secondary structure analysis. Some very small oligomers have been studied by other groups (Chernovitz *et al.*, 1987; Robert *et al.*, 1988; Diem *et al.*, 1984; Oboodi *et al.*, 1984; Lee *et al.*, 1989), but these studies have focused on vibrational spectral detail rather than on empirical correlation with structure. We have made studies of the length dependence of the VCD for oligopeptides forming β-sheets (Narayanan *et al.*, 1985, 1986) and 3_{10}-helical (Yasui *et al.*, 1986a,b) structures; these studies have been made in conjunction with Prof. Claudio Toniolo and co-workers of Padua, using small to medium oligomers of $n = 3$ to 8 in addition to the previously noted studies of the α-helical length dependence with $n = 18$ to 33 and the $(\text{Pro})_n$ helix length dependence with $n = 2$ to 12.

For the 3_{10}-helical studies, correlation of NMR, X-ray, and VCD data was used to show that a 3_{10}-helix dominated the conformations of Aib_8, $\text{Aib}_5\text{LeuAib}_2$, and $\text{Aib}_3\text{ValGlyLeu-Aib}_2$, all prepared as blocked oligomers (Aib is aminoisobutyric acid, a 3_{10}-promoter). This study showed that VCD could distinguish this short 3_{10}-helix from a right-handed α-helix by comparison of bandshapes as well as of magnitudes (Fig. 18).

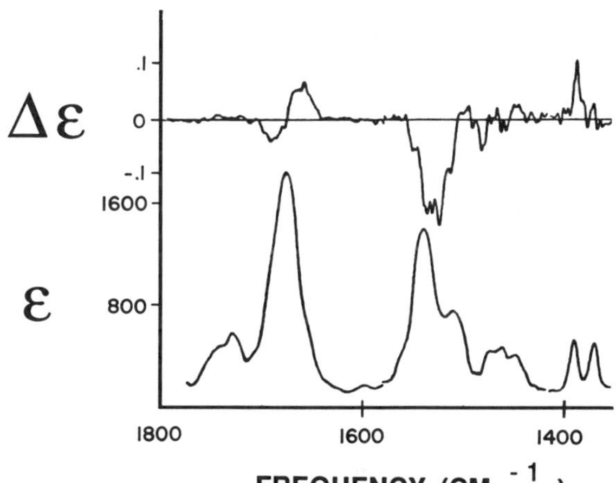

Fig. 18. VCD and absorption spectra of $\text{Aib}_5\text{LeuAib}_2$, a blocked 3_{10}-helical octamer, in the amide I and II regions. Note relative heights of amide I and II, compared with the *N*-protonated, α-helical spectra in Fig. 16 (upper right-hand corner).

In particular, the 3_{10} amide I and A VCD couplets are more conservative than found for the α-helix, and the intensity of the 3_{10}–amide I couplet is weaker than that of the strong negative amide II VCD, whereas the opposite relative intensities hold for the α-helix. While the real possibility exists that short α-helices give similar spectra and that the conservative nature of the couplets varies with the system studied (Yasui and Keiderling, 1986a), these results were encouraging in terms of expanded investigation of the detail of polypeptide structure with VCD, since the electronic CD of these 3_{10}-helices cannot be distinguished from that of α-helices. $(\text{Ala-Aib})_n$ oligomers, which were originally thought to vary from 3_{10}-helical to α-helical structure as chain length increased, do not yield a characteristic VCD that can differentiate between the two types (Yasui *et al.*, 1989a). This outcome is consistent with the recent X-ray results showing that several of these alternate oligopeptides are indeed mixtures of α- and 3_{10}-helical conformers (Toniolo *et al.*, 1989).

For length-dependence studies of the 3_{10}-helix alone, a series of blocked oligopeptides were synthesized, by the Padua group, with the general formula $\text{Aib}_n\text{LeuAib}_2$, $n = 0$ to 5, (Leu is included to establish a chiral preference). We measured the VCD of the amide A, I, and II bands of these oligomers as a function of n to determine at what level the characteristic 3_{10}-helical VCD develops (Yasui *et al.*, 1986a). For all three bands, a constant magnitude of $\Delta A/A$ develops by n = 3, which corresponds to almost two turns (three hydrogen bonds) of an ideal 3_{10}-helix. The spectra, in fact, evidence the same qualitative VCD even at $n = 1$, the tetramer level, where only a single type III β-turn can form. Such a result again is consistent with the dominance in VCD of relatively local interactions, as was noted earlier in conjunction with our observations of "random-coil" VCD spectra. If this proves to be consistently true, β-turn contributions to protein VCD will have a significant impact on the resultant VCD of globular proteins. Such effects are consistent with our preliminary analyses of amide I′ VCD for a series of 20 globular proteins (see next section). Thus, analysis of protein VCD could potentially lead to more precise structural differentiation than is possible with electronic CD. Furthermore, the variety and resolution of bands accessible with VCD as compared with electronic CD may also prove to be a decisive advantage for protein studies, especially if such analyses are done as a tandem study utilizing data from both techniques.

Regarding β-sheets, we were able to study the length dependence of the solid-state secondary structure, using film techniques, for oligopeptides with various aliphatic side chain oligopeptides (Ala, Val, Nva, and Leu) as prepared by the Padua group (Narayanan *et al.*, 1985, 1986). These studies indicated that there are three different conformational states of the oligomer as chain length is increased, whereas earlier studies, using amide I frequency shifts and electronic CD, could detect only two of these. The couplet band shapes (except Val_n) of these solid-state oligomer spectra for amide I bands are in contrast to the single-signed VCD found for the PLL antiparallel β-sheet result (Fig. 15).

This difference from PLL carries over to the solution phase. The Ala_7 and Val_7 oligomers in TFE yield bandshapes reflecting what is seen for the film samples but with much weaker intensities (Fig. 19). The Ala_7 and Val_7 oligomers had previously been assigned to antiparallel and parallel β-sheet conformations, respectively, based on both experimental and calculational results. Our VCD spectra can neither confirm nor deny this but do clearly show that they are different from each other *and* from the PLL antiparallel β-sheet result. Furthermore, they do *not* fit the dipolar-coupling- based theoretical calculations of VCD for finite sheets (Snir *et al.*, 1975). An alternate view of antiparallel β-sheet VCD can be obtained using gramicidin-S (Yasui and Keiderling, 1989), a cyclic decapeptide with four cross-ring β-type hydrogen bonds. The amide I VCD of this molecule in solution (see Fig. 7) is clearly a conservative couplet analogous to that seen for Ala_7 (Narayanan *et al.*, 1986) but not to the PLL high-pH result (Yasui and Keiderling, 1986b). Phosvitin at pD~2.5 yields VCD partially

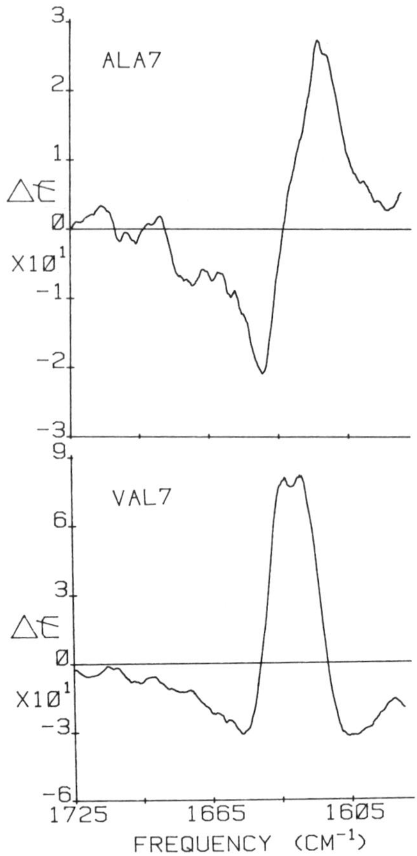

Fig. 19. VCD of Val_7 (bottom) and Ala_7 (top) oligomers in TFE solution (saturated) in the amide I region. Both are in a β conformation but give quite different bandshapes from each other and from that in Figs. 14 or 7, which also result from β structure.

comparable to that of β-form PLL but has an equilibrium between random-coil and β-sheet forms. At lower pD (~1.6) an alternative form for the antiparallel β-sheet VCD is found (Yasui, *et al.*, 1989b). Hence the question of the characteristic β-sheet VCD remains unsettled and may be another indication of the dominance of relatively short range, i.e., amide–amide coupling, effects in our studies.

C. Proteins

Our longer range goal in these particular studies is the development of a more detailed experimental picture of protein secondary structure in solution. To this end we have accumulated VCD data for a number of globular proteins for which crystallographic or other structural data is available and that have been previously characterized with near-ultraviolet CD. The near-ultraviolet CD of polypeptides and proteins is dominated by the contribution from the α-helical component, whereas all the "standard" polypeptide conformers have VCD bands with comparable intensities. The different secondary structures do have different frequencies and bandshapes in their characteristic VCD spectra, a factor that allows discrimination of conformation. For example, as might be expected, predominantly α-helical myoglobin gives a somewhat broadened amide I VCD in D_2O, with a dominant positive couplet and a very weak, low-energy negative band. Similarly, partly β-sheet β-lactoglobulin-A gives two predominantly negative VCD bands at roughly the expected frequencies. FT-IR VCD of these two proteins in D_2O are illustrated in Fig. 20. Vibrational circular dichroism of other proteins with more mixed structures are less clear when viewed in such simple terms, which is presumably due to the convolution of VCD components of differing bandshapes but similar intensities. Our measured protein spectra all correspond to samples systematically exchanged (N—D) prior to making solutions in D_2O for measurement, but the degree of deuteration will clearly depend on solvent accessibility of individual subunits. The strong variation in the VCD spectra among proteins that we have studied is indicative of a high conformational information content. It is a reasonable result in light of the intense VCD we have seen for turns and coils in oligopeptides, which imply that the VCD technique has a short-range sensitivity.

Denatured proteins still yield significant but altered VCD spectra (Yasui and Keiderling, 1988b) as would be consistent with the dominance of short-range effects in VCD. These

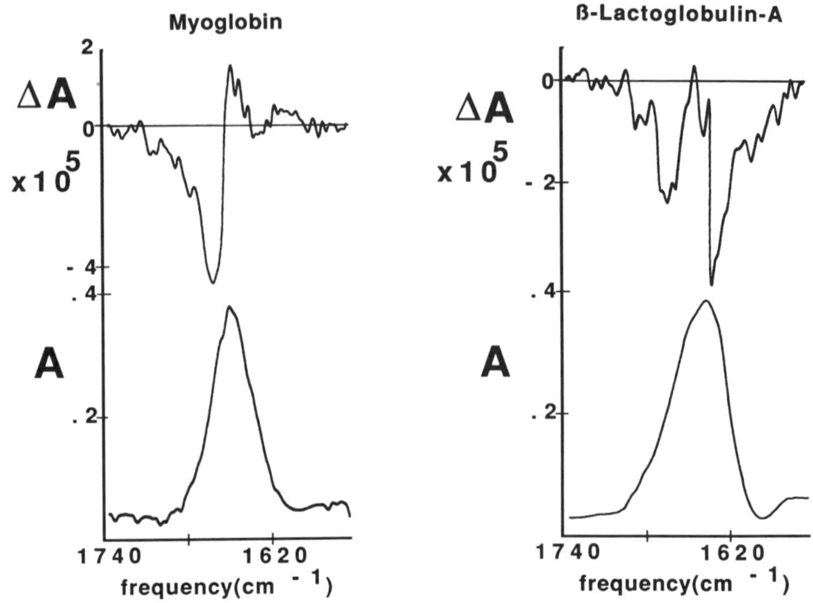

Fig. 20. FT-IRVCD of myoglobin and β-lactoglobulin-A in D_2O at neutral pD, measured on the lens setup at 4-cm^{-1} resolution (4 × 4096 scans) and corrected with a baseline obtained with a neutral D_2O solution of poly-DL-lysine.

latter experiments also demonstrate the sensitivity of VCD to conformational variation as compared with infrared absorption data alone. Dramatic line shape changes are seen in VCD upon denaturization, and only subtle shifts are evident in the infrared absorption spectrum as the protein is denatured.

That this information content available from the amide I′ VCD might have new features as compared with UV-CD results is clearly demonstrated in Fig. 21 (Pancoska *et al.*, 1989a). There UV-CD data for four globular proteins in the n—π^* and π—π^* regions are compared to the amide I′ VCD of these same proteins. While the UV-CD indicates significant variation in magnitude in this set of proteins, each has a bandshape composed of roughly the same three bands with the same sign pattern (− − +, as energy increases). On the other

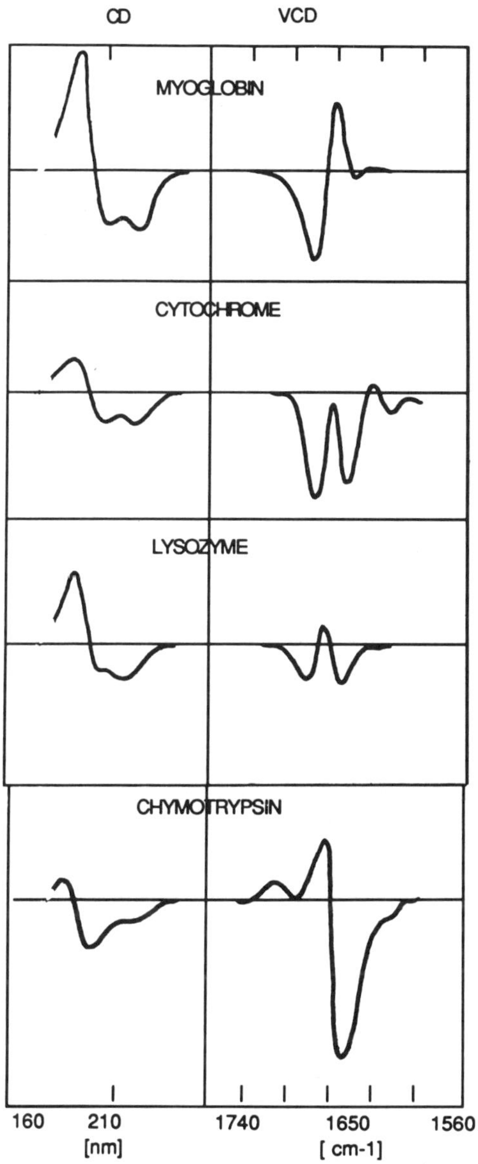

Fig. 21. Comparison of n-π^* and π-π^* UV-CD and amide I′ VCD for four globular proteins of differing secondary structure. The enhanced sensitivity of VCD to the variations in conformation is evident in the sign reversal between myoglobin and chymotrypsin. All spectra were redrawn and effectively smoothed for preparation of this figure. The scales are arbitrary but the same for all proteins.

hand, the VCD changes dramatically showing total sign pattern variation and large differences in bandshape among these four. It is clear from even such a qualitative study that VCD is more sensitive to whatever structural variations occur in this set than is UV-CD. The set was chosen to vary from a predominantly α-helical protein (myoglobin) to one with a relatively high degree of β-sheet character (chymotrypsin). The other two (cytochrome *c*, lysozyme) are more mixed structures. They give amide I′ VCD that can be shown to be a linear combination of the dominantly α- and β-forms. Our ability to extract a greater range of structural characteristics with improved reliability from this enhanced sensitivity remains to be proved, but the information content is certainly there.

In a preliminary effort to decompose the observed VCD into component bands, via an empirical bandshape analysis, we have found that the amide I data for 10 quite different proteins can be fit with four bands of equal half-width (15 cm^{-1}), constrained central frequency (± 5 cm^{-1}), but variable sign and magnitude (Pancoska *et al.*, 1989b). Fits to hemoglobin, cytochrome *c*, and casein dispersive VCD measured for the amide I′ band are shown in Fig. 22 as examples of three quite different bandshapes that can be shown to follow the above prescription. The relative consistency of the decomposed amide I′ VCD implies that the protein data set is well behaved and must be dominated by a few, analyzable components. Furthermore, analysis of the fits obtained shows that the proteins cluster into characteristic VCD types for the amide I′ when the component bands are considered, and that these types of modes bear a rough correlation to any dominance that can be found for α-helix, β-sheet, and "random-coil" components in the secondary structure.

Current work along this line is focused on using factor analysis as an objective method of determination of independent (orthogonal) subspectra that describe the VCD bandshape. This technique can also be used to generate a spectral representation for several of the structural components important for the overall protein VCD. These could then serve

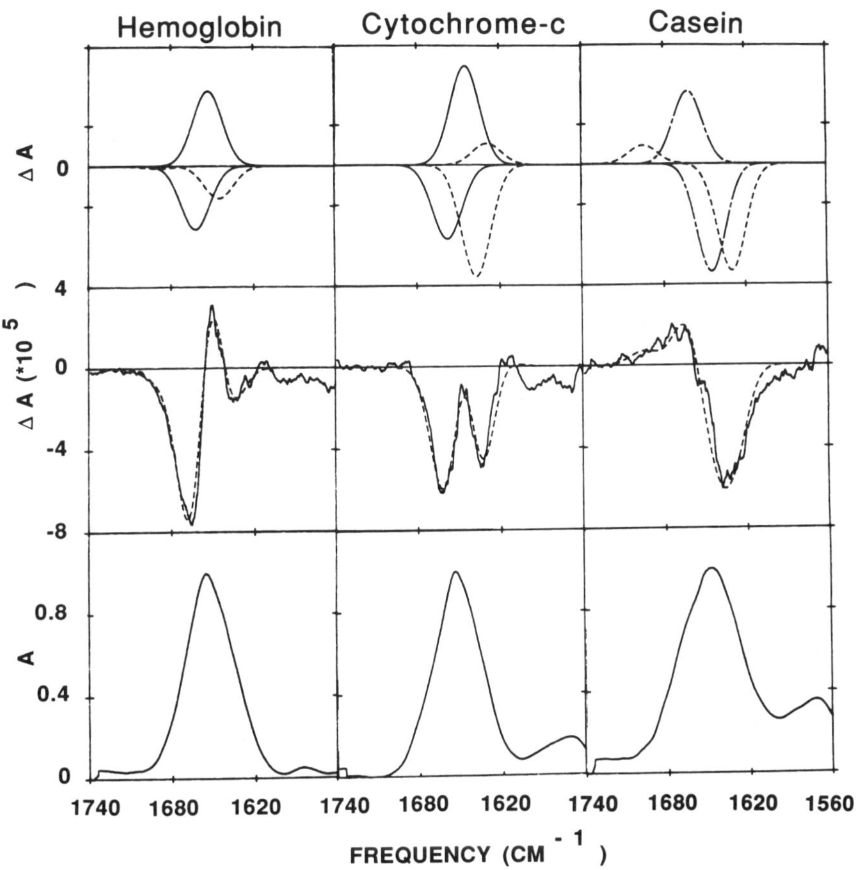

Fig. 22. Dispersive VCD of hemoglobin, cytochrome *c*, and casein illustrating the variation in bandshape that can be satisfactorily fit with four components of equal width (15 cm^{-1}), of constrained center frequency (± 5 cm^{-1}), but of variable sign and magnitude. It might be noted that these spectra have a somewhat better S/N ratio than those obtained on the FT-IR instrument (Fig. 19) but are of lower resolution (~ 12 cm^{-1}).

as a basis for bandshape deconvolution (Hennessy and Johnson 1981; Compton and Johnson, 1986; Pancoska *et al.*, 1979; Malon *et al.*, 1983). Initial results (Pancoska *et al.*, 1989b) show that the amide I′ VCD of an expanded set of 20 proteins can be fit with five orthogonal subspectra. The first subspectrum represents the most important common component of all the experimental spectra used in the analysis, the second contains the primary corrections to this common subspectrum, and so on. The fifth subspectrum is at the limits of meaningful contribution to the reconstruction of the protein VCD. It is of special interest to note that the most important subspectrum in the UV-CD factor analysis (Hennessy and Johnson, 1981) is effectively equivalent to the UV-CD spectrum of the α-helical factor. On the other hand, in VCD it is representative of a mixture of contributions, a finding implying that the different structural factors (α-helix, β-sheet, turns, etc.) must contribute to the overall VCD on an equivalent basis. In VCD, the second subspectrum appears to correct the common component for the α-helical contribution. It can, in fact, be shown that the coefficient of the second subspectrum for each protein is directly correlated to the α-helical content determined by X-ray crystallography (Pancoska *et al.*, 1989b). Additionally, this same coefficient is inversely correlated to the β-sheet content similarly determined. This behavior of the primary subspectra in the VCD factor analysis reinforces our contention that UV-CD of proteins is dominated by the CD of the α-helical component of the structure and that VCD, because of its different structural dependence, will offer an alternative view of solution secondary structure.

This initial effort at protein VCD analysis in the future will be coupled to efforts to extend the protein VCD data set to include other bands and encompass other data such as UV-CD and infrared absorption data. This should yield a more comprehensive secondary structure analysis capability for proteins than is now available. We and others are also studying nucleic acid systems (Annamalai and Keiderling, 1987; Yang *et al.*, 1989; Diem *et al.*, 1989) and multiple small molecule systems.

This "minireview" of our polypeptide and protein work was intended to give the reader an example of the type of practical problem to which VCD, in particular, could be addressed. We believe that vibrational optical activity as evidenced in VCD and ROA measurements offers the stereochemist a new, more precise spectroscopic tool as compared to "conventional" spectroscopies. By extension, it offers new benefit to that group of researchers whose analytical problems center around determination of various aspects of molecular structure. Bringing VCD and ROA into the multitechnique-oriented analytical effort that is characteristic of much modern biochemical and natural product chemical research should lead to further development of those fields. Extension to a greater number of practitioners will serve to better define the promise of vibrational optical activity measurement that is now evident through the explorations of a few research groups.

ACKNOWLEDGMENTS

Our work in vibrational optical activity has been funded by research and instrumentation grants from the National Science Foundation, the National Institutes of Health, and the donors of the Petroleum Research Fund for which we are most grateful. The VCD results from our laboratory are the product of the work of many graduate students and postdoctoral associates without whose efforts significant progress could not have been attained. These particularly include current and former students: C. N. Su, V. J. Heintz, T. R. Devine, U. Narayanan, A.-M. Sourigues, P. V. Croatto, L. Yang, R. K. Dukor, A. A. El-Azhary, and M. R. Vavra; postdoctorals: R. D. Singh, A. C. Sen, A. Annamalai, S. C. Yasui, and M.-C. Tissot; and senior collaborators temporarily at UIC: M. Pawlikowski, P. Malon, and P. Pancoska. We have also been fortunate to collaborate with a number of research groups around the world who have provided us both with research suggestions and samples. In particular, we have had a long-term co-operation with Claudio Toniolo at Padua and James Chickos at the University of Missouri, St. Louis.

REFERENCES

Abbate, S., Laux, L., Overend, J., and Moscowitz, A. (1981). *J. Chem. Phys.* **75,** 3161 [erratum (1983) **78,** 609].

Amos, R. D., Handy, N. C., Jalkanen, K. J., and Stephens, P. J. (1987). *Chem. Phys. Lett.* **133,** 21.

Annamalai, A., and Keiderling, T. A. (1987). *J. Am. Chem. Soc.* **109,** 3125.
Annamalai, A., Chickos, J. S., and Keiderling, T. A. (1984). *J. Am. Chem. Soc.* **106,** 6254.
Annamalai, A., Keiderling, T. A., and Chickos, J. S. (1985). *J. Am. Chem. Soc.* **107,** 2285.
Balasubramanian, D. (1974). *Biopolymers* **13,** 407.
Barnett, C. J., Drake, A. F., Kurada, R., and Mason, S. F. (1980a). *Mol. Phys.* **41,** 455.
Barnett, C. J., Drake, A. F., and Mason, S. F. (1980b). *Chem. Commun.* p. 43.
Barron, L. D. (1975). *Nature (London)* **257,** 372.
Barron, L. D. (1976). *In* "Molecular Spectroscopy" (R. F. Borrow, D. A. Long, and J. Sheriden, eds.), Vol. 4. Chem. Soc., London.
Barron, L. D. (1977). *Chem. Phys. Lett.* **46,** 579.
Barron, L. D. (1979). *In* "Optical Activity and Chiral Discrimination" (S. F. Mason, ed.). Reidel, Boston.
Barron, L. D. (1984). *Top. Curr. Chem.* **123,** 151.
Barron, L. D. (1987). *In* "Laser Scattering Spectroscopy of Biological Objects" J. Stepanek, D. Anzenbacher, and B. Sedlácek, eds.), p. 27. Elsevier, Amsterdam.
Barron, L. D. (1989). *Vib. Spectra Struct.* **17B,** 343.
Barron, L. D., and Buckingham, A. D. (1973). *J. Chem. Soc., Chem. Commun. p. 152.*
Barron, L. D., and Buckingham, A. D. (1975). *Annu. Rev. Phys. Chem.* **26,** 381.
Barron, L. D., and Clark, B. P. (1982). *Mol. Phys.* **46,** 839.
Barron, L. D., and Vrbancich, J. (1982). *Chem. Phys. Lett.* **92,** 466.
Barron, L. D., and Vrbancich, J. (1983). *J. Raman Spectrosc.* **14,** 118.
Barron, L. D., and Vrbancich, J. (1985). *Adv. Infrared Raman Spectrosc.* **12,** 215 (and references therein).
Barron, L. D., Bogaard, M. P., and Buckingham, A. D. (1973a). *J. Am. Chem. Soc.* **95,** 603.
Barron, L. D., Bogaard, M. P., and Buckingham, A. D. (1973b). *Nature (London)* **241,** 113.
Barron, L. D., Meehan, C., and Vrbancich, J. (1982). *J. Raman Spectrosc.* **12,** 251.
Blaha, K., Budnesinsky, M., Koblicora, Z., Malon, P., Tichy, M., Baker, J. R., Hossain, M. B., and van der Helm, D. (1982). *Collect. Czech. Chem. Commun.* **47,** 1000.
Bose, P. K., Barron, L. D., and Polavarapu, P. L. (1989). *Chem. Phys. Lett.* **155,** 423.
Boucher, M., Brocki, T. R., Moscovits, M., and Bosnich, B. (1977) *J. Am. Chem. Soc.* **99,** 6870.

Boyd, W. J., Jennings, D. E., Blass, W. E., and Gailar, N. M. (1974). *Rev. Sci. Instrum.* **45,** 1286.
Brocki, T. R., Moscovits, M., and Bosnich, B. (1980). *J. Am. Chem. Soc.* **102,** 495.
Chabay, I., and Holzwarth, G. (1975). *Appl. Opt.* **14,** 454.
Cheng, J. C., Nafie, L. A., and Stephens, P. J. (1975). *J. Opt. Soc. Am.* **65,** 1061.
Cheng, J. C., Nafie, L. A., Allen, S. D., and Braunstein, A. I. (1976). *Appl. Opt.* **15,** 1960.
Chernovitz, A. C., Freedman, T. B., and Nafie, L. A. (1987). *Biopolymers* **26,** 1879.
Chickos, J. S., Annamalai, A., and Keiderling, T. A. (1986). *J. Am. Chem. Soc.* **108,** 4398.
Compton, L., and Johnson, W. C. (1986). *Biophys. J.* **49,** 4940.
Croatto, P. V. (1989). Ph.D. thesis, University of Illinois at Chicago.
Croatto, P. V., and Keiderling, T. A. (1988). *Chem. Phys. Lett.* **144,** 455.
Croatto, P. V., and Keiderling, T. A. (1989). Manuscript in preparation and Proc. 7th Int. FTS Conf., Washington, D.C., in press.
Devine, T. R. (1984). Ph.D. thesis. University of Illinois at Chicago.
Devine, T. R., and Keiderling, T. A. (1983). *J. Chem. Phys.* **79,** 5796.
Devine, T. R., and Keiderling, T. A. (1984). *J. Phys. Chem.* **88,** 390.
Devine, T. R., and Keiderling, T. A. (1985). *J. Chem. Phys.* **83,** 3749 [erratum (1987) **87,** 5051].
Devine, T. R., and Keiderling, T. A. (1986). *Chem. Phys. Lett.* **124,** 341.
Devine, T. R., and Keiderling, T. A. (1987). *Spectro chim. Acta, Part A* **43A,** 627.
Devlin, F., and Stephens, P. J. (1987). *Appl. Spectrosc.* **41,** 1142.
Diem, M., Gotkin, P. J., Kupfer, J. M., and Nafie, L. A. (1978). *J. Am. Chem. Soc.* **100,** 5644.
Diem, M., Photos, E., Khouri, H., and Nafie, L. A. (1979). *J. Am. Chem. Soc.* **101,** 6829.
Diem, M., Oboodi, M. R., and Alva, C. (1984). *Biopolymers* **23,** 1917.
Diem, M., Roberts, G. M., Lee, O., and Barlow, A. (1988). *Appl. Spectrosc.* **42,** 20.
Diem, M., Gulotta, M., and Goss, D. (1989). *Biopolymers,* in press.
Dignam, M. J., and Baker, M. D. (1981). *Appl. Spectrosc.* **35,** 186.
Dukor, R., K., and Keiderling, T. A. (1989a). *Proc. Eur. Peptide Symp. 20th,* p. 519.
Dukor, R. K., and Keiderling, T. A. (1989b) Abstr. 7th Int. FTS Conf., Washington, D.C.
Escribano, J. R., Freedman, T. B., and Nafie, L. A. (1987). *J. Chem. Phys.* **87,** 3366.
Freedman, T. B., and Nafie, L. A. (1983a). *J. Chem. Phys.* **78,** 27.

Freedman, T. B., and Nafie, L. A. (1983b). *J. Chem. Phys.* **79,** 1104.
Freedman, T. B., and Nafie, L. A. (1987). *Top. Stereochem.* **17,** 113.
Freedman, T. B., Balukjian, G. A., and Nafie, L. A. (1985). *J. Am. Chem. Soc.* **107,** 6721.
Freedman, T. B., Young, D. A., Oboodi, M. R., and Nafie, L. A. (1987). *J. Am. Chem. Soc.* **109,** 1551.
Freedman, T. B., Kallmerten, J., Lipp, E. D., Young, D. A., and Nafie, L. A. (1988). *J. Am. Chem. Soc.* **110,** 689.
Greenfield, N. J., and Fasman, G. D. (1969). *Biochemistry* **8,** 4108.
Havel, H. A. (1981). Ph.D. thesis. University of Minnesota, Minneapolis.
Hecht, L., Barron, L. D., and Hug, W. (1989). *Chem. Phys. Lett.,* in press.
Heintz, V. J., and Keiderling, T. A. (1981). *J. Am. Chem. Soc.* **103,** 2395.
Henderson, D. O., and Polavarapu, P. L. (1986). *J. Amer. Chem. Soc.* **108,** 7110.
Hennessy, J. P., and Johnson, W. C. (1981). *Biochemistry* **20,** 1085.
Hoffman, G. G., Schrader, B., and Snatzke, G. (1987). *Rev. Sci. Instrum.* **58,** 1675.
Holzwarth, G., and Chabay, I. (1972). *J. Chem. Phys.* **57,** 1632.
Holzwarth, G., Hsu, E. C., Mosher, H. S., Faulkner, T. R., and Moscowitz, A. (1974). *J. Am. Chem. Soc.* **96,** 251.
Hsu, E. C., and Holzwarth, G. (1973). *J. Chem. Phys.* **59,** 4678.
Hug, W. (1981). *Appl. Spectrosc.* **35,** 115.
Hug, W. (1982). *In* "Raman Spectroscopy" (J. Lascombe and P. V. Huong, eds.), Wiley-Heyden, Chichester, England.
Hug, W., and Surbeck, H. (1979). *Chem. Phys. Lett.* **60,** 186.
Hug, W., and Surbeck, H. (1982). *J. Raman Spectrosc.* **13,** 1.
Jalkanen, K. J., Stephens, P. J., Amos, R. D., and Handy, N. C. (1987a). *J. Am. Chem. Soc.* **109,** 7193.
Jalkanen, K. J., Stephens, P. J., Amos, R. D., and Handy, N. C. (1987b). *Chem. Phys. Lett.* **142,** 153.
Jalkanen, K. J., Stephens, P. J., Amos, R. D., and Handy, N. C. (1988a). *J. Phys. Chem.* **92,** 1781.
Jalkanen, K. J., Stephens, P. J., Amos, R. D., and Handy, N. C. (1988b). *J. Am. Chem. Soc.* **110,** 2012.
Johnson, W. C. (1985). *Methods Biochem. Anal.* **31,** 61.
Kawiecki, R. W., Devlin, F., Stephens, P. J., Amos, R. D., and Handy, N. C. (1988). *Chem. Phys. Lett.* **145,** 411.
Keiderling, T. A. (1981a). *Appl. Spectosc. Rev.* **17,** 189.
Keiderling, T. A. (1981b). *J. Chem. Phys.* **75,** 3639.
Keiderling, T. A. (1986). *Nature (London)* **322,** 851.
Kobrinskaya, R., Yasui, S. C., and Keiderling, T. A. (1988). *Proc. Am. Pept. Symp., 10th* p. 65.
Lal, B. B., and Nafie, L. A. (1982). *Biopolymers* **21,** 216.

Laux, L., Pultz, V., Abbate, S., Havel, H. A., Overend, J., Moscowitz, A., and Lightner, D. A. (1982). *J. Am. Chem. Soc.* **104,** 4276.
Lee, O., Roberts, G. M., and Diem, M. (1989). *Biopolymers,* in press.
Lipp, E. D., and Nafie, L. A. (1984). *Appl. Spectrosc.* **38,** 20.
Lipp, E. D., and Nafie, L. A. (1985). *Biopolymers* **24,** 799.
Lipp, E. D., Zimba, G. G., and Nafie, L. A. (1982). *Chem. Phys. Lett.* **90,** 1.
Lowe, M. A., Segal, G. A., and Stephens, P. J. (1986a). *Chem. Phys. Lett.* **123,** 108.
Lowe, M. A., Segal, G. A., and Stephens, P. J. (1986b). *J. Am. Chem. Soc.* **108,** 248.
McCoy, C. A., and deHaseth, J. A. (1988). *Appl. Spectrosc.* **42,** 336.
Malon, P., and Keiderling, T. A. (1988). *Appl. Spectrosc.* **42,** 32.
Malon, P., Pancoska, P., Budesinsky, M., Hlavacek, J., Pospisek, J., and Blaha, K. (1983). *Collect. Czech. Chem. Commun.* **48,** 2844.
Malon, P., Kobrinskaya, R., and Keiderling, T. A. (1988a). *Biopolymers* **27,** 733.
Malon, P., Barness, C. L., Bodesinsky, M., Dukor, R. K., van der Helm, D., Keiderling, T. A., Koblicova, Z., Pavlikova, F., Tichy, M., and Blaha, K. (1988b). *Collect. Czech. Chem. Commun.* **53,** 2447.
Marcott, C. A. (1979). Ph.D. thesis. University of Minnesota, Minneapolis.
Marcott, C., Scanlon, K., Overend, J., and Moscowitz, A. (1981). *J. Am. Chem. Soc.* **103,** 483.
Martin, D. E., and Puplett, E. (1969). *Infrared Phys.* **10,** 105.
Moskovits, M., and Gohin, A. (1982). *J. Phys. Chem.* **86,** 3947.
Nafie, L. A. (1981). *Vib. Spectra Struct.* **10,** 153.
Nafie, L. A. (1984). *Adv. Infrared Raman Spectrosc.* **11,** 49.
Nafie, L. A., and Diem, M. (1979a). *Acc. Chem. Res.* **12,** 296.
Nafie, L. A., and Diem, M. (1979b). *Appl. Spectrosc.* **33,** 130.
Nafie, L. A., and Freedman, T. B. (1981). *J. Chem. Phys.* **75,** 4847.
Nafie, L. A., and Freedman, T. B. (1983). *J. Chem. Phys.* **78,** 7708.
Nafie, L. A., and Freedman, T. B. (1986). *J. Phys. Chem.* **90,** 763.
Nafie, L. A., and Freedman, T. B. (1987). *Chem. Phys. Lett.* **134,** 225.
Nafie, L. A., and Polavarapu, P. L. (1981a). *J. Chem. Phys.* **75,** 2935.
Nafie, L. A., and Polavarapu, P. L. (1981b). *J. Chem. Phys.* **75,** 2945.
Nafie, L. A., and Vidrine, D. W. (1979). *Proc. Soc. Photo-Opt. Instrum. Eng.* **56,** 191.
Nafie, L. A., and Vidrine, D. W. (1982). *In* Fourier Transform Infrared Spectroscopy (J. R. Ferraro and L. J. Basile, eds.) Vol. 3, p. 83. Academic Press, New York.
Nafie, L. A., and Walnut, T. H. (1977a). *J. Chem. Phys.* **67,** 1501.
Nafie, L. A., and Walnut, T. H. (1977b). *Chem. Phys. Lett.* **49,** 441.
Nafie, L. A., and Zimba, C. G. (1987). *In* "Biological Applications of Raman Spectroscopy" (T. G. Spiro, ed.), Vol. 1, p. 307. Wiley, New York.

Nafie, L. A., Keiderling, T. A., and Stephens, P. J. (1976). *J. Am. Chem. Soc.* **98,** 2715.

Nafie, L. A., Diem, M., and Vidrine, D. W. (1979). *J. Am. Chem. Soc.* **101,** 496.

Nafie, L. A., Lipp, E. D., and Zimba, C. G. (1981). *Proc. Int. Conf. Fourier Transform Infrared Spectrosc., SPIE* **289,** 82.

Nafie, L. A., Oboodi, M. R., and Freedman, T. B. (1983). *J. Am. Chem. Soc.* **105,** 7449.

Nafie, L. A., Lee, N. S., Paterlini, G., and Freedman, T. B. (1987). *Mikrochim. Acta [Wein]* **3,** 93.

Narayanan, U. (1987). Ph.D. thesis. University of Illinois at Chicago.

Narayanan, U., Keiderling, T. A., Bonora, G. M., and Toniolo, C. (1985). *Biopolymers* **24,** 1257.

Narayanan, U., Keiderling, T. A., Bonora, G. M., and Toniolo, C. (1986). *J. Am. Chem. Soc.* **108,** 2431.

Narayanan, U., Keiderling, T. A., Elsevier, C. J., Vermeers, P., and Runge, W. (1988). *J. Am. Chem. Soc.* **110,** 4133.

Oboodi, M. R., Alva, C., and Diem, M. (1984). *J. Chem. Phys.* **88,** 501.

Osborne, G. A., Cheng, J. C., and Stephens, P. J. (1973). *Rev. Sci. Instrum.* **44,** 10.

Pancoska, P., Fric, I., and Blaha, K. (1979). *Collect. Czech. Chem. Commun.* **44,** 1296.

Pancoska, P., Yasui, S. C., and Keiderling, T. A. (1989a). *Biochemistry,* **28,** 5917.

Pancoska, P., Yasui, S. C., and Keiderling, T. A. (1989b). *Biochemistry,* submitted.

Paterlini, M. G., Freedman, T. B., and Nafie, L. A. (1986). *Biopolymers* **25,** 1751.

Polavarapu, P. L. (1984a). *Vib. Spectra Struct.* **13,** 103.

Polavarapu, P. L. (1984b). *Appl. Spectrosc.* **38,** 26.

Polavarapu, P. L. (1985). *In "Fourier Transform Infrared Spectroscopy"* Vol. 4, (J. R. Ferraro and L. Basile, eds.), Academic Press, New York.

Polavarapu, P. L. (1988). *Infrared Phys.* **28,** 109.

Polavarapu, P. L., and Nafie, L. A. (1980). *J. Chem. Phys.* **73,** 1567.

Polavarapu, P. L., Michalska, D. F., and Back, P. M. (1984). *Appl. Spectrosc.* **38,** 438.

Prasad, P. L., and Nafie, L. A. (1979). *J. Chem. Phys.* **70,** 5582.

Robert, G. M., Lee, O., Calienni, J., and Diem, M. (1988). *J. Am. Chem. Soc.* **110,** 1749.

Schellman, J. A. (1973). *J. Chem. Phys.* **58,** 2882 [erratum (1974) **60,** 343].

Schlosser, D. W., Devlin, F., Jalkanen, K., and Stephens, P. J. (1982). *Chem. Phys. Lett.* **88,** 286.

Sen, A. C., and Keiderling, T. A. (1984a). *Biopolymers* **23,** 1533.

Sen, A. C., and Keiderling, T. A. (1984b). *Biopolymers* **23,** 1519.
Singh, R. D., and Keiderling, T. A. (1981a). *J. Chem. Phys.* **74,** 5347.
Singh, R. D., and Keiderling, T. A. (1981b). *J. Am. Chem. Soc.* **103,** 2387.
Singh, R. D., and Keiderling, T. A. (1981c). *Biopolymers* **20,** 237.
Smith, G. R., Fridovich, B., Ivancic, W., and Rao, K. N. (1978). *Rev. Sci. Instrum.* **49.** 1223.
Snir, J., Frankel, R. A., and Schellman, J. A. (1975). *Biopolymers* **14,** 173.
Stephens, P. J. (1974). *Annu. Rev. Phys. Chem.* **25,** 201.
Stephens, P. J. (1976). *Adv. Chem. Phys.* **35,** 197.
Stephens, P. J. (1985). *J. Phys. Chem.* **89,** 784.
Stephens, P. J. (1987). *J. Phys. Chem.* **91,** 1712.
Stephens, P. J., and Clark, R. (1979). *In* "Optical Activity and Chiral Discrimination" (S. F. Mason, ed.). Reidel, Boston.
Stephens, P. J., and Lowe, M. A. (1985). *Annu. Rev. Phys. Chem.* **36,** 213.
Su, C. N. (1982). Ph.D. thesis. University of Illinois at Chicago.
Su, C. N., and Keiderling, T. A. (1980). *J. Am. Chem. Soc.* **102,** 511.
Su, C. N., Heintz, V. J., and Keiderling, T. A. (1980). *Chem. Phys. Lett.* **73,** 157.
Tiffany, M. L., and Krimm, S. (1968a). *Biopolymers* **6,** 1767.
Tiffany, M. L., and Krimm, S. (1968b). *Biopolymers* **6,** 1379.
Tiffany, M. L., and Krimm, S. (1969). *Biopolymers* **8,** 347.
Toniolo, C., Benedetti, E., *et al.* (1989). *Proc. Eur. Peptide Symp. 20th,* p. 465.
Yang, L., Annamalai, A., and Keiderling, T. A. (1989). Manuscript in preparation.
Yang, Y. T., Wu, C. S. C., and Martinez, H. M. (1986). *Methods Enzymol.* **130,** 208.
Yasui, S. C., and Keiderling, T. A. (1986a). *Biopolymers* **25,** 5.
Yasui, S. C., and Keiderling, T. A. (1986b). *J. Am. Chem. Soc.* **108,** 5576.
Yasui, S. C., and Keiderling, T. A. (1987). *J. Am. Chem. Soc.* **109,** 2311.
Yasui, S. C., and Keiderling, T. A. (1988a). *Proc. Am. Pept. Symp., 10th* p. 90.
Yasui, S. C., and Keiderling, T. A. (1988b). *Mikrochim. Acta,* **II,** 325.
Yasui, S. C., and Keiderling, T. A. (1989). Unpublished observations.
Yasui, S. C., Keiderling, T. A., Formaggio, F., Bonora, G. M., and Toniolo, C. (1986a). *J. Am. Chem. Soc.* **108,** 4988.
Yasui, S. C., Keiderling, T. A., Bonora, G. M., and Toniolo, C. (1986b). *Biopolymers* **25,** 79.
Yasui, S. C., Keiderling, T. A., and Katakai, R. (1987a). *Biopolymers* **26,** 1407.
Yasui. S. C., Keiderling, T. A., and Sisido, M. (1987b). *Macromolecules* **20,** 2403.

Yasui, S. C., Keiderling, T. A., and Toniolo, C. (1989a). *Biopolymers,* submitted.

Yasui, S. C., Pancoska, P., Dukor, R. K., Keiderling, T. A., Renugopalakrishnan, V., Glimcher, M. J., and Clark, R. C. (1989b). *J. Biol. Chem.,* submitted.

Young, D. A., Freedman, T. B., Lipp, E. D., and Nafie, L. A. (1986). *J. Am. Chem. Soc.* **108,** 7255.

Young, D. A., Freedman, T. B., and Nafie, L. A. (1989). *J. Am. Chem. Soc.,* in press.

6

Characterization of Semiconductor Silicon Using Fourier Transform Infrared Spectrometry

K. Krishnan
P. J. Stout
BioRad Digilab Division
Cambridge, Massachusetts

Masaharu Watanabe
Toshiba ULSI Research Center
Kawasaki, Japan

Practical Fourier Transform Infrared Spectroscopy

I. INTRODUCTION

Infrared spectrometry is a powerful tool for the characterization of semiconductor materials and finds wide application in the silicon and Group III-V, Group II-VI compound semiconductor industry. The applications of the infrared technique deal with the basic understanding of the lattice structure of these materials, identification and quantification of impurities in the materials, and measurements of the thicknesses of thin films on semiconducting materials. However, this chapter only deals with the application of FT-IR techniques to the impurity and epitaxial thickness measurements in silicon. Some applications of FT-IR microsampling techniques to the study of semiconductor materials can be found in the chapter by Krishnan and Hill in this volume. Detailed discussions of multivariate quantitative analytical methods applied to silicate passivation layers on silicon can be found in the chapter by Haaland in this volume.

In 1947 the point-contact transistor was invented by Bardeen and Brattain (1949), and the junction transistor was invented in 1948 by Shokley (1949, 1976). Investigators now understand that the action of the transistor is mainly in the single crystalline regions of silicon and germanium. Consequently, the need for crystalline perfection and freedom from undesirable impurities in these materials has been well established. During the last few decades, the microelectronic industry has succeeded in building increasing amounts of computing power into smaller and smaller device geometries on integrating circuit chips (ICs). With few exceptions, these ICs have been built on pure, single-crystal silicon substrates. The substrate silicon could be doped with Group V elements (P, As, Sb, etc.) to produce excess negative carriers (n-type silicon) or with Group III elements (B, Al, Ga, etc.) to produce excess positive carriers (p-type silicon). Two types of ICs are commonly built over the doped substrates: bipolar and metal oxide semiconductor (MOS). In the bipolar technology, current amplifications are achieved over regions of silicon that are alternately doped n or p, that is, by applying appropriate voltages to npn- or pnp-type structures built on silicon. In the MOS technology, the ICs are built by using local structures known as gates, which are sandwiches of metal on silicon dioxide over adjacent n- or p-doped regions of silicon. Positive or negative electric charges accumulate in the areas between the neighboring doped regions. These active regions are then interconnected to each other or other components on the same chip by conducting gold or aluminum stripes, thereby forming the complete IC. The electrical characteristics (such as speed and amplification) of the IC are functions of applied voltages, doping levels, and the physical dimensions of the device. Over the past couple of decades, the ICs have evolved from large scale (LSI) to very large scale integrated (VLSI) structures, and the focus in the industry now is on ultralarge scale integration (ULSI). These structures involve device and metal interconnect features in the sub-micron (<1 μm) range.

As mentioned above, the bulk of ICs manufactured today are built using silicon as the semiconducting material. Single-crystal silicon is grown from melt using the Czochralski (CZ) technique. In the early years of the semiconductor industry, silicon crystals 1 to 3 in. in diameter were commonly grown using the CZ method. The technology has evolved over the years to the extent that silicon crystals 6 to 8 in. in diameter are now routinely grown. The silicon crystal is sectioned into thin (0.3 to 0.6 mm) [100] or [111] wafers. Numerous ICs are then built on a single wafer, which is sliced into individual ICs and appropriately packaged. Since the yield and the electrical performance of the finished device depends upon the material properties of the silicon used in device fabrication, numerous analytical techniques are commonly used during manufacture for material characterization. These techniques include scanning electron microscopy, X-ray, surface analysis, optical spectroscopy, wet chemistry, chromatography, and thermal analysis. Of the optical methods, Fourier transform infrared (FT-IR) spectroscopy is a routine, rapid, and nondestructive technique used widely throughout the semiconductor industry. The FT-IR technique can be used for the quantitative determination of impurity concentrations in silicon and other semiconducting materials, for the determination of the epitaxial thicknesses of the n- or p-doped layers on silicon, and for the determinations of the dopant concentrations in silicon dioxide. These and other FT-IR applications in the semiconductor industry are reviewed in this chapter.

II. CHARACTERIZATION OF INTERSTITIAL OXYGEN AND SUBSTITUTIONAL CARBON IN SILICON

A. General Considerations

During the crystal growing process, oxygen impurity is easily incorporated into the silicon crystal from the quartz crucibles that are generally used to hold the melt and also from the ambient air surrounding the melt. The oxygen atoms can occupy interstitial sites in the silicon unit cell (Fig. 1) and form two strong Si—O bonds with the nearest-neighbor silicon atoms. The oxygen impurity concentration in silicon can be as high as 2×10^{18} atoms/cm^3. During semiconductor device fabrication, the silicon is subjected to a number of thermal annealing steps. During these steps, some of the interstitial oxygen can be converted to SiO_2 precipitates. In smaller concentrations, these precipitates act as intrinsic or internal getters, removing harmful impurities in the silicon, and serve to strengthen the silicon substrate. However, excessively large amounts of oxygen precipitates will weaken the silicon substrate and introduce slip and warpage. Thus, depending upon its concentration, oxygen can act as either a beneficial or a destructive impurity in silicon. For improved silicon device manufacturing, it is therefore important to measure and control the amount of dissolved oxygen in the silicon crystal.

Carbon impurity can be introduced into the silicon crystal from the crucibles used in the crystal growth process; carbon, being tetravalent, can occupy lattice sites normally occupied by silicon. The location of the substitutional carbon is shown in Fig. 1. The concentration of the carbon impurity can

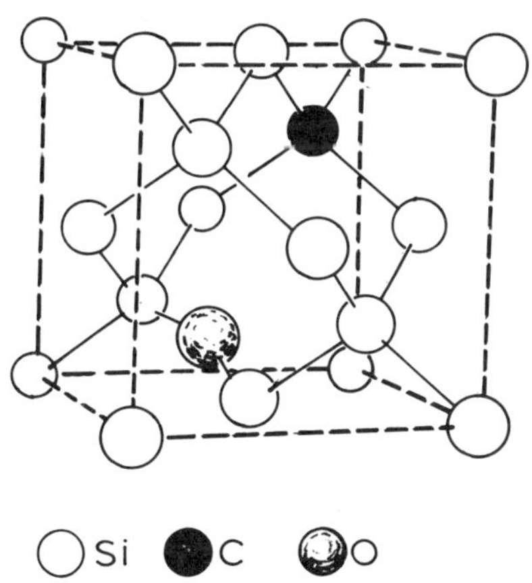

Fig. 1. Unit cell of silicon containing interstitial oxygen connected to two nearest-neighbor silicon atoms and substitutional carbon.

range from a few parts per billion in high-purity silicon to tens of parts per million in solar-grade silicon. Carbon impurity in silicon usually adversely affects the performance of the finished device, and its concentration must also be known precisely for efficient device manufacturing. FT-IR spectroscopy is currently the method of choice for the rapid and precise determination of these impurities.

Undoped or as-grown silicon is semi-insulating (resistivity on the order of 10 ohm-cm or higher) and is transparent to infrared radiation. The infrared transmission spectrum of such high-resistivity silicon will exhibit many absorption bands that are due to the vibrations of the silicon lattice: phonon bands. Inclusion of electrically inactive impurities such as oxygen, carbon, nitrogen, and hydrogen in the silicon lattice will lead to additional absorptions. Each impurity will have its own characteristic absorption band whose absorption coefficient can be used for the quantitative determination of the impurity. Highly doped silicon will have free carriers (n or p) and will exhibit free carrier absorption. At room temperature, the free carrier or plasma absorption will start at long wavelengths, and the absorption edge will move to shorter wavelengths with increasing concentrations of the free carriers. When the silicon resistivity is low (on the order of a few milliohms-cm), the free carrier absorption will dominate, and the absorptions due to the silicon phonons and the impurity vibrations may be impossible to detect.

B. Lattice Absorption Spectrum of Silicon

To identify and measure the absorption bands due to the impurities and defects in silicon, it is first of all necessary to measure the absorption spectrum of a high-purity silicon. High-purity, single-crystal silicon material could be produced by float zone (FZ) processing of the CZ single-crystal silicon. The lattice absorption spectrum of FZ silicon has been the subject of numerous studies (Johnson, 1959; Balkanski and Nusimovica, 1964; Abdullah *et al.*, 1984; Humlicek and Vojtechovski, 1985; Pradhan *et al.*, 1987a,b). Figure 2 shows the FT-IR absorption spectrum of a FZ silicon sample 2 mm thick. Table I lists the peak positions of the silicon lattice absorption bands. The strongest band seen in the spectrum at 610 cm^{-1} is assigned as the combination of a TO (transverse optical phonon) and TA (transverse acoustical phonon) and is commonly referred to as the "two-phonon" band. The assignments of the other bands in the silicon spectrum is open to discussion, and the calculation of the phonon spectrum of silicon is a current topic in solid-state physics. The assignments of the lattice absorption bands and absorption coefficients

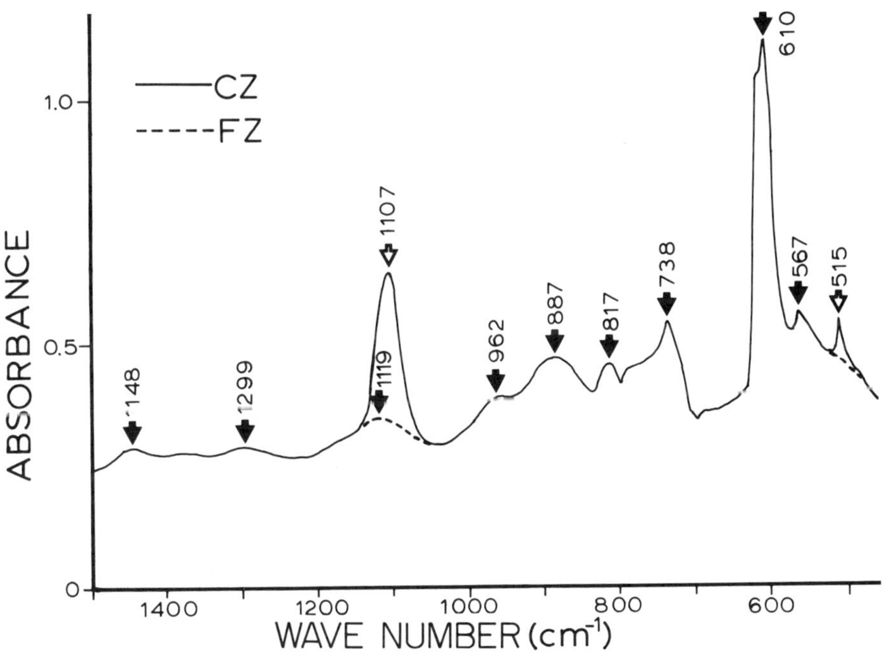

Fig. 2. Infrared absorption spectrum of a float zone silicon sample 2 mm thick showing the silicon phonon bands (closed arrows). CZ, Czochralski-process sample; FZ, float zone sample.

provided by Pradhan *et al.* (1987a,b), Johnson (1959), Balkanski and Nusimovika (1964), and Pajot (1977) are also included in Table I.

C. Optical Constants of Silicon

When infrared radiation is transmitted through a highly polished silicon sample, there will be multiple reflections of the infrared beam within the silicon sample and the incident beam

Table I Phonon Bands of Silicon

	Band position (cm^{-1}) for crystals[b]				
Absorption coefficient[a]	FZ(vac) (1)	FZ (2)	CZ(poly) (3)	(4)	FZ (5)
0.44	1448	1448.68	1446.61		1448
	1378		1385.17		
0.40	1302	1300	1296.16		1299
1.04					1118
		1107.50	1107.14		
1.35	964	968.51	968.55	965	960
				902	
2.17		887.33	887.31	880	886
1.93	819	823			819
	766			780	
2.84	740	739	736.81	743	739
	689			685	
	620				
9.29	610	609.50	609.50	608.5	610
2.79	566	567.11			566
			513.06	515	
	376				
	316				
	156				
	122				

[a]Absorption coefficients from Pajot (1977).
[b]Band assignments from (1) Johnson (1959); (2) Pradhan *et al.* (1987a); (3) Pradhan *et al.* (1987b); (4) Balkanski and Nusimovica (1964); (5) Pajot (1977).

will suffer reflection losses. These reflection losses will be dependent upon the refractive index of silicon. For accurate determination of the defect and impurity concentrations in silicon, it is thus necessary to have a good understanding of the optical constants of silicon. This property of silicon has been the subject of numerous investigations (Saltzberg and Villa, 1957; Loewenstein *et al.*, 1973; Passchier *et al.*, 1977; Edwards and Ochoa, 1980). The influence of the complex optical constants on the measurement of absorption coefficients, specifically that of the oxygen impurity, has been discussed by Engelbrecht and Lonbard (1986).

The reflectivity, R, is

$$R = [(\mathbf{n} - 1)^2 + k^2]/[(\mathbf{n} + 1)^2 + k^2] \tag{1}$$

where $\mathbf{n} = n + ik$ is the complex refractive index. The imaginary part k is given as

$$k = \alpha \lambda/4\pi \tag{2}$$

where α is the absorption coefficient and λ is the wavelength. For silicon in the mid-infrared wavelength range (2–50 μm), k is of the order of 10^{-3}, and its contribution to R is negligible. Figure 3 shows the refractive index and the reflectivity calculated with Eq. (1). It can be seen that for silicon the reflectivity is very close to 0.30 over the whole mid-infrared range.

Humlicek and Vojtechovski (1985) have fitted the following polynomial to the optical constants of silicon measured by Edwards and Ochoa (1980):

$$n = 3.41999 + 1.566 \times 10^{-9} \times \bar{\nu}^2 - 1.0386 \times 10^{-17} \times \bar{\nu}^4 \tag{3}$$

where $\bar{\nu} = 1/\lambda$ is the wavenumber (cm^{-1}).

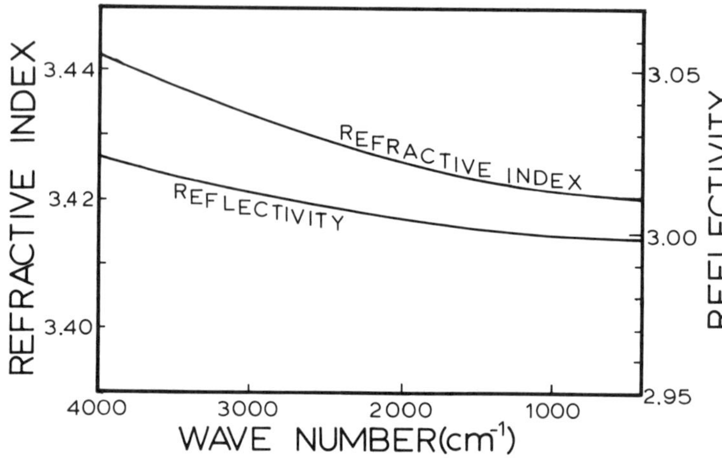

Fig. 3. The refractive index and reflectivity of silicon in the mid-infrared region.

D. Free Carrier Absorption

The effect of free carriers on the infrared absorption spectrum of silicon has been described earlier. For the case of the free carrier absorption, the complex refractive index for the free carrier absorption by the Drude model is given by the following expressions:

$$n^2 - k^2 = \epsilon\infty - \left(\frac{4\pi Ne^2}{\omega m^*}\right)\left[\frac{\omega\tau^2}{(1 + \omega^2\tau^2)}\right] \tag{4}$$

$$2nk = \left(\frac{4\pi Ne^2}{\omega m^*}\right)\left(\frac{\tau}{1 + \omega^2\tau^2}\right) \tag{5}$$

where $\epsilon\infty$ (= 11.7) is the high-frequency dielectric constant, e is the electronic charge, N is the free carrier density, m^* is the effective mass of the free carrier ($m^* = 0.26m_e$ for an electron), τ is the relaxation time of the free carrier scattering, and ω is the optical frequency. The dc conductivity, σ_0, is given by

$$\sigma_0 = \frac{Ne^2\tau}{m^*} \tag{6}$$

The free carrier absorption coefficient is then

$$\alpha = \left(\frac{4\pi Ne^2}{ncm^*}\right)\left(\frac{\tau}{1 + \omega^2\tau^2}\right) \tag{7}$$

where c is the speed of light. The free carrier absorption depends on the density of free carrier and increases gradually with infrared wavelength. As shown in Fig. 4, the free carrier absorption is not significant for carrier density less than 1×10^{16} cm^{-3}. Adequate measurement can be made for resistivity down to about 0.5 ohm-cm for p-type silicon and 0.05 ohm-cm for n-type silicon (Pidgeon, 1980).

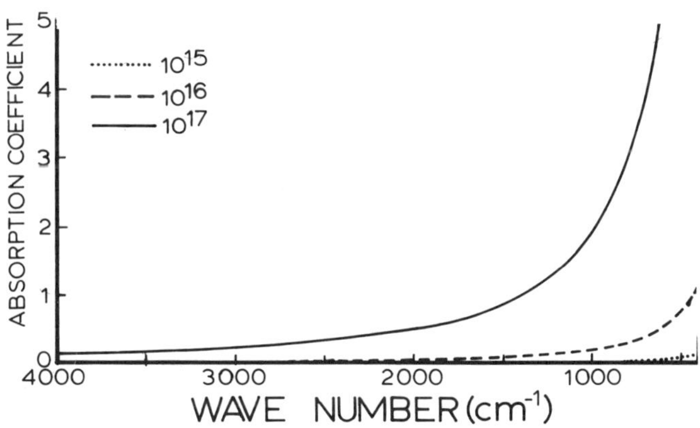

Fig. 4. Free carrier absorptions of silicon that contains 10^{15}, 10^{16}, and 10^{17} electrons.

E. Vibrational Modes Attributable to the Oxygen Impurity

Kaiser *et al.* (1956) were the first to observe the infrared absorption modes due to the interstitial oxygen impurity in silicon and germanium. Using a double-beam dispersive infrared spectrometer and a float zone reference sample, they observed for silicon an absorption band at 1107 cm^{-1} and assigned it to the Si—O stretching vibration. They also pointed out that this band could be used for the quantitative determination of the oxygen content in silicon.

Hrostowski and Kaiser (1957) carried out detailed study of this absorption as a function of temperature and oxygen isotopic ratios and assigned the 1107-cm^{-1} absorption to the antisymmetric stretching of a nonlinear Si_2O moiety. They also identified an absorption at 515 cm^{-1} as due to the bending vibration of the Si_2O. Since then numerous studies have been carried out, at both room and cryogenic temperatures, on the vibrational spectrum of the interstitial oxygen in silicon and the assignment of the observed absorption bands to the vibrations of the nonlinear Si_2O molecule (Hrostowski and Alder, 1960; Pajot, 1967; Pajot and Deltour, 1967; Bosomworth *et al.*, 1970; Krishnan and Hill, 1981; Vaikilova *et al.*, 1982; Stavola, 1984; Pajot *et al.*, 1985; Chen and Schroder, 1987).

Figure 5 shows the room-temperature FT-IR absorption spectrum of a polished CZ silicon sample 2 mm thick. On comparison with the silicon phonon spectrum (the FZ spectrum of Fig. 2), one can see the additional absorption bands due to impurities. These bands can be seen even more clearly in Fig. 6, which is the difference spectrum CZ − FZ. All of the bands seen in Fig. 6 are due to the interstitial oxygen impurity except for the one at 607 cm^{-1}, which is due to the carbon impurity. Table II lists the room-temperature absorption bands due to the oxygen impurity. Figure 7 shows an expanded view of the difference spectrum around the 1100-cm^{-1} region. The absorption band clearly visible at 1250–1310 cm^{-1}

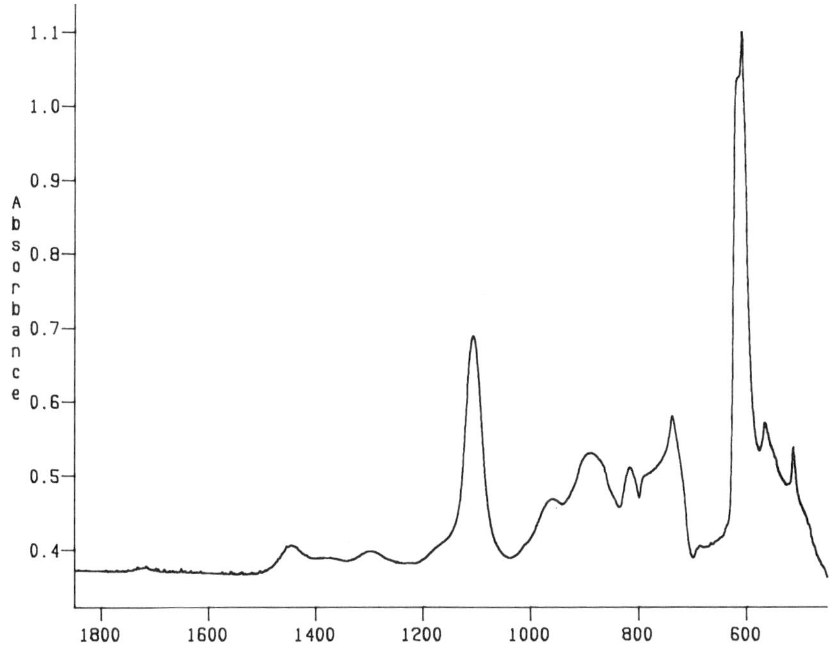

Fig. 5. Infrared absorption spectrum of a CZ silicon sample 2 mm thick showing the bands due to the silicon phonons and the oxygen and carbon impurities.

is due to oxygen precipitates. There are three major modes of vibration in the Si—O—Si quasi-molecule. The modes of this quasi-molecule, which has a bond angle of 164° are shown in Fig. 8. Bosomworth *et al.* (1970) were the first to calculate theoretically the vibrational spectrum of the Si_2O molecule, and Chen and Schroder (1987) have recently recalculated the spectrum using an *ab initio* Lowest Common Atomic Orbital-Molecular Orbital, (LCAO-MO) Si_2O model. The vibrational assignments based upon the preceding work as well as that of Pajot *et al.* (1985) and Stavola (1984) of the observed absorption bands due to the oxygen impurity are also included in Table II. There has been some controversy in the literature regarding the nature of the 515-cm^{-1} absorption. O'Mara

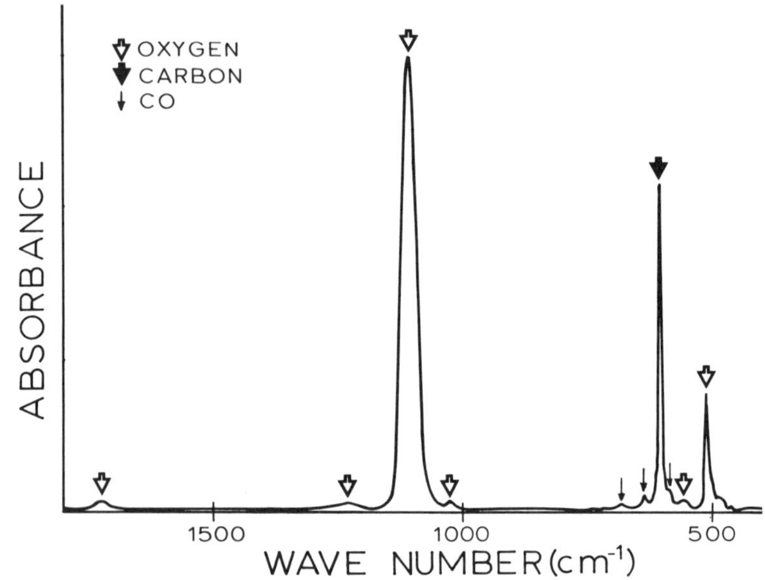

Fig. 6. The infrared difference spectrum CZ - FZ showing the absorption bands due to the oxygen and the carbon impurity.

Table II Infrared Absorption of Interstitial Oxygen in Silicon

Position (cm^{-1})	FWHH[a] (cm^{-1})	Relative intensity	Assignment
1720	31	0.016	$\nu_1 + \nu_2$
1226	22	0.011	
1107	33	1.000	ν_2 (asymmetric Si—O—Si stretch)
1059	—	—	
1013	8	0.006	$2 \times \nu_1$
560	—	—	
515	8	0.260	ν_1 (Symmetric Si—O—Si stretch)

[a]FWHH, full width half-height.

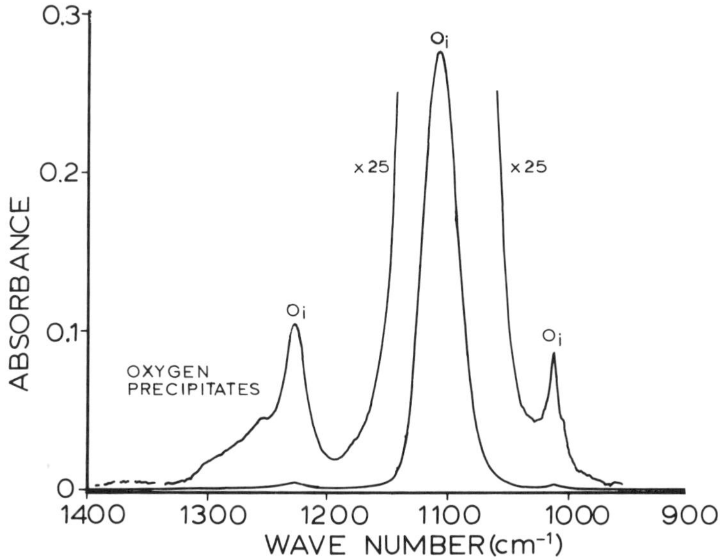

Fig. 7. Scale expansion of the oxygen-related absorption bands in the 1100-cm^{-1} region.

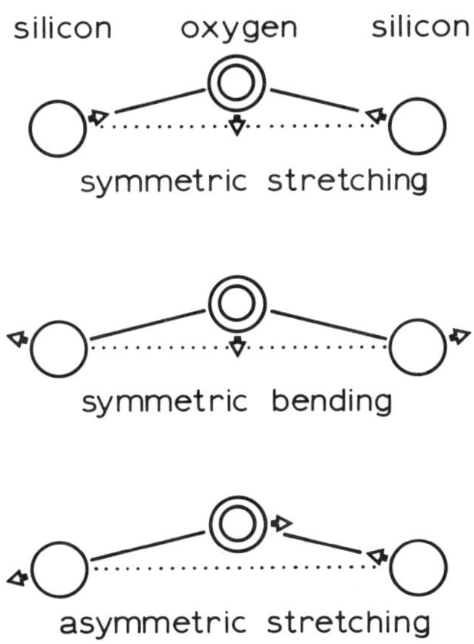

Fig. 8. Normal vibrational modes of Si—O—Si quasi-linear molecule.

(1983, 1985) had suggested that this band is not related at all to the Si_2O molecule but was due to a different oxygen species. However, Fig. 9 clearly shows that the intensity of the 515-cm^{-1} band linearly correlates to the intensity of the 1107-cm^{-1} absorption; so this band clearly is due to the Si_2O molecule. Ohnishi and Tsuya (1981) had suggested that the 515-cm^{-1} band arises from a Si_2O bending vibration. However, the recent investigation by Stavola (1984) would seem to suggest that this band is due to a symmetric O—Si—O stretching mode.

There is one band at 560 cm^{-1} in the spectra of the CZ silicon samples that, as far as the authors are aware, has not been reported earlier. This band has only been observed in CZ silicon crystals and seems to be independent of the carbon concentration. The intensity of this band seems to correlate linearly with the intensity of the 515-cm^{-1} band, as shown in Fig. 10. The assignment of this band is unclear, because there is no corresponding oxygen vibrational energy level in the quasi-linear or the LCAO-MO Si_2O model calculations.

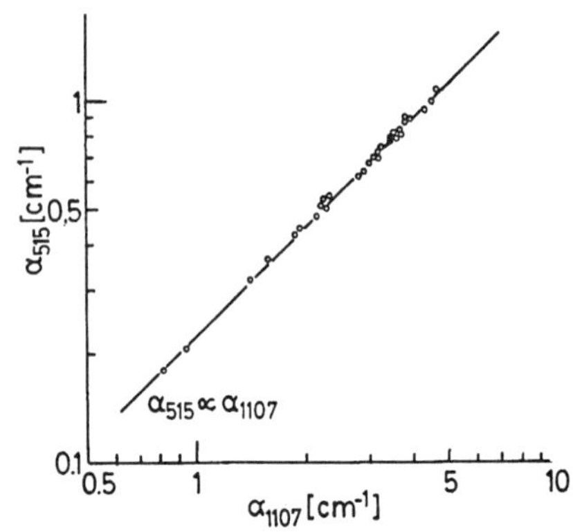

Fig. 9. Plot of the absorbance of the 515-cm^{-1} band against the absorbance of the 1107-cm^{-1} band. The plot shows that the two absorption bands are due to the same oxygen species.

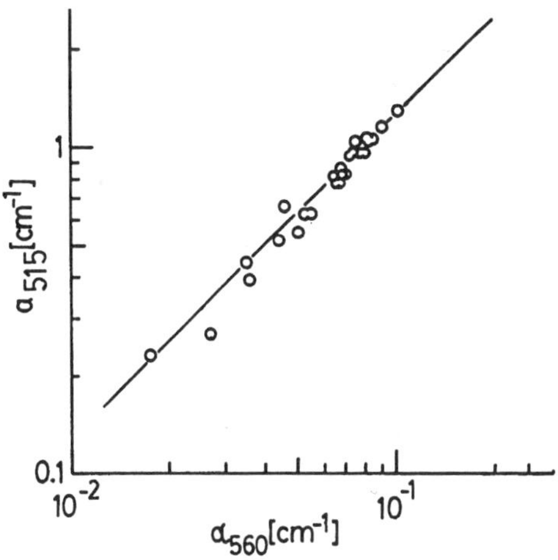

Fig. 10. Plot of the absorbance of the 560-cm^{-1} band versus that of the 515-cm^{-1} band.

On cooling the CZ silicon sample to liquid nitrogen or liquid helium temperatures, the room-temperature 1107-cm^{-1} absorption moves to higher frequencies, increases in intensity, becomes sharper, and splits into a number of components. The low-temperature behavior of the oxygen absorptions in silicon has been studied by a number of investigators (Graff *et al.*, 1973; Mead, 1980; Pradhan *et al.*, 1987b; Pajot, 1967; Hrostowski and Alder, 1960; Krishnan and Hill, 1981). Figure 11 shows the difference spectrum (after subtracting away the silicon phonon bands) in the 1130-cm^{-1} region of a CZ silicon sample 2 mm thick at 18 K. This spectrum was recorded at 0.5 cm^{-1} resolution. Figure 12 shows the same spectrum in the region 1250–1000 cm^{-1}. Some of the fine structure seen around the major bands in this region are due to various silicon isotopes. The reader is referred to the cited literature for a detailed analysis of the low-temperature silicon spectrum.

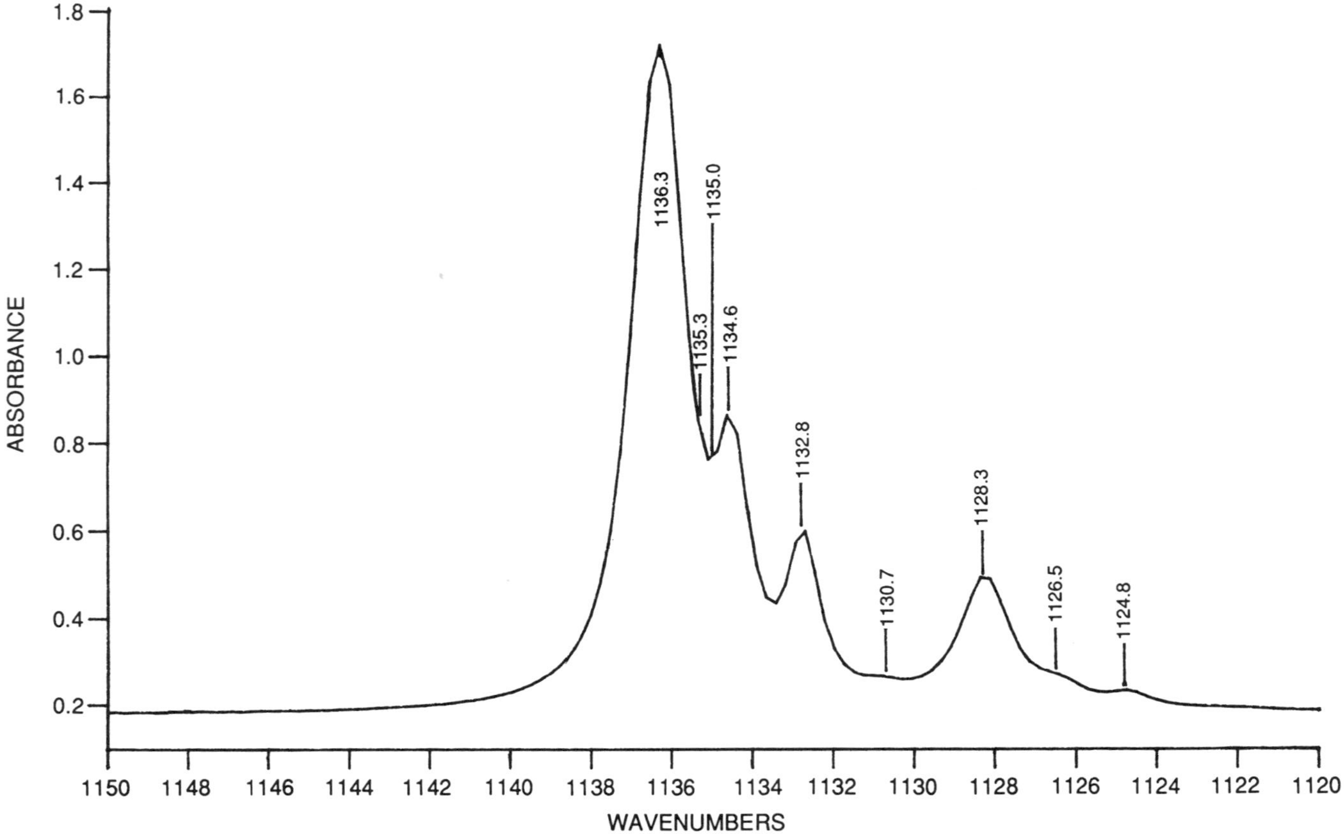

Fig. 11. Scale-expanded infrared absorption spectrum of a CZ silicon sample 2 mm thick held at 18 K. The silicon phonon bands have been subtracted away by using a FZ reference. The spectrum was recorded at a resolution of $0.5\ \text{cm}^{-1}$.

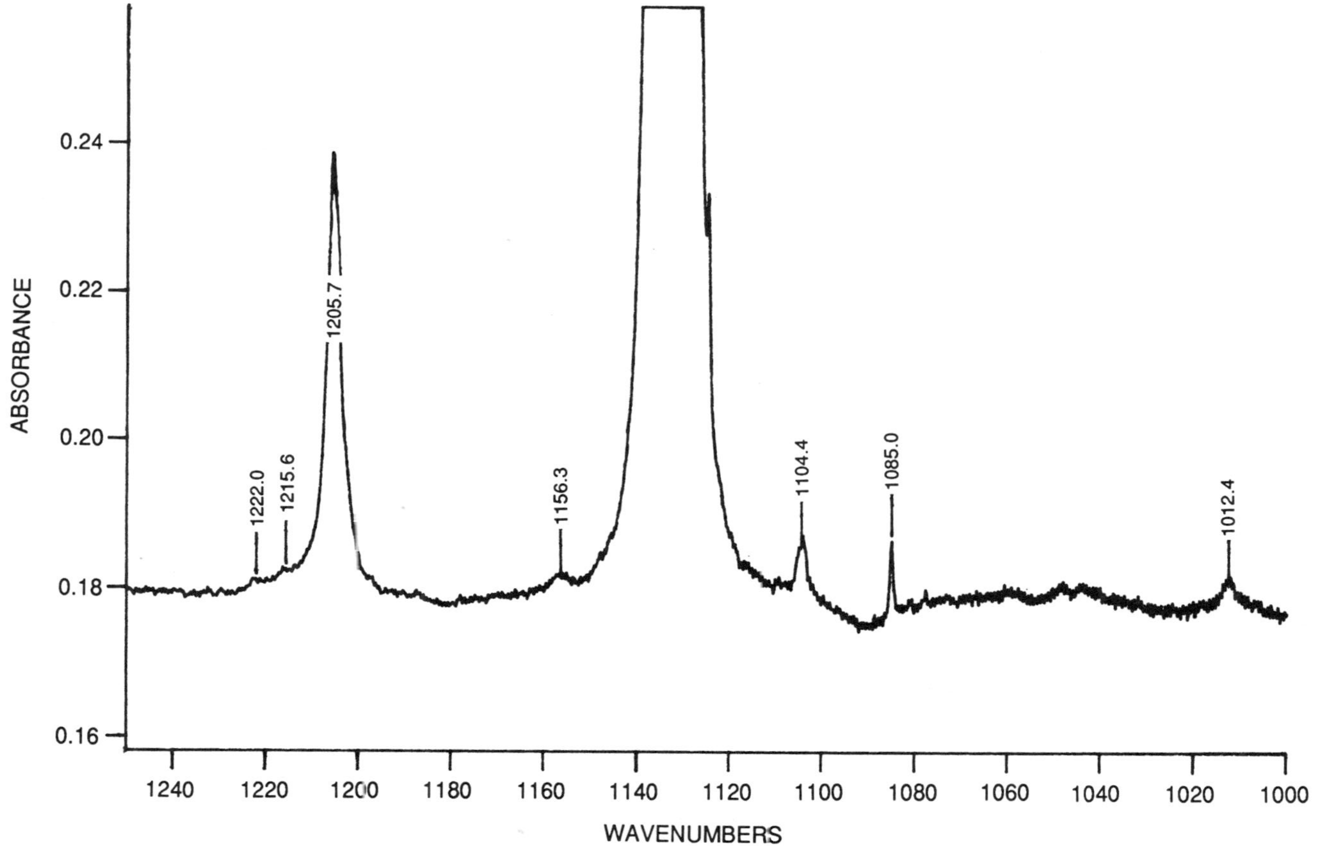

Fig. 12. Infrared absorption spectrum of a CZ silicon sample 2 mm thick held at 18 K; this plot shows the weaker bands. The silicon phonon bands have been subtracted away by using a FZ reference. The spectrum was recorded at a resolution of 0.5 cm^{-1}.

F. Quantitative Determination of the Interstitial Oxygen Concentration in Silicon

1. Conversion Factor

Regardless of the exact assignments of the oxygen-related absorptions, it is of paramount importance to the semiconductor process engineer that the oxygen concentrations in silicon be determined precisely, rapidly, and nondestructively. The infrared, particularly the FT-IR, technique offers these advantages and is used throughout the semiconductor industry as the method of choice for the oxygen measurements. Organizations such as ASTM (American Society for Testing Materials) in the United States, DIN (Deutsches Institut für Normung) in the Federal Republic of Germany, and JEIDA (Japan Electronics Industry Development Association) in Japan have established standard procedures for the measurement of the interstitial oxygen concentrations in silicon. All the methods depend upon the determination of the absorption coefficient of the 1107-cm^{-1} band, using appropriate local baseline points. The ASTM method (1983a), for instance, details the sample preparation, the measurement conditions, and the determination of the absorption coefficient from the peak height of the 1107-cm^{-1} band over local baseline points in the difference spectrum. The method also includes conversion factors that will convert the measured absorption coefficient into the oxygen concentrations expressed as parts per million atomic or atoms per cubic centimeter. The conversion factors have been arrived at by measuring the absorption coefficients of the 1107-cm^{-1} band in a number of silicon samples of known oxygen concentrations. The "known" oxygen concentration in silicon could be determined by vacuum fusion analysis, lithium diffusion, inert gas fusion analysis, charged particle activation analysis, and photon activation analysis. The conversion factors determined by various investigators vary significantly; they are listed in Table III. Even ASTM changed their factor for converting the absorption coefficient into atoms per cubic centimeter of oxygen from 4.81×10^{17} cm^{-2} in 1979 to 2.45×10^{17} cm^{-2} in 1983. This wide variation in the conversion factors is a source of confusion in the statement of the oxygen concentrations, both within the semiconductor industry and in the published literature.

The latest attempt to clarify this situation regarding the conversion factors was taken up by JEIDA, who undertook a round-robin measurement on a number of silicon samples in a number of laboratories in Japan and determined the conversion factor to be 3.03×10^{17} cm^{-2}. Subsequently, an international round-robin study on the same samples was organized involving laboratories in the United States, the Federal Republic of Germany, the United Kingdom, Japan, and the People's Republic of China. The details of this international

Table III Conversion Factors for Interstitial Oxygen

Reference	Factor[a]	No. of samples	Chemical analysis[b]
Kaiser and Keck (1957)	2.75	12	VFA
Iglitsyn *et al.* (1965)	6.00	8	LD
Rook and Schweikert (1969)	2.90	4	CPAA
Kim (1969)	3.80	6	CPAA
Baker (1970)	4.81	99	IGA
Gross *et al.* (1972)	3.85	4	CPAA
Yatsurugi *et al.* (1973)	3.00	10	CPAA
Graff *et al.* (1973)	2.45	12	VFA
He *et al.* (1983)	3.10	67	IGA
Iizuka *et al.* (1983)	3.03	22	CPAA
Rath *et al.* (1984)	3.00	8	PAA
Barraclough *et al.* (1986)	2.60	21	PAA
Regolini *et al.* (1986)	3.00	7	CPAA
Baghdadi *et al.* (1988)	3.14	—	CPAA
Chu *et al.* (1986)	3.45	1	SIMS

[a]Oxygen concentration = absorption coefficient × factor × 10^{17} atoms-cm^{-3}.
[b]VFA, vacuum fusion analysis; LD, lithium diffusion; IGA, inert gas fusion analysis; CPAA, charged particle activation analysis; PAA, photon activation analysis.

round-robin study will be published elsewhere, and the study yielded a conversion coefficient designated as IOC-87 (International Oxygen Coefficient 1987) with a value of 3.1×10^{17} cm^{-2}. During the international round-robin study, it was found that the various laboratories around the world were able to measure the absorption coefficient of the 1107-cm^{-1} oxygen band within a reproducibility of 1%. However, the measurement of the absolute oxygen concentration by nonspectroscopic methods exhibited very large scatter. In general, the charged particle activation analysis using 160 (^{3}He, p) ^{18}F gave the best interlaboratory reproducibility. A plot of the final oxygen content versus the infrared absorption coefficient is shown in Fig. 13.

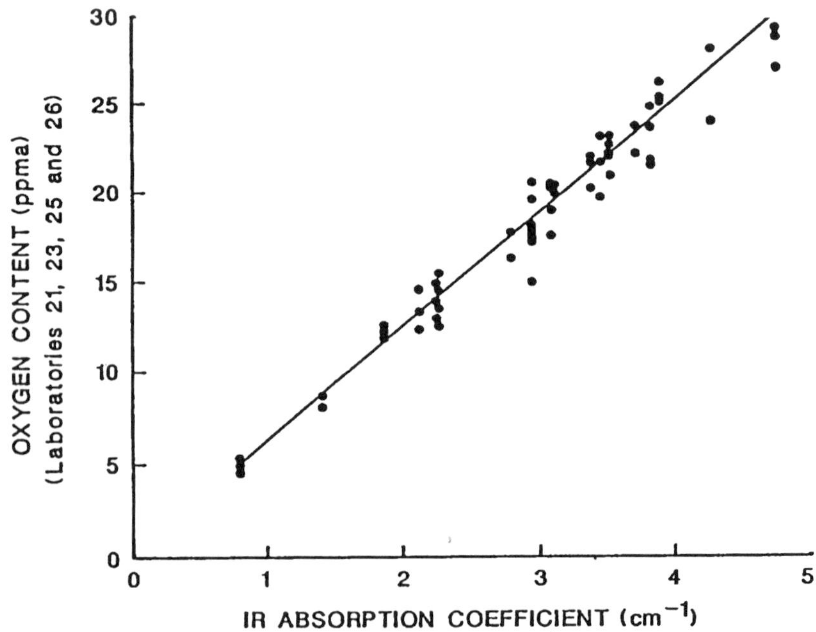

Fig. 13. Plot of the oxygen content against the absorption coefficient of the 1107-cm^{-1} band. The plot shows the typical result obtained in the JEIDA round-robin.

2. Corrections for Multiple Internal Reflections within Silicon

As mentioned earlier, Kaiser *et al.* (1956) were the first to measure the concentration of oxygen in silicon using infrared spectrometry. Subsequently, Kaiser and Keck (1957) and Thurber (1970) defined the procedure for the measurements. Baker (1970) reported on the measurement of the oxygen concentrations in the parts per billion range using infrared spectrometry. Kagel and Baker (1974) were the first to use FT-IR, and they reported the determination at room temperature of 122 parts per billion oxygen concentration in silicon using a sample and reference specimens 20 mm thick. Since then, numerous publications have appeared in the literature on the routine determination of the oxygen content in silicon (Liaw, 1979; Vidrine, 1980; Mead and Lowry, 1980; Pajot, 1977; Krishnan, 1983; Graff, 1983; Stallhofer and Huber, 1983; Engelbrecht and Lonbard, 1986; Weeks, 1984). Most of the measurements reported in these papers have been done using FT-IR spectrometers. Bullis and O'Mara (1983) have discussed the requirements of the FT-IR instrumental stability for precise oxygen measurements, and Bullis and Coates (1987) have compared the performances of two commercially available FT-IR spectrometers regarding oxygen measurements.

The standard methods for the oxygen measurements such as those published by ASTM (1983a) require a highly polished silicon sample 2 to 4 mm thick and FZ reference specimens of exactly matched thicknesses. In a dual-beam dispersive infrared spectrometer, the sample and the reference are placed in their respective beams, and the differential spectrum is recorded. With FT-IR instruments, separate absorbance spectra are created for the sample and the reference, and a difference spectrum is produced by absorbance subtraction of the silicon phonon bands from the sample spectrum. In either case, the absorption coefficient of the 1107-cm-1 oxygen band is determined by measuring the peak-height absorbance over a local baseline. An error in the absorbance measurement of 0.01 for a sample 2 mm thick would result in an error of 0.55 ppma in the determination of the oxygen concentration (using the 1983a ASTM conversion factor). Thus, it is of paramount importance that the absorption coefficient be measured precisely.

When the infrared beam is incident on a highly polished silicon sample, the beam will undergo multiple reflections within the sample, as shown in Fig. 14. In the absence of multiple reflections, the transmittance of the sample is defined as

$$T = \frac{I}{I_0} \tag{8}$$

where I_0 is the incident energy and I is the transmitted energy. When multiple reflections take place, the transmittance becomes

$$T = \frac{(1-R)^2\exp(-\alpha d)}{1 - R^2\exp(-2\alpha d)} \tag{9}$$

where R is the reflectivity of the sample, α is the absorption coefficient, and d is the sample thickness. The difference between the absorption coefficients of the sample and the reference is

$$\alpha_i = \alpha_t - \alpha_r \tag{10}$$

where α_i is the absorption coefficient of the impurity or defect under consideration, and α_t and α_r are the absorption coefficients of the test and reference specimens, respectively. The conventional absorbance subtraction of the test and the reference specimens is defined as

$$\begin{aligned}\ln T_t - \ln T_r &= 2.303\,(\log T_t - \log T_r) \qquad (11)\\ &= 2.303\left\{-(\alpha_t - \alpha_r)d + \log\left[\frac{(1-R^2\exp(-2\alpha_r d)}{(1 - R^2\exp(-2\alpha_t d)}\right]\right\}\end{aligned}$$

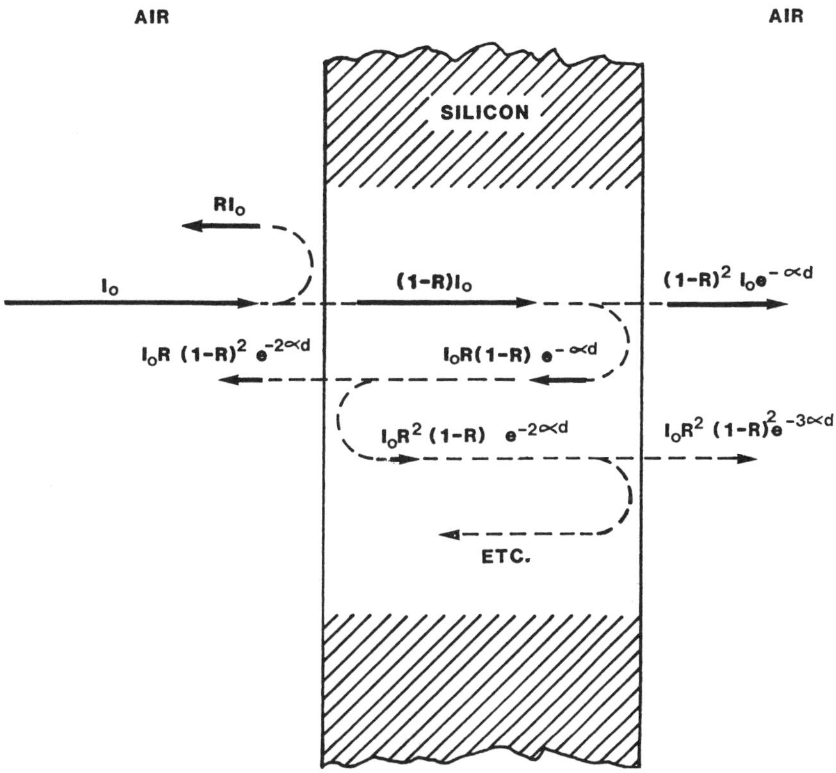

Fig. 14. Schematic representation of the multiple reflections of the infrared beam in a double-sided polished silicon sample.

where T_t and T_r are the transmittance of the test and reference specimens. The second term of Eq. (13) is not negligible, so the conventional absorbance subtraction technique is not quite accurate. Calculating the absorption coefficient from Eq. (10) as

$$\alpha = -\left(\frac{1}{d}\right)\ln\left[\frac{-(1-R)^2 + (1-R)^4 + 4R^2T^4}{2R^2T}\right] \tag{12}$$

and using Eq. (11) will result in a more accurate subtraction. In addition to this problem with the exact determination of the absorption coefficient, there are a number of other possible sources of error in the determination. These can be classified as those due to the test and the reference specimens or those related to the performance of the instrument used for the measurements. Among the specimen-related problems are the following differences between the test and the reference specimens: (1) thickness; (2) reflectivity; (3) free carrier absorption; (4) change in the optical constants; (5) interfering absorption bands; (6) interference fringes caused by multiple reflections within the sample; and (7) exact surface conditions. Among the instrument-related problems are (1) nonlinearity of the detector; (2) different apodization functions used by different FT-IR instruments—this problem will result in different absorbance peak heights for the same sample; (3) nonuniformity of the infrared beam intensity; (4) stray light; (5) purge variations; (6) interference between the sample or test specimen surface and the optical components of the instrument; and (7) the photometric accuracy.

The free carrier absorption results in the gradual increase of the absorption coefficients with increasing wavelengths. The free carrier effect is not very large, however, unless the resistivity is less than 0.5 ohm-cm for p-type materials and less than 0.05 ohm-cm for n-type materials. Oates and Lin (1988) have outlined a procedure for the determination of the oxygen concentrations in heavily doped (very low resistivity) silicon specimens.

The interference fringes due to multiple reflections within the sample usually appear in high-resolution measurements, as shown in Fig. 15. In this case, Eq. (8) leads to

$$T = \frac{(1-R)^2\exp(-\alpha d)}{1 + R^2\exp(-2\alpha d) - 2R\cos(4\pi n\bar{\nu}d)} \tag{13}$$

$$P = \frac{1}{2nd} \tag{14}$$

$$d\Delta\bar{\nu} < \frac{1}{2n} \tag{15}$$

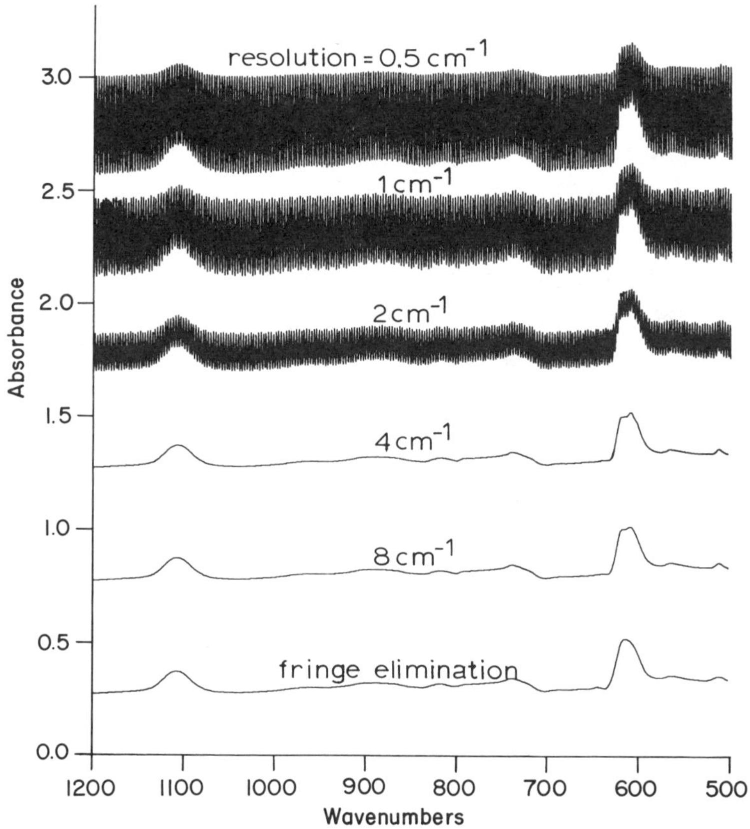

Fig. 15. Interference fringes of a silicon wafer measured at different resolutions. The bottom spectrum shows the effects of software fringe elimination.

The interference fringes can be removed by using smoothing or lower resolution measurements, by using very thick test and reference specimens, by using the Brewster angle of incidence method, or by using special software techniques (Krishnan and Ferraro, 1982; Hyland *et al.*, 1987). The fringe problems will always arise when one analyzes actual production wafers, which usually range in thickness between 0.3 and 0.6 mm, at very high resolution. Most of the preceding techniques may help in minimizing the fringe problem, but they may not be able to completely eliminate it.

3. Single-Sided Polished Samples

Whereas the referee methods defined by organizations such as ASTM, DIN, and JEIDA deal with silicon samples 2 mm thick and polished on both surfaces (hereafter called double-sided polished wafers), most silicon samples used in the semiconductor device manufacturing use wafers that range in thickness from 0.3 to 0.7 mm and are polished on one side only (hereafter called single-sided polished wafers). The semiconductor device under consideration is built on the polished surface, but the second or backside surface is "damaged" by various techniques. The reason for using backside-damaged wafers is as follows: The most important parameter affecting the performance of a semiconductor device is the contaminants introduced during silicon growth and device manufacturing. The backside damage provides a method of gettering or removing metallic impurities to the damaged side, and away from the polished side. This gettering effect is obtained by introducing controlled amounts of dislocations on the backside, to within a few micrometers of the back surface. This backside damage can be achieved by abrasion, by caustic, acid, or bright etching, and by high-power lasers. The dislocations act as sinks for the vacancies and impurities; and once these are removed from the polished side, they remain bound to the backside during the rest of the device manufacturing process. When the backside is damaged or roughened, the reflectivity of the backside will vary and become indeterminate, and the multiple reflection corrections discussed earlier will no longer apply. One can apply an empirical solution to the problem and define the transmittance as

$$T = \frac{(1-R)^2\exp(-\alpha df)}{1 - R^2\exp(2\alpha df)} \tag{16}$$

where f is the fraction of the transmitted and internally reflected light that is scattered out of the backside. The variable f is nothing but a fitting parameter, however, and there is no way to calculate it. Furthermore, the scattering of the infrared radiation by the roughened backside will result in distorted baselines in the recorded spectrum, making it very difficult to draw proper local baselines. Thus, there is no referee method to measure the oxygen content of a backside-damaged silicon wafer.

One of the possible ways to overcome the problem associated with the accurate measurement of the oxygen content of backside-damaged wafers is to measure the oxygen content of a double-sided polished specimen by the referee method outlined earlier, to damage one of its sides to different extents by different methods, and to measure the oxygen contents of these damaged specimens without applying any multiple reflection correction. It may then be possible to establish empirical correlations between the oxygen content of the double-sided polished wafers and the backside-damaged wafers, enabling one to measure the exact oxygen content of the backside-damaged wafers. Krishnan (1983) has outlined a different procedure for dealing with backside-damaged wafers. This technique involves the determination of the absorption coefficient of the 1107-cm^{-1} oxygen absorption band using a thickness of the silicon specimen calculated from measuring the absorbance of one of the silicon phonon bands itself. Since the absorbance of the silicon phonon band would itself reflect the effects of multiple reflections within the silicon specimen, this method should account for most of the errors introduced by the effects of multiple reflections regardless of the condition of the backside. Using this technique, Krishnan (1983) was able to demonstrate that accurate oxygen concentration measurements could be made regardless of the nature of the backside or the thickness of the samples. Bullis and Coates (1987) have independently verified that this method works satisfactorily and yields very good correlation in the measured oxygen concentrations between slices 2 mm thick and bright-etched product wafers.

In conclusion, it is now possible to routinely measure the oxygen content of any silicon sample precisely and rapidly using the FT-IR spectrometric technique.

G. Oxygen Precipitates in Silicon

The fabrication of ICs by the VLSI process is very complicated and involves a number of heat-treatment steps. Oxygen concentrations in the 10- to 20-ppma range are well above the solid solubility limit at these processing temperatures, and oxygen tends to precipitate out during the manufacturing process. The oxygen precipitation has two side effects. On the one hand, the precipitates degrade the performance of the device by increasing the leakage currents. On the other hand, the oxygen precipitates increase the manufacturing yield by gettering the contaminants, or by acting as the recombination sites for the excess carriers.

As mentioned in the previous sections, interstitial oxygen in silicon exhibits a strong absorption band at 1107 cm^{-1} and a medium absorption band at 515 cm^{-1}. When the silicon wafer containing interstitial oxygen is heat treated with the consequent formation of precipitates and related defects, additional infrared absorption bands are observed. The infrared absorption spectrum of the oxygen precipitates has been the subject of numerous studies (Tempelhoff *et al.*, 1977, 1979, 1981; Patel, 1981; Pajot and Bardeleben, 1984; Hu, 1977, 1986; Gaworzewski *et al.*, 1984). The strongest of these new absorptions are located near 1100 cm^{-1} and 1250 cm^{-1}. Quantitative estimates of the precipitate concentrations can be made up by using the absorption coefficients of the 1250-cm^{-1} band. Tempelhoff *et al.* (1977) have observed three stages of oxygen precipitation. At relatively low treatment temperatures (400° to 780°C), very small precipitates are produced that absorb between 1030 and 1075 cm^{-1}. Precipitates produced in the temperature range 870°–930°C produce an additional absorption near 1124 cm^{-1}. Tempelhoff *et al.* (1977) have assigned this latter absorption to cristobolite-type precipitates in the silicon crystal.

Gaworzewski *et al.* (1984) and Hu (1980) have undertaken a systematic study of the precipitates and have classified them, as shown in Fig. 16. They attributed the broad absorption at 1100 cm^{-1} to small, globular precipitates produced at lower treatment temperatures. The two absorption bands that appear at 1250 and 1100 cm^{-1} are attributed to platelike precipitates produced by medium temperature heat treatment. Higher temperature heat treatment produces precipitates exhibiting bulk silicate properties; when these precipitates have octahedral coordination, they lead to infrared absorbances at 1106 and 470 cm^{-1}. The authors have also shown that an amorphous, platelet-like precipitate is responsible for an absorption band around 1225 cm^{-1}. Most of the observed bands in the spectra of the various precipitates have been assigned to the various phonon modes in the silicon-defect lattice by the above authors.

Freeland (1980) has studied the problem of the oxygen precipitates generated by heat treatment of the crystal at 650°C. Jastrzbski *et al.* (1982) have proposed a method for measuring the precipitated and total oxygen content in silicon. They proposed heat treatment for 1 hr at 1300°C. After this treatment, all the oxygen in silicon is transformed into the interstitial form, whose concentration can easily be determined by FT-IR using the methods outlined earlier. If there is any difference between the interstitial oxygen concentrations before and after the heat treatment, this difference represents the amount of precipitated oxygen.

It should also be noted that heat treatment of oxygen-containing silicon at 450°C produces thermal donor states (Kaiser *et al.*, 1956; Hrostowski and Kaiser, 1958; Bean and Newman, 1972). When the silicon sample containing the thermal donors is cooled to near liquid helium temperatures, a progression of infrared absorption bands are found near 500 cm^{-1}. These bands are due to the electronic transitions

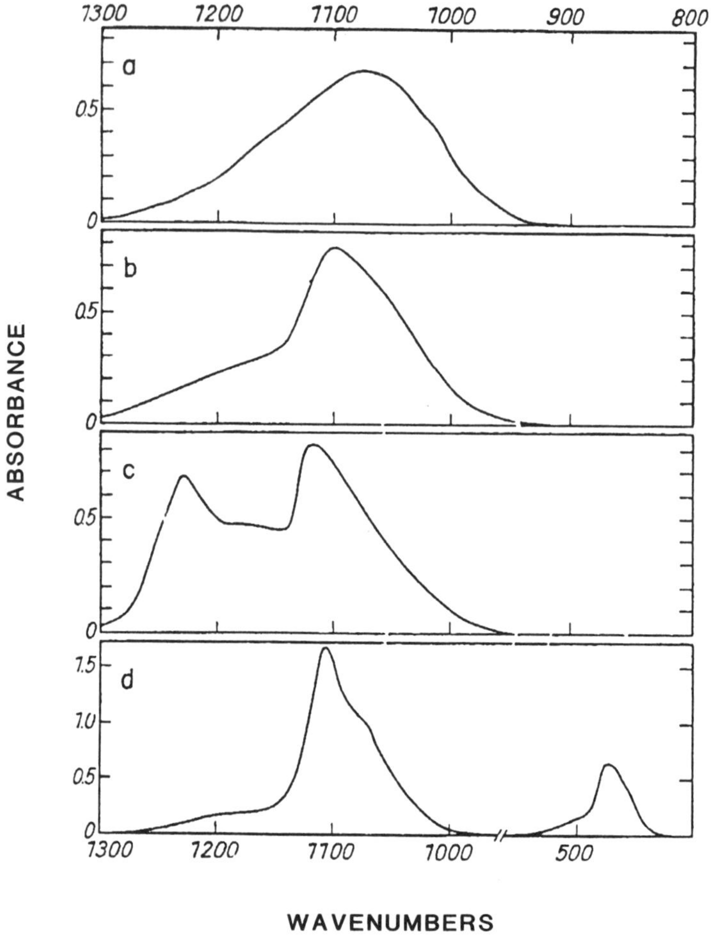

Fig. 16. Infrared absorption spectra due to different types of oxygen precipitates: (a) Sample a after heat treatment at 600°C for 240 hr; (b) sample c after heat treatment at 900°C for 64 hr; (c) sample b after heat treatment at 900°C for 64 hr; and (d) sample z after heat treatment at 1275°C for 2 hr. Reprinted with permission from Gaworzewski *et al.* (1984).

involving the thermal donor states and have been studied by Tempelhoff *et al.* (1977), Wruck and Gaworzewski (1979), and Pajot *et al.* (1983, 1985). The formation and the characterization of the thermal donors in silicon is a current topic in solid-state physics.

H. Substitutional Carbon in Silicon

The carbon impurity occupies substitutional sites in the silicon lattice, as shown in Fig. 1, and forms C—Si bonds. The stretching of this bond in the silicon lattice gives rise to an infrared absorption band at 607 cm^{-1}, first identified by Newman and Willis (1965). Since this band is completely overlapped by the strong two-phonon band of silicon (the band at 610 cm^{-1} in Fig. 2, silicon phonon spectrum), the 607-cm^{-1} band can only be seen in the difference spectrum obtained by subtracting away the silicon phonon bands by using a float zone reference. However, in practice, it is very difficult to perform a clean subtraction between the test and the float zone specimens over the whole two-phonon band. The quality of the subtracted spectrum depends upon, among other factors, the difference in surface finish, doping levels, and the thermal history between the test and the reference specimens. For the most accurate carbon measurements, it is best to use test and reference specimens that are polished on both sides and are exactly matched in thickness (about 2 mm). A number of papers that have been published in the literature deal with the routine measurement of the carbon impurity in silicon (Vidrine, 1980; Mead and Lowry, 1980; Engelbrecht, 1987). Leroueille (1982) has described a method to measure the carbon content of thin (approximately 400 μm thick) silicon ribbons, and Jaae (1986) has studied the distribution of carbon in thin silicon ribbons. Krishnan (1983) has demonstrated the possibility of improving the accuracy of the carbon determination by using the method of Fourier self-deconvolution (Kauppinen *et al.*, 1981).

The standard procedure for the determination of the carbon content in silicon has been described by a number of standards organizations (see, for instance, ASTM, 1983b). The measurement procedures are similar to those for the measurement of the interstitial oxygen concentrations in silicon. The factors for converting the measured absorption coefficients into atomic concentrations provided by different investigators are shown in Table IV. One can see that the scatter in the carbon conversion factor is much less than the one for oxygen. As has been done for oxygen, JEIDA had recently made round-robin tests in a number of laboratories in Japan for the carbon conversion factor. The absolute measurement of the carbon concentrations in the round-robin test samples was done by charged particle activation analysis using ^{12}C (p, B)F. The carbon conversion factor thus arrived at by JEIDA is 0.81×10^{17} cm^{-2}, which is not very different from the value of 1.0×10^{17} cm^{-2} used by ASTM (1983b). Figure 17 shows the plot of the absorption coefficient of the 607-cm^{-1} band against the carbon content.

Since the errors in the absorption coefficient measurements are very small for strong absorbances such as the silicon two-phonon band, and since the 607-cm^{-1} carbon absorption underlies the two-phonon band, it is not necessary to

Table IV Conversion Factors for Substitutional Carbon

Reference	Factor[a]	No. of samples	Chemical analysis
Newman and Willis (1965)	11.0	—	^{14}C
Endo *et al.* (1972)	8.3	7	^{3}He
Gross *et al.* (1972)	8.0	—	^{2}H
Spenke *et al.* (1973)	9.5	7	
ASTM (1983a)	10.0	—	
Regolini *et al.* (1986)	10.0	14	^{2}H
Inoue *et al.* (1986)	8.5	19	^{3}He

[a]Carbon concentration = absorption coefficient × factor × 10^{17} atoms-cm^{-3}.

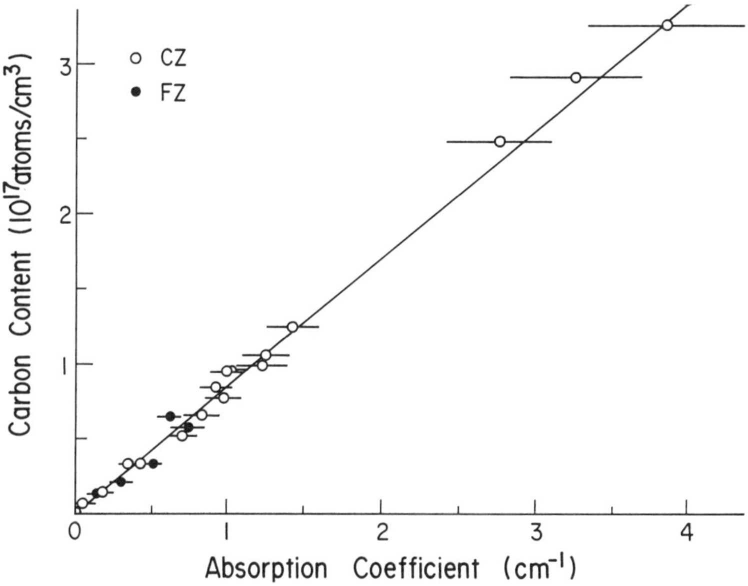

Fig. 17. Plot of the carbon concentration against the absorption coefficient of the 607-cm^{-1} band. The data shown in the figure formed part of the JEIDA round-robin results.

apply the corrections for multiple reflections during the carbon measurements. From a survey of the room-temperature carbon measurements reported in the literature on silicon samples 2 mm thick, it would seem that the minimum detectable concentration is of the order of 0.1 ppma. If the carbon-containing silicon sample is cooled down to near liquid helium temperatures, the 607-cm^{-1} band sharpens up somewhat and the sensitivity of the FT-IR method can be improved. However, the improvement in the sensitivity from cryogenic measurements for the carbon is not as large as that for the interstitial oxygen.

In silicon sample containing oxygen and relatively large amounts of carbon, carbon–oxygen complexes with their own characteristic infrared absorption bands are present. The presence of the carbon–oxygen complexes was first reported by Newman and Smith (1969) and has been reinvestigated by Pradhan *et al.* (1987b). Oehrlein *et al.* (1982) have investigated the carbon–oxygen complexes as nucleation centers for the precipitation of oxygen in CZ silicon. Table V lists the infrared absorption bands associated with the carbon-oxygen complexes in silicon. Figure 18 shows the room-temperature difference spectrum. (CZ − FZ) of a CZ silicon specimen 2 mm thick in the 700- to 500-cm^{-1} range. One can clearly see the

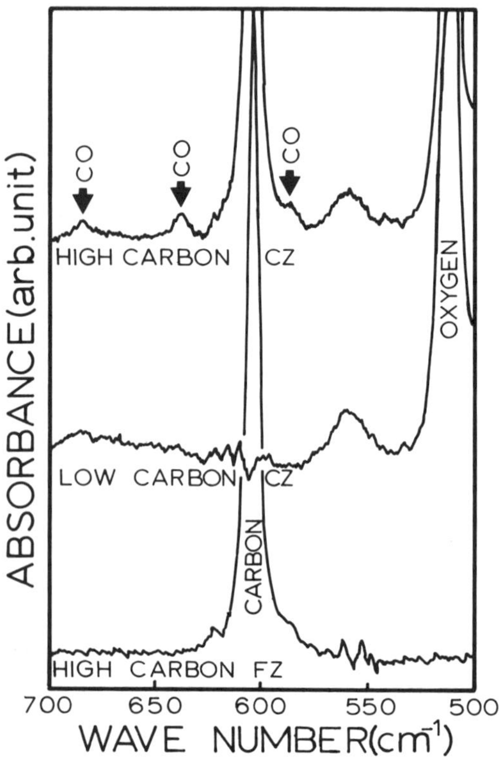

Fig. 18. Room-temperature infrared difference spectra between 700 and 500 cm^{-1} of (bottom) FZ silicon containing carbon; (middle) CZ silicon containing no detectable carbon; and (top) CZ silicon containing carbon. The bands due to the carbon–oxygen complexes are indicated by the arrows.

Table V Infrared Absorption Bands Due to Carbon–Oxygen Complexes

	Peak position (cm^{-1})			
Line	300 K	77 K	15 K	4.2 K
A	1093.6	1103.0	1111	1103.9
B	1047.5	1051.9	1106	1052.0
C			1086	1085.1
D		1098.4	1078	1099.0
			1060	669.3
			1040	693.3
X	586.2	588.5	1011	
Y	636.7	640.1		
Z	683.8	689.9		

three satellite peaks at 684, 637, and 586 cm^{-1} due to the carbon–oxygen complexes. These bands are only seen in CZ silicon crystals and have not been seen in FZ crystals, no matter how high the carbon content. These observations suggest that the carbon–oxygen complexes may exist in as-grown silicon single crystals and can be detected by room-temperature infrared spectroscopic measurements.

III. NITROGEN IN SILICON

The presence of nitrogen as an impurity in silicon has been reported to suppress the occurrence of thermal slip and to have a strong pinning effect on the dislocations in silicon (Abe *et al.*, 1981; Sumino *et al.*, 1984). Nitrogen, as a Group V impurity, would be expected to display donor properties similar to the other Group V dopants. However, the solubility of nitrogen in silicon is very low, and its donor activity is very different from the other Group V elements. In commercially grown CZ silicon, nitrogen is known to exist in concentrations of less than 2×10^{14} atoms-cm^{-3}. The infrared spectroscopic behavior of nitrogen in silicon has been studied by Stein (1985, 1986, 1987). The nitrogen impurity in silicon produces two infrared absorption bands at 963 and 766 cm^{-1}, as shown in Fig. 19. Because of the low concentrations of nitrogen in silicon, fairly thick (10 mm) specimens have to be used so that these bands can be seen clearly. Study of the isotope shift of the two bands suggest that they are due to an N—N pair near a silicon lattice site (Stein, 1987). The 963-cm^{-1} absorption band is similar in frequency to the main absorption band of silicon nitride, and trisilylamine, and can be used for the quantitative determination of nitrogen in silicon. The conversion factor for the quantitative determination of nitrogen in silicon has been estimated by Itoh *et al.* (1985) to be 1.83 ± 0.24 ×

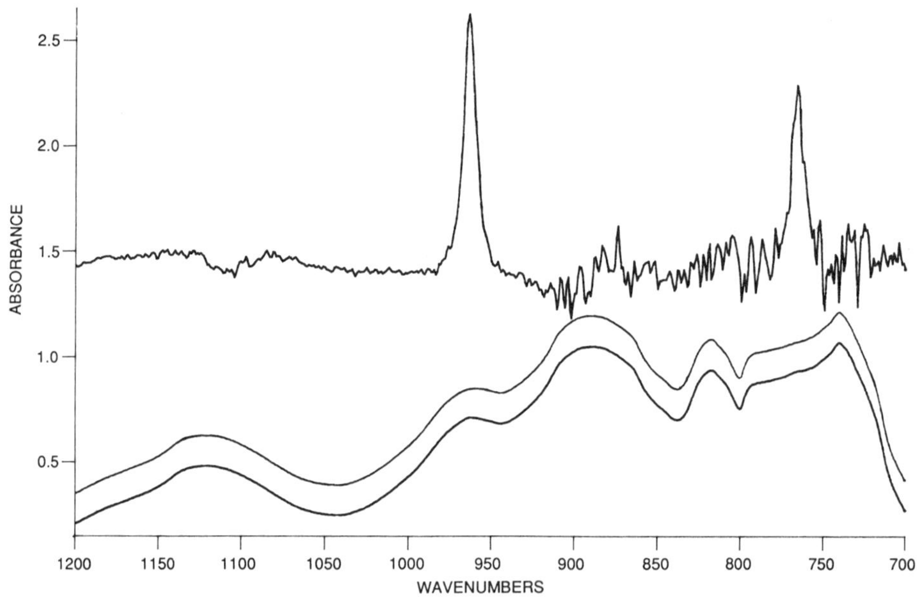

Fig. 19. Infrared absorption spectra of nitrogen in silicon. double-sided polished samples 10 mm thick were used. (Bottom) Silicon containing nitrogen; (middle) silicon without nitrogen; and (top) difference spectrum showing the 963- and the 766-cm^{-1} bands due to the nitrogen impurity.

10^{17} cm^{-2}. However, this calibration factor cannot be applied to nitrogen in CZ silicon to measure the total nitrogen content. A fraction of the nitrogen in CZ silicon is known to be infrared inactive.

Stein (1985) has identified an absorption band at 653 cm^{-1} due to off-center substitutional nitrogen in silicon; the nature of the off-center impurities has been studied theoretically by DeLeo *et al.* (1984). Nitrogen can, upon low-temperature annealing of the silicon, precipitate or form donor states similar to those produced by the oxygen precipitates. Suezawa *et al.* (1988) have studied the infrared absorption bands (in the region 400–200 cm^{-1}) arising from the electronic transitions of these donor states; and they have also identified transitions due to nitrogen–oxygen complexes.

IV. HYDROGEN IN SILICON

Atomic hydrogen could be introduced into silicon during crystal growth and VLSI processes such as exposure to hydrogen-containing plasma. Pearton *et al.* (1987) has reviewed the behavior of hydrogen in silicon. Hydrogen in silicon can exist either bound to the silicon in lattice sites or complexed to defect centers or residual impurities. The infrared spectrum of hydrogen in silicon exhibits strong absorbances at 2210 and 1946 cm^{-1} and weaker bands at 2191, 2178, 2048, 1994, and 1950 cm^{-1}. The band at 2210 cm^{-1} is assigned to defects containing four hydrogen atoms, namely, a silane molecule in a tetrahedral interstitial site—$(SiH_4)T$, or four hydrogen atoms at a vacancy—(V.4H). The 1946-cm^{-1} absorption band is usually assigned to defects containing only the Si–H groups. The 2191- and the 2123-cm^{-1} bands are thought to be oxygen related. In addition to the aforementioned bands, there are also two bands—at 812 and 791 cm^{-1}—due to the bending and wagging vibrations of these defects.

The foregoing discussion applies to the determination of the hydrogen content in crystalline silicon, which is mostly used in semiconductor device manufacturing. However, solar silicon technology uses amorphous silicon, usually in the form of sputtered films. These films normally contain alpha-Si:H bonds that give rise to infrared absorption bands around 2000 and 600 cm^{-1}, due to the stretching and wagging modes, respectively. The positions and assignments of the various infrared active bands due to the hydrogen vibrations in amorphous silicon have been well established (Brodsky *et al.*, 1977; Freeman and Paul, 1978), and the conversion factors for the hydrogen concentration measurements have been provided by Fang *et al.* (1979) and by Jeffrey *et al.* (1979). Ross *et al.* (1982) have found that for the quantification of the hydrogen content of amorphous silicon, infrared spectrometry, ^{15}N nuclear reaction, and SIMS all yielded similar results.

V. SHALLOW IMPURITIES IN SILICON

Even the purest semiconductor-grade silicon contains impurities in the 10^9–10^{14} atoms-cm^{-3} concentration range. Impurities such as the Group III (acceptor) and Group V (donor) elements in silicon are substitutional and have ionization potentials of around 0.1 eV and give rise to shallow acceptor and donor energy levels. Impurities such as boron, phosphorus, arsenic, and antimony are among the most common ones encountered in silicon. At room temperature, the free carriers introduced by the dopants have high mobility; however, when the silicon sample is cooled down to near liquid helium temperatures (less than 20 K), the electrons or holes are frozen out and become loosely bound to the defect centers. Under these conditions, the behavior of the ionized carriers can be described in terms of a pseudo-Bohr (hydrogen-like) model, and their transitions will give rise to characteristic absorptions in the mid- to the far-infrared. These absorptions can be seen in the infrared transmission spectra of silicon samples containing the impurities (sample thicknesses ranging from 2 to 10 mm) held in a low-temperature cryostat at temperatures less than 20 K. Ramdas and Rodriguez (1981) have reviewed the spectroscopy of these solid-state analogs of the hydrogen atom. A large number of papers have appeared in the literature on the low-temperature transmission infrared spectroscopy of silicon containing a variety of Group III and Group V impurities (Pajot *et al.*, 1979; Pajot, 1964; Tardella and Pajot, 1982; Colbow, 1963; Krishnan, 1981). Baber (1980) and Pajot and Debarre (1981) have addressed the issue of the quantitative determination of the shallow impurities in silicon. Table VI lists the major infrared absorption bands due to different impurities that can be used for the quantitative measurements. Figure 20 shows the infrared absorption spectrum of a typical silicon sample 2 mm thick held at 15 K. Baber (1980) has discussed in detail the factors for converting the measured infrared absorption coefficients to atomic concentrations.

Most of the impurity bands observed in the infrared spectrum of silicon at low temperatures are due to the transitions of the electron or the hole from the ground state of the neutral impurities to the pseudo-hydrogen levels lying below their respective band edges. Ionized centers cannot participate in

Table VI Major Analytical Absorption Bands of the Shallow Impurities in Silicon at 15 K

Impurity	Peak position (cm^{-1})
Phosphorus	316
Boron	320
Antimony	294
Arsenic	382
Aluminum	472
Gallium	548
Indium	1177

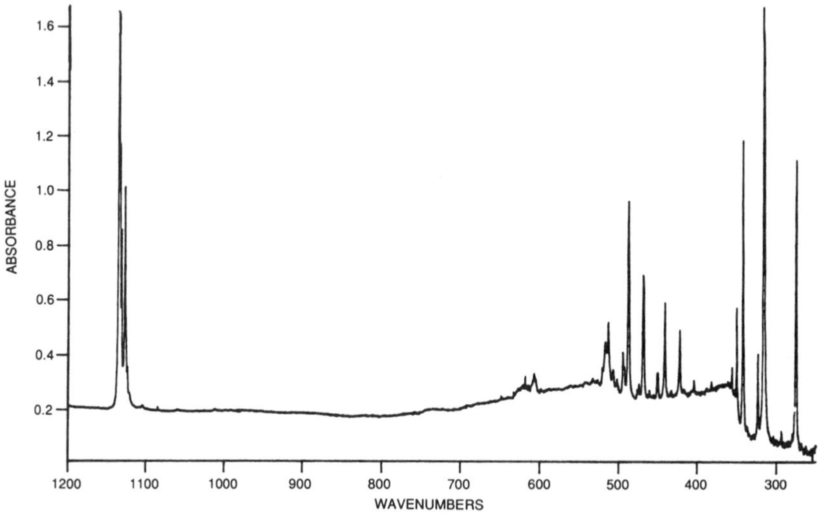

Fig. 20. Infrared absorption spectrum of a silicon sample 2 mm thick held at 15 K. The series of bands between 520 and 400 cm^{-1} are due to the thermal donor. The bands at 278 and 320 cm^{-1} are due to the boron impurity. The strong band near 1130 cm^{-1} is due to the oxygen impurity.

such transitions, so the absorption bands seen are due to the net impurity concentrations. Compensated impurities will not be detected by this method. However, White (1967) and Kolbesen (1975) have shown that by illuminating the silicon sample held at low temperatures with photons having energy higher than the band gap of silicon (1.11 eV), excess electron–hole pairs can be generated. These excess carriers neutralize the ionized centers, and the infrared transmission spectrum recorded under these conditions will show absorption bands corresponding to the total impurity content of the sample. Using this so-called Kolbesen technique, Baber was able to demonstrate the net and total impurity concentration determination in silicon. When used with this technique, the infrared transmission method can detect most of the common impurities in silicon in the low parts per billion concentration range.

The infrared photoluminescence (PL) technique—

whereby the near-infrared emission spectrum of silicon held at cryogenic temperatures and illuminated by powerful radiation, such as from an argon ion laser, is recorded—is capable of much lower detection limits for these impurities. There are several FT-IR instruments currently available that can perform these FT-PL measurements.

VI. RADIATION DAMAGE

When silicon is exposed to energetic particles, radiation damage may be introduced. In today's VLSI process, there are many opportunities for radiation damage such as ion implantation, reaction ion etching, plasma ashing of photo resist, and neutron transmutation doping. Devices used in the space must be able to withstand a heavy radiation environment. Behavior of defects can be traced by a variety of experimental techniques. Most of the defects are point defects and their clusters, so techniques that are sensitive to small defects are suitable for radiation damage studies. Infrared absorption is one of the most powerful techniques for this purpose; it has been reviewed by Newman (1969, 1974, 1982).

A wide variety of defects are introduced by different irradiation conditions. Figure 21 shows a typical spectrum of neutron-irradiated silicon. Two distinct features are in the spectrum. One is the absorption band at 830 cm^{-1}. This is an A center absorption, which is an oxygen trapped in a single vacancy $[VO]^0$. The 830-cm^{-1} absorption band is a neutral A center A^0. A negatively charged A center, A^- ($[VO]^-$), is located at 835 cm^{-1}; it is also observed in the spectrum. Another feature is a broad absorption band at 488 cm^{-1}. This band is assigned to a single phonon absorption band. The single absorption is usually infrared inactive, but disturbance of the silicon lattice causes breakdown of its symmetry and makes the single phonon band infrared active. Another feature is the small absorption at 1030–850 cm^{-1}. The 870-cm^{-1}

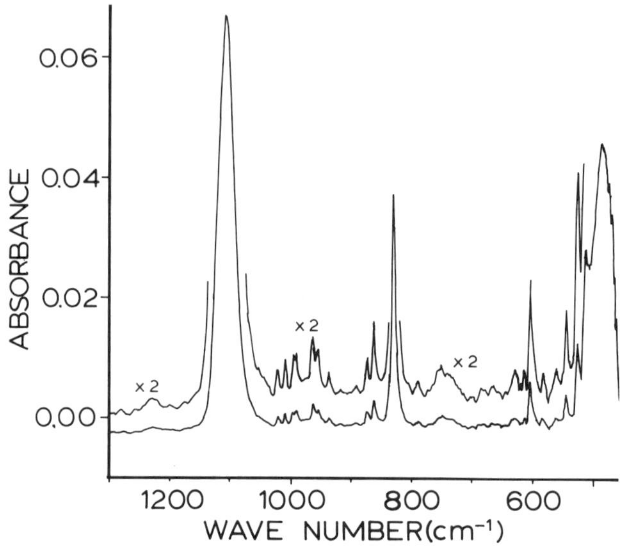

Fig. 21. The as-irradiated spectrum of neutron transmutation doping crystal made by light water reactor.

absorption band has been attributed to a substitutional oxygen pair [O_2], but it requires further study. Many absorptions due to radiation damage have been observed, but only a few are tentatively assigned. The A center absorption is well characterized, and the concentration of A centers can be estimated by

$$[\mathrm{VO}] = 6.1 \times 10^{16} \times \alpha_{830} \tag{17}$$

The radiation damage affects the silicon absorption edge. The transmittance around the absorption edge is reduced. This effect is called the Fan effect.

The complicated configuration of radiation damage causes a variety of absorption bands. Studies on the bulk radiation damage have been established to some extent. However, radiation damage caused by low-energy particles is not studied so much because the damage is located near the surface and inevitably causes weak absorptions. Making use of dry processes and ion implantation in VLSI processes is dependent on damage and damage control. This is an area of research in which FT-IR should be employed.

VII. EPITAXIAL THICKNESS MEASUREMENTS

As mentioned in the Introduction, silicon bipolar devices are built using npn- or pnp-type structures. These structures are usually fabricated using the epitaxial technology. Figure 22 shows the schematic structure of a typical silicon bipolar device. The silicon epitaxial technology has been reviewed by Bean *et al.* (1985). The epitaxial technique involves the growth of a thin film of a material on the surface of a single crystal substrate such that the epitaxial layer has the same crystallographic orientation as the substrate. In other words, the epitaxial layer becomes an extension of the substrate. However, the epitaxial layer can be doped to a different extent than the substrate. The epitaxial layer can be

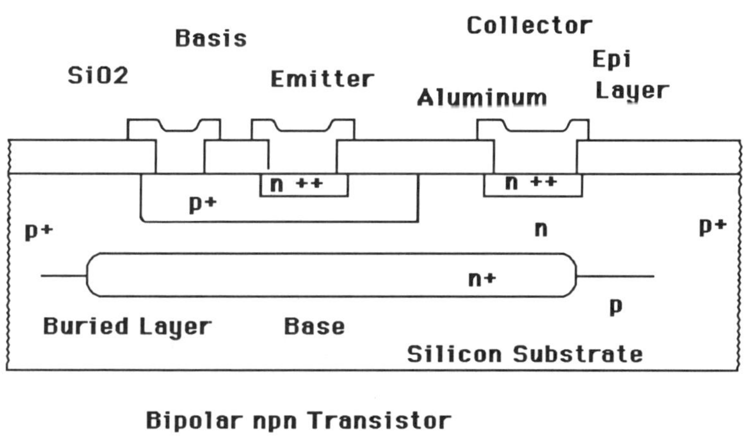

Fig. 22. Schematic representation of a typical silicon bipolar device.

homoepitaxy, or simply epi (the epi and the substrate being the same material, such as silicon on silicon), or heteroepitaxy (the epi and the substrate being different materials such as silicon on sapphire or mercury-cadmium-telluride on cadmium telluride).

Most ICs built using the epitaxial technology have high-resistivity epi layers built on low-resistivity substrates. In the case of silicon, the epi layer could be grown by molecular beam epitaxy (MBE), or by the thermal decomposition of silane gas, or by the thermal reaction between silane halides (silcon tetrachloride, dichloro silane, etc.) and hydrogen. The MBE technique is used mostly in research and development environments. The silane gas can be used only for producing very thin epi layers. Thus, most of the epi layers used in the production environment are produced by chemical vapor deposition processes involving silane halides. As mentioned earlier, the epitaxial layers of high resistivity can be built on highly and uniformly doped substrates. Alternately, the high-resistivity epitaxial layer can be built on high-resistivity substrates; islands of high doping levels could then be diffused into the substrate (buried layer epitaxy). Factors such as current amplification, leakage, and breakdown voltage of a finished IC depend upon the exact thickness of the epitaxial layer used in the process. Thus, for the process engineer, it is very important to measure the thickness of the epitaxial layer precisely. A number of techniques can be used for the measurement: angle lapping and staining (ASTM, 1983c) three-probe voltage breakdown method (ASTM, 1983d), spreading resistance using four-point probe measurements with mercury probes on freshly etched n-type silicon (Severin and Bulle, 1975a,b), and infrared. Since the highly doped regions of the substrate will not transmit infrared radiation (due to free carrier absorption), epitaxial thickness measurements are made using the infrared reflectance technique (ASTM, 1983e).

Figure 23 shows a schematic representation of an epitaxial layer on silicon; the figure shows the concentrations of the

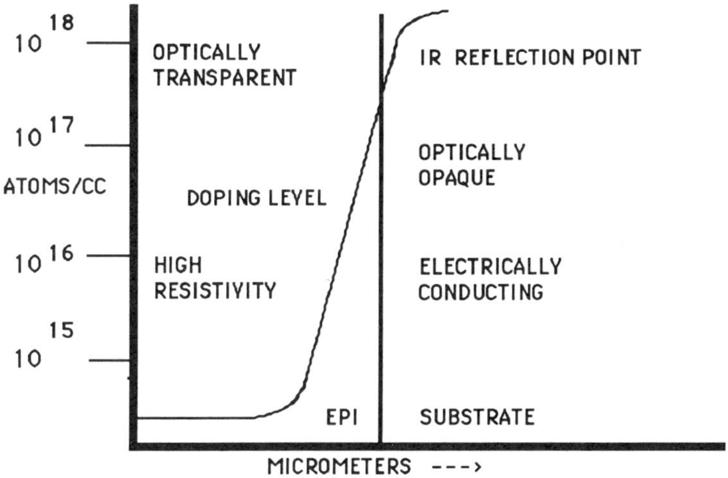

Fig. 23. Schematic representation of an epitaxial layer with low-resistivity substrate.

dopant as a function of depth away from the epitaxial surface. The epi is lightly doped and will transmit the infrared radiation in the wavelength range 2 to 50 μm; the substrate is heavily doped and will reflect the infrared radiation. If the heavily doped substrate contains buried layers, the infrared beam will be reflected by the buried layers and transmitted by the rest of the substrate. Figure 24 shows a schematic representation of the reflection of the infrared radiation from the epi–air and the epi–substrate interfaces. Depending upon the difference in doping levels between the epi and the substrate, multiple-reflected radiation may reach the detector. At each reflection, the infrared beam will undergo a phase shift, and this phase shift for the air–epi interface ($\phi1$) will be different from that of the epi–substrate interface ($\phi2$). In any event, since the detector senses the radiation reflected from the epi–air and the epi–substrate interfaces, the resulting spectrum will show interference fringes. The period of the interference fringes is given by Eq. (16) described earlier. As the epi layer increases

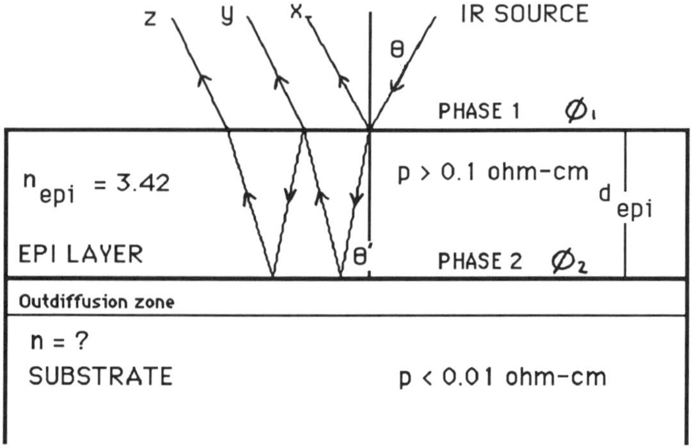

$$d = t \cos \theta$$

Fig. 24. Reflection of infrared radiation from an epi structure.

in thickness, the period of the fringes decrease. Figure 25 shows the reflectance spectrum of three silicon (n on n^+) epitaxial samples with thicknesses of 99, 25, and 3 μm.

Schumann *et al.* (1966) and Schumann and Schneider (1970) have studied the measurement of silicon epitaxial thickness from the infrared interference spectra such as those shown in Fig. 25, assuming an abrupt junction model. The reflectivity of the air-epi-substrate is given as (Schumann and Schneider, 1970)

$$R = \frac{r_1^2 + r_2^2 - 2r_1r_2\cos(\partial - \phi)}{1 + r_1^2r_2^2 - 2r_1r_2\cos(\partial - \phi)} \tag{18}$$

$$r_1^2 = \left(\frac{n_1 - 1}{n_1 + 1}\right)^2 \tag{19}$$

$$r_2^2 = \frac{(n_2 - n_1)^2 + k_2^2}{(n_2 + n_1)^2 + k_2^2} \tag{20}$$

$$\partial = (4\,\pi\, n_1 h/\lambda)\,[1 - \left(\frac{1}{n_1^2}\right) \sin^2\theta]^{\frac{1}{2}} \tag{21}$$

$$\phi = \tan^{-1}\left(\frac{2n_1k_2}{n_1^2 - n_2^2 - k_2^2}\right) \tag{22}$$

(n_1, k_1) and (n_2, k_2) are the optical constants of the epi and the substrate, respectively; θ is the angle of incidence of the infrared beam; ϕ is the phase shift at the epi–substrate interface; and h is the thickness of the epitaxial layer. Figure 26 shows the variation of the optical constants of silicon as a

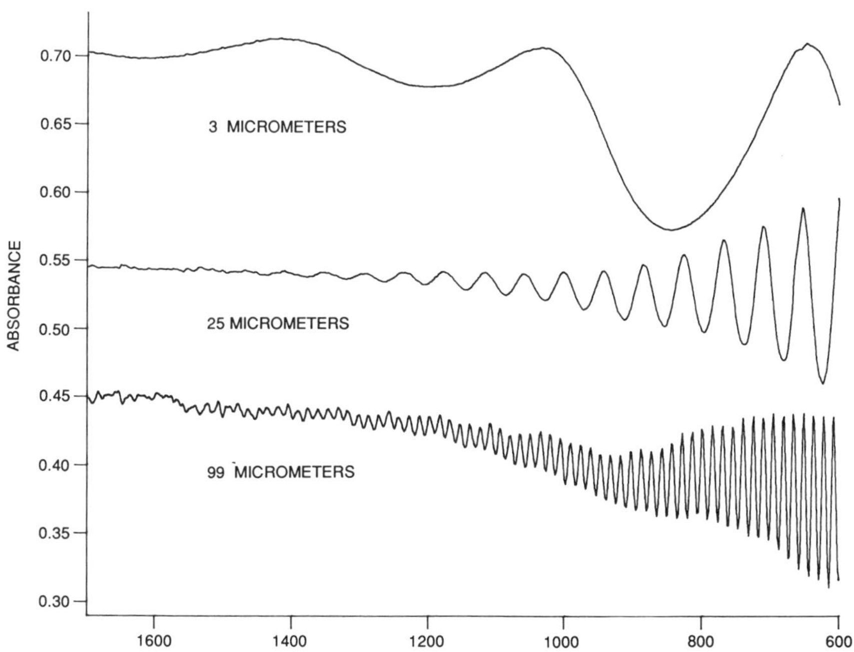

Fig. 25. Infrared reflectance spectra from (n/n$^+$) silicon epitaxies of thicknesses (bottom) 99, (middle) 25, and (top) 3 μm.

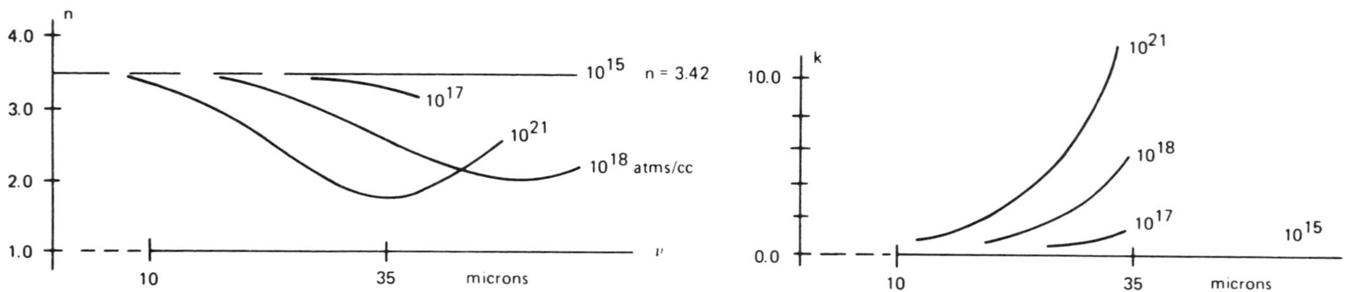

Fig. 26. The optical constants of silicon as a function of wavelength and doping levels.

function of dopant concentrations. From a measurement of the resistivity of the substrate and from known relationships between the silicon resistively and the dopant concentration given by Irvin (1962), Schumann and Schneider (1970) were able to establish correlations between the infrared reflectivity as a function of wavelength, substrate doping level, and the thickness of the epitaxial layer. Using these correlations, the authors were able to measure silicon epitaxial thicknesses on silicon. Severin and Eversteyn (1975) have critiqued the Schumann method and have outlined corrections needed to be applied for proper epi thickness measurements. The Schumann method, of course, requires a knowledge of the substrate dopant concentrations, which may not always be readily available.

The epitaxial thickness can be measured rapidly using FT-IR instruments (Fluornoy *et al.*, 1972; Foteva *et al.*, 1975). Rather than using the reflectance spectrum computed from the recorded interferograms, the technique uses the information in the interferogram itself. Figure 27 represents schematically the epitaxial measurement using the interferometric technique. If there is only one reflecting surface on the sample, the recorded interferogram will exhibit a maximum signal (known as the center burst) at the point of zero retardation corresponding to the position of the moving mirror of the Michelson interferometer, where the infrared beams through the two arms of the interferometer have traveled equal distances. However, when the reflection takes place from an epitaxial sample, two secondary maxima (known as side bursts) will be seen in the interferogram. These side bursts will occur for the equivalent positions of the moving mirror before and after the point of zero retardation. In each of these positions, the moving mirror is located at a point separated from the point of zero retardation by a distance that corresponds to twice the distance the infrared beam travels through the epi before being reflected from the epi–substrate interface. Since the position of the moving mirror can be calculated precisely by counting the He-Ne laser (which is used in most FT-IR

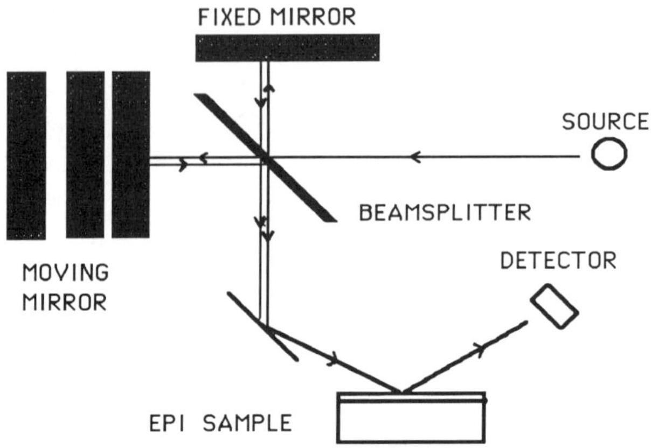

Fig. 27. Schematic representation of the epitaxial thickness measurements made with an FT-IR instrument. Additional constructive interferences are produced by the infrared beams reflected from the fixed and the moving mirrors of the interferometer reflecting from the air–epi and the epi–substrate interfaces.

instruments to aid in sampling the interferogram) fringes, the epitaxial thickness can be calculated easily from the location of the side bursts.

Figure 28 (top) shows the reflected interferogram from a silicon (n on n^+) epitaxial layer 8 μm thick. One can see the center burst and the two side bursts. As the epitaxial layer gets thinner, the side bursts move closer to the center burst and may not be distinguishable in the recorded interferogram. In this case, one needs to collect a reference interferogram, preferably from a similar epi, but of a much different thickness. The reference interferogram can then be subtracted away from the sample interferogram to clearly bring out the side bursts. Figure 28 shows this subtraction process for the epi 8 μm thick. The reference used in this case was an epi of similar structure and had a thickness of 2 μm (Fig. 28, middle). In Fig. 28 (bottom), the center burst has almost completely been canceled away, and the side bursts due to the 2-μm reference as well as the 8-μm sample could easily be seen.

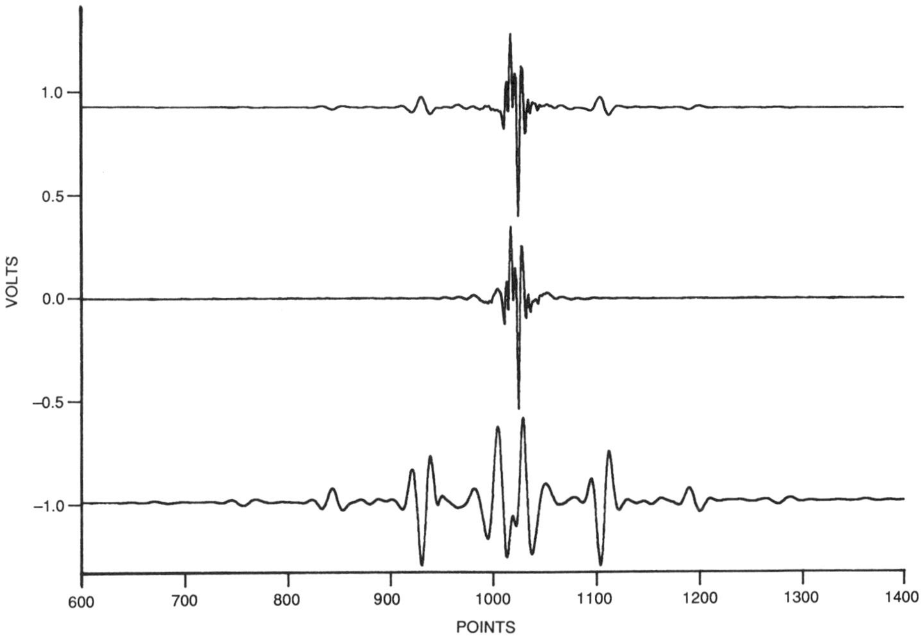

Fig. 28. Reflectance interferograms from (n/n$^+$) silicon epitaxies. (Top) Reflectance interferogram from an epi 8 μm thick. One can clearly see the two side bursts spaced symmetrically around the center burst. (Middle) Reflectance interferogram from an epi 2 μm thick. (Middle) The difference.

The difference interferogram is replotted on an expanded scale in Fig. 29. In this figure, one can also see side bursts of progressively decreasing amplitudes corresponding to thicknesses of 16, 24, and 32 μm. These are due to the infrared beam undergoing multiple reflections within the epitaxial layer.

When the epi thickness is very small, less than 1 μm, the interferogram subtraction method outlined in the preceding paragraph may not work satisfactorily when there is no perfect match between the sample and the reference epitaxies. When the epitaxial layer is thick, the nature of the reflected radiation is dominated by the characteristics of the epi layer

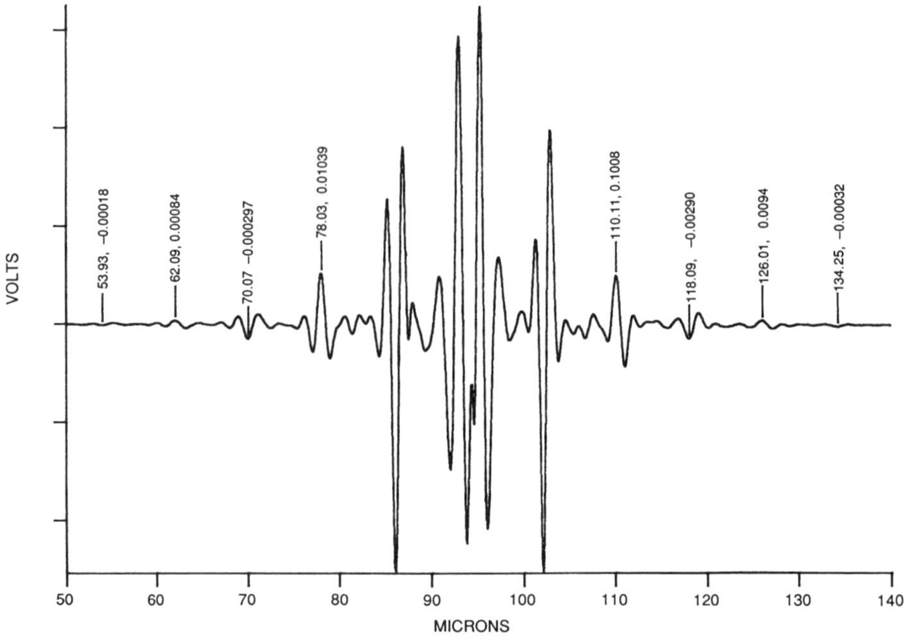

Fig. 29. Scale expansion of the difference interferogram of Fig. 28. The multiple reflections within, and the phase changes introduced by reflection can be clearly seen.

and the difference in the doping levels between the epi and the substrate. However, when the epi layer is very thin, the reflection is mostly influenced by the substrate properties. Since the interferogram center burst is influenced by these factors, complete cancellation of the center burst may not be possible, and the side bursts may be obscured by the residual modulations of the center burst. In such cases, one can Fourier transform the interferogram from the sample and the reference epitaxies and apply digital filtering techniques to the transformed spectrum to enhance the features in the spectrum due to reflections within the epi layer, and reverse Fourier transform the processed spectra. The subsequent sample and the reference interferograms can then be subtracted to highlight the side burst due to the epi thickness. This process is usually referred to as the "thin epi mode" in the FT-IR industry. Figure 30 shows the thin epi mode difference interferogram for the 8-μm epitaxy, using the 2-μm epi as the reference. In this case, one can see that the result is very similar to that shown in Fig. 29.

One can see in Figs. 29 and 30 that the shape or the phase of the side burst changes with number of reflections; for example, the 16-μm sideburst (arising from two reflections within the epi) is 180° out of phase from the main 8-μm side burst (arising from a single reflection within the epi). The phase of the side burst can also change with the thickness and the substrate doping level. Figure 31 shows the side bursts from an epi layer 6 μm thick built on silicon substrates doped to different extents with arsenic. Since the location of the side burst could be determined as the point of maximum, minimum, or

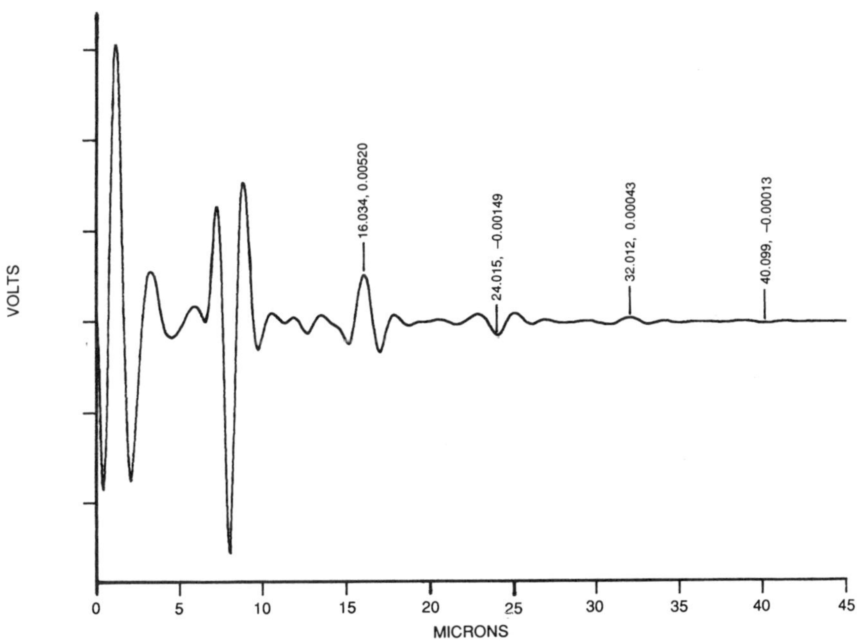

Fig. 30. The difference interferogram of the epi 8 μm thick obtained by using the "thin mode" algorithm.

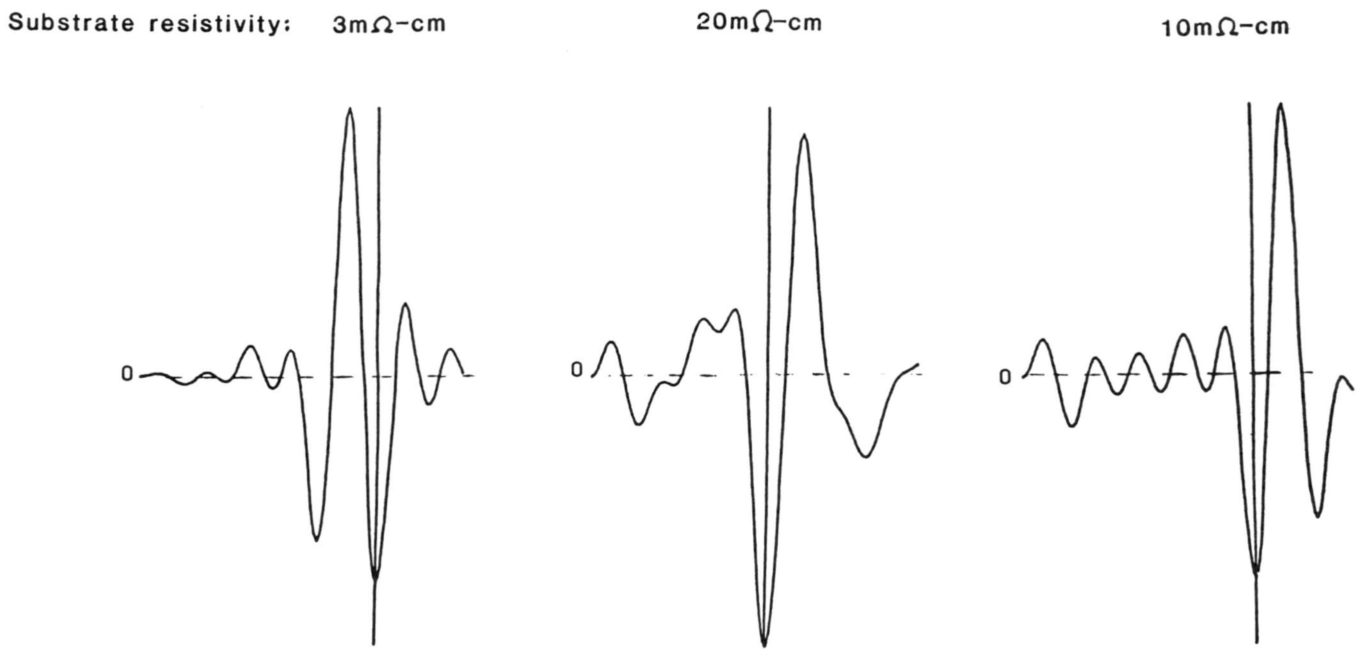

Fig. 31. Side burst shapes from a epi 6 μm thick for different levels of arsenic doping of the substrate. Note the change in measurement point within the side burst profile as the substrate doping level changes.

the various zero-crossings, precise measurements of the epi thicknesses depend upon the correct measurement location within the center burst. The choice of the measurement point is usually empirically determined using prior knowledge of the phases of the side bursts from epitaxies of known characteristics such as substrate doping level and the epi thickness determined by other techniques. Thus, the FT-IR reflectance can be used as a rapid and routine method for the measurement of epitaxial thicknesses in the semiconductor around the world. Most FT-IR instruments used for this purpose can perform the thickness measurement from a single-scan, low-resolution interferogram in a total measurement time, including the necessary computations, of 5 to 10 sec.

While most of the preceding discussion have been about silicon on silicon epitaxies, the FT-IR technique can also be used effectively for the determination of the epitaxial thicknesses on other systems such as Group III-V superlattice structures. Figure 32 shows the difference interferogram from an InGaAs/InP/InP structure. The thickness of the top InGaAs layer was 4.0 μm and that of the lower InP layer 2.5 μm. In Fig. 32 one can see side bursts corresponding to 4.0 and 6.5 μm corresponding to reflections from the InGaAs − InP, and the InGaAs + InP − InP interfaces.

VIII. QUANTITATIVE MEASUREMENTS ON PASSIVATION LAYERS

As we mentioned earlier, the silicon manufacturing technology uses passivation layers during the device fabrication process. Until recently, phosphosilicate glass films on silicon used to be the only passivation layer encountered in the industry. But VLSI and ULSI with their lower processing temperatures have necessitated the use of passivation layers (of thicknesses ranging from 50 to 150 nm) with lower reflow temperatures. Thus, the use of borophosphosilicate (BPSG)

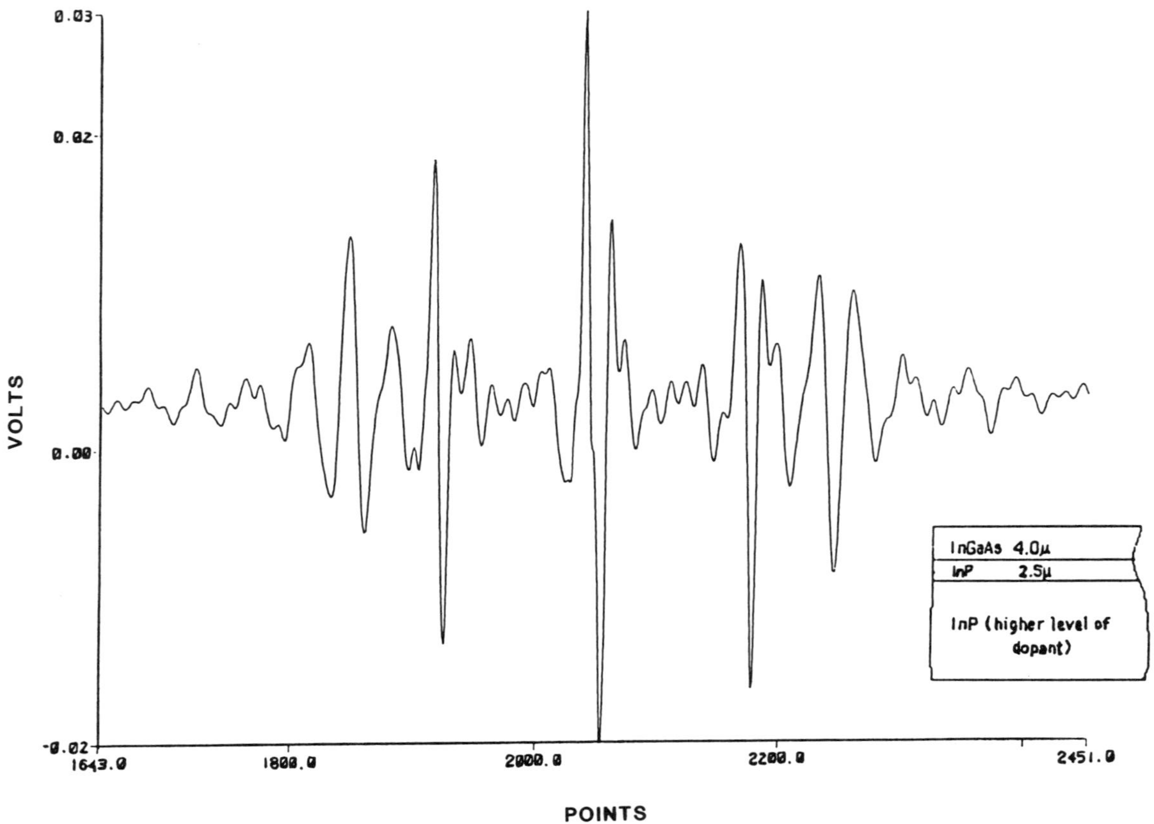

Fig. 32. Subtracted interferogram from a superlattice structure: InGaAs/InP/InP. The side bursts correspond to thicknesses of 4.0 and 6.5 μm.

and silicon nitride passivation layers are currently used. The infrared spectrometric methods can readily and rapidly be used for the characterization and quantitative measurements of these films. The mechanical and electrical characteristics of the silicon passivation layers are a function of the atomic hydrogen concentration present in these films.

A. Silicon Nitride Films on Silicon

The characterization of plasma-deposited silicon nitride films has been discussed by Kember and Liddell (1985). The hydrogen atoms in silicon nitride could be bound to the silicon or the nitrogen sites and give rise to infrared absorptions around 2160 cm^{-1} (due to Si—H stretching vibration) and around 3350 cm^{-1} (due to the N—H stretching vibration). In addition, the infrared spectrum exhibits a very strong band at 875 cm^{-1} due to the Si—N stretching vibration. The amount of Si—H bonds and the N—H bonds in the film can be measured quantitatively from the infrared absorption coefficients using the following conversion factors given by Lanford and Rand (1978).

$$\text{Number of Si—H bonds-cm}^{-2} = \text{area of the 2160-cm}^{-1} \text{ absorbance band} \times 1.36 \times 10^{17}$$

$$\text{Number of N—H bonds-cm}^{-2} = \text{area of the 3350-cm}^{-1} \text{ absorbance band} \times 1.90 \times 10^{17}$$

Using an FT-IR spectrometer, Knolle (1984) has correlated the infrared absorbances with the film thickness and refractive index. According to him,

$$\text{Silicon nitride film thickness} = \text{area of 875-cm}^{-1} \text{ absorbance band} / (243 - 8.2 \times K)$$

$$\text{Refractive index of the film} = 1.95 + 0.0418 \times K$$

In the above relations, K is the ratio of the absorbances of the Si—H to the N—H bands.

B. Doped Silicon Dioxide Films on Silicon

The characteristics of BPSG films for use in IC manufacturing have been discussed by Kern and Smeltzer (1985) and by Becker *et al.* (1986). FT-IR spectroscopy can be used for the routine measurement of the concentrations of phosphorus and boron in PSG and BPSG films (Krishnan, 1984). In these cases, the best quantitative results are obtained by using statistical multivariate techniques such as the partial least squares or the principal component regression analysis methods. These methods are reviewed elsewhere in this book (see the chapter by Haaland) and will not be discussed here.

IX. CONCLUSION

We have presented a number of applications of the FT-IR spectrometric technique for the characterization of silicon during the various stages of semiconductor device manufacturing. FT-IR has a number of other applications in the semiconductor industry, for example, the determination of the concentrations of impurities such as carbon, boron, and silicon in gallium arsenide; the characterization of defect centers in Group II-VI materials such as ZnSe and MCT; and the determination of dopant concentrations on highly doped semiconductors by plasma resonance measurements. Some of these measurements are routine and can easily be carried out at room temperature; others may require cryogenic temperatures and the use of special techniques such as photoionization spectroscopy. Unfortunately, these measurements cannot be described here. The present authors are convinced, however, that these and other future applications will make FT-IR spectrometry one of the most powerful techniques for semiconductor material characterization.

REFERENCES

Abdullah, A. K. W., Maslin, K. A., and Parker, T. J. (1984). *Infrared Phys.* **24,** 185.

Abe, T., Kikuchi, K., Shirai, S., and Muraoka, S. (1981). *In* "Semiconductor Silicon, 1981" (H. R. Huff, R. J. Kriegler, and Y. Takeishi, eds.), p. 189. Electrochem. Soc., Pennington, New Jersey.

American Society for Testing Materials (1983a). *Annu. Book ASTM Stand.* F121-83.

American Society for Testing Materials (1983b). *Annu. Book ASTM Stand.* F123-83.

American Society for Testing Materials (1983c). *Annu. Book ASTM Stand.* F110-72.

American Society for Testing Materials (1983d). *Annu. Book ASTM Stand.* F108-72.

American Society for Testing Materials (1983e). *Annu. Book ASTM Stand.* F95-76.

Baber, S. C. (1980). *Thin Solid Films* **12,** 201.

Baghdadi, A., Bullis, W. M., Croakin, M. C., Yue-zhen, Li, Scace, R. I., Series, R. W., Stallhofer, P., and Watanabe, M. (1988). *J. Electrochem. Soc.,* submitted for publication.

Baker, J. A. (1970). *Solid State Electron.* **13,** 1431.

Balkanski, M., and Nusimovica, M. (1964). *Phys. Status Solidi* **5,** 635.

Bardeen, J., and Brattain, W. (1949). *Phys. Rev.* **75,** 1208.

Barraclough, K. G., Series, R. W., Hilsop, J. S., and Wood, D. A. (1986). *J. Electrochem. Soc.* **133,** 187.

Bean, A. R., and Newman, R. C. (1972). *J. Phys. Chem. Solids* **33,** 255.

Bean, K. E., Runyan, W. R., and Massey, R. G. (1985). *Semicond. Int.* **May,** 136.

Becker, F. S., Pawlik, D., Schafer, H., and Staudigl, G. (1986). *J. Vac. Sci. Technol. B* **4,** 732.

Bosomworth, D. R., Hayes, W., Spray, A. R. L., and Watkins, G. D. (1970). *Proc. R. Soc. London, Ser. A* **317,** 133.

Brodsky, M. H., Cardona, M., and Cuomo, J. J. (1977). *Phys. Rev. B* **16,** 3556.

Bullis, W. M., and Coates, L. B. (1987). *Solid State Technol.* **March,** 69.

Bullis, W. M., and O'Mara, W. C. (1983). *Proc. Symp. Defects Silicon* **83-9,** 275.

Chen, C. S., and Schroder, D. K. (1987). *Appl. Phys. A* **42,** 257.

Chu, P. K., Hockett, R. S., and Wilson, R. G. (1986). *Proc. Mater. Res. Soc. Symp.* **59,** 67.

Colbow, K. (1963). *Can. J. Phys.* **41,** 1801

DeLeo, G. G., Fowler, W. B., and Watkins, G. D. (1984). *Phys. Rev. B* **29,** 3193.

Edwards, D. F., and Ochoa, E. (1980). *Appl. Opt.* **19,** 4130.

Endo, Y., Yatsurugi, Y., Akiyama, N., and Nazaki, T. (1972). *Anal. Chem.* **44,** 2258.

Engelbrecht, J. A. A. (1987). *Infrared Phys.* **27,** 203.

Engelbrecht, J. A. A., and Lonbard, O. J. (1986). *Infrared Phys.* **26,** 75.

Fang, C. J., Gruntz, K. J., Ley, L., Cardona, M., Demond, F. J., Muller, G., and Kalbitzer, S. (1979). *J. Non-Cryst. Solids* **35/36,** 255.

Flournoy, P. A., McClure, R. W., and Wyntjes, G. (1972). *Appl. Opt.* **11,** 1907.

Foteva, I. I., Gorbunov, G. G., and Vol'kenshtein, I. A. (1975). *Sov. J. Opt. Technol.* **42,** 46.

Freeland, P. E. (1980). *J. Electrochem. Soc.* **127,** 754.

Freeman, E. C., and Paul, W. (1978). *Phys. Rev. B* **18,** 4288.

Gaworzewski, P., Hill, E., Kirochi, F. G., and Vecserny'es, L. (1984). *Phys. Status Solidi* **85,** 133.

Graff, K. (1983). *J. Electrochem. Soc.* **130,** 1378.

Graff, K., Grallath, E., Ades, S., Goldbach, G., and Tolg, G. (1973). *Solid State Electron.* **16,** 887.

Gross, C., Gaetano, G., Tucker, T. N., and Baker, J. A. (1972). *J. Electrochem. Soc.* **119,** 926.

He, H.-N., Li, Y.-Z., Xhao, G.-D., Yan, R.-H., Lu, Q.-R., and Qi, M.-W. (1983). *Talanta* **30,** 7661.

Hrostowski, H. J., and Alder, B. J. (1960). *J. Chem. Phys.* **33,** 980.

Hrostowski, H. J., and Kaiser, R. H. (1957). *Phys. Rev.* **107,** 966.

Hrostowski, H. J., and Kaiser, R. H. (1958). *Phys. Rev. Lett.* **1,** 199.

Hu, S. M. (1977). *Appl. Phys. Lett.* **31,** 139.

Hu, S. M. (1980). *J. Appl. Phys.* **51,** 5945.

Hu, S. M. (1986). *Proc. Mater. Res. Soc. Symp.* **59,** 249.

Humlicek, J., and Vojtechovski, K. (1985). *Phys. Status Solidi A* **92A,** 249.

Hyland, S. L., Ast, D. G., and Baghdadi, A. (1987). *J. Cryst. Growth* **82,** 191.

Iglytsin, M. I., Kekelidze, G. P., and Lazareva, G. V. (1965). *Sov. Phys. Solid State* **6,** 2508.

Iizuka, T., Takasu, S., Tajima, M., Arai, T., Nozaki, T., Inoue, N., and Watanabe, M. (1983). *Proc. Symp. Defects Silicon* **83-9,** 265.

Inoue, N., Arai, T., Nozaki, T., Endo, K., and Mizuma, K. (1986). *ASTM Spec. Tech. Publ.* **960.**

Irvin, J. C. (1962). *Bell Syst. Tech. J.* **41,** 387.

Itoh, Y., Nozaki, T., Masui, T., and Abe, T. (1985). *Appl. Phys. Lett.* **47,** 488.

Jaae. (1986). *Infrared Phys.* **26,** 243.

Jastrzbski, L., Zanzucchi, P., Thebault, D., and Lagowski, J. (1982). *J. Electrochem. Soc.* **129,** 1638.

Jeffrey, F. R., Shanks, H. R., and Danielson, G. C. (1979). *J. Appl. Phys.* **50,** 7034.
Johnson, F. A. (1959). *Proc. Phys. Soc.* **73,** 265.
Kagel, R. O., and Baker, J. A. (1974). *Appl. Spectrosc.* **28,** 65.
Kaiser, W., and Keck, P. H. (1957). *J. Appl. Phys.* **28,** 882.
Kaiser, W., Keck, P. H., and Lange, C. F. (1956). *Phys. Rev.* **101,** 1264.
Kauppinen, J. K., Moffat, D. J., Mantsch, H. H., and Cameron, D. G. (1981). *Appl. Spectrosc.* **35,** 271.
Kember, P. N., and Liddell, S. C. (1985). *Semicond. Int.* **August,** 158.
Kern, W., and Smeltzer, R. K. (1985). *Solid State Technol.* **June,** 171.
Kim, C. K. (1969). *Radiochem. Radioanal. Lett.* **2,** 53.
Knolle, W. R. (1984). *Proc. Adv. Plasma Technol. Conf. Semin., 5th* p. 1.
Kolbesen, B. O. (1975). *Appl. Phys. Lett.* **27,** 353.
Krishnan, K. (1981). "Digilab FTS/IR Note 39." BioRad Digilab Div., Cambridge, Massachusetts.
Krishnan, K. (1983). *Proc. Symp. Defects Silicon* **83-9,** 285.
Krishnan, K. (1984). *ASTM Spec. Tech. Publ.* **850,** 358.
Krishnan, K., and Ferraro, J. R. (1982). *In* "Fourier Transform Infrared Spectroscopy" (J. R. Ferraro and L. J. Basile, eds.) Vol. 3, p. 188. Academic Press, New York.
Krishnan, K., and Hill, S. L. (1981). *Proc. Soc. Photo-Opt. Instrum. Eng.* **289,** 27.
Lanford, W. A., and Rand, M. J. (1978). *J. Appl. Phys.* **49,** 2473.
Leroueille, J. (1982). *Appl. Spectrosc.* **36,** 153.
Liaw, H. M. (1979). *Semicond. Int.* **October,** 71.
Loewenstein, E. V., Smith, D. R., and Morgan, R. L. (1973). *Appl. Opt.* **12,** 398.
Mead, D. G. (1980). *Appl. Spectrosc.* **34,** 171.
Mead, D. G., and Lowry, S. R. (1980). *Appl. Spectrosc.* **34,** 167.
Newman, R. C. (1969). *Adv. Phys.* **18,** 545.
Newman, R. C. (1974). "Infrared Studies of Crystal Defects." Tailor & Francis, London.
Newman, R. C. (1982). *Rep. Prog. Phys.* **45,** 1163.
Newman, R. C., and Smith, R. S. (1969). *J. Phys. Chem. Solids* **30,** 1493.
Newman, R. C., and Willis, J. B. (1965). *J. Phys. Chem. Solids* **26,** 373.
Oates, A. S., and Lin, W. (1988). *J. Cryst. Growth* **89,** 117.
Oehrlein, G. S., Lindstrom, J. L., and Corbett, J. W. (1982). *Appl. Phys. Lett.* **40,** 241.
Ohnishi, Y., and Tsuya, H. (1981). *Appl. Phys. Lett.* **38,** 867.
O'Mara, W. C. (1983). *Proc. Symp. Defects Silicon* **83-9,** 120.
O'Mara, W. C. (1985). *Proc. Soc. Photo-Opt. Instrum. Eng.* **524,** 61.
Pajot, B. (1964). *J. Phys. Chem. Solids* **25,** 613.
Pajot, B. (1967). *J. Phys. Chem. Solids* **28,** 73.
Pajot, B. (1977). *Analysis,* **5,** 293.
Pajot, B., and Bardeleben, J. (1984). *In* "Defects in Semiconductors,"p. 685.

Pajot, B., and Debarre, D. (1981). *In* "Neutron-Transmuted-Doped Silicon" (J. Guldberg, ed.), p. 423. Plenum, New York.
Pajot, B., and Deltour, J. P. (1967). *Infrared Phys.* **7,** 195.
Pajot, B., Kauppinen, J., and Anttila, R. (1979). *Solid State Commun.* **31,** 759.
Pajot, B., Compain, H., Lerouille, J., and Clerjaud, B. (1983). *Physica. B* **117/118,** 110.
Pajot, B., Stein, H. J., Cales, B., and Noud, C. (1985). *J. Electrochem. Soc.* **132,** 3034.
Passchier, W. F., Honlik, D. D., Mandel, M., and Afsar, M. N. (1977). *J. Phys. D: Appl. Phys.* **10,** 509.
Patel, J. R. (1981). *In* "Semiconductor Silicon, 1981" (H. R. Huff, R. J. Kriegler, and Y. Takeishi, eds.), p. 189. Electrochem. Soc., Pennington, New Jersey.
Pearton, S. J., Corbett, J. W., and Shi, T. S. (1987). *Appl. Phys. A* **43,** 153.
Pidgeon, C. R. (1980). *Handb. Semicond.* **2,** 233.
Pradhan, M. M., Garg, R. K., and Arora, M. (1987A). *Infrared Phys.* **27,** 25.
Pradhan, M. M., Garg, R. K., and Arora, M. (1987B). *Infrared Phys.* **27,** 207.
Ramdas, A. K., and Rodriguez, S. (1981). *Rep. Prog. Phys.* **44,** 1297.
Rath, H. J., Stallhofer, P., Huber, D., and Schmitt, B. F. (1984). *J. Electrochem. Soc.* **131,** 1920.
Regolini, J. L., Stoquert, J. P., Ganter, C., and Siffert, P. (1986). *J. Electrochem. Soc.* **133,** 2165.
Rook, H. L., and Schweikert, E. A. (1969). *Anal. Chem.* **41,** 958.
Ross, R. C., Tsong, I. S. T., and Messier, R. (1982). *J. Vac. Sci. Technol.* **20,** 406.
Saltzberg, C. D., and Villa, J. J. (1957). *J. Opt. Soc. Am.* **47,** 244.
Schumann, P. A., Jr., and Schneider, C. P. (1970). *J. Appl. Phys.* **41,** 3532.
Schumann, P. A., Jr., Phillips, R. P., and Olshefski, P. J. (1966). *J. Electrochem. Soc.* **113,** 368.
Severin, P. J., and Bulle, H. (1975a). *J. Electrochem. Soc.* **122,** 133.
Severin, P. J., and Bulle, H. (1975b). *J. Electrochem. Soc.* **122,** 137.
Severin, P. J., and Eversteyn, F. C. (1975). *J. Electrochem. Soc.* **122,** 962.
Shokley, W. (1949). *Bell Syst. Tech. J.* **28,** 435.
Shokley, W. (1976). *IEEE Trans. Electron. Devices* **23,** 597.
Spenke, E., *et al.* (1973). "Research Report on Project 74." Ministry of Research and Technology, FRG.
Stallhofer, P., and Huber, D. (1983). *Solid State Technol.* **August,** 233.
Stavola, M. (1984). *Appl. Phys. Lett.* **44,** 514.
Stein, H. J. (1985). *Appl. Phys. Lett.* **47,** 1339.
Stein, H. J. (1986). *Proc. Mater. Res. Soc. Symp.* **59,** 523.
Stein, H. J. (1987), *J. Electrochem. Soc.* **134,** 2592.

Suezawa, M., Sumino, K., Harada, H., and Abe, T. (1988). *Jpn. J. Appl. Phys.* **27,** 62.

Sumino, K, Yonenaga, I., Imai, M., and Abe, T. (1984). *J. Appl. Phys.* **54,** 5016.

Tardella, A., and Pajot, B. (1982). *J. Phys. (Orsay, Fr.)* **43,** 1789.

Tempelhoff, K., Spiegelberg, F., and Gleichmann, R. (1977). *In* "Semiconductor Silicon, 1977," p. 585. Electrochem. Soc. Pennington, New Jersey.

Tempelhoff, K., Spielberg, E., Gleichmann, R, and Wruck, D. (1979). *Phys. Status Solidi* **56,** 213.

Tempelhoff, K., Hahn, B., and Gleichmann, R. (1981). *In* "Semiconductor Silicon, 1981" (H. R. Huff, R. J. Kriegler, and Y. Takeishi, eds.), p. 244. Electrochem. Soc., Pennington, New Jersey.

Thurber, W. R. (1970). *National Bureau of Standards. Tech. Note,* **529.**

Vaikilova, G., Vitman, R. F., Lebedev, A. A., and Mukhammedov, S. (1982). *Sov. Phys. Semicond.* **16,** 1426.

Vidrine, D. W. (1980). *Anal. Chem.* **52,** 92.

Weeks, S. P. (1984). *ASTM Spec. Tech. Publ.* **850,** 335.

White, J. J. (1967). *Can. J. Phys.* **45,** 2797.

Wruck, D., and Gaworzewski, P. (1979). *Phys. Status Solidi A* **56A,** 557.

Yatsurugi, Y., Akiyama, N., Endo, Y., and Nazaki, T. (1973). *J. Electrochem. Soc.* **120,** 975.

7

Industrial Applications of FT-IR

H. Ishida
A. Ishitani
Material Science Laboratories
Toray Research Center, Inc.
Otsu, Shiga 520, Japan

Practical Fourier Transform Infrared Spectroscopy

I. INTRODUCTION

The position of FT-IR as an useful technique for characterization of industrial materials has been firmly established within the last decade. FT-IR has brought additional merits such as high sensitivity, high precision, quickness of measurement, and extensive data processing capability besides the intrinsic advantages of infrared spectroscopy such as wide applicability, nondestructiveness, measurement under ambient atmosphere, capability of providing detailed structural information, and a huge data base. The introduction of FT-IR has changed infrared spectroscopy from a technique of identification of bulk materials to a comprehensive system of material characterization and has opened up a new field especially in surface studies and microanalyses of industrial materials.

We started analytical use of FT-IR in 1977 and have been developing various kinds of application techniques. Five instruments are being used in our laboratory at present, mainly for characterization of industrial materials (Ishitani, 1985).

Most of the applications described in this chapter cover the analytical studies done recently in our laboratory. Throughout the chapter, we will attempt to point out the usefulness and the position of FT-IR as an analytical technique in bulk and surface studies and microanalyses of industrial materials.

II. MICROANALYSIS BY FT-IR MICROSCOPY

Introduction of a microscope as an accessory of FT-IR has opened up a new field in microanalysis of industrial materials (Messerschmidt and Harthcock, 1988). FT-IR microspectroscopy provides qualitative information about the chemical structure of small specimens or of a small area of samples down to 10–15 μm; this information was impossible to obtain with earlier techniques. Figure 1 illustrates the position of FT-IR microscopy in microanalytical methods from the

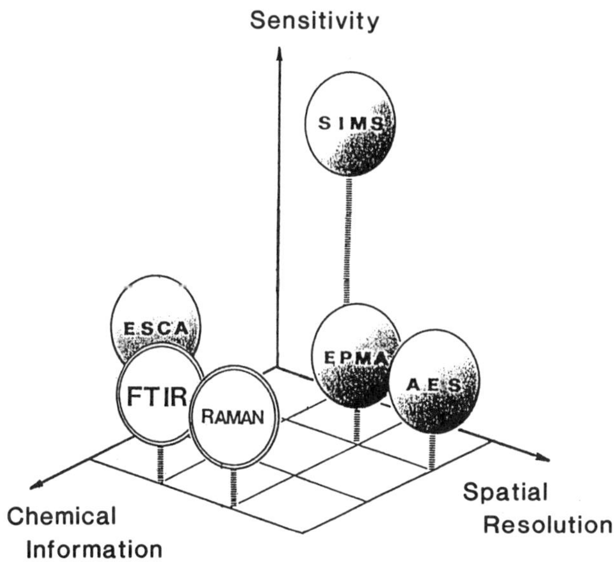

Fig. 1. Tentative illustration of the position of FT-IR microscope among other microanalytical techniques.

viewpoints of sensitivity, spatial resolution, and chemical information in comparison to other techniques. Among the micro-analytical methods in the figure, such as Raman microprobe, FT-IR microscopy, micro-ESCA (electron spectroscopy for chemical analysis), AES (Auger electron spectroscopy), EPMA (electron probe microanalysis), and SIMS (secondary ion mass spectrometry), methods based on vibrational spectroscopies possess outstanding advantages for chemical analysis. Raman microprobe, although having higher spatial resolution than FT-IR microscopy, suffers often from problems such as fluorescence and thermal decomposition by a condensed laser beam in real samples encountered in industrial analysis laboratories. FT-IR microscopy has much wider applicability than Raman microprobe. They are also mutually complementary in the intensity selection rule. We can therefore choose the technique most appropriate for the compound being studied.

Table I shows typical examples of industrial applications of FT-IR microscopy. Identification of small inclusions and contaminants is frequently required in quality control and trouble shooting. These applications are described in the chapter by Krishnan and Hill in this volume. Small area analy-

Table I Industrial Applications of FT-IR Microscopy

Objects	Applications (example)
Small inclusion	Plastics, fibers, films, ceramics
Impurity	Carbon and oxygen in Si, carbon in GaAs
Contaminant (deposit)	Magnetic tapes, hard disk, electrical contacts, LSI
Microstructure	orientation and crystallinity, crystal form
Composite material	Laminated films, polymer alloy, interface of composites
Biological material	Gallstone, calculus, tissue section

sis is also needed to clarify compositions of composite materials and biological materials. A cross-sectioning technique by a microtome is indispensable for these applications. One of important advantages of FT-IR microscopy is convenience of measurement under ambient atmosphere. In practical analyses, the most fruitful combination is the one with EPMA, which gives the elemental information in small particles and areas.

In the present section, we mainly focus our interest on the spatial resolution of the FT-IR microscope, which is important for the practical analysis of small areas and small samples. Applications are limited to microanalysis of two kinds of polymeric systems, multilayer films and treeing of polyethylene cable.

A. Experimental Evaluation of Practical Spatial Resolution of an FT-IR Microscope

A FT-IR microscope uses a remote aperture inserted at the focal plane of an objective (Cassegrainian) lens. The aperture is used to mask the area surrounding the image of a sample to be analyzed in order to improve the spatial resolution of measurement. The following experiments were done with an UMA-300 microscope attached to a Digilab FTS-60 spectrometer. Figure 2 shows the single beam intensities monitored at 1000 and 2000 cm^{-1} as a function of the aperture size. A good linearity with slope of 2 holds down to 20 μm. This result indicates that the single beam intensity passing through the aperture is proportional to the area (L^2) limited by the aperture. Below 20 μm, the intensity largely deviates from linearity downward as a result of the effect of diffraction of an infrared beam that has a wavelength comparable with the aperture size. It has a significant effect upon the spatial resolution of an FT-IR microscope used with small apertures.

We prepared special samples for evaluation of practical

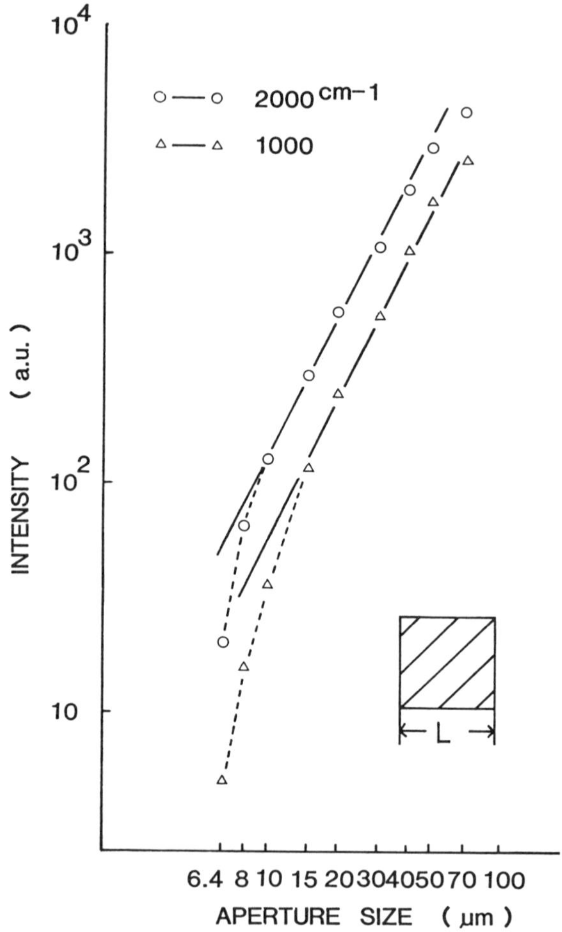

Fig. 2. Dependence of single beam intensities at 1000 and 2000 cm^{-1} on aperture size. a.u., arbitrary units.

spatial resolution as shown in Fig. 3 (Ishitani *et al.*, 1988a). Polyethylene terephthalate (PET) filaments with three different diameters were embedded in an epoxy resin matrix and sliced to thin sections of about 1 μm thickness. An infrared absorption spectrum of each section was measured as a function of aperture size as shown in Fig. 4 for the PET filament of 10 μm diameter. The relative absorption intensity of PET

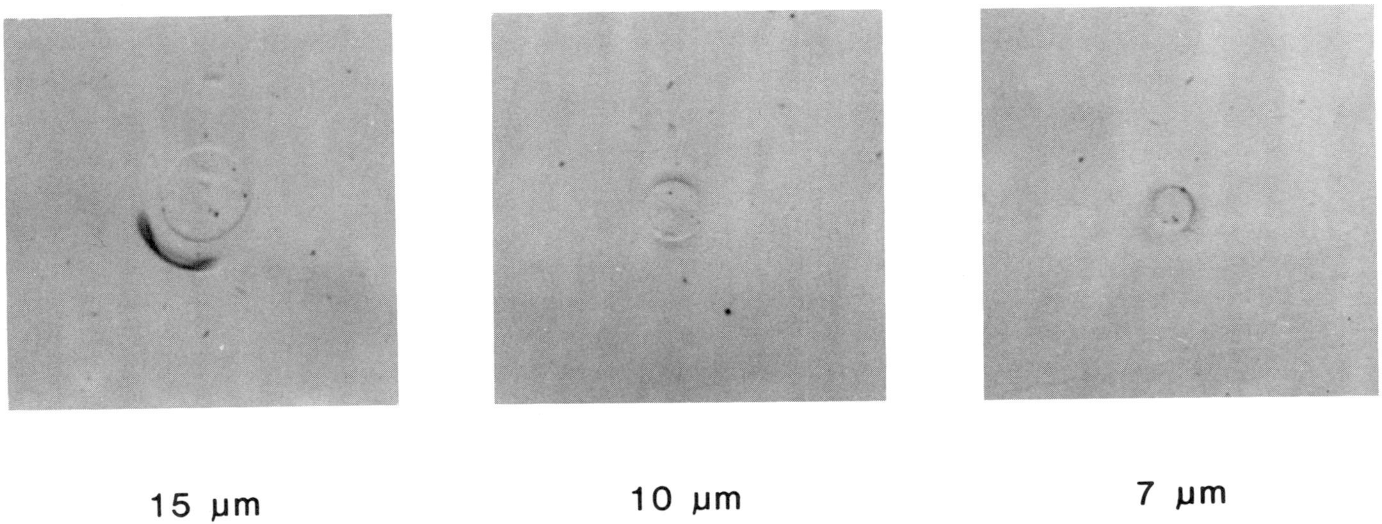

Fig. 3. Special samples used for evaluation of the spatial resolution of the FT-IR microscope. Polyethylene terephthalate (PET) filaments of various diameters embedded in an epoxy resin matrix.

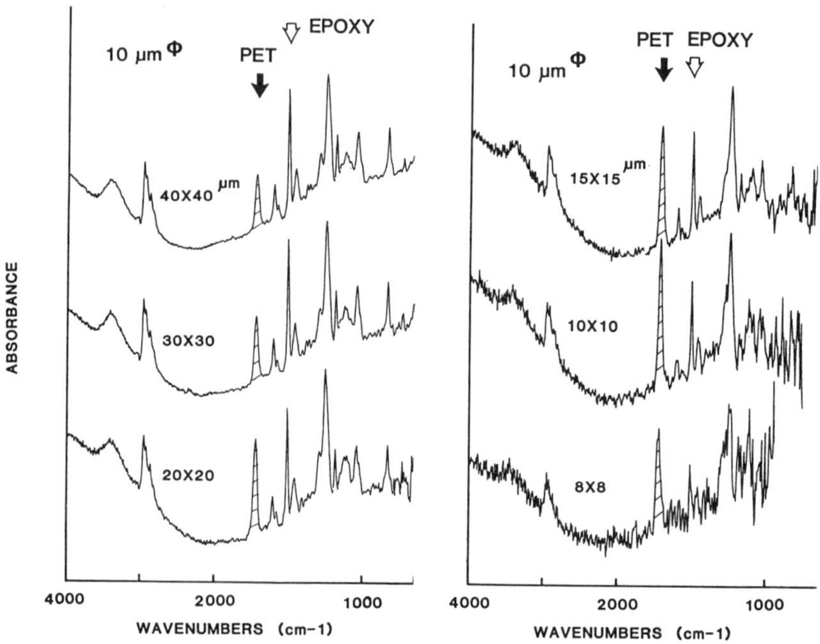

Fig. 4. Infrared absorption spectra of PET (10 μm)/epoxy cross section measured with different aperture sizes.

signal increases with a decrease of the aperture size. The background with a large slope is due to the interference fringe of the thin cross section. It is worth noting that the characteristic absorption bands of epoxy are still observed even in the spectrum obtained with a 8-μm^2 aperture which is smaller than the diameter (10 μm) of the PET filament.

The quantitative intensity ratios between characteristic PET (1720 cm^{-1}) and epoxy (1510 cm^{-1}) absorption bands are plotted as a function of the aperture size (Fig. 5). A theoretical curve is calculated from the employed aperture size and the area of each component within the square aperture. A major cause for the large difference between the theoretical and experimental curves below 20 μm is due to stray light arriving at the detector from the masked area. The stray light in the present optics was experimentally evaluated by using gold

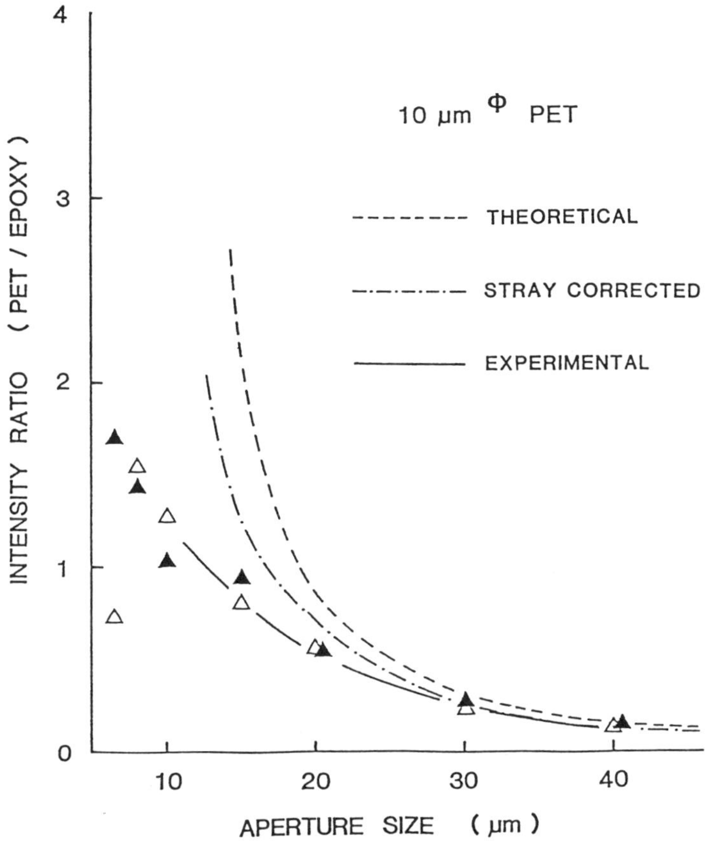

Fig. 5. Aperture size dependence of relative absorption intensity between PET and epoxy bands for a PET filament 10 μm in diameter.

wires whose diameters ranged from 10 to 50 μm. A corrected curve is calculated by taking the stray effect into consideration. Although small improvement is obtained down to 15 μm, a large deviation under 10 μm cannot be explained by the correction.

Figure 6 summarizes the experimental data for three kinds of PET filaments with diameters of 7, 10, and 15 μm. Below the aperture size of 10 μm, the relative intensity shows a large fluctuation. In order to explain the results obtained by use of extremely small-sized apertures, it is necessary to

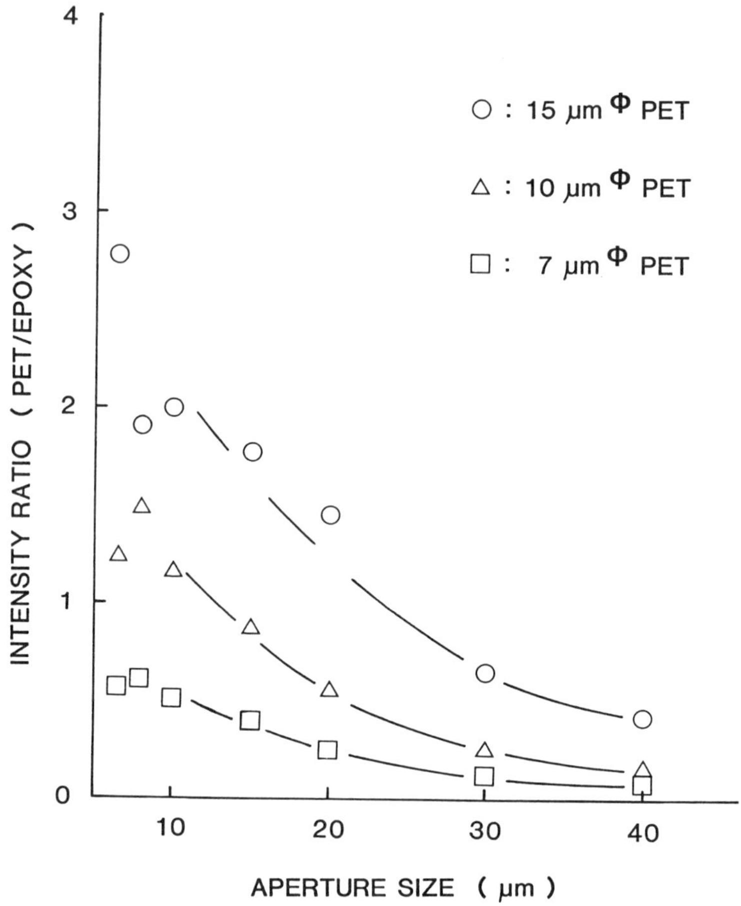

Fig. 6. Aperture size dependence of relative absorption intensity (PET/epoxy) for three sizes of PET filaments.

consider many other optical phenomena associated with a microscope and a sample, for example, reflection and refraction effects by the sample, optical matching between visible light and infrared beam optics of the microscope besides the inherent diffraction effect.

B. Applications to Polymeric Materials

1. Compositions of Multilayered, Laminated Films

Many applications have been reported for compositional analysis of multilayered polymer films (Harthcock *et al.*, 1986; Reffner *et al.*, 1987). In this section, we describe problems in compositional analysis of thin multilayered films on the basis of the spatial resolution discussed in the preceding section. Figure 7 shows infrared absorption spectra of a cross section of a three-layered film with 30 × 30 μm aperture size. The composition of each layer is A, polyamide; B, polyethylene (main component) and polyamide; C, polyethylene, $CaCO_3$, and polyamide. In the spectrum of the middle layer (B), weak absorption bands due to polyamide are observed in the window region of polyethylene. Spectral contributions from layer A due to the stray light can be eliminated by measurement using smaller aperture. As shown in Fig. 8, the relative absorption intensity between polyamide and polyethylene is found to be independent of the aperture size. The result indicates that polyamide is really a minor component of layer B, rather than coming from the neighboring layer.

Figure 9 shows the infrared absorption spectra of another three layered film having a thinner middle layer (B). Layer B originally consisted of polyethylene and a small amount of carboxylic acid, a film given the commercial name Sarlin. In this case, the spectral contributions of polyamide from the neighboring layer (A) should be taken into consideration.

2. Study of Trees Formed in Cable Insulation Polyethylene

The treeing phenomenon of cross-linked polyethylene (XLPE) used as an electrical insulation is attracting much

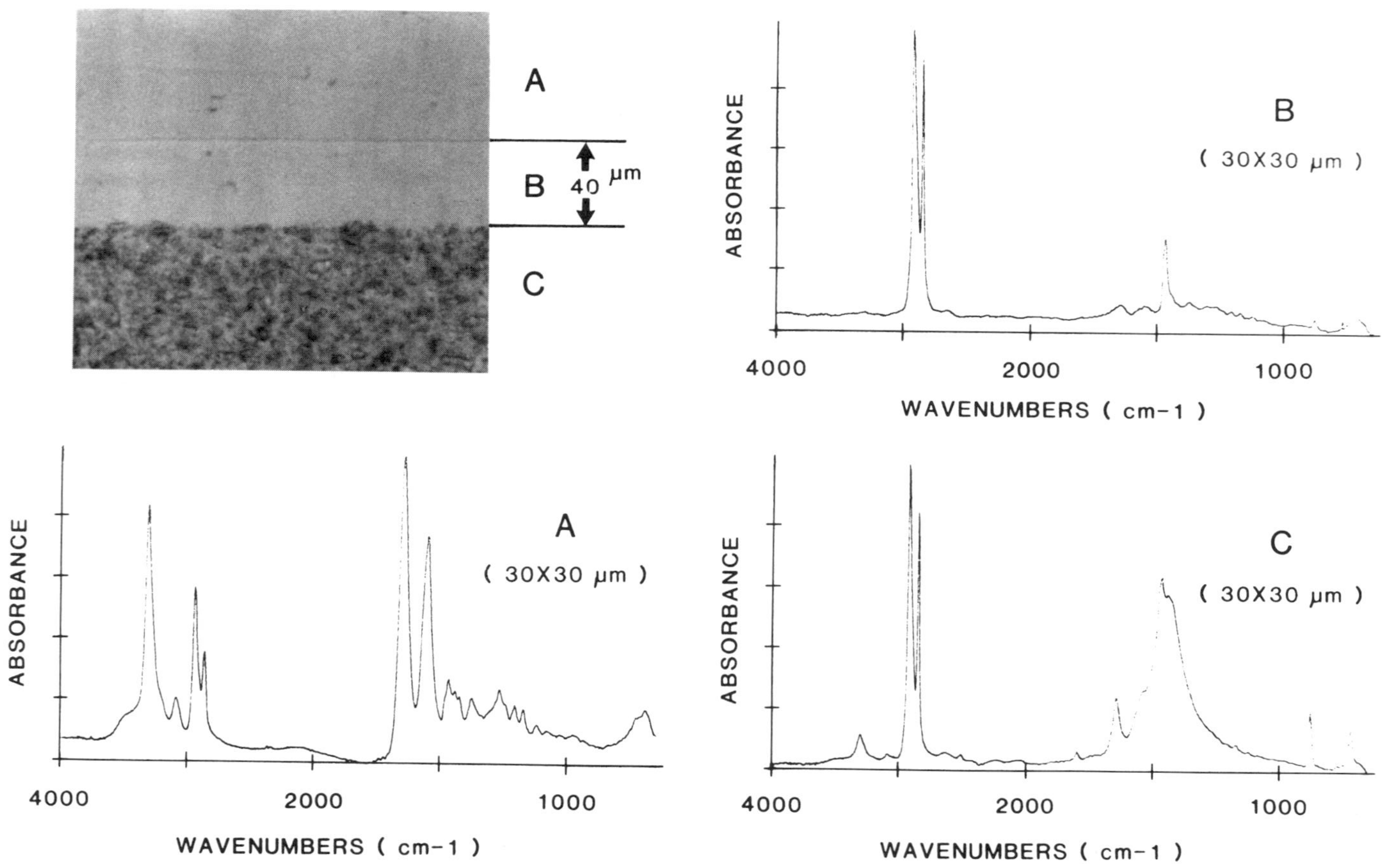

Fig. 7. Infrared absorption spectra of a cross section of a three-layered, laminated film measured by a FT-IR microscope with an aperture size of 30 × 30 µm. Compositions of layers: A, polyamide; B, polyethylene (main component) and polyamide; C, polyethylene, $CaCO_3$, and polyamide.

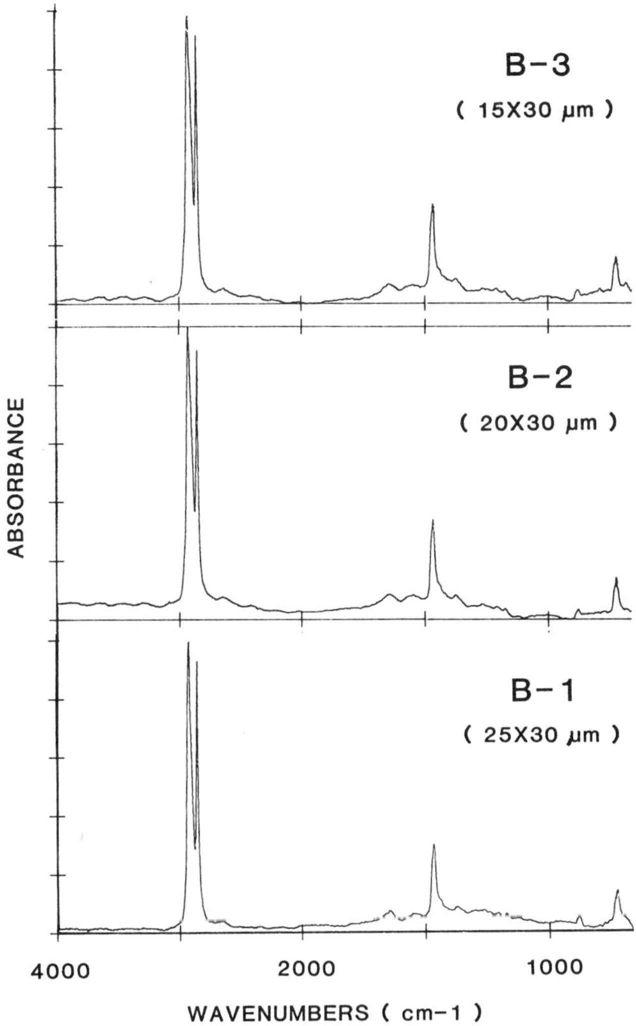

Fig. 8. Aperture size dependence of infrared absorption spectrum of the middle layer (B) in the three-layered film shown in Fig. 7.

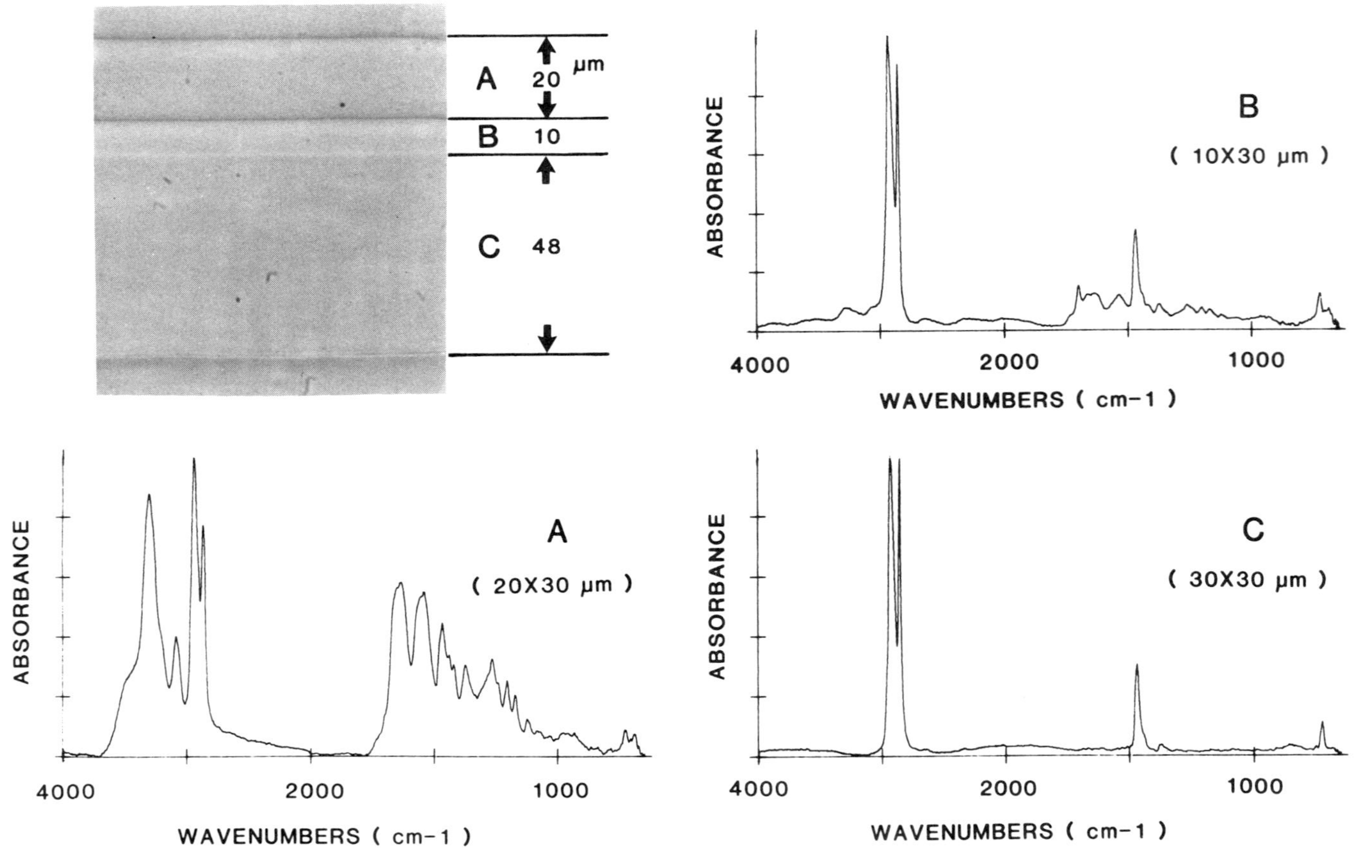

Fig. 9. Infrared absorption spectra of a three-layered, laminated film having a thin middle layer (10 μm). Compositions of layers: A, polyamide; B, polyethylene (main component) and carboxylic acid (Sarlin); C, polyethylene, $CaCO_3$, and polyamide.

interest (Garton *et al.*, 1987). FT-IR microscopy is expected to provide useful information about the microstructure of trees, which is needed if we are to understand the mechanism of tree formation. XLPE insulation cables were kept in water at 60°C under constant electrical stress of 50 kV for about 1 year. Thin sections of 10 μm thickness were carefully prepared by a microtome to include nuclei and tree areas. Scanning electron microscopy and EPMA were also used for morphological studies and elemental analysis, respectively (Kamoto *et al.*, 1988).

Figure 10 shows a microscopic image of a typical water tree (so-called bow-tie tree A). A nucleus of the tree is indicated by arrow "a". FT-IR microscopic measurement was done on five different tree areas (Figs. 11 and 12). Small chemical changes are detected in the spectra obtained in the area around the nucleus. In the expanded spectrum (a), $CaCO_3$ is clearly identified as a component of the nucleus. Elemental analysis by EPMA also detected calcium (Table II). Another major spectral change is due to formation of carboxylate groups generated by oxidative degradation of XLPE. An apparent spectral change is not observed outside the tree area (e). The oxidative reaction seems to increase in the vicinity of the nucleus. A small additional absorption band at 1740 cm^{-1} is considered to come from an ester additive in XLPE probably condensed in the tree areas.

Another type of water tree (D) is shown in Fig. 13, together with a microscopic image of the section. In this case, the nucleus is a bubble, indicating no other spectral change than additional ester (1740 cm^{-1}) and carboxylate (1610 cm^{-1}) absorption bands. Other water trees and an electrical tree were also examined by FT-IR-microscopy and EPMA. The determined compositions of the trees are summarized in Table II. There exist various kinds of contaminants, such as $CaCO_3$, SiO_2, polyamide, and cellulose, in the nuclei of these trees. These species are regarded as impurities possibly contained in manufacturing processes of XLPE cables. Oxidative degradation of XLPE leading to the formation of carboxylate is a

Fig. 10. Microscopic image of a typical water tree formed in cross-linked polyethylene cable insulation. Arrows indicate the areas analyzed by FT-IR microscopy. a–e, Areas of the tree.

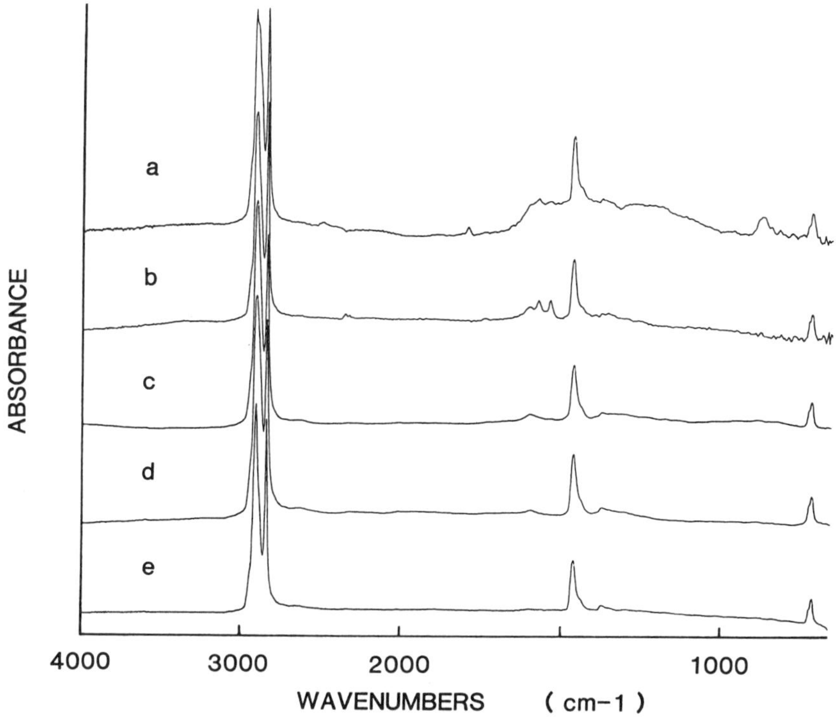

Fig. 11. Infrared absorption spectra of water tree A measured by FT-IR microscopy. Letters correspond to those in Fig. 10.

major chemical process observed in water trees. FT-IR microscopy is found to be very useful for elucidation of treeing phenomena. Compositional mapping with a computer-controlled *x–y* stage (Harthcock and Atkin, 1988) will furnish a more detailed picture of tree formation. Our study along these lines is now in progress.

III. SURFACE ANALYSIS

Surface analysis with vibrational spectroscopy techniques are becoming more and more important, because of the necessity of acquiring chemical information more detailed than that

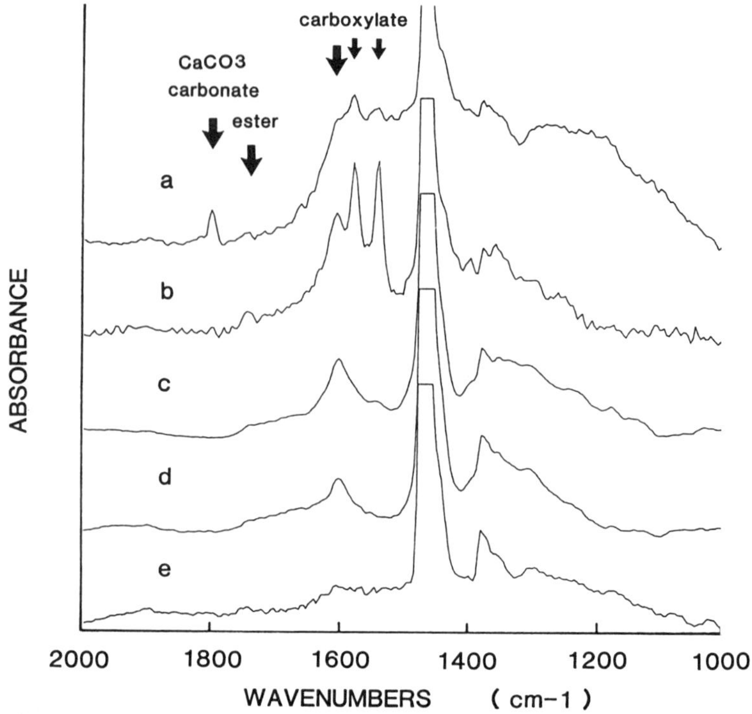

Fig. 12. Infrared absorption spectra of water tree A expanded in the 1000- to 2000-cm^{-1} region.

Table II Chemical Structure of Various Types of Trees

Type	Nucleus: Composition	Nucleus: Element	Tree Area Functional Group
Electrical Tree	Cellulose	—	None
Bow-tie tree A (Water tree)	$CaCO_3$	Ca, S	Carboxylate (COO^-) ester
B	SiO_2	S, Si	Carboxylate (COO^-) ester
C	Polyamide	S, Si	Carboxylate (COO^-) ester
D	Bubble	S	Carboxylate (COO^-) ester

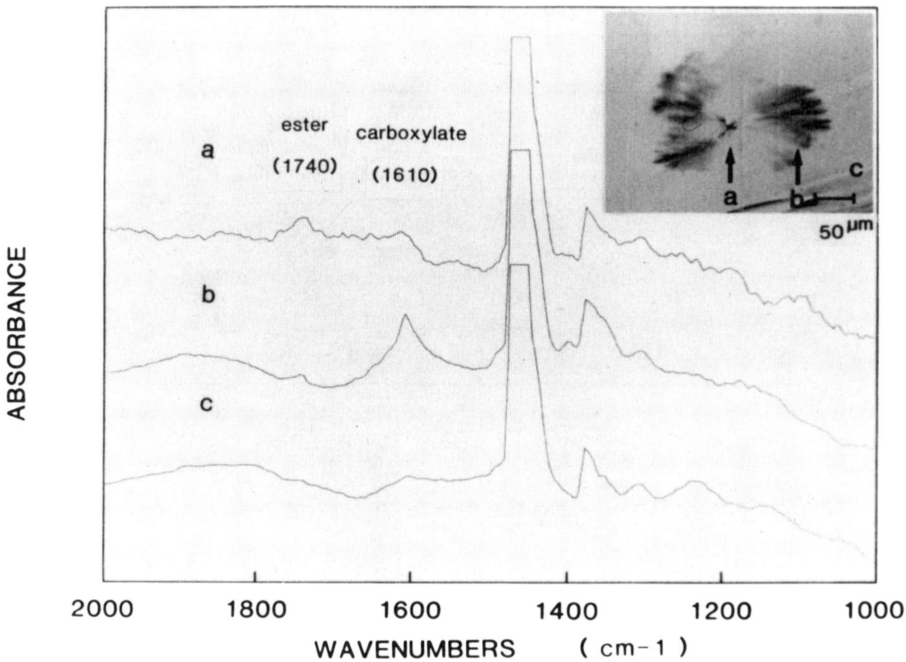

Fig. 13. Infrared absorption spectra of water tree D. Letters denoting the spectra correspond to those in the photograph.

provided by mere elemental analysis by conventional surface analytical techniques. Surface analysis by FT-IR is acquiring an important and indispensable position among other techniques like AES, ESCA, and SIMS. Many kinds of measuring modes, summarized in Table III, are used in surface analysis by FT-IR. Major advantages of the surface analysis by FT-IR are its ability to give molecular information including orientation, sample measurability under ambient atmosphere, its nondestructiveness, and the ease with which insulating samples can be measured, because of the absence of a charging problem. In practical terms, a combination of FT-IR and elemental analyses by ESCA and AES is most fruitful. Surface sensitivity, still much lower than that of X-ray photoelectron spectroscopy (XPS) and AES, has been improved to the

Table III Surface Analysis by FT-IR

Method[a]	Sensitivity	Application
ATR	Monolayer	Polymer films, fibers, glass, paper, cloth (surface treatment, deposits, orientation, crystallinity, etc.)
DRIFTS	~10 Å	Powders (ceramics, carbon, catalyst, glass, etc.) TLC-IR, LC-IR
RAS	Monolayer	Metal (smooth surface)
EMS	~100 Å	Metal (wire, rough surface, etc.) catalyst
PAS	~1 μm	Carbon, coal, ceramics
SEWS	~10 Å	Polymer films, metal, organic thin film

[a]ATR, attentuated total reflection; DRIFTS, diffuse reflectance infrared Fourier transform spectroscopy RAS, reflection absorption spectroscopy; EMS, emission spectroscopy; PAS, photoacoustic spectroscopy; SEWS, surface electromagnetic wave spectroscopy.

monomolecular level by attenuated total reflection (ATR) and reflection absorption spectroscopy (RAS) methods (Ohnishi *et al.*, 1978; Allara and Swalen, 1982).

In the present section, we describe briefly applications of ATR and photoacoustic spectroscopy (PAS) methods for the surface analysis of typical industrial materials like oxide layers on Si, ion-implanted polyethylene films, and multilayered polymer films. The latter two topics are related to a depth-profile study down to a subsurface layer, which is very important in the surface characterization of real samples.

A. Surface Analysis of the Thin Oxide Layer on a Si Wafer

Charges trapped within an oxide layer (SiO_2) and at its interface with the substrate are known to have a large effect on certain electric properties, such as the breakdown and the threshold voltages of a silicon-based metal-oxide semiconductor (MOS). These electric traps, present at the level of an impurity, have been suggested to be generated from SiOH and SiH groups in the oxide film (Fig. 14). FT-IR ATR technique was used to determine SiOH and SiH contents in thermal oxide films grown on Si wafers (Nagasawa *et al.*, 1988).

Si wafers (0.5 mm thick) were treated in a pyrogenic steam atmosphere at 850°C (wet oxidation) and annealed at 850°C and 1000°C in dry N_2. The thickness of the oxide films depends on the oxidation time. After the oxidation treatment, the wafers were cut to trapezoids to be used as internal reflection elements of ATR measurement as shown in Fig. 15. The absorption band of the thin film on the Si wafer is remarkably enhanced by the multiple reflections (100 times) of the infrared beam through the Si wafer itself (Harrick, 1967). Figure 16 shows a typical ATR spectrum of an oxide layer (815 Å) and a transmission spectrum of the same wafer. This method enables us to detect a small number of SiOH and SiH groups in thin oxide films.

Figure 17 shows ATR spectra of a thick oxide film (815 Å) in the SiOH stretching region. Figure 18 shows those of a thin oxide film (160 Å) in the same region. The SiOH content in the thick oxide film decreases markedly by N_2 annealing at 850°C, whereas a much smaller decrease is observed for the thin oxide film. On the other hand, annealing at 1000°C

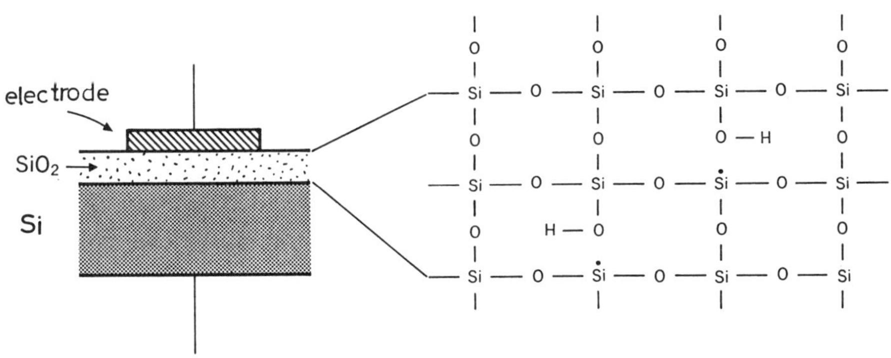

Fig. 14. Schematic illustration of a metal-oxide semiconductor and the chemical structure of its oxide layer.

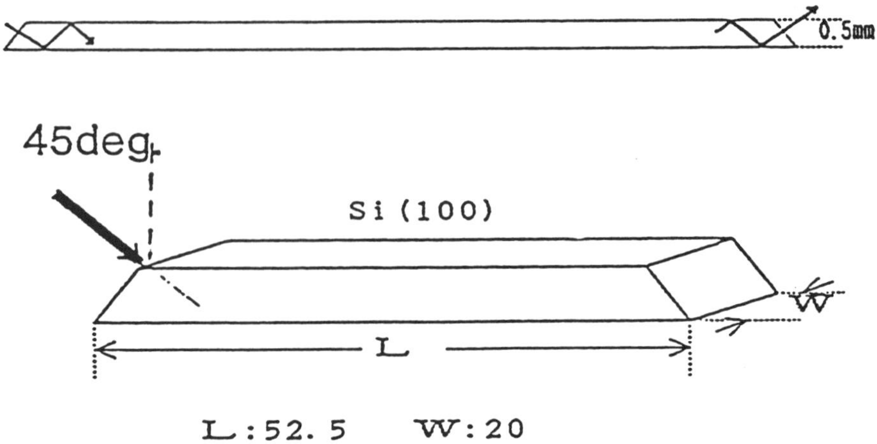

Fig. 15. Set up of ATR measurement using the Si wafer itself as an internal reflection element.

decreases the SiOH content in both films. These results indicate that annealing at 850°C is effective in removing the bulk of SiOH, and annealing at 1000°C is necessary to remove SiOH at the SiO_2–Si interface.

Figure 19 shows also ATR absorption spectra of an oxide film (815 Å) in the wavenumber region of SiH absorption. A band around 2310 cm^{-1} is assigned as the SiH stretching mode in amorphous SiO_2, and a band around 2210 cm^{-1} could be attributed to the SiH_2 group, which has two neighboring oxygen atoms (Lucovsky, 1979). Annealing at both 850°C and 1000°C does not decrease the number of SiH groups in the oxide film.

The present method can also be used to study oxide films on commercial wafers 400–500 μm thick by guiding an infrared beam into Si wafers with an optical flat edge of the appropriate angle. The method is also very useful for studying hydrogen incorporated into amorphous silicon and silicon nitride films prepared on Si substrates.

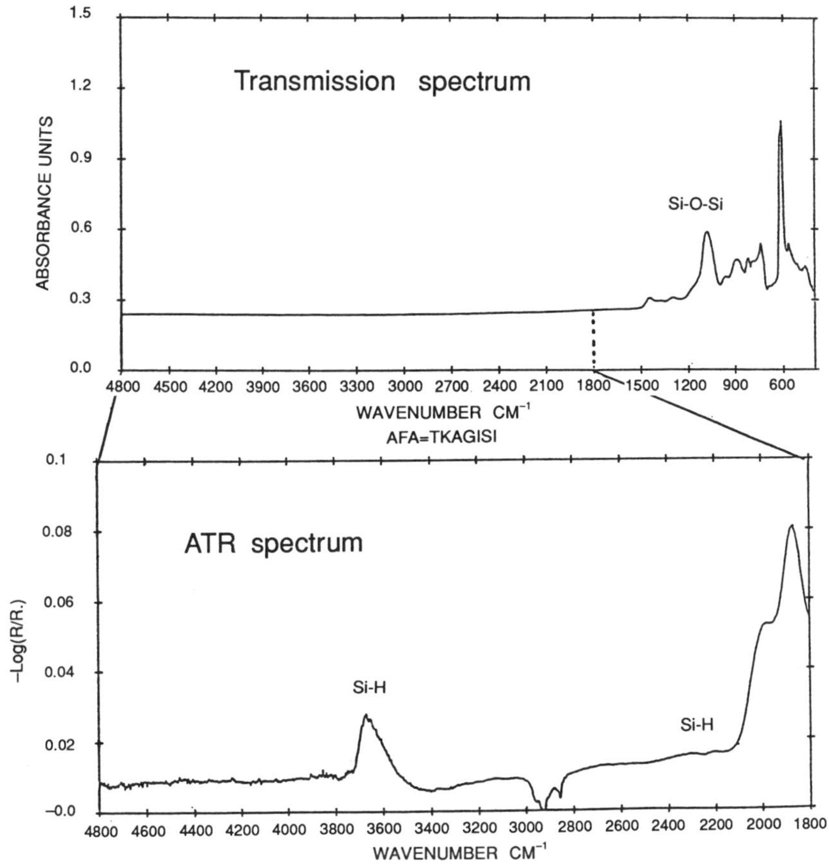

Fig. 16. ATR spectrum (bottom) of a SiO_2 layer (815 Å) on a Si wafer and a transmission spectrum (top) of the same wafer.

B. Depth-Profile Study of Ion-Implanted Polyethylene Films by the ATR Method

Surface modification of polymer materials by ion implantation has been attracting much interest recently (Venkatesan, 1985). It is well known that ion implantation produces

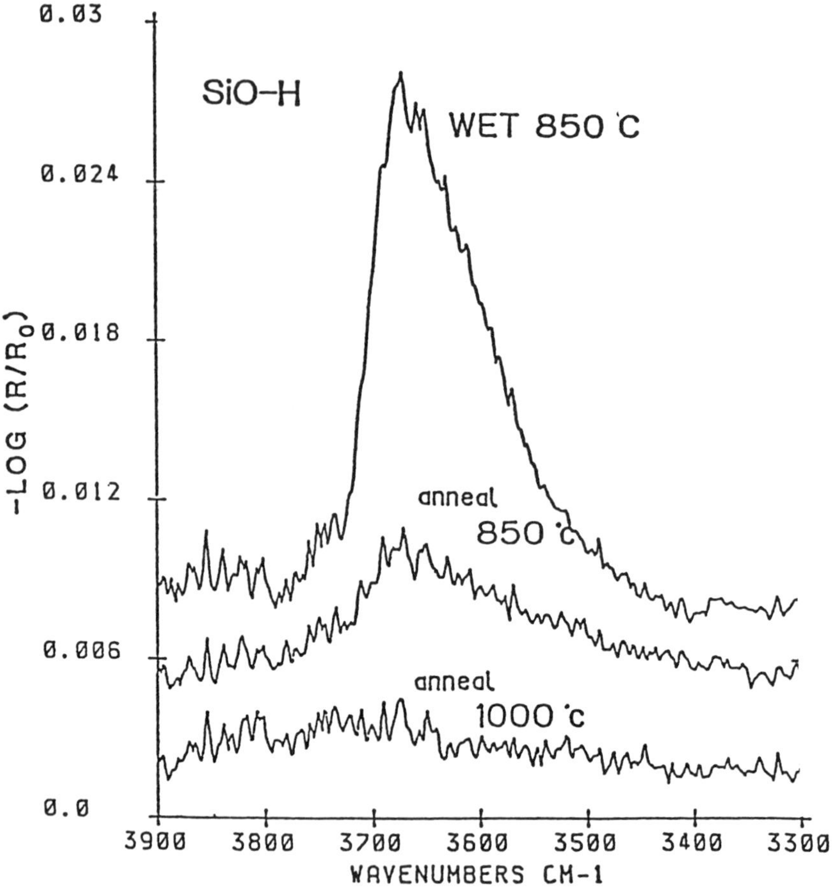

Fig. 17. ATR absorption spectra of a thick oxide layer (815 Å) in the SiOH stretching region.

electrically conducting layers in the subsurface of implanted polymers. However, detailed knowledge of the composition and structure of the modified layer is lacking. Extensive and irreversible decomposition accompanied with various chemical reactions is expected to occur in implanted polymer materials. We have recently undertaken a systematic approach to the problem by characterizing oxygen-implanted polyethylene (PE) films by combined techniques of FT-IR, SIMS, Raman

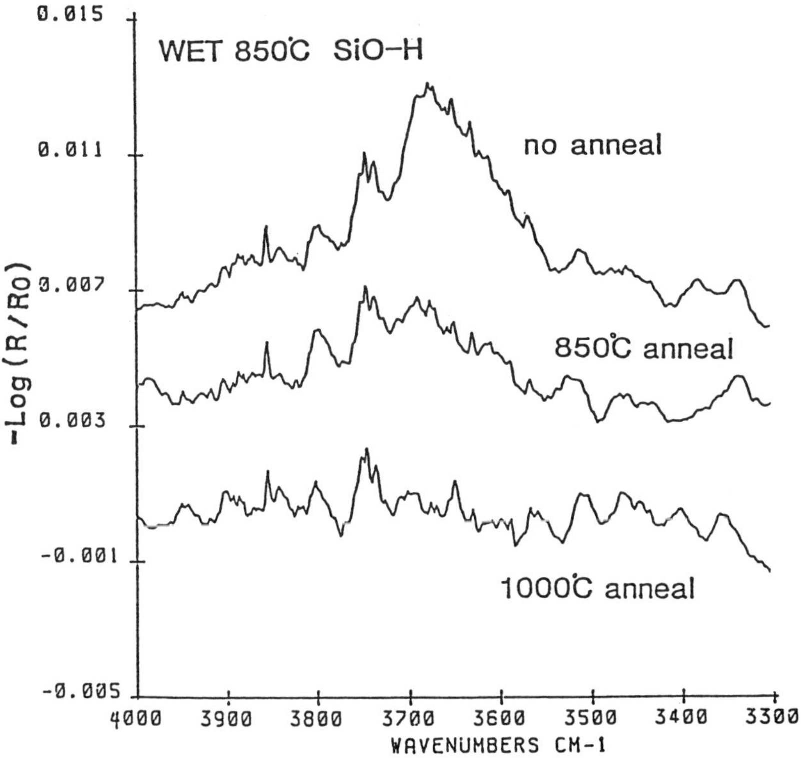

Fig. 18. ATR absorption spectra of a thin oxide layer (160 Å) in the SiOH stretching region.

spectroscopy, and XPS, as schematically illustrated in Fig. 20 (Ishitani *et al.*, 1988b).

The ATR measurement using both Ge and KRS-5 crystals as internal reflection elements (IRE) with different incidence angles can provide nondestructive depth profiles of various chemical species generated in the subsurface layers of PE films (Briggs *et al.*, 1980). As shown in Fig. 21, the sampling depth (roughly, penetration depth, d_p) depends on wavenumber, incidence angle (θ), and refractive index of IRE (Harrick, 1967).

Oxygen ions (O_2^+) were implanted into PE films of 190 μm thickness with energy of 60 keV at doses ranging from

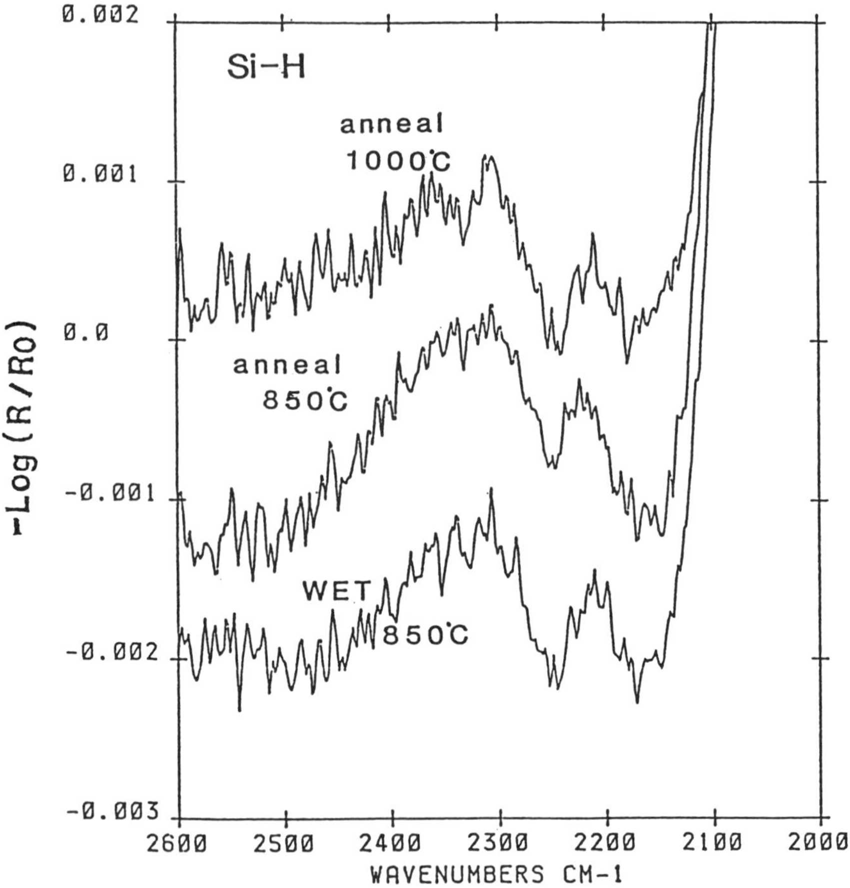

Fig. 19. ATR absorption spectra of a thick oxide layer (815 Å) in the SiH stretching region.

5×10^{14} (5E14) to 5×10^{16} (5E16) ions/cm^2. The implanted film surface was the color of metallic gold. Typical ATR spectra of the implanted films are shown in Fig. 22. A difference spectrum between an implanted film and the control clearly indicates various kinds of functional groups generated by implantation. They are hydroxyl, carbonyl, acid anhydride, carboxylic acid, and unsaturated hydrocarbon groups, including aromatic rings.

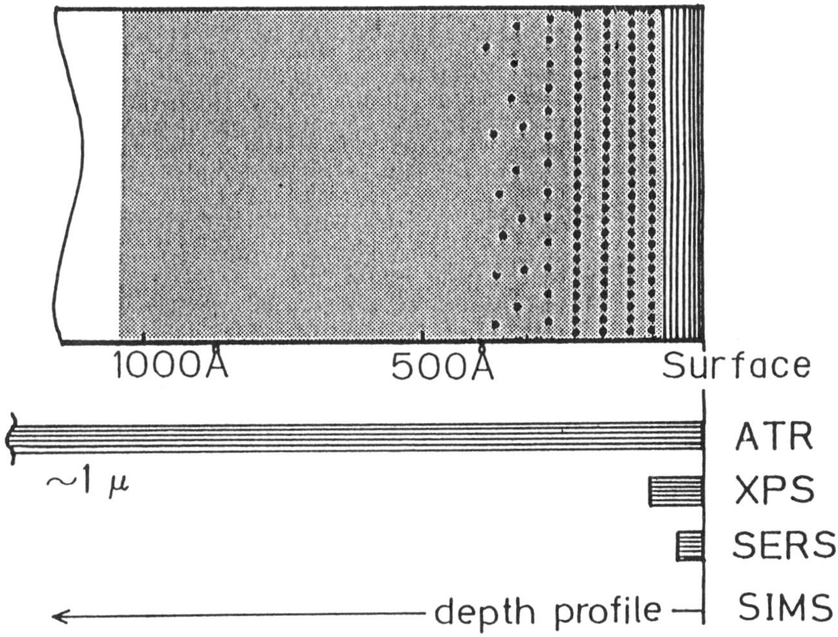

Fig. 20. Schematic illustration of probing depths of various kinds of analytical methods used for the surface characterization of ion-implanted films. ATR, attenuated total reflectance; XPS, X-ray photoelectron spectroscopy; SERS, surface-enhanced Raman scattering; SIMS, secondary ion mass spectrometry.

ATR measurement was done by use of Ge and KRS-5 IRE crystals with two incidence angles, 45° and 60° (d_p, 0.3–1.25 μm at 1700 cm^{-1}). The relative intensity of absorption bands between 1500 and 1850 cm^{-1} that are due mainly to carboxyl, carboxylic acid, and unsaturated hydrocarbon groups are plotted in Fig. 23 for three different doses as a function of d_p. The depth profiles of these functional groups show maxima in the subsurface region, a result in agreement with those expected from the projected range theory. The other functional groups also show profiles similar to those in Fig. 23.

It has also been confirmed that these profiles agree with an oxygen depth profile determined by SIMS analysis. Raman

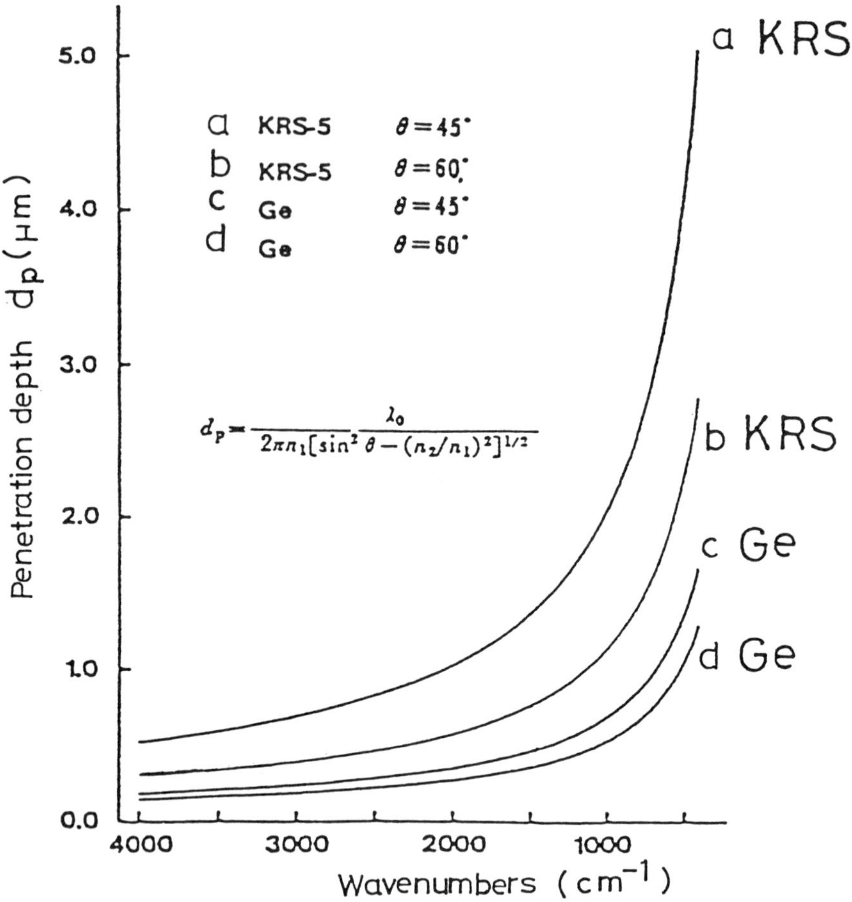

Fig. 21. Penetration depths in the ATR measurement of polymer films (n = 1.50) by use of Ge and KRS-5 with the incidence angles of 45° and 60° plotted as a function of wavenumber.

spectroscopy showed formation of amorphous carbon in the subsurface layer. The present results indicate that implanted oxygen induces condensation reactions, and a fair amount of oxygen is fixed in the damaged layer as oxygen-containing chemical species. ATR depth-profiling is a technique indispensable to our understanding of the mechanism of complicated chemical reactions of ion-implanted polymer materials.

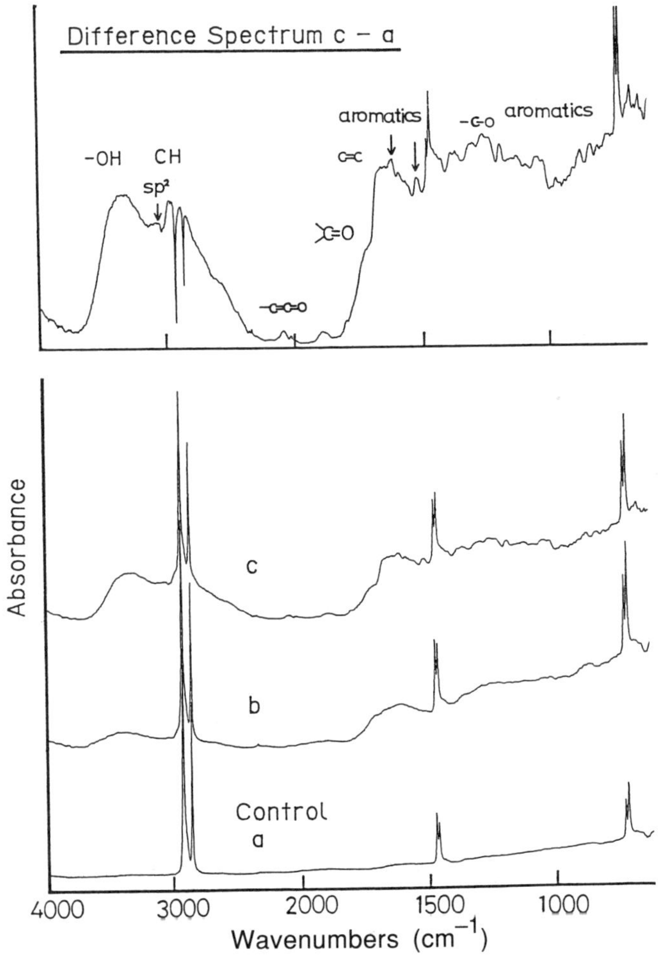

Fig. 22. FT-IR ATR spectra of ion-implanted polyethylene films obtained by KRS-5 with a 45° incidence angle. a, Control; b, 2.5×10^{14} ions/cm^2; c, 5×10^{15} ions/cm^2.

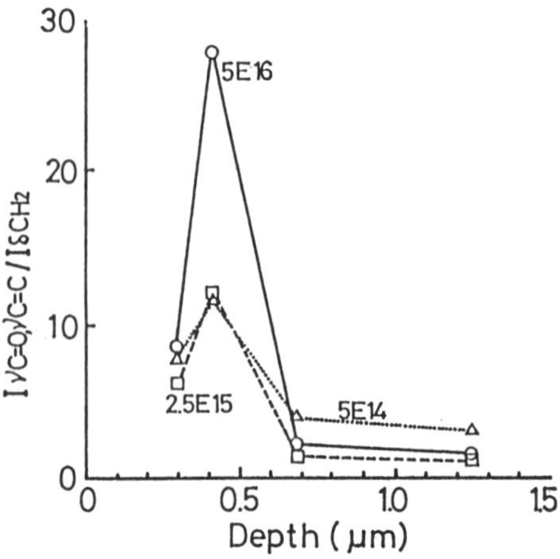

Fig. 23. Depth profiles of relative absorption intensity of functional groups generated by oxygen ion implantation plotted as a function of penetration depth.

C. PAS Study of Multilayered Polymer Films

We are finding various kinds of interesting applications of PAS, notwithstanding its low surface sensitivity. PAS is convenient for nondestructive infrared analysis of irregular shaped samples without any pretreatment. Another interesting application of PAS is nondestructive sampling of underlying layers. The sampling depth of PAS is known to extend even to 15–20 μm (Saucy *et al.*, 1985; Teramae and Tanaka, 1985). The depth is dependent on optical and thermal properties of a sample and also on instrumental parameters. In the present section, we will show an application of PAS to analysis of underlying layers of polymeric laminated materials (Ishitani, 1985).

Figure 24 shows the photoacoustic spectra of a three-layered film. PAS signals from the underlayers were examined as a function of thickness of a polypropylene upper layer. The PAS signal of the underlying polyester layer comes from a depth of about 18 μm, much deeper than the conventional

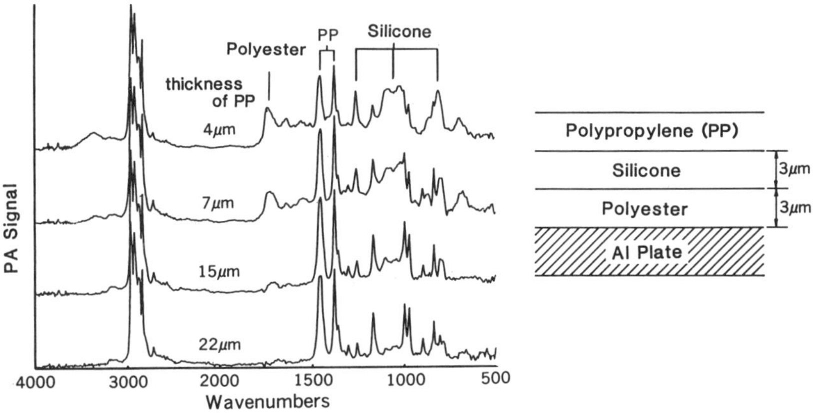

Fig. 24. FT-IR PAS spectra of three-layered polymer films on an aluminum plate.

ATR method (see Section III,B). The deep sampling depth is useful in many practical problems. The PAS spectrum of an individual layer can be separated by subtracting those of the other layers.

Figure 25 shows a PAS study of a photochemical reaction occurring in an underlying photopolymer layer. A polymer layer containing a photodecomposable diazo group is laminated on an aluminum plate and covered with a thin silicon rubber film. The photochemical reaction of the diazo compound is shown by the following reaction scheme. A photoacoustic band characteristic of the diazo group clearly

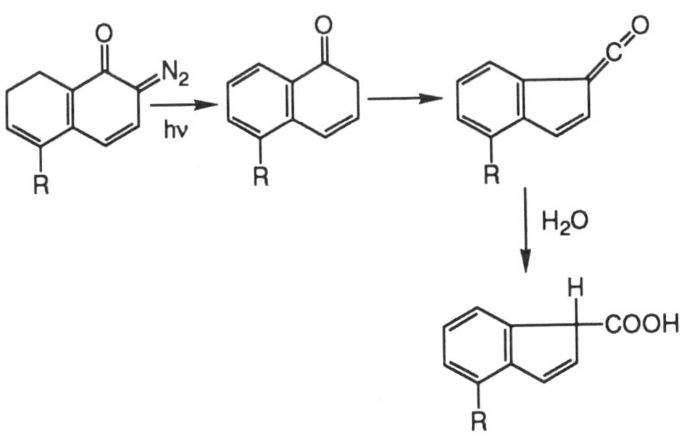

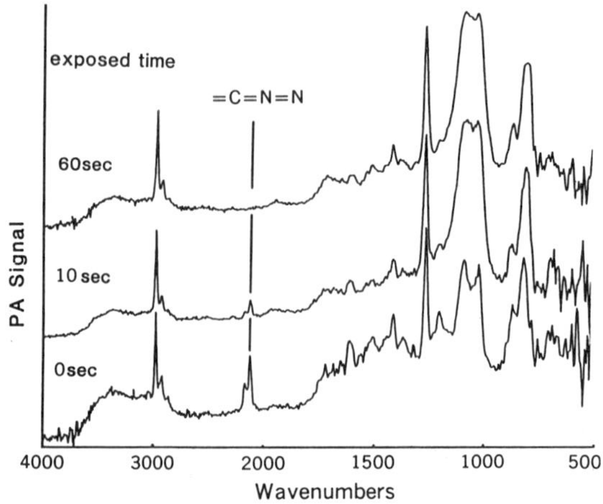

Exp. Time \ Method	ATR	PAS
10 Sec.	60	66
20	80	83
40	96	95
60	100	100

(%)

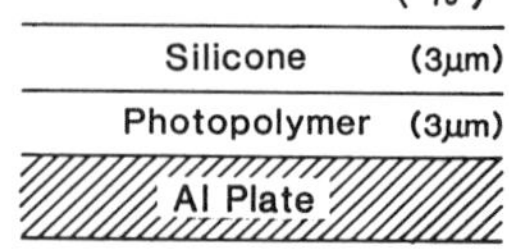

Fig. 25. FT-IR PAS study of a photochemical reaction in the underlayer of a two-layered polymer film. The table indicates the photochemical decomposition reaction behavior of the diazo group detected by PAS compared with the ATR measurement of the bare photopolymer layer.

decreases during ultraviolet illumination and completely disappears after 60 sec of illumination. Photochemical reaction behavior of the diazo group is also shown in the table of Fig. 25 as a function of exposure time compared with a direct ATR measurement of a bare photopolymer layer. A good agreement is observed between them. Nondestructive reaction analysis of a deep underlying layer, as presented here, becomes possible only by use of the FT-IR PAS method.

IV. BULK ANALYSIS

FT-IR is one of the most widely used analytical techniques for bulk analysis of materials. It can be used for all kinds of materials, organic or inorganic, crystalline or amorphous. FT-IR analysis can provide the most comprehensive information on chemical composition and structure, crystallinity, molecular and lattice orientation, and even subtle intermolecular interactions. FT-IR is the first technique to be used for bulk analysis; it was followed by X-ray diffraction, NMR (nuclear magnetic resonance), and others. The transmission mode coupled with the digital difference-spectrum technique is frequently used in qualitative and quantitative analyses of materials in industrial applications. Far-infrared absorption spectra also give important information on chemical and crystal structures, particularly those of inorganic materials.

In the following sections, we describe our recent quantitative analysis of carbon impurity in GaAs by high-resolution and low-temperature measurements. We then describe a FT-IR study of bonding states of hydrogen in diamond-like amorphous carbon films.

A. Quantitative Analysis of Carbon Impurity in GaAs by a High-Resolution Measurement

Carbon is one of the most important residual impurities in an undoped GaAs single crystal grown by the liquid-encapsulated Czochralski (LEC) method. It is important to determine the concentration of carbon in GaAs, because the semi-insulating property of GaAs crystals is realized by the balance between carbon acceptors and the EL2 deep donors. The infrared absorption of the localized vibrational mode (LVM) of carbon atoms at arsenic lattice sites was studied to determine carbon concentration (Brozel *et al.*, 1978; Homma *et al.*, 1985; Hunter *et al.*, 1984). We have recently carried out a high-resolution measurement of the carbon LVM at low temperatures down to liquid helium temperature; and we have found a new, improved calibration factor for the quantitative determination of carbon concentration. We have applied the new calibration factor to determine the carbon concentration in commercial GaAs wafers.

A Bruker IFS-113V Fourier transform spectrometer

equipped with a HELI-TRAN LT-3-110 cryostat of Air Products was used for the present study. A Mylar beamsplitter 3.5 μm thick and a DTGS detector with a polyethylene window were used for measurements, and the triangle apodization function was employed for Fourier transformation.

Figure 26 shows dependence of the carbon LVM absorption band on resolution at liquid helium temperature. The absorption coefficient increases with a decrease of the full width at half-maximum by increasing the resolution. Fine structure appears at the resolution of 0.06 cm^{-1}. It is well known that the fine structure originates from various distributions of ^{69}Ga and ^{71}Ga isotopes on the nearest-neighbor sites (Theis *et al.*, 1982). Figure 27 shows a temperature dependence of the carbon LVM measured with a high-resolution condition (0.06 cm^{-1}). Because the population of excited states of carbon LVM depends on the Bose-Einstein distribution, the absorption coefficient of LVM is expected to decrease with an increase of the sample temperature. As shown in Figure 28, the absorption coefficient reaches a maximum around liquid nitrogen temperature. One possible explanation for this result is that the intrinsic full width at half-maximum of the carbon LVM lines is narrower than the employed instrumental resolution (0.06 cm^{-1}). The results indicate that liquid nitrogen temperature is optimal for the determination of carbon concentration by infrared absorption spectrometry with a 0.06 cm^{-1} resolution.

We have used the highest peak located at 582.7 cm^{-1} to obtain a calibration factor from the measurement of round-robin samples prepared by the Japan Electron Industry Development Association. A calibration factor of 8.0×10^{15} (atoms/cm^2) has been newly determined under these conditions. Our result indicates that the higher resolution measurement increases sensitivity approximately 2.5 times compared with the conventional medium resolution analysis (0.5 cm^{-1}) at liquid nitrogen temperature, although it requires more data collection time.

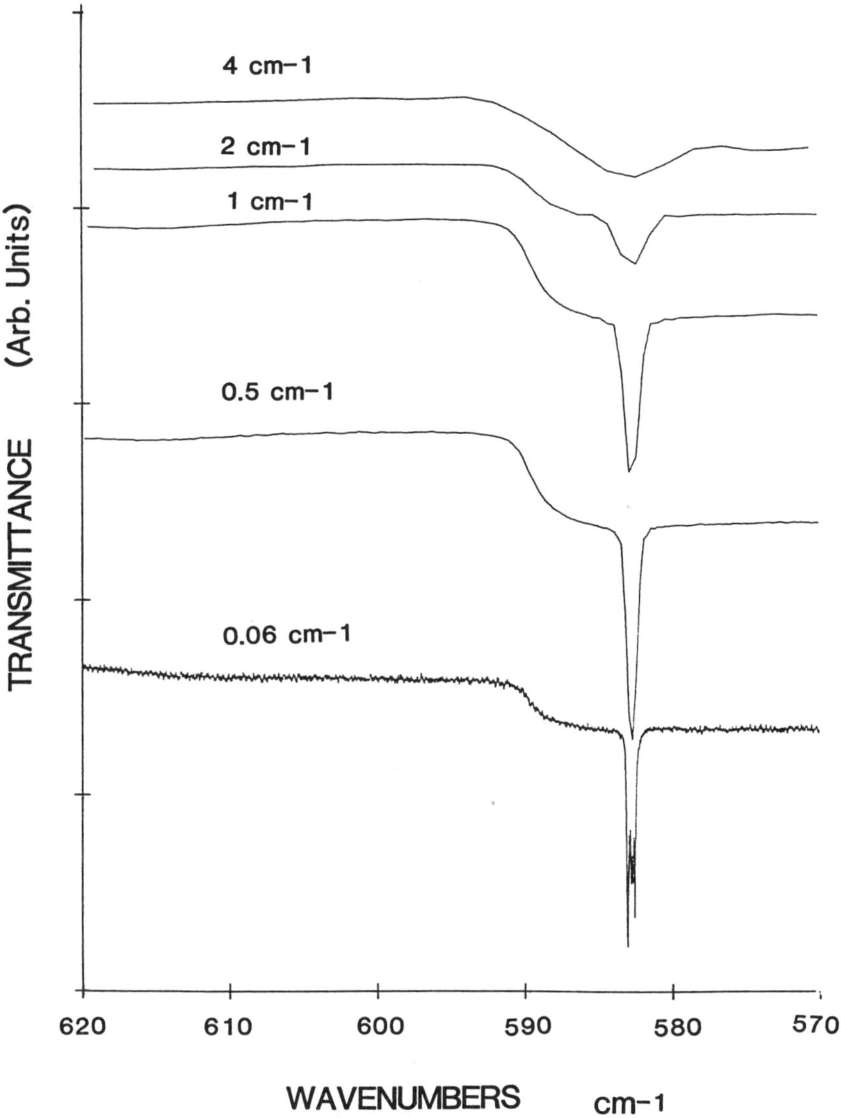

Fig. 26. LVM infrared absorption band of the carbon impurity in GaAs, measured as a function of resolution at liquid helium temperature.

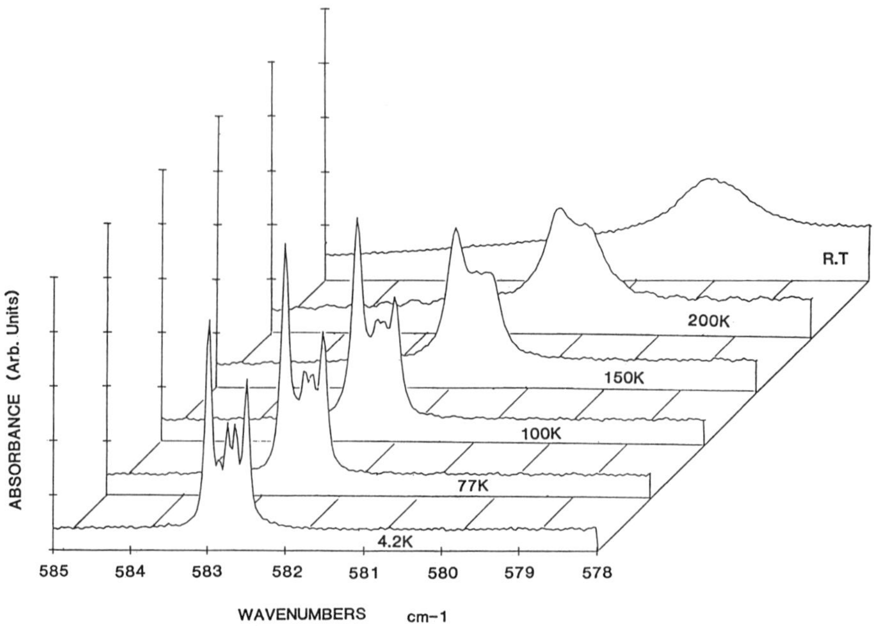

Fig. 27. Temperature dependence of the carbon LVM absorption band measured with a resolution of 0.06 cm^{-1}. R. T, room temperature.

In a real sample analysis, quantitative determination of the carbon concentration has to be performed in a very thin wafer (~500 μm thick), where multiple reflections produce an undesirable interference pattern superimposed on the spectrum. Under the high-resolution condition at low temperature, the full width at half-maximum of LVM becomes narrower than the half period of the interference fringe. We can therefore discriminate the carbon LVM from the interference pattern under these conditions. In addition, a simple spectral manipulation technique can eliminate the interference fringe pattern in the spectrum. Figure 29 shows an example of subtracting a spectrum shifted in the lower frequency direction by one period of the interference fringe from the original one. This technique enables quantitative analysis of carbon in a thin GaAs wafer using the aforementioned calibration factor. The detection limit of carbon in GaAs wafer by our method

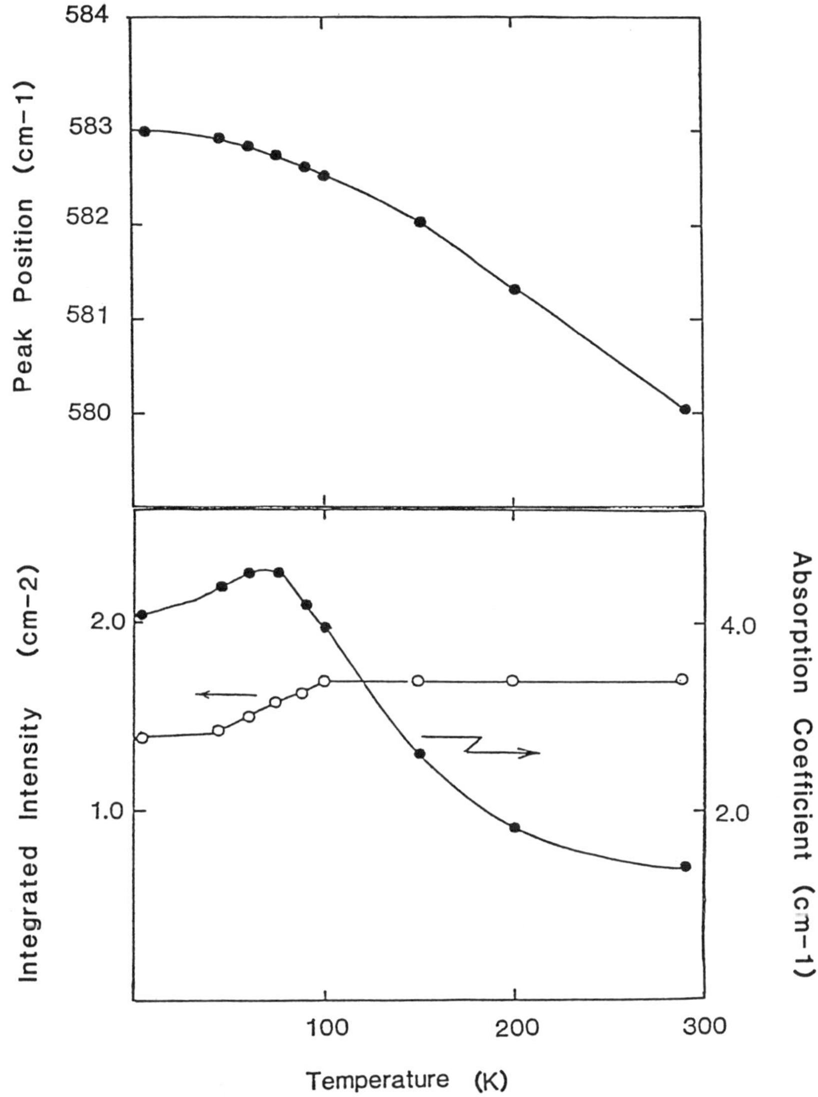

Fig. 28. Temperature dependence of the peak position, absorption coefficient, and integrated intensity of the carbon LVM absorption band.

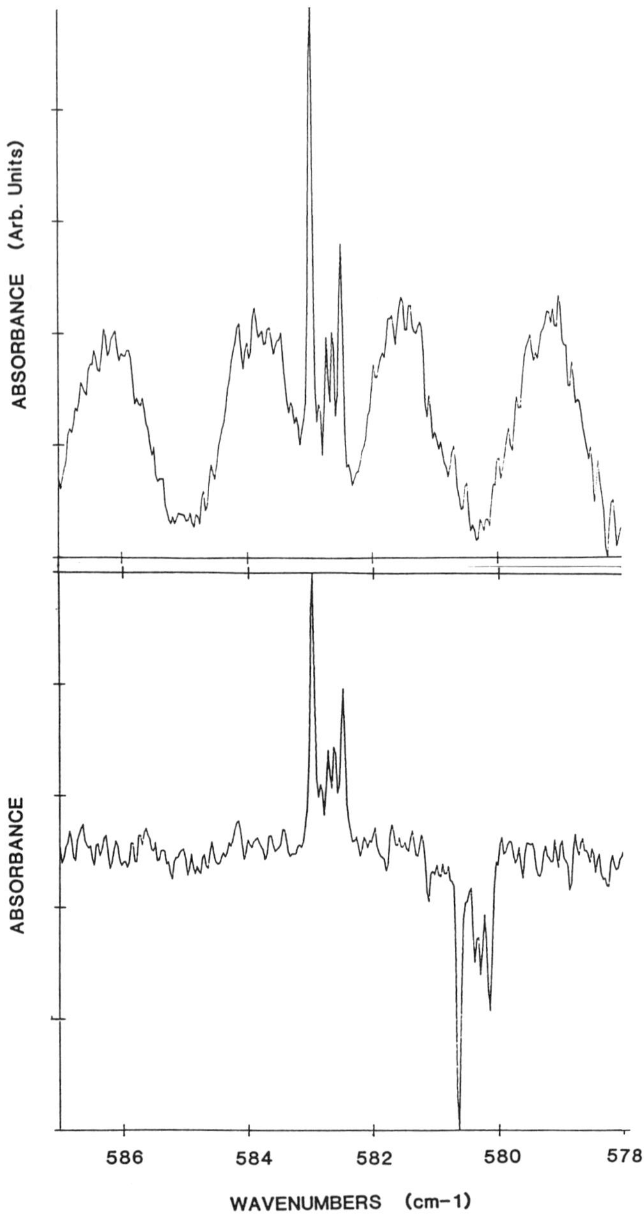

Fig. 29. Carbon LVM absorption band of a thin commercial GaAs wafer measured at 4 K with a resolution of 0.06 cm^{-1} (upper). Difference spectrum between the original and the shifted spectra (lower).

is about 2.0 × 10^{15} atoms/cm^3, which is higher than that of conventional SIMS bulk analysis.

B. Microstructure of Diamond-like Amorphous Carbon Films

Diamond-like amorphous carbon (DLC) films have a number of attractive properties such as hardness, electrical insulation, chemical inertness, and infrared transparency (Savvides, 1986). DLC films are believed to consist of a mixture of sp^3 (diamond-like) and sp^2 (graphitic) bonding structures. Raman spectroscopy is well known to be powerful for characterization of carbon materials. We have recently studied the microstructure of DLC films prepared by plasma chemical vapor deposition (CVD) and sputtering methods by Raman spectroscopy; we have proposed that the observed Raman bands of a DLC film originate from carbon clusters with a sp^2 configuration (Yoshikawa *et al.*, 1988a). It is very important to characterize bonding states of hydrogen incorporated into the DLC films. We have investigated infrared absorption spectra of DLC films prepared by the plasma CVD method, using a mixture of H_2 and C_2H_4 (Table IV) in the wavenumber region of

Table IV Hydrogen Bonding States Determined by FT-IR Analysis of DLC Films

				Relative hydrogen content (%)[b]			
				A		B	
Sample	Substrate temperature	Gas composition (C_2H_4:H_2)	Sample thickness (Å)	sp^3	sp^2	sp^3	sp^2
R1	R.T.[a]	1:0	1250	63	37	65	35
R2	R.T.	1:50	910	62	38	64	36
T1	120°C	1:0	1410	67	33	70	30
T2	120°C	1:50	920	64	36	67	33

[a]R.T., room temperature.
[b]A, area ratios; B, peak ratios.

the C—H stretching vibration (3200–2700 cm^{-1}). Infrared spectroscopy can easily distinguish between C—H sp^2 and sp^3 bonding sites by peak positions of characteristic absorption bands. Table V shows C—H stretching absorption bands and their assignment observed for the DLC films (Dischler *et al.*, 1983). The high sensitivity of FT-IR measurement is useful for the present study, because the total hydrogen content in DLC films is low and the thickness of DLC films is less than 1 μm. Figure 30 shows transmission infrared absorption spectra of the DLC films prepared on silicon substrates held at 120°C (samples T_1 and T_2). A broad C—H absorption band can be resolved to six components with Gaussian line shape by the self-deconvolution method. The hydrogen concentration estimated from the absorption bands ranges from 3×10^{22} to 1×10^{23} atoms/cm^3. Among six bands, the two bands located above 3000 cm^{-1} correspond to hydrogen bound to sp^2 carbon atoms. The relative hydrogen content of sp^2 and sp^3 bonding sites can be estimated from peak heights or peak areas, as summarized in Table IV (Yoshikawa *et al.*, 1988b). These results agree well with the Raman results and indicate that sp^3 hydrogen content increases with a decrease of the content of the sp^2 carbon cluster. Properties of DLC films depend greatly on the content of the sp^2 carbon cluster thus determined. Microstructural characterization by FT-IR as well as Raman spectroscopy will contribute greatly to a improvement of the

Table V C—H Stretching Absorption Bands Observed for DLC Films

Band	Peak Position (cm^{-1})	Assignment	Absorption coefficient ($mole^{-1}$ cm^{-1})
1	3050	CH sp^2 (aromatic)	30
		CH_2 sp^2 (olefin)	30
2	3000	CH sp^2 (olefin)	30
3	2970	CH_3 sp^3 (asymmetrical)	70
4	2920	CH_2 sp^3 (asymmetrical)	75
		CH sp^3	
5	2870	CH_3 sp^3 (symmetrical)	30
6	2850	CH_2 sp^3 (symmetrical)	45

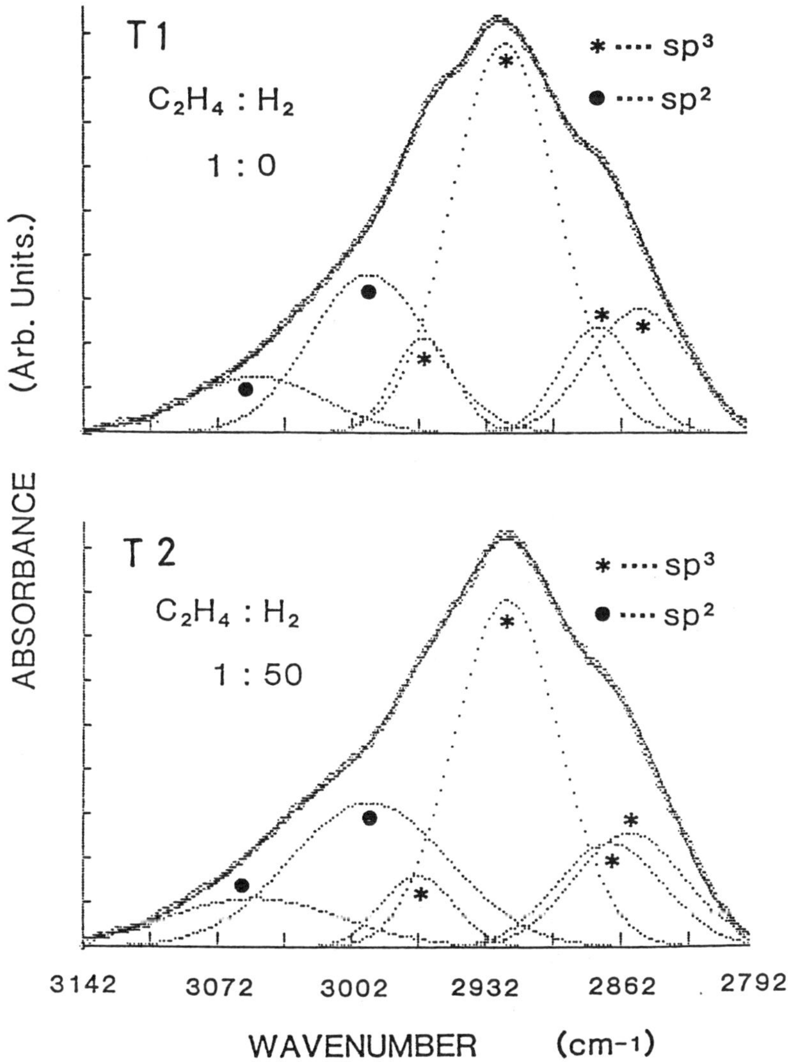

Fig. 30. Infrared absorption spectra of diamond-like amorphous carbon films prepared by the plasma CVD method in the C—H stretching vibrational region.

properties of DLC films. The ATR method is also very useful for thin DLC films with low hydrogen content, as shown in the study of thin oxide layer on Si (Section III,A). A silicon substrate itself also can be used as an internal reflection element of ATR measurement. Figure 31 shows both ATR and transmission spectra for sample R1. A remarkable absorption intensity enhancement is attained by the ATR method as a result of the multiple reflections (25 times) of the infrared beam through the Si substrate. It also enables quantitative analysis of hydrogen in thin DLC films that are used for surface coating of hard disks and various tools.

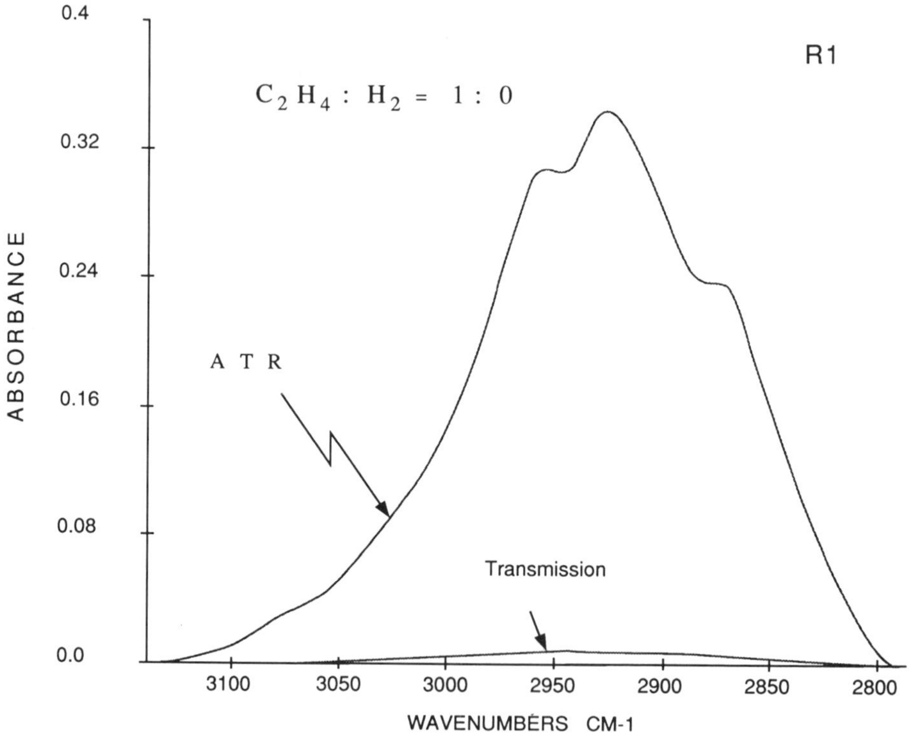

Fig. 31. Comparison of infrared absorption spectra of a diamond-like amorphous carbon film on a Si crystal between transmission and ATR measurement modes.

V. SUMMARY

Recent industrial applications of FT-IR as an analytical tool for materials characterization have been described. The applications introduced here are limited to the latest work done in our laboratory. The surface and microanalyses of industrial materials by FT-IR seem to have established a firm and important position among other conventional analytical techniques.

In closing this chapter, we would like to mention briefly the future advancement of FT-IR expected in industrial materials characterization. Compositional mapping by a FT-IR microscope will become of practical importance not only for industrial materials but also for biological materials. Combination of FT-IR microscopy with other classic analytical techniques such as DSC, TLC, and LC will become important. FT-Raman spectroscopy will accelerate advancement of Fourier Transform spectroscopy in the near-infrared and visible regions. Compact and low-cost FT-IR instruments will rapidly replace the dispersive infrared spectrometers widely used in industrial laboratories. In any event, the versatile capabilities of FT-IR will be widely recognized and utilized in many analytical fields, not only by instrumental advancements but also by developments of new measurement and sample preparation techniques.

ACKNOWLEDGMENTS

The authors wish to express our thanks to Dr. Y. Nagasawa, Mr. R. Kamoto, Mr. K. Shoda, Mr. N. Nagai, and Dr. M. Yoshikawa for their experimental support and helpful discussions.

REFERENCES

Allara, D. L., and Swalen, J. D. (1982). *J. Phys. Chem.* **86,** 2700.

Briggs, D., Zichy, J. I., Brewis, D. M., Comyn, J., Dahm, R. H., Green, M. A., and Konieczko, M. B. (1980). *Surf. Interface Anal.* **2,** 107.

Brozel, M. R., Clegg, J. B., and Newman, R. C. (1978). *J. Phys. D* **11,** 1331.
Dischler, B., Bubenzer, A., and Koidl, P. (1983). *Solid State Commun.* **48,** 105.
Garton, A., Bamji, S., Bulinski, A., and Densley, J. (1987). *IEEE Trans. Elect. Insulation* **EI-22,** 405.
Harrick, N. J. (1967). "Internal Reflection Spectroscopy." Wiley (Interscience), New York.
Harthcock, M. A., and Atkin, S. C. (1988). *Appl. Spectrosc.* **42,** 449.
Harthcock, M. A., Lentz, L. A., Davis, B. L., and Krishnan, K. (1986). *Appl. Spectrosc.* **40,** 210.
Hatta, A., Ohnishi, T., and Suetaka, W. (1982). *Appl. Phys. A* **29,** 71.
Homma, Y., Ishii, Y., Kobayashi, T., and Osaka, J. (1985). *J. Appl. Phys.* **57,** 2931.
Hunter, A. T., Kimura, H., Bukus, J. P., Winston, H. V., and Marsh, O. J. (1984). *Appl. Phys. Lett.* **44,** 74.
Ishida, H., Kawai, T., and Ishitani, A. (1986). "The Pittsburgh Conference, Atlantic City."
Ishitani, A. (1985). *Proc. Soc. Photo-Opt. Instrum. Eng.* **553,** 25.
Ishitani, A., Kamoto, R., and Ishida, H. (1988a). *Microbeam Anal.* 221.
Ishitani, A., Shoda, K., Ishida, H., Watanabe, T., Yoshida, K., and Iwaki, M. (1988b). *Proc. Int. Conf. Ion Beam Modif. Mater., 6th,* in press.
Kamoto, R., Takeda, S., Ishida, H., Ishitani, A., Endou, T., and Seki, Y. (1988). *Proc. FACSS Annu. Meet., 15th* p. 178.
Lucovsky, G. (1979). *Solid State Commun.* **29,** 571.
Messerschmidt, R. G., and Harthcock, M. A. (eds.) (1988). "Infrared Microspectroscopy: Theory and Applications." Dekker, New York.
Nagasawa, Y., Ishida, H., Soeda, F., Ishitani, A., Yoshii, I., and Yamamoto, K. (1988). *Microchim. Acta* **1,** 431.
Ohnishi, T., Ishitani, A., Ishida, H., Yamamoto, N., and Tsubomura, H. (1978). *J. Phys. Chem.* **82,** 1989.
Reffner, J. A., Messerschmidt, R. G., and Coates, J. P. (1987). *Microbeam Anal.* 180.
Saucy, D. A., Smiko, S. J., and Linton, R. W. (1985). *Anal. Chem.* **57,** 871.
Savvides, N. (1986). *J. Appl. Phys.* **59,** 4133.
Teramae, N., and Tanaka, S. (1985). *Appl. Spectrosc.* **39,** 797.
Theis, W. M., Bajaj, K. K., Litton, C. W., and Spitzer, W. G. (1982). *Appl. Phys. Lett.* **41,** 70.
Venkatesan, T. (1985). *Nucl. Instrum. Methods* **B7,** 461.
Yoshikawa, M., Katagiri, G., Ishida, H., and Ishitani, A. (1988a). *Solid State Commun.* **66,** 1177.
Yoshikawa, M., Nagai, N., Katagiri, G., Ishida, H., and Ishitani, A. (1988b). *Proc. Int. Conf. New Diamond Sci. Technol., 1st* p. 222.

8

Multivariate Calibration Methods Applied to Quantitative FT-IR Analyses

David M. Haaland
Sandia National Laboratories
Albuquerque, New Mexico

Practical Fourier Transform Infrared Spectroscopy

I. INTRODUCTION

The significant improvements in infrared spectroscopy brought about by the introduction of computerized Fourier transform infrared (FT-IR) spectrometers have resulted in a dramatic revitalization and expansion in quantitative infrared spectroscopy. The availability of digitized spectra produced by spectrometers having high signal-to-noise ratios (S/N) and linearity over a wide dynamic range in a spectral region containing an extremely high information content has led to an ever-increasing use of modern statistical methods in the analysis of infrared spectra. The application of multivariate statistical calibration and prediction methods to quantitative FT-IR spectroscopy has extended the potential of quantitative infrared spectroscopy and has allowed quantitative problems to be solved by FT-IR analyses that only a few years ago had not been thought possible. These statistical methods are just a few of the many tools available to chemists in a field that has come to be known as chemometrics. Chemometrics can be broadly defined as the application of statistical and mathematical methods for the design or optimization of chemical experiments and for the efficient extraction of information from chemical data. Since computerized FT-IR spectrometers can quickly generate vast quantities of spectral data, spectroscopists can rapidly become swamped in a sea of data. Therefore, efficient chemometric methods for converting spectral data into chemical information has made chemometrics one of the more rapidly growing areas of chemistry. Reviews of a variety of chemometric methods have been published recently (Sharaf *et al.,* 1986; Deming and Morgan, 1987; Massart *et al.,* 1988; Ramos *et al.,* 1986; Brown *et al.,* 1988; Brereton, 1987; Reyment, 1987; Meglen, 1988; Berridge, 1987). For spectroscopists unfamiliar with the more sophisticated chemometric methods, their significant power when applied to quantitative spectral analyses has at times made these methods appear magical. However, when used without an understanding or appreciation of their capabilities, assumptions, and limitations, these methods are quite capable of "magically" producing errors or meaningless results. Therefore, it is important that the various chemometric tools used in quantitative FT-IR analyses be understood by the analyst, and these methods should be approached with realistic expectations for their powerful capabilities.

Although there are numerous experimental aspects to quantitative FT-IR spectroscopy that must be considered, this chapter will focus on understanding and applying the various multivariate data analysis methods that have been published for use with FT-IR spectral data. These methods include classical least squares (CLS), inverse least squares (ILS), **q**-matrix, partial least squares (PLS), and principal component regression (PCR). Correlation and Kalman filter methods have received less attention in the infrared literature and therefore will only be briefly mentioned. Classical least-squares and inverse least-squares methods have often been referred to as **K**-matrix and **P**-matrix methods, respectively, by infrared spectroscopists. However, since these latter names are neither descriptive nor used when they are applied in other fields to different types of spectroscopy, it is probably more appropriate to use the more general statistical classifications of classical and inverse least-squares methods. The term *classical least-squares method* describes the procedure where the calibration model represents the physical law that describes the variation in the spectra with composition, i.e., Beer's law where spectral absorbance is represented as a linear function of component concentrations. Therefore, CLS explicitly uses the Beer's law model and is sometimes referred to as a hard model or explicit model method (Geladi, 1988).

Some of the other methods described here use linear models with concentrations represented as linear functions of either spectral intensities or linear combinations of intensities. These alternate calibration methods are not restricted to use the same number of components as the individual chemical components present in the spectral region being analyzed. They can use fewer or more linear components as required to achieve the best fit to the data. These statistical models are often described as "implicit" or "soft" models since the data are empirically modeled without using a direct physical model. Therefore, these methods can provide more flexibility in modeling the data. However, this greater flexibility is often at the expense of useful qualitative information that might have been available had the physical or "hard" model been used. Because they are less restrictive, soft models often permit better fits to real data that may experience deviations in Beer's law. These deviations in Beer's law are usually described as nonlinearities (Maris *et al.,* 1983; Haaland *et al.,* 1985; Haaland, 1987a), but often they can be represented by mathematically linear models (e.g., linear in the coefficients associated with higher order polynomials) that cause curvature in the Beer's law type plots of absorbance versus concentration. True mathematical nonlinearities cannot be completely modeled by these linear methods; but as described by Cahn and Compton (1988), they can approximate true nonlinearities if the nonlinearities can be precisely expressed as a power series expansion.

Inverse least squares is the simplest implicit or soft model method. However, when solved using the normal equations, the ILS method is limited in the number of spectral intensities that can be included in the analysis, while CLS is capable of being a full-spectrum method (i.e., with CLS, all spectral intensities can be included in the analysis without increasing the complexity of the analysis and without need for the inclusion of additional calibration samples). The ILS method uses the inverted form of Beer's law, in which concentration is modeled as a linear function of spectral absorbances. It will be shown that when the normal equations are solved in CLS and ILS (i.e., when the standard least-squares analysis is performed without any effort to eliminate collinearities in the data), these two methods employ different model assumptions, and they exhibit quite different properties. Multiple linear regression (MLR) is a term that is often used in the near-infrared literature to refer to the ILS method, since CLS methods are rarely used for near-infrared analyses. However, since both CLS and ILS use MLR techniques, the more general MLR designation could lead to some confusion, and its use in describing multivariate methods applied to infrared spectroscopy is not recommended.

Principal component regression and partial least squares

are two factor analysis-based methods that employ soft modeling. Factor analysis methods are used to factor the spectral data matrix into the product of two smaller matrices in order to simplify the representation of the data and to achieve a more stable solution to the ILS problem. Often PCR has been described simply as factor analysis, but PCR is only one of many possible factor analysis methods. The PCR method has also been called target factor analysis (Gemperline *et al.*, 1987), where the analyte concentrations are used as the target for obtaining the relationship between spectral data and concentrations. However, this term could be confused with a method by the same name that uses spectra rather than component concentrations as the target (Malinowski and Howery, 1980; McCue and Malinowski, 1981). In this latter case, a spectrum of the analyte is used as the target to determine whether the analyte is contained in the sample spectrum. Because PLS and PCR are capable of being full-spectrum methods, they have some of the increased flexibility of CLS as well as the advantages of the soft model methods.

A number of reviews of quantitative infrared spectroscopy are available (Smith, 1979; Brown and Obremski, 1984; Gillette *et al.*, 1985; Marcott, 1986; Griffiths and de Haseth, 1986; McClure, 1987, 1988). Some of these include a discussion of multivariate methods, but they generally do not review the newer factor-based methods of PLS or PCR, which are rapidly becoming more popular in the quantitative analysis of infrared spectra. In this chapter, the methods discussed in the preceding paragraphs are presented and compared in some depth, and their application to laboratory and industrial problems is reviewed. Yet, it must be emphasized that, although the multivariate statistical methods have been accurately described as being capable of extracting information from "dirty" (i.e., poor quality) data, starting with calibration data that are as high as possible in quality and that represent an adequate sample distribution should always be considered to be of primary importance. Planning and care at the beginning of a quantitative analysis problem will always be more valuable than trying to utilize the power of chemometrics to salvage data from a inadequately planned and poorly executed experiment. Therefore, a section of this chapter outlines the planning and care that should be the initial focus of all quantitative FT-IR analyses.

In this chapter, no attempt was made to present an exhaustive review of the literature. Rather, representative papers that illustrate the use of multivariate calibration and prediction methods for the quantitative analysis of infrared spectral data are referenced.

II. EXPERIMENTAL ASPECTS OF QUANTITATIVE FT-IR ANALYSES

A. Methods for Obtaining Spectra for Quantitative Analyses

Methods for collecting infrared spectra, sample preparation techniques, and potential pitfalls and errors have all been reviewed numerous times and are only mentioned here (Smith, 1979; Hirschfeld, 1978, 1979a; Krishnan and Ferraro, 1982; Gillette *et al.,* 1985; Marcott, 1986; Griffiths and de Haseth, 1986; Willis *et al.,* 1987). Transmission methods for solutions and gases are probably the most common, but quantitative transmission studies have also been made using KBr wafers, mulls, cast films, free-standing thin films, and bulk solid materials. Attenuated total reflection (ATR) methods, especially those using the CIRCLE cell (Sperline *et al.,* 1986; Braue and Pannella, 1987a,b; Miller *et al.,* 1988), are becoming popular for solutions, but selective adsorption of solutes on the ATR element is a constant concern. The ATR method can also be used for powders, textiles, rubbers, and plastics, but one must be careful to ensure that reproducible contact with the ATR element is achieved (Harrick, 1967). Quantitative specular reflection from thin films on a reflecting surface is possible

(Walder *et al.*, 1984), but highly nonlinear effects between absorption bands and interference fringes become important if the thin films have a thickness similar to the wavelength of the infrared radiation (Swallow and Allen, 1982; Allen and Swallow, 1982). Diffuse reflection methods, which are very popular in near-infrared spectroscopy (Wetzel, 1983; Martens, 1985), are starting to be used for quantitative FT-IR analyses of solid samples (Fuller and Griffiths, 1978; Hamadeh *et al.*, 1984; McKenzie and Koenig, 1985; Fredericks *et al.*, 1985a–c; Blitz *et al.*, 1986; Murthy and Leyden, 1986; Murthy *et al.*, 1986; Reinecke *et al.*, 1987; Thompson and Palmer, 1988; Brimmer and Griffiths, 1988). Photoacoustic spectroscopy can be used for the quantitative analysis of solids (Highfield and Moffat, 1985; Thompson and Palmer, 1988; Tsuge *et al.*, 1988). Infrared emission has been used in the analysis of hot gases (Herget, 1979), and emission methods are finding use in the quantitative analysis of condensed phases (Lauer *et al.*, 1985). With the introduction of redundant apertures on infrared microscopes, quantitative FT-IR microscopy is becoming a possibility, but stray light effects must still be evaluated for their presence, since these can cause significant nonlinear behavior (Friedman *et al.*, 1987; Messerschmidt, 1988).

B. Sample Preparation and Design

When obtaining or preparing samples for analysis, it is crucial that the samples be well mixed, as homogeneous as possible, and representative of the entire chemical sample or mixture from which the sample was obtained. Particle sizes should be uniform or smaller than the wavelength of the infrared radiation used in order to eliminate the large nonlinearities in band heights and shapes due to scattering and dispersion in refractive index (Duyckaerts, 1959). Interference fringes due to plane parallel reflecting surfaces must be avoided or corrected (Hawranek *et al.*, 1976; Hirschfeld and Mantz, 1976; Clark and Moffatt, 1978). Samples or sample cells that are wedged can cause quantitative errors (Koenig, 1964; Hirschfeld, 1979b), and errors will occur if any part of the beam bypasses part of the sample (e.g., pin holes, cracks, bubbles, nonuniformly distributed or sized particles, or the infrared beam extending beyond the edges of the sample) (Jones, 1952; Hirschfeld and Cody, 1977). Packing density, particle size, temperature, humidity, and sample height are important for quantitative diffuse reflection (Murthy and Leyden, 1986; Murthy *et al.*, 1986) and probably photoacoustic spectroscopy as well. Sample temperatures should be constant, especially for gases or samples with components that experience molecular associations. Since the infrared beam can cause a significant temperature rise, equilibration of the sample temperature in the spectrometer beam can be important. For gases, pressures need to be controlled unless pressures are above a limit where pressure broadening is no longer causing changes in the natural line widths of the spectral bands. For very dilute gases and solutions, care must be taken to assure that neither adsorption nor reaction with the walls of the infrared cell is taking place.

The linearity of Beer's law is affected by the aforementioned factors as well as by molecular or chemical interactions. These can often be minimized by using diluted or buffered solutions. Dispersion in refractive index occurring at strong absorptions causes nonlinearities due to nonconstant sample reflectance that results in intensity-dependent band distortions. Multiple reflections can cause nonlinear behavior, which is especially important in materials that have a high index of refraction, e.g., Ge or Si (Leroueille, 1982). Particle size-dependent scattering causes distortion (Christiansen effects) and frequency-dependent changes in effective pathlengths. Unexpected components in the samples with overlapping spectral features can introduce large errors (C. W. Brown, 1986) and should be included in the calibration or at least detected so that results from these samples can be checked by independent methods.

The selection of the calibration samples is crucial to the success of the quantitative analysis, especially when the number of components present in the sample becomes large. Often times, the calibration samples can be independently prepared. In these cases, efficient statistical experimental designs such as factorial designs (Box *et al.*, 1978) and mixture designs (Scheffe, 1958; Cornell, 1981; Haaland and Thomas, 1988a; Cahn and Compton, 1988) can be employed to select the composition of the calibration samples. In general, component concentrations are varied in a systematic and orthogonal fashion to obtain the greatest amount of information from a restricted number of samples. These designs can allow interaction and nonlinear effects to be detected and included in the analysis. When samples cannot be independently prepared, random sampling from a large population of chemical samples may be required. In this case, lower precision is generally achieved from a given number of samples relative to the controlled and designed calibrations (Haaland and Thomas, 1988b). In addition, it is more difficult in a random calibration to assure that all sources of variation and the entire range of variation expected in the unknown samples are included in the calibration. If all the sources of independent variation in the spectra that might be present in an unknown sample are not included in the calibration, then quantitative errors can be expected for unknown samples that contain additional sources of variation. In general, extrapolation beyond the range of variation in the calibration samples is subject to greater uncertainties and errors than are obtained by interpolations. Hruschka and Norris (1982) and Honigs *et al.* (1985) have listed methods for selecting samples to include in the calibration if spectra from all available samples are measured. Samples selected for inclusion in the calibration set will then be those representing the greatest sources of variation in the spectra. These methods will only be appropriate if a separate and slow reference method must be used to obtain the concentrations or properties necessary for the calibration.

Calibration samples should also be selected to ensure

that the range of variation in component concentrations or properties is large relative to the precision to which the calibration samples can be made or relative to the precision of the reference method used to determine the concentrations or properties of the calibration samples. In any quantitative analysis, the signal-to-noise ratios of both the spectra and the analyte concentrations will affect the precision of the estimated concentrations of unknown samples (Haaland and Thomas, 1988c). As an example, a S/N of 1000 in the spectra cannot be fully used if the range of analyte concentrations is only 10 times the precision of the reference method (i.e., concentration S/N = 10).

In designed calibrations, the target composition of a quality control application should be the center of the design. The range of concentrations should be much greater than the precision of making or independently analyzing the concentration. These features of experimental design are required both to improve the precision of the FT-IR analysis and to prevent the need for recalibration when the target value changes. The trade-off with increased range of calibration concentrations is that nonlinearities and model complexities increase as the concentration range is increased. As an extreme example, pure components could be used, but they are usually not representative of the actual sample mixtures, which generally contain a myriad of sources of nonlinearities. However, it should be stressed that careful planning of the calibration samples is the most important aspect of the quantitative FT-IR analysis problem.

C. Spectrometer Considerations

Once the calibration samples have been prepared, focus shifts to the spectrometer. For quantitative analysis, it is important that the resolution of the spectrometer be high enough to ensure that resolution-related deviations in Beer's law do not degrade the results beyond the required precision of the analysis. Ramsey (1952) has shown that the errors due to inadequate spectrometer resolution increase as the absorption intensity increases. For highest accuracy, the spectrometer resolution should be much less than the natural line width of the analyte. Ramsey has shown that for Lorentzian bands at an absorbance of 1.0, there is a 3% difference in measured and true absorbances when the instrument band-pass is one-fifth the width of the band, and a 24% difference exists when the instrument band pass is only one-half the bandwidth. Nonlinearities in the spectrometer response are also usually present, and most become more important as the absorbance increases. In particular, detector nonlinearities and saturation (Chase, 1984) along with nonlinearities in the detector preamplifier output can become major sources of nonlinearity above one absorbance unit. New advances in detector preamplifiers have recently reduced the severity of these problems. Digitizer errors (Baghdadi *et al.*, 1986) can also affect spectrometer nonlinearities. Other problems with spectrometer nonlinearities have been identified and reviewed in the excellent chapter by Hirschfeld (1979a).

Although the signal-to-noise ratios of the spectra are important in a quantitative experiment, it is often found that spectral variations from other sources can greatly exceed the detector-related noise sources (Bertie *et al.*, 1986). Removing and replacing the same sample in the spectrometer often results in greater spectral variations than expected from detector noise alone. Small variations and inhomogeneities in the sample, its pathlength, and its surface as well as focal shifts, temperature variations, and reflection differences can all contribute to this sample-in, sample-out variation. Some of these can be minimized by taking efforts to reproducibly introduce samples in the same position every time a spectrum is measured. For liquids in liquid cells or ATR cells or gases in gas cells, these problems are reduced by not removing the cells from the spectrometer when samples are changed. These sample-in, sample-out effects might also be minimized by averaging multiple spectra taken during the course of repeated loadings of the sample into the spectrometer.

D. Selection of the Multivariate Calibration Method to Use in the Analysis

Once the calibration spectra are obtained, one must select an appropriate calibration method. Unfortunately, selection of the method to use has often been limited by the software supplied with the spectrometer. However, this limitation is rapidly becoming less important as more software is made available by the manufacturers of FT-IR spectrometers and by third-party sources. The number of methods currently available and their variations can make the selection of software quite difficult. Haaland and Thomas (1988c) have performed Monte Carlo calculations using various multivariate calibration and prediction methods applied to simulated data that exhibit many of the parameters affecting spectral calibration data (e.g., amount of spectral noise or reference concentration errors, numbers of spectral data points and numbers of calibration samples, design of the calibration set, the presence or absence of spectral baselines, resolution of overlapping peaks, relative intensities of spectral peaks, and the presence or absence of deviations in Beer's law). They concluded that no single calibration method is best but that each calibration technique has relative advantages and disadvantages that are dependent on the numerous factors influencing the spectral and concentration data. Selection also depends on the type of information required and how important qualitative information and outlier detection is to the quantitative analysis problem. Both computer simulations and experience with real data will generate the most detailed comparisons with the various multivariate calibration methods. However, by understanding the details of the concepts behind the methods, educated decisions about which method to use in a given analysis can be made. It is clear from the aforementioned simulation studies that errors in calibration concentrations can often dominate other factors and can be most detrimental to the results of any calibration method. Thus, the precision and accuracy of the reference calibration methods is an important determinant for the selection of the appropriate calibration method.

III. UNIVARIATE VERSUS MULTIVARIATE CALIBRATION

To enhance our understanding of the revolution occurring in quantitative FT-IR spectroscopy, it is useful to compare the older experimental methods with the new statistical methods presented here. In addition to the improvements in ease of use of ratio-recording spectrometers over optical null spectrometer methods, FT-IR spectrometers ushered in the era of computerized infrared spectroscopy, high frequency precision, large optical throughputs, tremendous S/N enhancements, and speed advantages. Because the spectral data generated by an FT-IR spectrometer are digitized, many statistical methods can be readily applied to quantitative analysis problems. In the past, it was necessary to locate an isolated or nearly isolated spectral band in order to quantify the analyte. Also, baseline variations often required subjective baseline corrections. Unfortunately, many samples did not conform to the isolated band requirement. Therefore, methods that provided chemical separations such as chromatographic or mass spectral methods replaced infrared spectroscopy as the method of choice in performing many quantitative analyses. However, the ability to use many spectral frequencies or even the entire spectral range in the analysis has greatly increased the types of samples that can now be quantitatively analyzed by infrared spectroscopy. It is no longer necessary to identify isolated analyte bands, to perform subjective baseline corrections, or even to be restricted to a determination of component concentrations. Now one can even estimate chemical and physical properties from the infrared spectra of samples.

The one-spectral-band-at-a-time analyses are known as univariate analyses (i.e., only the absorbance at one frequency is related to concentration). Methods that simultaneously use information at two or more frequencies are known as multivariate methods. Not only do multivariate statistical methods provide enhanced analysis of component concentrations, but these multivariate methods have also recently made possible the estimation of physical and chemical properties of materials from their infrared spectra (Fredericks *et al.*, 1985a–c; Haaland and Thomas, 1986). A simple illustration of the increased capability of multivariate methods for component concentration determination is provided by the comparison of the data presented in Fig. 1. Figure 1A shows that an impurity component whose spectrum overlaps that of the analyte can affect the spectrum of the analytic band; therefore, the accuracy of the analysis will suffer when the analysis is performed at a single frequency ν_1. The measured absorbance, A_m, at the analysis frequency, ν_1, for a sample containing the impurity is different from the true absorbance A_t of the analyte at that frequency. If one had obtained the calibration curve shown in Fig. 1B from the spectra of samples containing no impurity, then the presence of the impurity in the sample would yield an apparent concentration that may be quite different from the true concentration. This error will remain undetected if the intensity was measured at only one frequency. If the impurity is included in the calibration samples but varies randomly in concentration in the samples, a calibration plot similar to that in Fig. 1B will exhibit large scatter among the data, and the result will be both a poor calibration curve and concentration estimates that are in error for the unknown samples. But with analysis at more than one frequency, not only can the presence of the impurity be detected (Fig. 1C), but, if its presence is included in the calibration, quantitative analysis is possible with multivariate calibration methods even when the impurity and its concentration are unknown. An

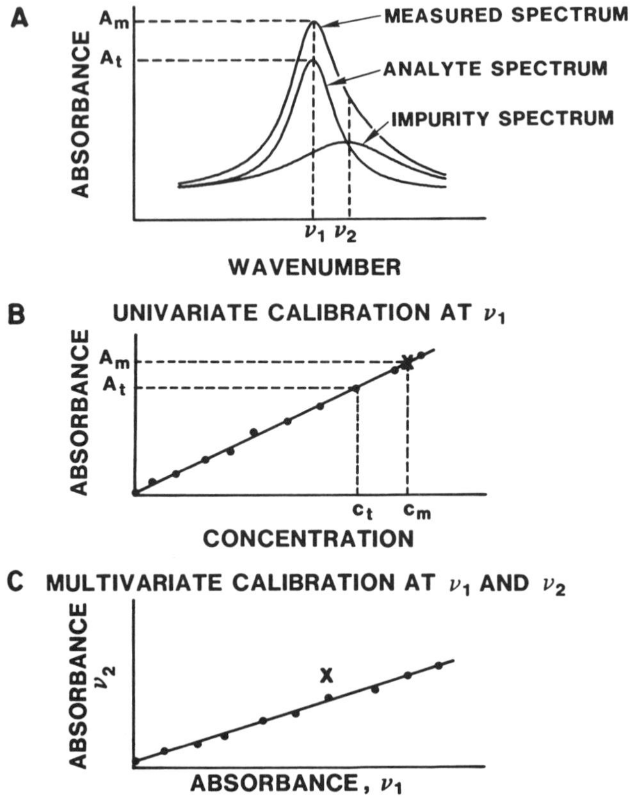

Fig. 1. An advantage of multivariate calibration over univariate calibration. The X in the two lower graphs indicates data obtained from a sample containing the analyte and the impurity.

indication that the unknown sample is different from the set of calibration samples not containing the impurity is obtained by plotting the absorbances of the calibration and unknown sample spectra at the two frequencies selected for analysis. As exhibited in Fig. 1C, the spectrum of the sample containing the impurity is obviously different from that of the calibration spectra (i.e., the unknown sample is an outlier whose analysis result must be suspect). The sensitivity in detecting outliers is generally increased by increasing the number of frequencies included in the analysis. The number of independently varying impurities that can be accounted for in the analysis is also increased by increasing the number of frequencies used. Accurate univariate methods are dependent on the ability to identify a unique, isolated band for each analyte. Multivariate methods can be used even when there is overlap of spectral information from the various components over all measured spectral regions. Unlike univariate methods, multivariate techniques can achieve increased precision from the redundant information in the spectra, can account for baseline variations, and can provide outlier detection.

When statistical calibration methods are applied to quantitative spectroscopy, separate calibration and prediction steps are generally required. These are illustrated in Fig. 2 for

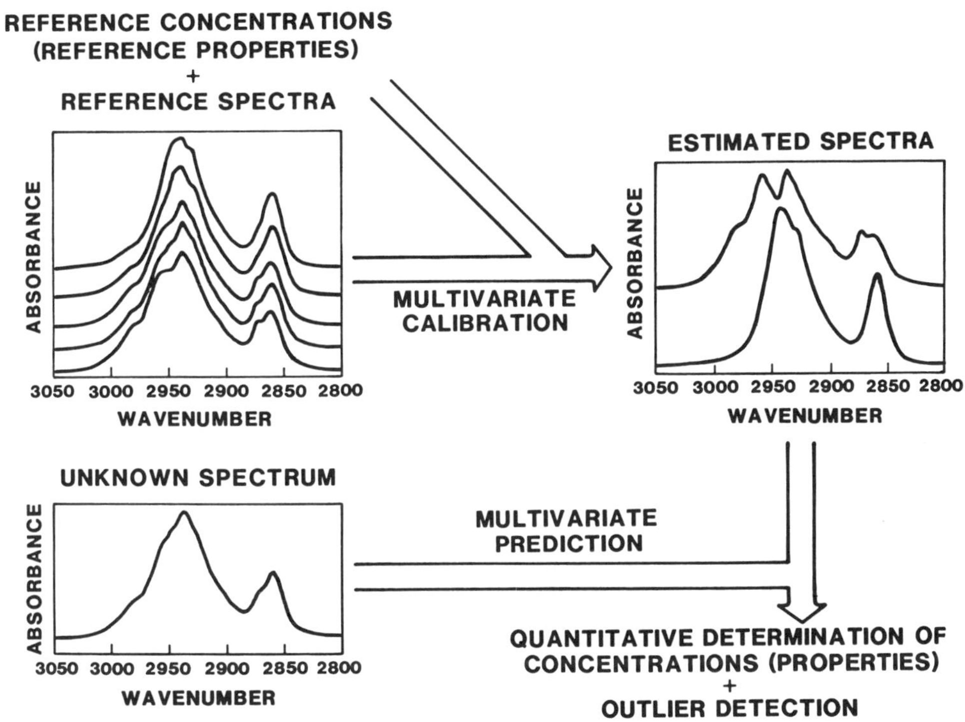

Fig. 2. Multivariate calibration and prediction applied to the analysis of FT-IR spectra.

multivariate calibration. One obtains the infrared spectra of a series of calibration standards that span the range of variation of all the factors influencing the spectra that are expected in the unknown samples to be analyzed. (Figure 2 illustrates the calibration for component concentrations from the infrared spectra of the C—H stretching bands of mixtures of two esters in a 15-μm liquid cell.) Infrared spectra contain information about the composition and molecular structure of the material. So, in theory, a chemical or physical property that is related to the composition and structure represented in the infrared spectra can be estimated directly from the infrared spectrum. Because one usually does not know *a priori* how composition or properties are reflected in the infrared spectra, one performs the empirical calibration illustrated in the top half of Fig. 2. When one assumes that the calibration uses samples that contain all the components expected in the unknown samples and spans their expected range of variation, the calibration will be able to empirically account for (or at least approximate) nonideal behavior in Beer's law in the spectra independent of their source. The result of the calibration illustrated in Fig. 2 (classical least-squares calibration) is the estimated spectra of the two components in solution as they exist in the series of calibration mixtures. These estimated spectra can be compared with the measured spectra of the pure components, if available. Differences in the measured and estimated pure-component, spectra indicate deviations in Beer's law if the linear Beer's law was the model used in the calibration (Haaland *et al.*, 1985; Haaland and Barbour, 1985). Other multivariate calibration methods produce either abstract spectra (PCR or PLS) or simply vectors of intensities if data from isolated frequencies are used in the calibration (ILS). As will be described in more detail later, much useful information can be obtained about the calibration samples during the calibration phase of the analysis.

Once the empirical calibration relating spectra and component concentrations or properties has been performed, then the spectrum of the unknown sample can be used in the multivariate prediction step to estimate the component concentrations and/or properties that were included in the calibration. If the calibration samples were truly representative of the unknown sample, then the result of the analysis will be an estimate that should have a precision similar to that found in the set of calibration samples. In addition, spectral residuals (i.e., the difference between measured and estimated spectra) can be used to determine whether the unknown sample is similar to the calibration samples. If the unknown sample is not representative of the calibration samples, spectroscopic interpretation of the spectral residuals can often be made to determine the sources of any differences between unknown and calibration samples.

IV. CLASSICAL LEAST-SQUARES CALIBRATION

To obtain a detailed understanding of the various multivariate statistical methods, it is instructive to demonstrate one of the methods with a simple two-component sample system that can be solved exactly. Since classical least squares (CLS) is one of the easiest methods to understand conceptually, it will be illustrated in this example. The other multivariate methods use different models, but they can often be understood by using variations or abstract extensions of the CLS calibration and prediction models. The basis of the CLS quantitative spectral analysis is the linear Beer's law, which states that the absorbance A_i at each frequency i is equal to the absorptivity–pathlength product ($a_i b$ or k_i) times the concentration, c. For samples containing m components, Beer's law is

$$A_i - \sum_{j=l}^{m} k_{ji} c_j \tag{1}$$

The left side of Fig. 3 shows the spectra of two pure components made from computer-generated overlapping Lorentzian bands. The center portion of Fig. 3 illustrates two

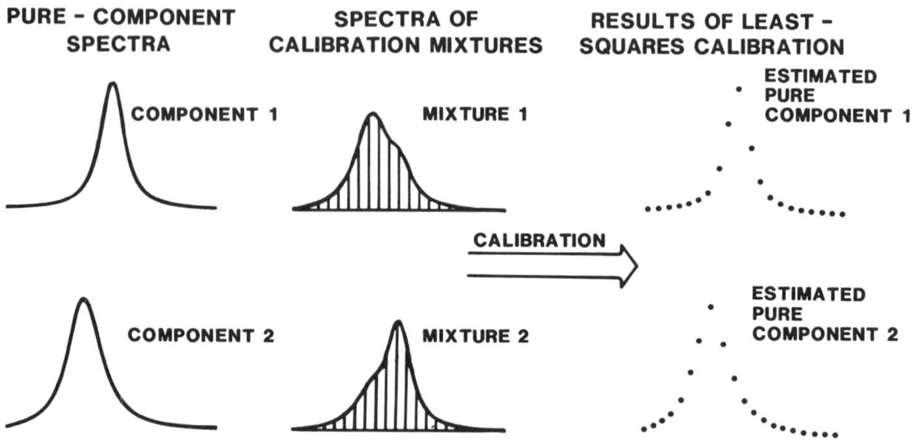

Fig. 3. Details of the classical least squares calibration.

calibration spectra generated by adding the appropriate amounts of the pure-component spectra according to Beer's law. At each frequency we can write two equations representing the total measured absorbance corresponding to the contributions from each of the two components. That is, at each frequency *i*,

$$
\begin{array}{lcccccr}
 & & \text{Component 1} & & \text{Component 2} & & \\
A_{1i} & = & k_{1i}c_{11} & + & k_{2i}c_{12} & \text{for mixture 1} & (2) \\
A_{2i} & = & k_{1i}c_{21} & + & k_{2i}c_{22} & \text{for mixture 2} & (3)
\end{array}
$$

Since these are calibration samples, the concentrations in Eqs. (2) and (3) are known and their absorbances have been measured. Therefore, at each digitized frequency in the calibration spectra, we have a system of two simultaneous equations in the two unknown quantities k_{1i} and k_{2i} that can be solved exactly. If this procedure is repeated at each frequency, then a plot of the k_{1i} and k_{2i} quantities determined from the solution of these simultaneous equations reconstructs the pure-component spectra at unit concentration and unit pathlength for each component, as shown at the right side of Fig. 3. If this procedure is performed at all digitized frequencies, then the entire spectrum of each component is reconstructed, albeit one frequency at a time. If more than two calibration samples are available and their spectra are measured, then there will be more than two equations in two unknowns. The general method to solve this overdetermined case is a least-squares analysis where the sum of the squares of the differences in the measured and estimated spectral intensities are separately minimized at each frequency. In addition, the analysis can be extended to include mixtures with more than two components. The least-squares estimated pure-component spectra generated during the calibration can then be used in the prediction phase of the analysis. Since the spectra of the mixture samples may contain deviations in Beer's law, the estimated spectra represent the linear least-squares estimate of the deviations over the concentration and spectral range contained in the calibration samples. Thus, they will more accurately represent the pure-component spectra as they exist in the mixtures. They will also yield better predictions when analyzing the spectra of the unknown samples.

In the prediction phase of the analysis, the estimated pure-component spectra are used to analyze the unknown sample spectrum, as illustrated in Fig. 4 for a sample that is a 50:50 mixture of the two pure components. One again performs a least-squares analysis, but now the analysis can be accomplished at all frequencies simultaneously, i.e., one uses a least-squares curve-fitting procedure. In Case I, the least-squares analysis is providing an answer to the question: "What is the best fit, in a least-squares sense, of the estimated pure-component spectra to the spectrum of the unknown sample?" That is, what linear combination of the estimated pure-component spectra gives the fit to the sample spectrum that minimizes the sum of squared differences between the sample spectrum and this linear combination of estimated pure-component spectra. Since the amount of the pure-component spectrum of each component in a mixture spectrum represents

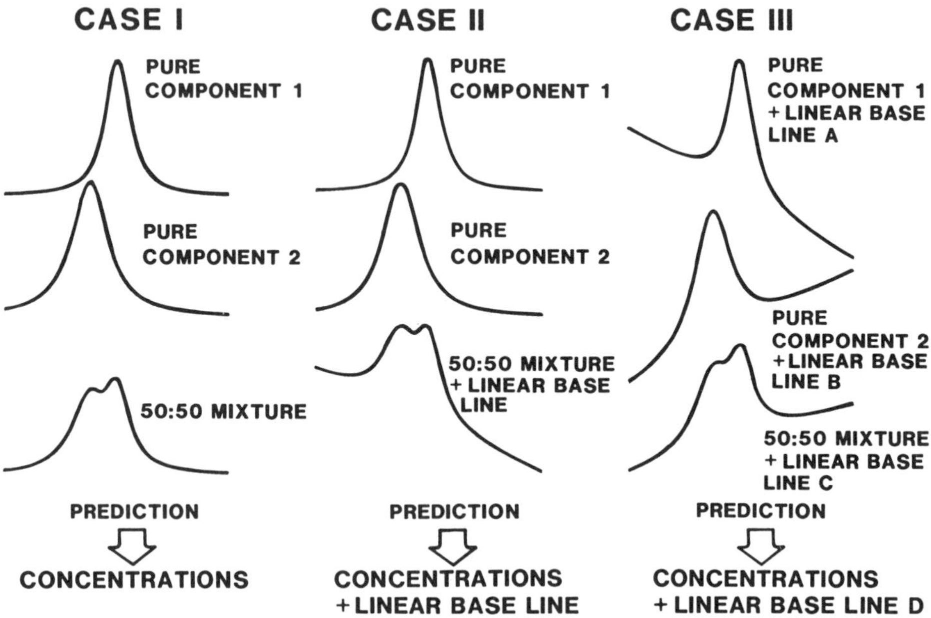

Fig. 4. Demonstration of classical least squares prediction. Case I: Calibration and unknown sample spectra all have the same spectral baseline. Case II: Calibration spectra have a constant spectral baseline and the unknown sample spectrum has a random linear baseline. Case III: All calibration and unknown sample spectra have random linear baselines.

its concentration in the mixture, the quantitative analysis of the unknown sample is complete. Again the prediction can be extended to samples containing more than two components.

Often it is found in real samples that baseline variations unrelated to sample component concentrations are present in the spectra. This situation is illustrated in Fig. 4, Case II, where a linear baseline has been added to the spectrum of the unknown sample. Assuming a linear model for the baseline, one can fit the sample spectra with a linear combination of the two pure-component spectra, a baseline offset and a linear slope. The baseline offset and slope can be treated simply as additional components whose "concentrations" (i.e., their magnitudes) are generally not of interest. In fact, even if there are random linear baselines superimposed upon the calibration and unknown sample spectra (see Fig. 4, Case III), this same approach can be used to eliminate the effects of random linear baselines in all spectra. This task can be accomplished because any linear combination of linear baselines still results in a linear baseline. The only difference in the results of the least-squares analyses from Fig. 4, Cases II and III, is that the estimated baseline offsets and slopes will be changed. The least-squares estimated concentrations of the analytes are identical in all cases presented in Fig. 4. The baseline fitting is not limited to linear baselines. Any functional form of the baseline (e.g., higher order polynomials, sines and cosines, exponentials, or linear combinations of these) may be used. The assumption that is made in the analysis is that the functional form of the baseline chosen is similar for both the unknown and the calibration samples. Better approximations of complex baselines can also be obtained by separately fitting simpler baseline models to each spectral band or to individual spectral regions and then pooling the results obtained from all spectral regions as described by Haaland and Easterling (1980, 1982). If the pooling of results is performed according to how well each spectral region is fit (i.e., inversely proportional to the estimated variances of each component in each band), then this latter method allows efficient deemphasis in the quantitative analysis of those spectral regions containing unexpected components or regions where Beer's law is not followed. Further improvements in CLS methods that have been presented previously include weighted least-squares calibrations and predictions, introducing nonzero intercept terms into Beer's law, accounting for pathlength variations, and using models other than the linear Beer's law model (i.e., models that include concentration cross-product terms, quadratic terms, or higher order terms) (see Haaland and Easterling, 1982; Haaland *et al.*, 1985; Haaland, 1986, 1987a). It should be noted that by centering the spectral and concentration data (i.e., removing the average spectrum and concentration from the calibration data), problems previously discussed (Brown *et al.*, 1982; Crocombe *et al.*, 1987) concerning the inclusion of nonzero intercepts in the spectral data during calibration are eliminated. Nonzero intercepts are useful for eliminating constant spectral backgrounds and for achieving better approximations of deviations in Beer's law (Haaland *et al.*, 1985).

These concepts are most readily presented in matrix form. In the remainder of the chapter, I use the convention that uppercase bold letters represent matrices, lowercase bold letters represent vectors, and letters in italics represent scalar quantities. Primes are used for transposed matrices and vectors. Vectors are expressed as column vectors, with row vectors being written as transposed column vectors. I follow the convention used in PLS and many PCR methods that the spectra included in **A** are presented as rows rather than as columns, as is commonly done with CLS and ILS. Therefore, Beer's law can be written for multiple samples at many frequencies as

$$\mathbf{A} = \mathbf{CK} + \mathbf{E_A} \tag{4}$$

where **A** is the $m \times n$ matrix consisting of the absorbances of each of the m samples at n frequencies. **C** is the $m \times l$ matrix of the l component concentrations in the m samples, and **K** represents the $l \times n$ matrix of pure-component spectra. $\mathbf{E_A}$ is the $m \times n$ matrix of spectral noise (or model error) present in the spectra. During calibration, the least-squares solution for the pure-component spectra **K** in Eq. (4) is simply

$$\hat{\mathbf{K}} = (\mathbf{C'C})^{-1}\mathbf{C'A} \tag{5}$$

where the $\hat{\mathbf{K}}$ is used to represent the least-squares estimate of the matrix **K** that minimizes the sum of squared spectral errors in Eq. (4). During prediction, the unknown sample spectrum **a** is also modeled with Beer's law. That is,

$$\mathbf{a} = \mathbf{K}'\mathbf{c} + \mathbf{e_a} \tag{6}$$

where the order of the matrix multiplication is reversed from that in Eq. (4) as a result of the convention that vectors are represented as column vectors. The least-squares solution for the component concentrations of the unknown is obtained by using $\hat{\mathbf{K}}$ from Eq. (5).

$$\hat{\mathbf{c}} = (\hat{\mathbf{K}}\hat{\mathbf{K}}')^{-1}\,\hat{\mathbf{K}}\mathbf{a} \tag{7}$$

To understand in more detail the significant advantages of CLS multivariate methods over univariate methods, it is useful to enumerate and explain some of the important advantages. As pointed out earlier, greater precision is possible with CLS relative to univariate methods. This can be illustrated by the simple spectrum presented in Fig. 5. This spectrum contains a single spectral band that has a rectangular line function containing n digitized intensities of equal intensity between ν_1

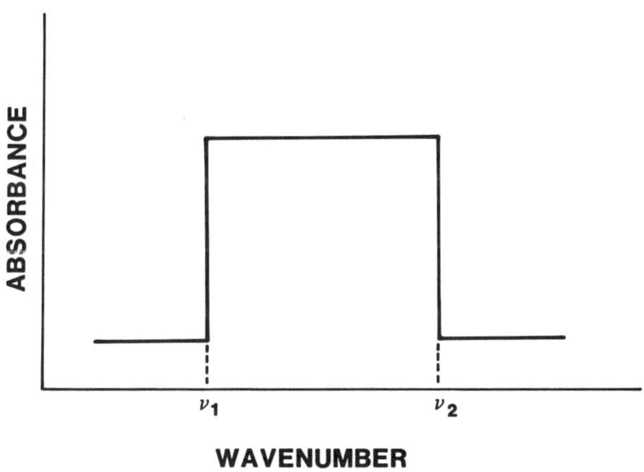

Fig. 5. Spectral band used to demonstrate improved precision when full-spectrum multivariate prediction methods are used in the analysis of spectral data.

and ν_2. Each digitized intensity contains noise plus an analyte signal that is the same at all n frequencies. If the baseline intensity is zero, if the noise at each frequency is independent, and if the variance of the noise is constant at all frequencies included in the analysis, then performing the quantitative analysis at all frequencies simultaneously during CLS prediction causes the signal from each frequency to add linearly while the noise at each frequency contributes as the square root of the number of frequencies. Thus, the improvement in S/N for quantitative analysis of the spectrum illustrated in Fig. 5 is simply $n^{1/2}$ relative to that which would be obtained by univariate methods. This result is similar to the S/N enhancement found when signal averaging n interferograms rather than one interferogram during data collection. Of course, real spectra do not normally have signal intensities that are equal (within the noise) at all frequencies in the band nor do they always have independent errors. Therefore, the $n^{1/2}$ improvement in S/N is not generally achieved, yet the improvement in S/N is still significant, especially if there are many data points with spectral intensity from the analyte. Haaland and Easterling (1980) have shown that for the analysis of the rotation–vibration band of N_2O centered at 2225 cm^{-1} for spectra taken at 0.06 cm^{-1} resolution, the improvement in quantitative precision relative to univariate methods is a factor of 12 for a given measurement time. If it had been desired to achieve the same precision as obtained from an analysis at a single frequency, a N_2O spectrum analyzed by CLS could have been obtained using a factor of 144 (i.e., 12^2) less time for signal averaging. If baselines are neither zero nor constant, then the baselines should be fitted simultaneously, and the improvement in concentration precision for a given measurement time falls to 9.6 or 4.8 for the N_2O example, depending on the baseline assumptions required for the analysis. Haaland and Easterling (1980) showed that the improved precision of CLS predictions coupled with baseline fitting permitted the quantitative analysis of N_2O even in the case where $S/N < 1$ at each frequency in the spectrum. This improved precision is demonstrated in Fig. 6 for one of the rotation–vibration bands of N_2O. Figure 6A shows the spectrum of 962 ppm (parts per million) N_2O in a N_2/O_2 mixture taken with 200 interferograms signal averaged. The spectrum in Fig. 6B is that of 3.5 ppm N_2O with the same amount of signal averaging but with the absorbance scale expanded by a factor of 10 so that the spectral features can be seen just above the noise level. The lower spectrum, in Fig. 6C, is that of the same 3.5-ppm sample but

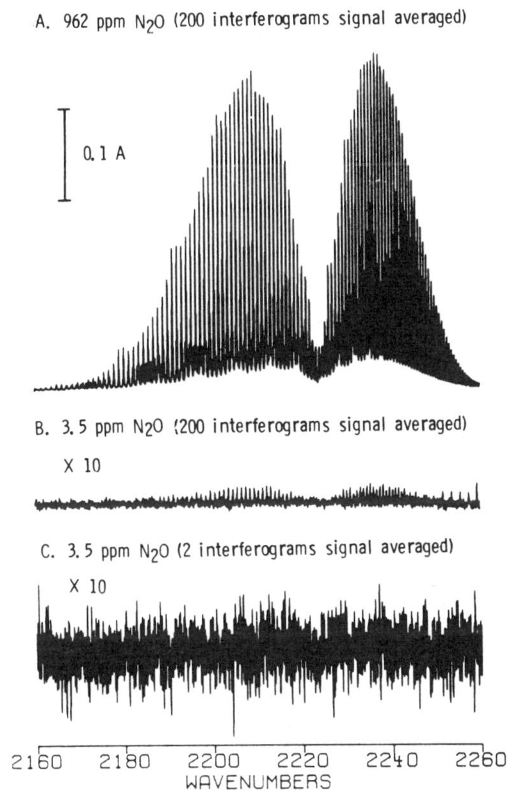

Fig. 6. High-resolution infrared spectra of N_2O in 80:20 mixtures of N_2/O_2 at 640 Torr in a cell with a 10-cm pathlength. (From Haaland and Easterling, 1980 with permission. Copyright Society for Applied Spectroscopy.)

with 2 orders of magnitude less signal averaging, thereby reducing the S/N by a factor of 10 and producing a signal that is clearly below the noise level. Yet the application of CLS prediction, which included the fit of separate linear baselines under each of the nearly 100 spectral peaks, and the use of the Fig. 6A spectrum as a reference resulted in a determination of the N_2O concentration as 4.2 ppm, with an estimated precision of ± 1.5 ppm at the 95% confidence level. It was shown that these methods improve the precision of the N_2O determination by a factor of 7 when compared to the determination using the intensity of the strongest feature in the spectrum (Haaland and Easterling, 1980).

In addition to the significant improvement in precision often obtained with the full-spectrum CLS method relative to that obtained with univariate or frequency-limited methods, CLS yields significant qualitative information that is not possible from either univariate or multivariate methods that are limited to small numbers of frequencies. For example, Haaland *et al.* (1985) showed that deviations in Beer's law could be identified by comparing the measured pure-component spectra with the estimated pure-component spectra. Any differences beyond the noise level represent deviations in the Beer's law model. Deviations caused by spectrometer-related nonlinearities or effects due to refractive index dispersion were separated from molecular interaction effects by performing studies as a function of pathlength, since only molecular interactions are independent of pathlength. Also, the presence, identity, and even concentration of unexpected components that are present in the unknown sample but not present in the calibration samples can sometimes be determined from an examination of the spectral residuals. For example, Haaland and Barbour (1985) in their FT-IR determination of polychlorinated biphenyls (PCBs represented as commercial mixtures of the PCB isomers called Aroclors) in transformer oils demonstrated the significant advantages of examining full-spectrum residuals. This work is represented in Fig. 7, where the measured and CLS-estimated sample spectra are com-

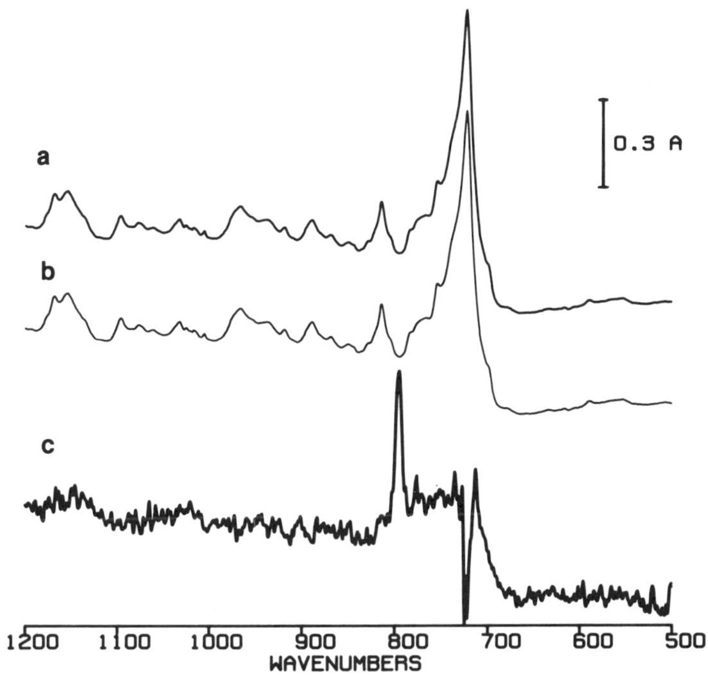

Fig. 7. (a) Measured spectrum of 0.5246% Aroclor 1242 and 0.4755% Aroclor 1254 in transformer oil in a cell with a 200-μm pathlength. (b) Spectrum generated from a linear combination of CLS-estimated pure Aroclor PCBs and transformer oil spectra using CLS-estimated concentrations. (c) Difference (a − b) scale expanded by a factor of 100. (From Haaland and Barbour, Copyright 1985 by International Scientific Communications, Inc.)

pared. Within the width of the lines of the two spectra displayed in Fig. 7a and b, the two spectra are the same. However, by expanding the spectral residual by a factor of 100, differences beyond the noise are observed as shown in Fig. 7c. These differences can be interpreted as the presence of an unexpected impurity with a single band at 790 cm^{-1}. This impurity caused a displacement of a corresponding amount of the sample, and this displacement is represented primarily by the negative deviation of the major 720 cm^{-1} band of the transformer oil. The impurity was identified as CCl_4 at a concentration of approximately 10 ppm based on the intensity of the 790 cm^{-1} band. Carbon tetrachloride was the last solvent used in cleaning the cell, and apparently it was not completely removed from the cell before the introduction of the sample. This level of CCl_4 was not detected in the FT-IR spectrum of the empty cell because of the presence of interference fringes in the spectrum of the empty cell. This experiment demonstrates both the potential power of examining residuals obtained from full-spectrum multivariate methods and attests to the high S/N possible with FT-IR spectra.

Finally, Ward *et al.* (1989) have shown that the CLS calibration applied to a reacting system of organometallic compounds and solvents could be used to identify which components in the 6-component mixture were reacting. These same methods were used to determine the nature of the reaction products. They compared measured pure-component spectra of all the initial components with CLS-estimated spectra obtained from a series of calibration mixtures with known amounts of each of the starting materials. Differences between these measured and estimated spectra indicated a failure of Beer's law. In reacting systems, these differences can be used to identify those components that react, and they can yield information about the products of reactions.

Even with all these advantages, there are some disadvantages to the CLS method compared with other multivariate methods that are available for quantitative infrared analysis. For example, in general, all components that have spectral features in the spectral region analyzed should be included in the analysis. This requirement can be relaxed significantly if the band-by-band analysis discussed earlier is performed with baseline fitting and pooling of the results from all bands. Thus, only those components that overlap all analyte spectral features and that cannot be fit by the functional form of the baseline model for each spectral band must be included in the analysis. Nevertheless, there are a significant number of real samples that fall into this latter category and for which all overlapping components are not known. In addition, molecular interactions, spectrometer nonlinearities, and poor baseline modeling can all lead to deviations in Beer's law that are only approximated as the linear estimates of the nonlinearities over the range of the calibration samples. Finally, CLS methods are not generally useful for estimating chemical and physical properties of samples from their spectra.

V. q-MATRIX METHOD

Some of the other multivariate statistical techniques do not have the aforementioned limitations of the CLS method. For example, the **q**-matrix method described by McClure *et al.* (1987) and McClure (1987) is quite similar to CLS without the limitations outlined earlier. This method basically follows the CLS prediction except that, rather than using measured or least-squares-estimated pure-component spectra, the spectra of mixtures are used as reference spectra. The **q**-matrix method is simply a least-squares curve-fitting approach to quantitative spectral analysis. The spectrum of an unknown sample is assumed to be a linear combination of the calibration mixture spectra. The **q**-matrix method can be illustrated with portions of Figs. 3 and 4. The calibration mixture spectra shown in the center of Fig. 3 are initially treated by the **q**-matrix model as though they were pure-component spectra. These calibration spectra are then used instead of estimated pure-component spectra in the CLS prediction of the 50 : 50 "unknown" spectrum in Fig. 4. The linear combination of these mixture spectra giving the least-squares fit to the unknown sample spectrum yields the amount of each calibration spectrum in the sample spectrum. However, the calibration spectra are not pure-component spectra but rather known mixtures of the pure components. The **q**-matrix estimated concentration of the analyte in the unknown sample is then determined from the concentration of the analyte in each calibration sample weighted by the amount of each calibration spectrum found in the least-squares fit of the unknown sample spectrum. Thus, the model for the unknown sample spectrum in the **q**-matrix least-squares analysis is

$$\mathbf{a} = \mathbf{A}'\mathbf{q} + \mathbf{e_a} \tag{8}$$

where the elements of **q** represent the amount of each calibration spectrum contributing to the fit of the unknown sample. The least-squares solution to Eq. (8) is

$$\hat{\mathbf{q}} = (\mathbf{AA}')^{-1}\,\mathbf{Aa} \tag{9}$$

which is analogous to Eq. (7) but with the measured calibration spectra replacing the estimated pure-component spectra. The component concentrations, **c,** in the unknown sample are derived from the **q**-matrix by summing the concentration contributions from each analyte in each calibration sample, i.e.,

$$\mathbf{c} = \mathbf{C}'\hat{\mathbf{q}} \tag{10}$$

where **C** represents the matrix of calibration concentrations. Although not stated explicitly by McClure *et al.* (1987), the analysis can actually be performed one component at a time if the variation of the component concentrations in Eq. (10) are uncorrelated. In addition, any chemical or physical property that is linearly related to the absorbances can be substituted for concentration in Eq. (10) to achieve a least-squares estimate of the property of the unknown sample. Thus, some of the limitations of the CLS method are overcome by the **q**-matrix method.

There are, however, difficulties with the **q**-matrix method. For example, there is no minimization of spectral errors in the calibration spectra. In addition, if Beer's law is followed for the calibration and unknown samples, then the **q**-matrix method will yield poorer results than the CLS method when there are more calibration samples than chemical components (i.e., the overdetermined case). In this case, the **q**-matrix method relies on spectral noise to keep the matrix to be inverted in Eq. (9) from becoming singular. That is, without the presence of noise in this ideal system, there will be only as many sources of variation in the data as there are components with spectral intensities at the frequencies analyzed. If additional spectra beyond the number of components are used, they will necessarily be linear combinations of the other calibration spectra within the spectral noise. Since the addition of each spectrum adds an unknown to Eq. (9) of the **q**-matrix method, the result is that there will be more unknowns than linearly independent equations. The presence of noise will keep the matrices from being truly singular, but the noise will make the solution to the matrix inversions unstable, and the precision of the analysis will be reduced. Therefore, the **q**-matrix method overfits the data since noise and model error in the calibration spectra are not minimized in any step of the analysis as they are in CLS method with its separate least squares calibration phase of the analysis. (Overfitting the data implies that in addition to reproducible sources of variation in the data, noise and model error are included in the calibration model. This overfitting then increases prediction errors.) In addition, the often valuable qualitative information that is available in the CLS calibration (i.e., the direct estimates of the pure-component spectra) is not obtained with the **q**-matrix approach. If many calibration samples are available for infrared analysis but slow reference methods of analysis are required to calibrate the infrared spectra, then the method of Hruschka and Norris (1982) for selecting spectra to use in the calibration can help reduce the overfitting problem of the **q**-matrix method. They obtained the spectra of all available samples and use only those spectra representing the greatest variance in their calibration. These fewer numbers of samples are than calibrated by the slow reference method and the remaining samples are analyzed by the calibrated infrared analysis.

Nyden and co-workers (Nyden and Babrauskas, 1987; Nyden *et al.*, 1988) also have presented several papers describing a **q**-matrix approach. However, they used numerically stable methods of calibration that circumvent some of the problems with overfitting described earlier. In effect, they eliminated spectra in the calibration set that, within the spectral noise level, can be represented as linear combinations of the other calibration spectra. This procedure makes their method closer to the principal component regression method (described later) than to the **q**-matrix method described by McClure *et al.* (1987).

VI. INVERSE LEAST-SQUARES METHOD

Another multivariate method that has been used with infrared spectra and that predates (Barnett and Bartoli, 1960; Sternberg *et al.,* 1960; see also Brown, 1982, 1986a; Brown *et al.,* 1982; Kisner *et al.,* 1983; Haaland, 1987b) the application of either CLS or the **q**-matrix techniques to infrared spectra is inverse least-squares (ILS). The multivariate ILS method can be illustrated most easily by comparing CLS and ILS multivariate methods when they are limited to the univariate case. The center portion of Fig. 1 illustrates the univariate classical least-squares calibration with absorbances at a single frequency expressed as a linear function of concentration. In this univariate calibration, the least-squares fit to the calibration data represents the line that minimizes the sum of squared vertical distances of the data to the line. This model assumes the error is exclusively in spectral intensities with no error in reference concentrations. The ILS method models concentration as a function of absorbance and assumes errors are exclusively in the reference concentrations. Thus, ILS interchanges dependent and independent variables in the Beer's law model. In the ILS model, the least-squares line fitted to the data in the center of Fig. 1 would be that line which minimized the sum of squared horizontal distances of the data to the line. In general, the ILS and CLS solutions to the calibration data will result in different calibration lines. These different solutions also mean that predictions of unknown sample concentrations will be dependent on the model used in the calibration. In the extension to multivariate analyses, additional differences become important, as discussed later.

The multivariate ILS method starts with the inverse of the Beer's law model such that component concentrations are described as a linear function of spectral absorbances, i.e.,

$$\mathbf{C} = \mathbf{A}\mathbf{P} + \mathbf{E}_c \tag{11}$$

where **C** and **A** are as defined previously, **P** is the $n \times l$ matrix of coefficients relating the l component concentrations to the spectral intensities, and $\mathbf{E}_c$ is the $m \times l$ matrix of concentration noise or errors that are not fit by the model. If the l concentration errors in the rows of **C** are uncorrelated, then Eq. (11) can be written for each component separately as

$$\mathbf{c} = \mathbf{A}\mathbf{p} + \mathbf{e}_c \tag{12}$$

During calibration, the least-squares solution for **p** is

$$\hat{\mathbf{p}} = (\mathbf{A}'\mathbf{A})^{-1}\,\mathbf{A}'\mathbf{c} \tag{13}$$

Unlike the CLS model, the number of frequencies that can be included in Eq. (13) is restricted. Therefore, the **A** that is used in the ILS model in Eq. (11) is generally much smaller than the **A** used in the CLS and **q**-matrix models in Eqs. (4) and (8). Thus, ILS does not use the curve-fitting procedures of the CLS or **q**-matrix methods. During ILS prediction, the single analyte concentration, c, is then determined by applying the model in Eq. (12) to the unknown sample spectrum.

$$\hat{c} = \mathbf{a}'\hat{\mathbf{p}} \tag{14}$$

When the normal equations are solved in Eq. (13) (i.e., the straight least-squares analysis is performed by inverting the full $(\mathbf{A}'\mathbf{A})$ matrix), the ILS method is quite restricted in the number of frequencies that should be included in the analysis. One practical limitation to the number of frequencies is that there can be no more frequencies added than calibration samples available for analysis. This limitation is imposed because each frequency added to Eq. (13) adds another unknown, so an additional equation (calibration spectrum plus analyte concentration) must be added to prevent Eq. (13) from being underdetermined. In addition, it can be shown that if more frequencies are included than true sources of variation in the data other than noise (e.g., variations due to component concentrations, baselines, and nonlinearities), then the ILS model will be overfitted and poorer results are to be expected (Haaland and Thomas, 1988c).

A significant difficulty with the ILS method that often occurs is the selection of how many frequencies and which frequencies are to be used in the analysis. Until very recently, statistically based frequency selection methods had not been employed in the analysis of infrared spectral data [see Sasaki *et al.* (1986) and Donahue *et al.* (1988) for some of the first applications of frequency selection methods to ILS analyses of mid-infrared data]. However, a variety of frequency selection methods have routinely been used in the quantitative analysis of near-infrared spectra, and these could be readily adapted to infrared analyses [see Honigs *et al.* (1983) for a review of some frequency selection methods]. Unfortunately, for spectra containing large numbers of spectral intensities, algorithms for frequency selection and the determination of optimal numbers of frequencies to include in the analysis can be extremely computationally time-consuming. Another area of deficiency in the ILS method is that full-spectrum residuals are not available for spectral analysis and interpretation. Spectral residuals are available at each of the small number of frequencies used in the analysis, and these can be used to determine the quality of the least-squares fits, but they are not generally useful for spectral interpretation. Finally, because of the limitation in number of frequencies used in the analysis, ILS does not provide the improvement in analysis precision experienced by CLS [see Haaland (1985) for a comparison of CLS and ILS methods]. Nevertheless, ILS methods have been successfully applied to many quantitative infrared analyses.

VII. FACTOR ANALYSIS METHODS

A. Partial Least Squares and Principal Component Regression

Factor analysis-based methods can overcome some of the limitations of the CLS, **q**-matrix, and ILS methods. In fact, factor analysis methods are needed to obtain full-spectrum solutions to the ILS model. There are a variety of factor analysis methods available for quantitative infrared analysis, but only variations of two main methods have been presented in the infrared literature. These are principle component regression (PCR) (Fredericks *et al.*, 1985a–c; Haaland and Thomas, 1988a,b; Haaland, 1988; Malinowski and Howery, 1980; Jolliffe, 1986; Wold *et al.*, 1987; Cahn and Compton, 1988) and partial least squares (PLS) analyses (Fuller *et al.*, 1988a,b; Haaland and Thomas, 1988a,b; Haaland, 1988; Cahn and Compton, 1988; Lindberg *et al.*, 1983; Martens, 1985; Hoskuldsson, 1988; Geladi, 1988). Often in the chemical literature, factor analysis has been equated with only principle component analysis (PCA), which is the first step of PCR. Factor analysis, however, is more general; and as applied to quantitative spectroscopy, it attempts by a variety of methods to decompose the calibration spectra into a series of factors that represent the spectra to within the noise level. Thus, the spectra are decomposed into the product of two matrices as

$$\mathbf{A} = \mathbf{TB} + \mathbf{E}_\mathrm{A} \tag{15}$$

where $\mathbf{T}$ is an $m \times r$ matrix of scores, $\mathbf{B}$ is the $r \times n$ matrix of loading vectors, and $\mathbf{E}_\mathrm{A}$ is the matrix of spectral noise and model error. As before, m is the number of samples in the calibration model, and n is the number of frequencies used. The dimension r of the model generally is chosen to represent the rank of **A,** such that optimal prediction properties are achieved. This dimension might be considered as the number of spectral components required to obtain optimal predictions. Spectral components can arise from true chemical components, chemical interactions, baseline variations, or nonlinearities.

When the factor analysis method uses principal component analysis as the method to decompose **A,** then the rows of **B** are the eigenvectors (sometimes called eigenspectra) of the product matrix **(AA′).** The eigenvectors represent those vectors that account for the maximum variance in the calibration spectra. The factor analysis presented in Eq. (15) is simply a transformation of the coordinate system that compresses the intensities at n frequencies to the r intensities in the new full-spectrum coordinate system of the r eigenvectors. In fact, it can be seen that the CLS model given in Eq. (4) is also a factor analysis model where the factors are constrained to follow Beer's law, with one factor representing each analyte that is present. Thus, the intensities in the CLS factor analysis are the component concentrations, and the new full-spectrum coordinates are the pure-component spectra. The CLS-based factor analysis is just a restatement of Beer's law that the sample spectra are linear combinations of the pure-component spectra. With CLS, the amounts (or intensities) of each pure-component spectrum in the sample spectrum are the concentrations of the pure components.

Unlike the CLS method, Eq. (15) is not restricted to the same number of factors as known chemical components. Therefore, as explained in more detail by Cahn and Compton (1988), apparent Beer's law nonlinearities can be modeled by PLS and PCR factor analysis methods. For example, if there are chemical interactions (which, in the extreme case, might be chemical reactions), then oftentimes more than the original number of chemical components will be required to represent the variations in the calibration spectra. If the concentration of the analyte is measured by a reference method after the reaction has taken place, then PLS or PCR factor analysis methods will require that more factors be added to represent the components generated during the reaction. Thus, the sources of variation in the spectral data are implicitly modeled by these factor analysis methods while CLS has imposed an explicit model (Beer's law) on the data. True mathematical nonlinearities cannot be modeled with these linear factor analysis models, but they may often be approximated by them (Cahn and Compton, 1988).

Once the factor analysis of the calibration spectra has been obtained, then the scores (or intensities in the new full-spectrum coordinate system of the loading vectors) are related to the concentrations with a model similar to that used in the ILS model:

$$\mathbf{c} = \mathbf{T}\mathbf{v} + \mathbf{e}_c \tag{16}$$

where $\mathbf{v}$ is the $r \times 1$ vector of coefficients relating scores to concentrations, and $\mathbf{T}$ is the matrix of scores (intensities in the new coordinate system) determined during the PLS or PCA decomposition in Eq. (15). The analogy to the ILS model in Eq. (12) is straightforward. The details of the methods used to factor analyze the calibration spectra are given elsewhere (Martens, 1985; Lindberg *et al.*, 1983; Geladi and Kowalski, 1986a,b; Haaland and Thomas, 1988a). It should be noted that the CLS "factor analysis" method does not require relating scores (i.e., intensities in the full-spectrum coordinate system where the CLS-estimated pure-component spectra are the new basis vectors) to component concentrations since the scores have been generated in CLS in such a way that they already represent the linear least-squares fit to the component concentrations.

The spectral decomposition of PCR and PLS will result in the generation of "abstract" spectra rather than the estimated pure-component spectra determined by CLS (see Fig.

2 for CLS calibration results). Therefore, some qualitative information will be lost during PCR or PLS calibration relative to CLS calibration. Haaland and Thomas (1988a) have shown that the calibration results of PLS generate qualitative spectral information that is generally poorer than that of CLS but better than that possible with PCR. The difference between PCR and PLS factor analysis methods lies in the fact that PCR performs the factor analysis in such a way that the factors describe only the variance in the calibration spectra. Variations in the spectra that are not related to the concentration of the analyte may receive large weight in the analysis, if, for example, baseline variations or major interfering components dominate spectral variations due to the analyte concentration differences. The PLS method, however, has been designed to extract factors that simultaneously account for the variance in the spectral data and correlate with analyte concentrations. Thus, better concentration predictive ability is expected for PLS relative to PCR when a portion of the variation in the spectral data does not correlate well with concentration variations. This improved prediction ability of PLS has been demonstrated through computer simulations (Haaland and Thomas, 1988c) at the expense of poorer fits of the spectral data than obtained with PCR. Since prediction of analyte concentrations is usually our primary goal, this poorer spectral fit should not be a major concern. During the PLS or PCR prediction phase, a least-squares curve-fitting procedure is used to find the best fit of the PLS or PCR loading vectors to the unknown sample spectrum. The amounts (scores or intensities) of each loading vector yielding the linear least-squares fit to the unknown spectrum are then related to concentrations. This relation is based on the calibration coefficients which relate scores to concentration (Eq. 16). Thus, PLS and PCR predictions are analogous to CLS prediction, but the abstract PLS or PCR loading vectors are used in the prediction rather than estimated pure-component spectra. However, an additional inverse least-squares relation between intensities (scores) and concentrations is required for PLS and PCR.

It should be emphasized that if noise is not present and Beer's law is followed, all full-spectrum methods discussed here will yield identical results, given sufficient precision in the computer. The major differences between the methods are how they account for noise and model error. The CLS and the **q**-matrix methods minimize spectral noise, ILS minimizes errors in concentration, while PCR and PLS contain steps that separately minimize both spectral errors (using the model in Eq. 15) and concentration errors (see Eq. 16). It can also be shown that when the **q**-matrix method and ILS are solved using the pseudoinverse [see Lorber *et al.* (1987) for a discussion of the psuedoinverse as it applies to PLS and PCR] and the full-rank singular value decomposition (SVD) retaining all eigenvectors with nonzero eigenvalues, then these two methods are identical. The SVD analysis is a numerically stable method of performing PCA on the calibration data. It removes all collinearities in the calibration concentrations and spectra. These collinearities are not removed when the "normal" equations are used to obtain the **q**-matrix and ILS solutions (Eqs. 9 and 13, respectively). When solved using SVD methods, **q**-matrix, ILS, and PCR all become equivalent. It should be stressed that, as generally implemented, these methods do not use SVD because it is computationally expensive. But in the limit of numerically stable computations, these three methods become equivalent. The only difference between the **q**-matrix method and the ILS method when the pseudoinverse is obtained from SVD is whether concentrations or spectra are considered the independent variables. Differences in this selection of independent variables have caused some confusion in the literature because the labels "scores" and "loading vectors" become interchanged when the calibration spectra are written as rows or columns of the **A** matrix. Fredericks *et al.* (1985a,b) have described the **A** matrix with the columns being the calibration spectra, while Haaland and Thomas (1988a,b) have followed the near-infrared and PLS literature and have written the **A** matrix with the spectra in the rows, a practice that has resulted in differences in the labels of the scores and loading vectors in these papers.

B. Selection of the Optimal Number of Factors in PLS and PCR

One difficulty with PLS and PCR methods is how to choose the optimal number of factors to include in the PLS or PCR model. As illustrated in Fig. 8, adding more factors increases the complexity of the model and allows a better fit of the calibration data to be obtained. However, the addition of more factors also increases the amount of noise and model errors that are included in the PLS or PCR model. The inclusion of too many factors results in overfitting the data, and poorer predictions on independent unknown samples will be obtained. At some number of factors, the better fit of the calibration data obtained by increasing model complexity is negated by the increased prediction error that results from overfitting of the data. The optimal number of factors to be used in the model will be that number of factors that corresponds to the

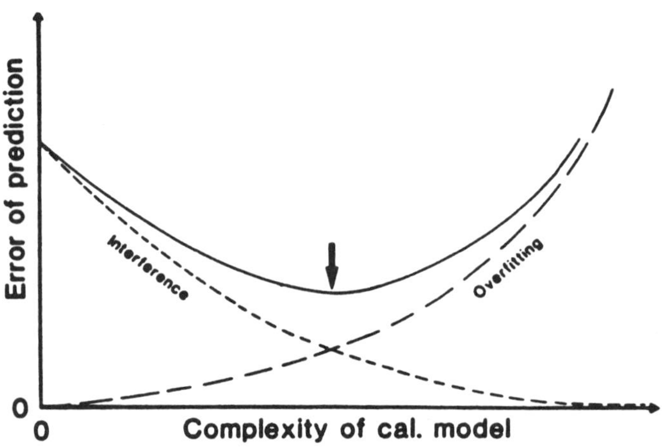

Fig. 8. Selecting the optimal model during calibration. The optimal model in factor analysis methods occurs when the improved performance of using more factors to model the interferences and other sources of spectral variation is countered by the poorer performance due to overfitting the calibration data. (From Martens and Naes, 1984 with permission. Copyright Elsevier Science Publishers.)

point at which decreasing error from increased model complexity is compensated by the increasing error from overfitting, and at which a minimum in prediction error is produced (see Fig. 8). This trade-off between model complexity and overfitting can be demonstrated with a simple situation where three calibration spectra have been measured at a single frequency that contains a single component known to follow the linear Beer's law model. However, because noise in the spectra or concentration errors are present, the data will generally not fall on a straight line. By employing a quadratic model, we could fit the three calibration data without error. However, at this point, we have overfit the calibration data, and we expect poorer prediction results on independent unknown samples than we would obtain if the calibration data had been modeled with a least-squares fit to a straight line, which is the correct model in this example.

There are many methods for selecting the optimal model, but cross-validation is one of the best in utilizing the available calibration data (Stone, 1974; Wold, 1978; Haaland and Thomas, 1988a,b; Osten, 1988). However, cross-validation can require extensive computer resources. Cross-validation is accomplished by removing a fraction of samples from the calibration set and calibrating on the remaining samples, with prediction performed on the samples left out of the calibration. The process is repeated with the same fraction of different samples held out of the calibration each time. This procedure continues until all samples have been held out once. Leaving one sample out at a time is preferred but requires more computing time than leaving out a larger fraction of the samples. A measure of total prediction error is used to determine the optimal number of factors to include in the analysis. Often the sum of squared prediction errors (PRESS, or prediction error sum of squares) is used as the optimal number of factors. However, Haaland and Thomas (1988a) have recommended using the number of factors that are less than that required for the minimum PRESS if the smaller number of factors yields a PRESS that is not statistically different at the 0.5 level of significance. This latter method often provides a simpler model that further guards against overfitting the calibration data. Cross-validation also allows more efficient outlier detection among the calibration samples (Lindberg *et al.*, 1983; Haaland and Thomas, 1988a,b). Other methods to select numbers of factors tend to be less expensive computationally and many have been described by Malinowski and Howery (1980) (see also Donahue *et al.*, 1988). These latter methods use only spectral information, and selection of factors is independent of calibration concentrations. Often, many sources of variation in the spectral data are not useful for prediction, and inclusion of these can decrease prediction precision. In addition, it is often empirically found that different numbers of factors give optimal predictions for different components or properties (Fuller *et al.*, 1988a,b; Haaland, 1988; Fredericks *et al.*, 1985a–c). Thus, factor selection methods that rely solely on the spectral data are best used when computer resources are limited or if computation speed is an important consideration. However, it is this author's opinion that the amount of information achieved by cross-validation leaving out one sample at a time is usually worth the extra computation time given the large expenditure in time often required to obtain good calibration standards.

VIII. CROSS-CORRELATION AND KALMAN FILTER METHODS

Cross-correlation and Kalman filter methods are two additional multivariate analysis methods that have been used in spectral analyses. However, these methods are not currently available in the software supplied by FT-IR instrument manufacturers, and very little has been published using these methods for infrared spectral data analysis. Therefore, these methods will only be briefly described here, and the cited literature should be consulted for details of these two methods. It is expected that of the two methods, the Kalman filter will have the larger impact on the analysis of infrared spectral data in the future.

A. Cross-Correlation Method

Mann and co-workers have presented the bulk of the application of cross-correlation methods to infrared and Raman spectroscopy (Mann *et al.*, 1982; Marley *et al.*, 1984; Tyson *et al.*, 1984a,b; Lin *et al.*, 1987; Shope *et al.*, 1987). In this method, the spectrum of the analyte to be quantified is cross-correlated with the unknown sample spectrum containing the analyte. The reader is referred to these references for a detailed description of the cross-correlation method applied to spectral data. However, it should be noted that when data are not filtered in the Fourier transform phase of the cross-correlation calculation but are processed in the same manner as in CLS, cross-correlation methods are mathematically equivalent to CLS prediction methods (Lam, 1983). The mathematical procedures of performing Fourier transforms on the spectral data to accomplish the cross correlation provide for filtering the data, and this often causes different results between the two methods. Tyson *et al.* (1984a) found that cross correlation yielded better results than CLS predictions in the infrared determination of $^{13}CO_2$. However, in the infrared analysis of lipid mixtures, CLS predictions yielded better results than the cross-correlation analysis.

B. Kalman Filter Method

To date, only one application of the Kalman filter to infrared spectroscopy has been published in the literature (Monfre and Brown, 1988). Because of this limited application to infrared spectroscopy, the Kalman filter will only be given cursory treatment in this chapter. The use of the Kalman filter in spectroscopic analyses has been reviewed several times (S. D. Brown, 1986; Cooper, 1986; Rutan, 1987). The Kalman filter is an efficient, recursive method for analyzing multivariate data. It has the advantage that it can be continually updated to modify the model by using the spectral residuals (called innovations) to correct for model error. It also can be used for studying kinetic data since it has the ability to fit and predict time-varying spectral data. In their pioneering work, Monfre and Brown (1988) showed that the extended Kalman filter applied to the infrared spectra obtained during the hydrolysis of two esters in aqueous solutions can yield reaction rates and initial reactant concentrations in the presence of interactions between the solvent, reactant, and products. Additional results of the Kalman filter applied to FT-IR spectral data will be expected from this and other groups in the future.

IX. VARIATIONS IN MULTIVARIATE CALIBRATION METHODS

There are many variations of the multivariate methods presented earlier, and it is instructive to outline some of them here. Haaland (1986, 1987a) has introduced the use of either cross product or quadratic concentration terms in the CLS model in order to better account for deviations in Beer's law. This modification improved the quantitative FT-IR determination of esters in binary solutions of similar esters, N_2O in samples whose spectra were obtained at relatively low resolution, and boron in a series of seven-component bulk glasses. Brown and co-workers (Brown, 1982; Maris *et al.*, 1983) also introduced higher order polynomial terms in absorbance intensities to better model deviations in Beer's law when applying ILS methods. They demonstrated improved quantitative precision for the analysis of various gas mixtures of methane, ethane, and propane when the spectra were obtained at a low resolution (25 cm^{-1}) to induce Beer's law deviations in the spectra. However, evidence of overfitting of the data sometimes occurred when too many fitting parameters were added, for example, the standard errors of prediction increased with the addition of more parameters. In these cases, the simpler models more correctly modeled the available data. Kisner *et al.* (1983) showed that a quadratic term added to the ILS infrared analysis of serum lipids generally resulted in a poorer analysis precision. This result suggested that the serum lipids more closely followed the linear Beer's law model over the concentration range used for the calibration samples.

Brown and co-workers (Brown *et al.,* 1986a,b; Donahue *et al.,* 1988) have also recommended performing multivariate analyses in the Fourier domain. They take the Fourier transform of the spectral absorbance data, which results in the formation of "Fourier vectors." The transformed data are then used directly in the multivariate analysis rather than the original absorbance spectra. When spectral data are transformed in this fashion, broad baseline variation information is concentrated in the early elements of the vectors while only high-resolution information with lower S/N is contained in the later elements. Therefore, they propose that a small number of data points starting shortly after the first few elements in the Fourier vector can often be used more effectively in the analysis. These selected data in the Fourier vectors are less contaminated by baseline variations and noise. It has been stated that this processing of the data can serve to improve the precision of the analysis as well as minimize computation times by reducing the amount of data used in the quantitative analysis. Brown *et al.* (1986a) and Donahue *et al.* (1988) have used both Gram-Schmidt orthogonalization and principal component analysis to perform factor analysis in the Fourier domain with considerable success. They compare CLS and ILS methods in the spectral domain to factor analysis methods in the Fourier domain and find in all cases that the analysis in the Fourier domain resulted in better precision (or in one case the same precision) as the methods applied in the spectral domain. However, because different methods were tested in the spectral and Fourier domains, it is not clear whether the improvements noted were due to the application of the more powerful factor analysis methods or because the data were processed in the Fourier domain. More work needs to be performed in this area to clarify the advantages of analyzing spectra in the Fourier domain. One might expect that analyzing data in the Fourier domain will be most efficient for moderately overlapped spectral features with spectra that have spectral noise as their dominant source of model error. In other cases, one might expect poorer results by processing data in the Fourier domain. For example, in the case of isolated bands, overlap is increased in the Fourier domain. In the case of high overlap, a truncated Fourier vector may not provide adequate resolution to resolve components for quantitation. Finally, if spectral model error is not dominated by spectral noise, low-resolution spectral information will likely dominate the model error, and useful spectral information may be lost in the truncation. In each of these cases, lower precision might be expected with an analysis in the Fourier domain. However, since this type of analysis in the Fourier domain results in an analysis of low-resolution spectral features, it more closely approximates a band area analysis and has the effect of deemphasizing the importance of sharp spectral features in the analysis. Often band areas follow Beer's law better than peak heights, and strong, sharp spectral bands experience more deviations in Beer's law than broad bands. Thus, in these cases, analysis in the Fourier domain may yield better analysis results. Careful systematic evaluations will be required to identify more precisely the advantages of processing data in the Fourier domain.

In principal component regression, several variations have been presented. Haaland and Thomas (1988a,b) have followed the near infrared literature and used factors in the order in which they contribute to the spectral variance. They eliminated only those factors representing smaller variance if they did not improve predictions. On the other hand, Fredericks *et al.* (1985a,b) calculate most of the principal components and use only those that are statistically significant in influencing the prediction results. A trade-off in this latter case represents a compromise between reduced precision of the estimate and increased bias. Major sources of variation in the spectra that are not important for prediction (e.g., baseline variations) might be more effectively excluded by the methods of Fredericks *et al.* (1985a,b), while they might be included in the methods of Haaland and Thomas (1988a,b). Thus, much like the PLS model, correlation with concentration may be improved with the PCR method of Fredericks *et al.* (1985a,b). However, Fredericks *et al.* did not give the significance level required before factors are eliminated. If the significance level is set too high, then fewer factors than necessary will be included in the model and underfitting will be the result. The net result would be a lower prediction precision for unknown samples. In addition, Fredericks *et al.* did not use cross validation in the selection of factors. Outlier detection can be less sensitive without the use of cross validation. Thus, the relative advantages of these methods awaits further study.

There are also several versions of PLS that have been described. A PLS analysis may be performed one component at a time, in which case it has been described as PLS1 (Martens, 1985). If the calibration is performed using two or more chemical components simultaneously, then the method involves a more complex interative procedure that is called PLS2 (Martens, 1985). In general, PLS1 has better predictive properties unless component concentrations are correlated in both calibration and unknown samples beyond the normal correlation that all components sum to a mole fraction of one. However, more data storage space is required for PLS1 to save the calibration results used for prediction if many components are to be quantified. In addition, different PLS algorithms for producing either orthogonal loading vectors or orthogonal scores are available, but these have been shown to yield identical results (Manne, 1987).

X. ERROR ANALYSIS, DIAGNOSTICS, AND OUTLIER DETECTION

A question that should always be asked when performing any analysis is, "How good are the results of the analysis?" This question should be asked both of the calibration results and the analysis of the unknown samples. This question is more readily answered for the calibration and validation samples, because concentrations have been determined for these samples. Therefore, relative errors or standard errors (either standard error of estimation, SEE, for the calibration samples or standard error of prediction, SEP, for the validation samples) can be used to estimate the quality of the calibration and to determine the precision that might be expected for unknown samples that are representative of the calibration samples. Methods to calculate these errors are given in Table I. Fuller *et al.* (1988a) recommended that the concentrations of the various components be plotted against each other to assure that the concentrations of various components in the calibration or validation samples are not correlated. Correlated concentrations can reduce the precision of the analysis or even prevent an analysis from being performed if the correlation is quite high. Properly designed calibrations and even random calibrations can reduce the concentration correlations in the calibration. Fuller *et al.* (1988 a,b) have shown that plots of estimated concentrations (or concentration errors) of the calibration samples versus reference concentrations are useful in spotting significant outliers. Fredericks *et al.* (1985b); Haaland and Barbour (1985); Fuller *et al.* (1988a); and Haaland and Thomas (1988b) have all shown that spectral outliers can be identified from their spectral residuals. Others (Lindberg *et al.,* 1983; Haaland and Thomas, 1988a,b; Haaland, 1988)

Table I Measures of Error[a]

Ave. % relative error $= \sum_{i=1}^{m} \frac{\lvert \hat{c}_i - c_i \rvert}{m}$	
$\mathrm{SEE}^b = \left[\sum_{i=1}^{m} \frac{(\hat{c}_i - c_i)^2}{m - df} \right]^{1/2}$	where df = number of components, frequencies, or factors used in the analysis for CLS, ILS, and PLS or PCR, respectively, and is increased by 1 if the data are centered
$\mathrm{SEP}^c = \left[\sum_{i=1}^{m} \frac{(\hat{c}_i - c_i)^2}{m - df} \right]^{1/2}$	where $df = 0$ if the date are not centered or 1 if they are centered

[a] c_i and $\hat{c}_i$ are the reference and estimated concentrations for the ith sample, m is the number of calibration or prediction samples included in the analysis, and df is the number of degrees of freedom.

[b] SEE has also been called internal standard error of prediction and root mean square error of estimation.

[c] SEP has also been called external standard error of prediction and root mean square error of prediction.

have shown that the use of F-statistics for both concentration and spectral residuals can be quite effective at identifying outliers both in the calibration samples and the unknown samples. These statistics essentially compare the results of a single calibration sample or unknown sample to that of the other calibration samples. If the results are unusual compared to that of the other calibration samples (i.e., relatively large F-statistic), then that particular sample can be flagged as an outlier. The outlier sample can be studied further for the cause of its being an outlier (e.g., incorrect analyte determination by the reference method, unique type of sample, unexpected components, unusual baseline, incorrectly labeled or documented sample). During calibration, F-statistics are most useful if cross validation is performed, because a single outlier sample might be able to be accurately fit by a factor analysis-type method using all calibration samples. The unique sample will often be fit by the addition of a single factor that represents the new features of the outlier sample spectrum. Thus, F-ratios can be small for an outlier sample when all calibration samples are included in the calibration while they will generally be large during cross-validated calibrations.

The precision of the analysis can also be estimated in the calibration and prediction for any of the methods presented. For example, Haaland and Easterling (1980, 1982) have given a procedure to estimate the precision of the CLS analyte determination based on the spectral residuals. However, this estimated precision tends to be better than the actual precision, because it is based on the assumption that the spectral errors are independent and normally distributed and that there are no errors in the calibration spectra or their concentrations. Similarly, Fredericks *et al.* (1985a,b) and Fuller *et al.* (1988a,b) have used the Mahalanobis distance to estimate errors when performing PCR and PLS, respectively. The Mahalanobis distance indicates how far the reduced spectrum (scores) of a particular sample is from the center of the reduced spectral space of the calibration samples for normally distributed multivariate data. Since errors are greater for samples further away from the center of the calibration, the Mahalanobis distance gives an indication of the concentration precision. The Mahalanobis estimation of precision is overly optimistic, because it is based on the assumption that the concentration errors in the unknown samples are independent and normally distributed and that there are no spectral errors in either the calibration spectra or the unknown sample spectrum. The Mahalanobis estimate of precision is especially poor for true outlier samples. Nevertheless, these measures of precision can be used as guidelines for the relative quality of the analysis of any given sample.

In the case of well-designed experiments, the standard error of prediction for a series of unknown samples that are not found to be outliers should be bounded by the SEP found during cross validation of the calibration samples and the SEE for the model with all calibration samples included. A sample's leverage or influence in the calibration can also be investigated to identify samples that are relatively unique and therefore heavily influencing a single factor in factor analysis therefore heavily influencing a single factor in factor analysis methods. A more complete discussion of outlier detection and influence measures is described by Chatterjee and Hadi (1986) in their excellent review of outlier detection and influence measures. However, the methods summarized by Chatterjee and Hadi are strictly appropriate only when errors are predominately in the dependent variable. Additional developments need to be made to address the case where errors are present in both the dependent and independent variables.

XI. APPLICATIONS

In the following sections, I present a collection of representative applications of multivariate calibration methods to the analysis of infrared spectra. It is not meant to be an exhaustive review of the literature. Also, the discussion is generally restricted to applications involving analyses of mid-infrared data. Therefore, many important applications involving the analysis of spectra in other regions of the electromagnetic spectrum have been excluded from this discussion.

A. Gas Analyses

The quantitative FT-IR analysis of gas-phase samples has a relatively long history of use with multivariate calibration. Maker *et al.* (1979) used an iterative scheme employing spectral intensities at large numbers of frequencies for the FT-IR analysis of gases of interest in automotive exhaust gases. Maker *et al.* used an iterative subtraction method that minimized the point-to-point variation in the sample absorbance spectrum during the subtraction of a reference spectrum of the analyte from the unknown sample spectrum. This method was used to find the subtraction scale factor required to remove the spectrum of the analyte. This scale factor is then proportional to the concentration of the analyte gas in the unknown sample. This method could be used even in the presence of overlapping components. Rivord (1979) in our laboratory used this same method for the FT-IR analysis of CO, CO_2, and N_2O in N_2/O_2 mixtures. However, the long computation times and lack of precision estimates for the final results caused Haaland and Easterling (1980) to develop CLS prediction methods with simultaneous baseline fitting to speed computation times, to provide more precise concentrations, and to obtain estimates of precision of the final concentration results. They demonstrated that these CLS methods could improve detection limits for CO, CO_2, and N_2O in N_2/O_2 mixtures. Detection limits were less than 1 ppm for each of the three gases studied in cells with a 10-cm pathlength, using spectra obtained at 0.06-cm^{-1} resolution. The improved precision of the multivariate prediction combined with simultaneous baseline fitting allowed these gases to be detected even when the S/N was less than one at any frequency. Later Haaland (1987a) used models that included quadratic terms in concentration to account for resolution-related deviations in Beer's law when spectrometer bandwidth was increased to 0.5 cm^{-1} in the determination of N_2O in air samples.

Brown (1982) and Maris *et al.* (1983) used ILS methods applied to the infrared spectra of the C—H stretching bands to determine gas-phase concentrations of light alkanes. They each introduced deviations in Beer's law by degrading the spectrometer resolution to 25 cm^{-1}. By adding terms quadratic in absorbance to the least-squares analysis, they were able to better model the data, and they achieved more accurate results than possible with the linear model.

Strang and Levine (1989) and Strang *et al.* (1989) have used CLS with simultaneous baseline fitting to monitor air quality for eleven gases important in the semiconductor industry (e.g., arsine, borane, phosphine, freons). They found that the CLS methods always yielded better limits of detection than methods using single frequency peak heights. Their potential improvements in the limits of detection were limited, however, by the necessity to limit the spectral region analyzed to maintain linearity and to avoid interference with common infrared-active atmospheric components. Ying and Levine (1989) have also used CLS with simultaneous baseline fitting of FT-IR spectra to provide air monitoring for multicomponent gas mixtures for gases important in environmental and industrial hygiene air monitoring. Their work indicates that CLS applied to the FT-IR monitoring of air is a viable method for detecting gases below the parts-per-million level in many cases.

Nyden and Babrauskas (1987) have proposed using the **q**-matrix method coupled with FT-IR to analyze gases generated during combustion of materials. Although they do not present experimental results, their methods have been tested on simulated mixtures containing 50 components, and the **q**-matrix method gave accurate quantitations down to 10 ppm.

Tyson *et al.* (1984b) have used both cross-correlation and least-squares techniques applied to dispersive infrared spectra to determine the ^{13}CO-to- ^{12}CO ratios, which are important in breath tests as a noninvasive method of monitoring metabolic disorders. They found that the cross-correlation method gave better results than the CLS prediction. However, the lack of frequency precision of dispersive infrared and the absence of baseline fitting coupled with digital filtering of the data when cross correlation was used may have influenced the relative performance of these two methods. Tyson *et al.* concluded that by using multivariate analyses, infrared determination of ^{13}C enrichment was adequate for this application.

B. Analyses of Organic Liquid Mixtures

Antoon *et al.* (1977) were probably the first to use CLS prediction methods applied to the quantitative analysis of infrared spectra. They adapted the methods that Blackburn (1965) developed for the quantitative analysis of gamma ray spectra. They applied these methods to the quantitative analysis of the FT-IR spectra of mixtures of the three xylene isomers. This pioneering work showed that accurate quantitative determinations of components with highly overlapping spectra could be achieved using multivariate curve fitting methods. Haaland and Easterling (1982) extended these CLS methods to include simultaneous baseline fitting and least-squares weighting appropriate for IR spectra. Pooled results from band-by-band analyses achieved greater precision and accuracy. Pooled results also reduced effects due to Beer's law deviations and the presence of unexpected components in the unknown sample. Mann *et al.* (1982) used cross-correlation methods to analyze mixtures of *o*- and *m*-xylene and *p*-anisaldehyde in *m*-xylene. Brown and co-workers on several occasions (Brown *et al.*, 1982, 1986a,b; Kisner *et al.*, 1982, 1983; C. W. Brown, 1986; Donahue *et al.*, 1988) have applied a variety of methods to the quantitative infrared analysis of xylenes and related substituted phenyl groups and mixtures of other organic liquids. Brown *et al.* (1982) used ILS to analyze mixtures of *p*-xylene and pseudocumene. Kisner *et al.* (1982) used a frequency-limited CLS prediction method to analyze chloroform solutions of triglyceride, phospholipid, and cholesteryl ester. Since a multicomponent CLS calibration step was not used, a one-component-at-a-time calibration was used with all other components approximated as a constant solvent to be fit by the inclusion of a nonzero intercept. Kisner *et al.* (1983) extended the latter analyses using ILS calibration and prediction with a nonzero intercept to improve the analysis precision. When introducing multivariate analysis in the Fourier domain, Brown *et al.* (1986a,b) used either principal component regression or Gram-Schmidt orthogonalization factor analysis methods to analyze mixtures of 2-butanone and acetonitrile in CCl_4 with and without additional impurities as well as binary mixtures of 2-heptanone and 2-octanone. C. W. Brown (1986) used the latter binary mixtures and ternary mixtures of *m*- and *p*-xylene and pseudocumene to compare ILS and CLS methods [see Haaland (1987b) for further discussion of this comparison].

Recently, Cahn and Compton (1988) applied PCR and PLS to the quantitative FT-IR analysis of the xylene isomers. They pointed out the importance of using statistically efficient calibration mixture designs for minimizing prediction error. Both PLS and PCR yielded similar results, and errors were within 0.002 weight fraction.

Haaland *et al.* (1985) were the first to combine multivariate CLS calibration and prediction methods to the quantitative analysis of FT-IR spectra. They showed that the accuracy of the analysis of binary mixtures of esters could be improved using CLS calibration applied to mixtures of the esters rather than using the measured pure-component spectra as references. Haaland (1987a) showed that more accurate analyses could be obtained from this same data by adding terms containing the cross product of the two ester concentrations to the normal linear Beer's law relationship. They also showed that useful qualitative information could be obtained from these full-spectrum CLS quantitative analyses (Haaland *et al.*, 1985). For example, Fig. 9A shows the spectrum of a two-component mixture of esters. Figure 9B shows the spectral residual after obtaining the linear least-squares fit to this sample spectrum when pure-component spectra were used as references. Residuals above the noise indicate in which regions of the spectra Beer's law is failing. Studies as a function of pathlength showed that the primary source of this failure in Beer's law was due to spectrometer nonlinearities or dispersion in refractive index effects rather than to molecular interactions. However, these model errors were reproducible, as evidenced by the lower residuals (Fig. 9C) found when the spectra of mixtures of the two esters were used as references in the CLS calibration. By modeling these Beer's law deviations over a narrower concentration range, lower residuals and more precise concentration estimates were obtained. Thus, average percent relative error was reduced from 3.8% when pure-component spectra were used to 1.7% when a se-

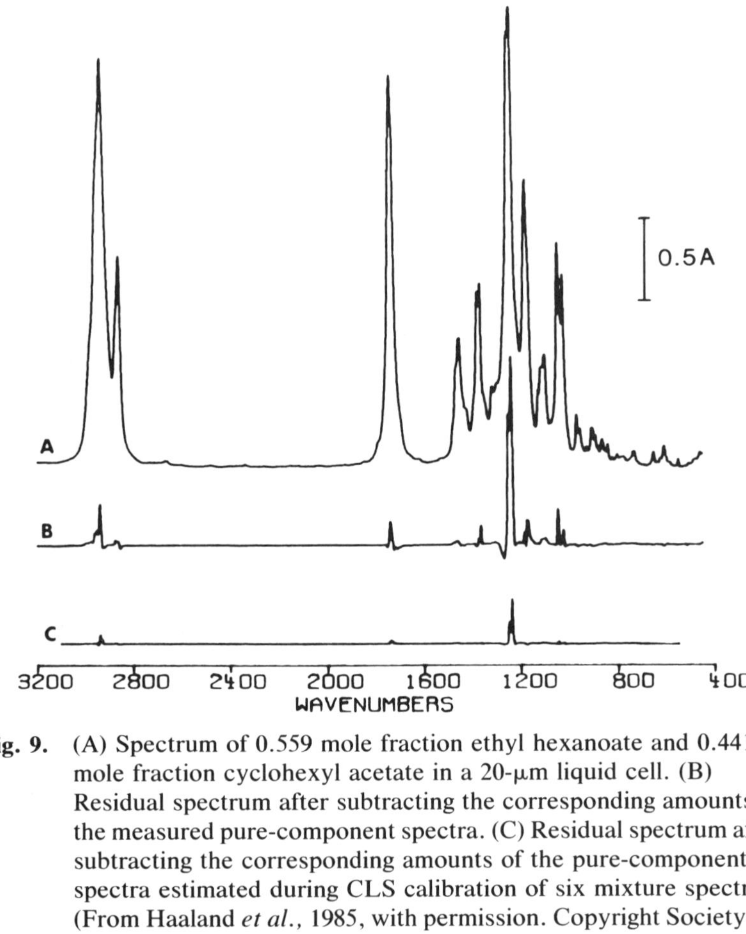

Fig. 9. (A) Spectrum of 0.559 mole fraction ethyl hexanoate and 0.441 mole fraction cyclohexyl acetate in a 20-μm liquid cell. (B) Residual spectrum after subtracting the corresponding amounts of the measured pure-component spectra. (C) Residual spectrum after subtracting the corresponding amounts of the pure-component spectra estimated during CLS calibration of six mixture spectra. (From Haaland *et al.*, 1985, with permission. Copyright Society for Applied Spectroscopy.)

ries of mixture spectra were used as references. Further improvements in precision were obtained by restricting the calibration mixtures used in the calibration to those that were closer in concentration to that of the unknown sample.

Haaland and Barbour (1985) also showed that the application of CLS calibration and prediction applied to the quantitative FT-IR analysis of PCBs in transformer oils could be used under laboratory conditions to detect the PCBs down to the 50-ppm level. This level of sensitivity was achieved despite the significant overlap in spectral features between the various classes of commercial PCB mixtures and the transformer oil.

Rao and Zerbi (1984) have used principal component analysis to determine the number of components during phase transitions of nonadecanes and $C_{24}H_{49}COOH$. Least-squares curve-fitting coefficients were used to monitor the concentration of the various species as a function of temperature.

C. Analyses of Coal and Minerals

Painter *et al.* (1981) have used CLS prediction methods to quantitatively determine the mineral matter in coal. Fredericks *et al.* (1985a–c) showed in their excellent papers on the application of factor analysis (in this case, PCR) to quantitative FT-IR that the analyses of coals and minerals for their constituents and properties could be realized using diffuse reflectance spectroscopy. They found that the carbon, organic hydrogen, oxygen, nitrogen, sulfur, and ash content of coals could be determined from their infrared spectra. Properties of coal that could be estimated from their infrared spectra included the amount of volatile matter, mineral matter, fluidity, grindability, specific energy, vitrinite reflectance, and volume percentage of vitrinite, inertinite, and exinite. Proper selection of range of variation in the calibration samples and spectral range were found to affect the precision of the results. A plot of specific energy determined by PCR applied to the diffuse reflection spectra of 20 unknown coal samples is shown in Fig. 10 as a function of specific energy measured by calorimetry. Correlations were also found for the constituents and/or properties of magnesium dioxide ore, bauxite, diesel fuel, and iron ore. This wide range of constituents and properties that could be determined from the infrared spectra of coals and minerals coupled with the relative speed of FT-IR

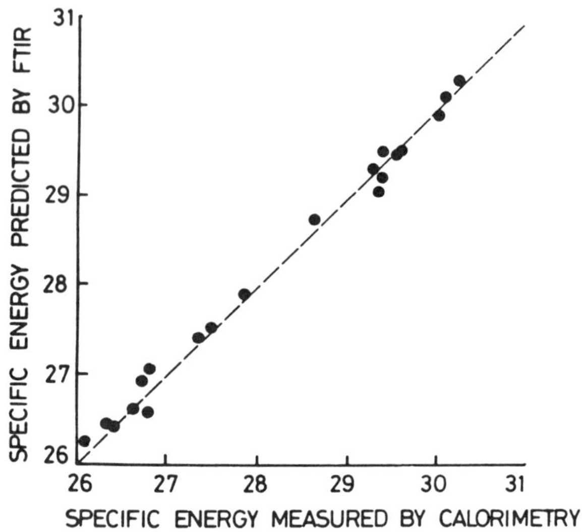

Fig. 10. Plot of measured specific energy (MJ/kg) against specific energy predicted from the diffuse FT-IR spectra using PCR for 20 coals; 60 similar coals were used as a calibration set. (From Fredericks *et al.*, 1985a, with permission. Copyright Society for Applied Spectroscopy.)

makes the use of infrared a tremendously powerful analytical tool.

Christy *et al.* (1987) used PLS coupled with diffuse reflection FT-IR to estimate the vitrinite reflectance of coals. Again by reducing the range of the calibration standards and eliminating outliers, the quantitative results were improved.

D. Infrared Analyses for the Microelectronics Industry

As previously mentioned, the use of CLS for rapid infrared analysis of air for components of environmental interest in the microelectronics industry have been successfully demonstrated (Strang and Levine, 1989; Strang *et al.*, 1989). Krishnan and Mundhe (1983) and Krishnan (1984) have demonstrated the use of CLS methods for the analysis of phosphosilicate glass (PSG), borosilicate glass (BSG), and borophosphosilicate glass (BPSG) thin films on Si wafers. These thin-film glasses are used as dielectric films for a variety of purposes in the processing of integrated circuits (ICs). Infrared spectroscopy serves as a rapid, nondestructive method required for the quality control of these dielectric thin films. Hurley (1987) showed that these CLS methods could be used for monitoring the affect of annealing and moisture sensitivity on the BPSG thin films. Haaland (1988) compared CLS, PLS, and PCR methods applied to the FT-IR analysis of BPSG thin films over a wide range of boron and phosphorus concentrations and film thicknesses. He found that transmission measurements taken at a 60° incident angle were superior to those taken at 0° for quantitative analysis when film thicknesses are not available. The importance of using statistically designed calibration sets was emphasized. All multivariate methods were found to be superior to the more traditional peak-height methods. In addition, PLS and PCR significantly outperformed CLS. Although PLS and PCR performed comparably, Haaland preferred PLS to PCR because of the shorter computation time, less complex model, and greater qualitative information content available with PLS. He also found that the most precise predictions were made when the spectra were not corrected for film thickness. Apparently, the interaction between absorption bands and the thickness-dependent interference fringes could be modeled by the PLS and PCR methods. The spectra of four BPSG samples showing the greatest range of spectral variation in the 44 samples used for calibration are presented in Fig. 11, along with assignments of the vibrational bands. Figure 12 shows an attempt to correlate the peak height of the P═O stretching band with the concentration of phosphorus determined by the reference method of inductively coupled plasma (ICP) emission spectroscopy. A positive correlation is not found in this case because of the difficulty in defining baselines, the overlap with the B—O stretching vibration, and the changing molecular interactions with P and B contents. Figure 13 shows the fit found using

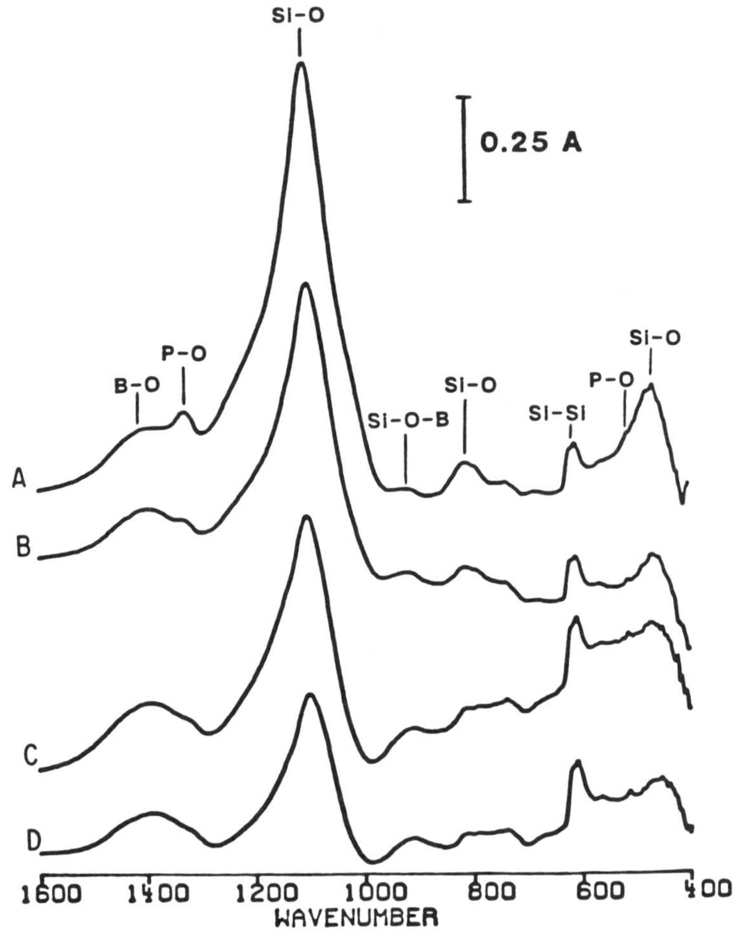

Fig. 11. Infrared spectra of representative BPSG thin films on Si wafers at 0° incident angle. (Reprinted with permission from Haaland (1988). Copyright American Chemical Society.)

PLS applied to the same calibration spectra but now the entire spectral range exhibited in Fig. 11. This dramatic improvement is a result of using all spectral information and the ability of PLS to model baselines and deviations in Beer's law. The determination of boron was a factor of two more precise than that found for P, and results for boron determinations can be

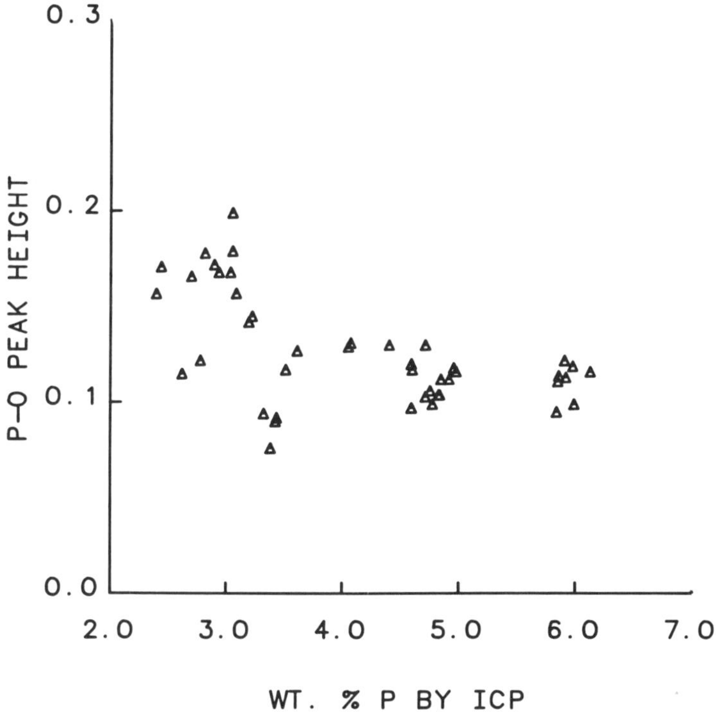

Fig. 12. Baseline-corrected infrared peak height of the P═O stretching vibration in BPSG versus the weight percent of P determined by ICP analyses for the 44 BPSG thin-film spectra taken at a 60° incident angle.

found in Haaland (1988). In addition, thickness could be determined from the infrared spectra. The qualitative interpretation and assignment of spectral bands was greatly facilitated by the analysis of the pure-component estimated spectra obtained by either CLS or PLS. Finally, the use of outlier detection methods was demonstrated, and these could be used to identify unusual samples and to improve the accuracy and precision of the calibrations and prediction. Recent unpublished analyses from our laboratory using more uniform BPSG thin-film samples have improved the phosphorus analysis precision by

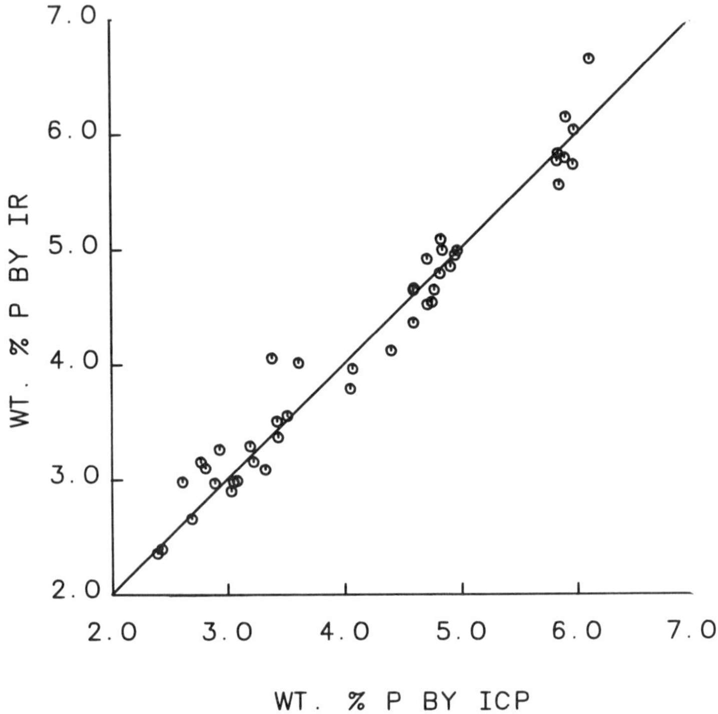

Fig. 13. Weight percent P determined during cross-validated PLS analyses of the uncorrected infrared spectra of 44 BPSG thin-film samples on Si at a 60° incident angle as a function of the weight percent P determined by ICP analyses. The solid line represents the ideal correlation with a slope of 1.0 and an intercept of 0. (Reprinted with permission from Haaland (1988). Copyright 1988 American Chemical Society.)

more than a factor of 2 relative to that demonstrated in Fig. 13.

E. Analyses of Glasses

In addition to the analysis of thin-film glasses, the FT-IR analysis of bulk glasses has been performed using multivariate calibration. Haaland (1986) used CLS with a model including quadratic terms in concentration to more accurately model the concentration-dependent frequency shifts in a band related to B_2O_3 content for bulk samples of a seven-component borosilicate glass. Later Haaland and Thomas (1988b) applied CLS, PLS, and PCR to the entire spectral range of the FT-IR data obtained from the same glasses to determine both OH and B_2O_3 concentrations in the glass. Both PLS and PCR outperformed linear CLS models, and both yielded results that were similar to the previous quadratic model CLS analyses that were restricted to a small spectral range. The ability to perform infrared analyses of bulk glasses in transmission mode significantly reduces sample preparation time. Spectral residuals from CLS, PLS, and PCR all were shown to be useful for outlier detection and for detecting the presence and the identity of unexpected components in the sample spectra.

Haaland and Thomas (1986) have also applied PLS to the quantitative analysis of five-component phosphate-based glasses to estimate component concentrations and glass properties. Because two of the oxide components (BaO and Na_2O) did not have spectral features of their own in the spectral range studied, their determinations were dependent on the perturbations of the spectral features of the other components. However, since both BaO and Na_2O modify the glass in a similar manner, their determination along with that of most of the glass properties were poor. However, by including density and index of refraction with the spectral data, the analysis of all component concentrations and properties of the glass (glass transition temperature, softening temperature, crystallization temperature, melting temperature, and thermal expansion coefficient) could be determined to the precision required for quality control of the glasses. Thus, a single infrared measurement coupled with rapid determinations of density and index of refraction could replace more traditional methods of determining composition and properties of the glasses that required approximately 1 day per sample to perform.

In a creative application of factor analysis methods, Culler *et al.* (1984) used principal component analysis combined with constraints to extract the pure-component spectra of a silane coupling agent on E-glass. The pure-component spectra estimated using their factor analysis methods were then used in a CLS prediction step to estimate the concentrations of the silane coupling agent on the E-glass.

F. Analyses of Detergents

Fuller *et al.* (1988b) were successful at using PLS to analyze detergents for six components. They used separate calibration and validation sets of samples to determine the number of factors to include in their PLS model. They showed that uncertainty estimates and spectral residuals could be used to determine the presence of outlier samples. They also showed that poorer but still adequate analysis precision could be obtained by performing the analysis on a small portion of the single-beam interferograms. This later finding may serve to make the quantitative FT-IR analysis of detergents even faster when a penalty in analysis precision can be tolerated.

G. Medical and Biological Applications

A number of groups have applied multivariate statistical calibrations to the infrared analysis of components in blood. Brown and co-workers first applied CLS (Kisner *et al.*, 1982) and then ILS methods (Kisner *et al.*, 1983) to the analysis of cholesterol and lipid blood serum components. McClure *et al.* (1987) demonstrated better precision by applying the **q**-matrix method to spectra of similar samples. Tyson *et al.* (1984a) also analyzed lipid components and found that CLS prediction methods performed somewhat better than cross-correlation methods. Using CLS methods applied to infrared spectra, McClure (1987) successfully determined the amounts of aspirin, acetaminophen, and caffeine in solution with relative errors of under 1%. Kuehl and Crocombe (1984) have used CLS prediction methods applied to the ATR infrared spectra of samples related to the analysis of fermentation broths. They were able to determine methanol, ethanol, and acetone concentrations in water solutions using intensities at a small number of frequencies.

Therrien *et al.* (1985) and Nyden *et al.* (1988) showed that protein content and conformation could be quantitated using multivariate calibrations applied to the infrared spectra of proteins in solution. They used CLS calibrations and predictions to determine quantitatively the amount of three protein conformations in proteins from the infrared spectra of the amide I and II bands of proteins. Nyden *et al.* (1988) showed that because there are deviations in Beer's law with mixtures of proteins, the **q**-matrix method yielded more accurate quantitative results than they were able to obtain using the CLS method. As mentioned previously, they used a numerically robust algorithm to reduce collinearity, which has the net effect of making this **q**-matrix method more similar to PCR than to the **q**-matrix method described by McClure *et al.* (1987). Nyden *et al.* (1988) used the spectral noise level as their criterion for selection of the spectra to include in the calibration. This criterion may lead to some overfitting since sample-in, sample-out effects, model error, and concentration errors may be limiting factors in the analysis precision rather than spectral noise. Also, spectral noise in absorbance units is not constant with changes in band intensities, a finding that makes selection of a single noise level, based on the spectral noise when no sample is in the infrared beam, much less than ideal. Nevertheless, the results of the application of the modified **q**-matrix method of Nyden *et al.* show it to be a very powerful quantitative tool in the case where molecular interactions are present and to provide better estimates than possible with CLS methods.

H. Analyses of Polymers

Antoon *et al.* (1977) used CLS prediction methods to determine the antioxidant content of butadiene rubbers from their infrared spectra. In this case, spectral residuals were used to determine the spectrum of antioxidant interaction with butadiene rubber. They used similar analyses to successfully analyze blends of isotactic polystyrene and polypropylene oxide. Koenig and Rodriquez (1981) used principal component analysis applied to infrared thin-film spectra to determine the number of components in blends of polyphenylene oxides and polystyrene. Spectral subtractions were used to obtain spectra to be used in CLS predictions. In this manner, concentrations of interaction components could be determined as a function of original polymer composition. Fenner (1984) has used ATR infrared spectra and CLS prediction methods to quantify the constituents in poly(vinyl acetate) adhesives. The base polymer spectrum was subtracted from the polymer with each additive in order to obtain "pure" component spectra of the additives including interactions with the base polymer. Miller and Obremski (1987) have used ILS methods applied to the Fourier-transformed infrared absorbance spectra of thin films of poly(vinyl acetate) and poly(vinyl chloride). The results of these analyses were generally within the precision of the reference values of poly(vinyl acetate) in the films. A matrix adjust method was implemented so that film thickness determinations were not necessary.

XII. SUMMARY

Chemometric methods in general and multivariate calibration methods in particular are starting to become widely available and to be used in the quantitative analysis of FT-IR spectral data. The potential of these methods is just now beginning to be widely tapped by the spectroscopic community. It is important to keep in mind that these methods should never be used to compensate for poor experimental technique or poorly prepared samples. Complete understanding of the methods and concepts involved in multivariate calibration and prediction can help to eliminate errors and prevent inappropriate application of the various available methods. When coupled with carefully prepared calibration samples and when close attention is paid to spectroscopic methods, multivariate calibration methods yield improvements in analysis precision. They also provide greater understanding of the individual calibration and unknown samples and thus they can allow greater quantitative accuracy to be achieved. In addition, a large range of new problems can now be solved that only a few short years ago were impossible to solve with conventional infrared spectroscopic methods.

REFERENCES

Allen, G. C., and Swallow, G. A. (1982). *Oxid. Met.* **17,** 157.

Antoon, M. K., Koenig, J. H., and Koenig, J. L. (1977). *Appl. Spectrosc.* **31,** 518.

Baghdadi, A., Gladden, W. K., and Flach, D. R. (1986). *Appl. Spectrosc.* **40,** 617.

Barnett, H. A., and Bartoli, A. (1960). *Anal. Chem.* **32,** 1153.

Beebe, K. R., and Kowalski, B. R. (1987). *Anal. Chem.* **2,** 187.

Berridge, J. C. (1987). *Analyst* **112,** 385.

Bertie, J. E., Jones, R. N., and Behnam, V. (1986). *Appl. Spectrosc.* **40,** 427.

Blackburn, J. A. (1965). *Anal. Chem.* **37,** 1000.

Blitz, J. P., Murthy, R. S. S., and Leyden, D. E. (1986). *Appl. Spectrosc.* **40,** 829.

Box, G. E. P., Hunter, W. G., and Hunter, J. S. (1978). "Statistics for Experimenters: An Introduction to Design, Data Analysis, and Model Building." Wiley, New York.

Braue, E. H., Jr., and Pannella, M. G. (1987a). *Appl. Spectrosc.* **41,** 1057.

Braue, E. H., Jr., and Pannella, M. G. (1987b). *Appl. Spectrosc.* **41,** 1213.

Brereton, R. G. (1987). *Analyst* **112,** 1635.
Brimmer, P. J., and Griffiths, P. R. (1988). *Appl. Spectrosc.* **42,** 242.
Brown, C. W. (1982). *J. Test. Eval.* **12,** 86.
Brown, C. W. (1986). *Spectroscopy* **1,** 4.
Brown, C. W., and Obremski, R. J. (1984). *Appl. Spectrosc. Rev.* **20,** 373.
Brown, C. W., Lynch, P. F., Obremski, R. J., and Lavery, D. S. (1982). *Anal. Chem.* **54,** 1472.
Brown, C. W., Obremski, R. J., and Anderson, P. (1986a). *Appl. Spectrosc.* **40,** 734.
Brown, C. W., Bump, E. A., and Obremski, R. J. (1986b). *Appl. Spectrosc.* **40,** 1023.
Brown, S. D. (1986). *Anal. Chim. Acta* **181,** 1.
Brown, S. D., Barker, T. Q., Larivee, R. J., Monfre, S. L., and Wilk, H. R. (1988). *Anal. Chem.* **60,** 252R.
Cahn, F., and Compton, S. (1988). *Appl. Spectrosc.* **42,** 865.
Chase, D. B. (1984). *Appl. Spectrosc.* **38,** 491.
Chatterjee, S., and Hadi, A. S. (1986). *Stat. Sci.* **1,** 379.
Christy, A. A., Velapoldi, R. A., Karstang, T. V., Kvalheim, O. M., Sletten, E., Telnacs, N. (1987). *Chemom. Intell. Lab. Syst.* **2,** 199.
Clark, F. R. S., and Moffatt, D. J. (1978). *Appl. Spectrosc.* **32,** 547.
Cooper, W. S. (1986). *Rev. Sci. Instrum.* **57,** 2862.
Cornell, J. A. (1981). "Experiments with Mixtures." Wiley, New York.
Crocombe, R. A., Olsen, M. L., and Hill, S. L. (1987). *ASTM Spec. Tech. Publ.* **934,** 95.
Culler, S. R., Gillette, P. C., Ishida, H., and Koenig, J. L. (1984). *Appl. Spectrosc.* **38,** 495.
Deming, S. N., and Morgan, S. L. (1987). "Experimental Design: A Chemometric Approach." Elsevier, New York.
Donahue, S. M., Brown, C. W., and Obremski, R. J. (1988). *Appl. Spectrosc.* **42,** 353.
Duyckaerts, G. (1959). *Analyst* **84,** 201.
Fenner, R. A. (1984). *Appl. Spectrosc.* **38,** 84.
Fredericks, P. M., Lee, J. B., Osborn, P. R., and Swinkels, D. A. (1985a). *Appl. Spectrosc.* **39,** 303.
Fredericks, P. M., Lee, J. B., Osborn, P. R., and Swinkels, D. A. (1985b). *Appl. Spectrosc.* **39,** 311.
Fredericks, P. M., Osborn, P. R., and Swinkels, D. A. (1985c). *Anal. Chem.* **57,** 1947.
Friedman, C. R., D'Agostino, J. M., and Messerschmidt, R. G. (1987). Paper 352, presented at the Pittsburgh Conference and Exposition on Analytical Chemistry and Applied Spectroscopy, March 1987.
Fuller, M. P., and Griffiths, P. R. (1978). *Anal. Chem.* **50,** 1906.
Fuller, M. P., Ritter, G. L., and Draper, C. S. (1988a). *Appl. Spectrosc.* **42,** 217.

Fuller, M. P., Ritter, G. L., and Draper, C. S. (1988b). *Appl. Spectrosc.* **42,** 228.
Geladi, P. (1988). *J. Chemom.* **2,** 231.
Geladi, P., and Kowalski, B. R. (1986a). *Anal. Chim. Acta* **185,** 1.
Geladi, P., and Kowalski, B. R. (1986b). *Anal. Chim. Acta* **185,** 19.
Gemperline, P. J., Boyette, S. E., and Tyndall, K. (1987). *Appl. Spectrosc.* **41,** 454.
Gillette, P. C., Lando, J. B., and Koenig, J. L. (1985). *In* "Fourier Transform Infrared Spectroscopy" (J. R. Ferraro and L. J. Basile, eds.) Vol. 4, p. 1. Academic Press, Orlando, Fl.
Griffiths, P. R., and de Haseth, J. A. (1986). "Fourier Transform Infrared Spectrometry." Wiley, New York.
Haaland, D. M. (1985). *Proc. Soc. Photo-Opt. Instrum. Eng.* **553,** 241.
Haaland, D. M. (1986). *Appl. Spectrosc.* **40,** 1152.
Haaland, D. M. (1987a). *ASTM Spec. Tech. Publ.* **934,** 78.
Haaland, D. M. (1987b). *Spectroscopy* **2** (6), 56.
Haaland, D. M. (1988). *Anal. Chem.* **60,** 1208.
Haaland, D. M., and Barbour, R. L. (1985). *Am. Lab.* **17** (7), 14.
Haaland, D. M., and Easterling, R. G. (1980). *Appl. Spectrosc.* **34,** 539.
Haaland, D. M., and Easterling, R. G. (1982). *Appl. Spectrosc.* **36,** 665.
Haaland, D. M., and Thomas, E. V. (1986). Paper 86, presented at the 1986 Eastern Analytical Symposium, New York.
Haaland, D. M., and Thomas, E. V. (1988a). *Anal. Chem.* **60,** 1193.
Haaland, D. M., and Thomas, E. V. (1988b). *Anal. Chem.* **60,** 1202.
Haaland, D. M., and Thomas, E. V. (1988c). Paper K05, presented at the meeting of the Federation of Analytical Chemistry and Spectroscopy Societies, Boston, November 1988.
Haaland, D. M., Easterling, R. G., and Vopicka, D. A. (1985). *Appl. Spectrosc.* **39,** 73.
Hamadeh, I. M., Yeboah, A. S., Trumbull, K. A., and Griffiths, P. R. (1984). *Appl. Spectrosc.* **38,** 486.
Harrick, N. J. (1967). "Internal Reflection Spectroscopy." Wiley (Interscience), New York.
Hawranek, J. P., Neelakantan, P., Young, R. P., and Jones, R. N. (1976). *Spectrochim. Acta, Part A* **32A,** 75.
Herget, W. F. (1979). *In* "Fourier Transform Infrared Spectroscopy" (J. R. Ferraro and L. J. Basile, eds.), Vol. 2, p. 111, Academic Press, New York.
Highfield, J. G., and Moffat, J. B. (1985). *Appl. Spectrosc.* **39,** 550.
Hirschfeld, T. (1978). *Appl. Opt.* **17,** 1400.
Hirschfeld, T. (1979a). *In* "Fourier Transform Infrared Spectroscopy" (J. R. Ferraro and L. J. Basile, eds.), Vol. 2, p. 193. Academic Press, New York.
Hirschfeld, T. (1979b). *Anal. Chem.* **51,** 495.

Hirschfeld, T., and Cody, C. (1977). *Appl. Spectrosc.* **31,** 551.
Hirschfeld, T., and Mantz, A. W. (1976). *Appl. Spectrosc.* **30,** 552.
Honigs, D. E., Freelin, J. M., Hieftje, G. M., and Hirschfeld, T. B. (1983). *Appl. Spectrosc.* **37,** 491.
Honigs, D. E., Hieftje, G. M., Mark, H. L., and Hirschfeld, T. B. (1985). *Anal. Chem.* **57,** 2299.
Hoskuldsson, A. (1988). *J. Chemom.* **2,** 211.
Hruschka, W. R., and Norris, K. H. (1982). *Appl. Spectrosc.* **36,** 261.
Hurley, K. H. (1987). *Solid State Technol.* **30** (3), 103.
Jolliffe, I. T. (1986). "Principal Component Analysis." Springer-Verlag, New York.
Jones, R. N. (1952). *J. Am. Chem. Soc.* **74,** 2681.
Kisner, H. J., Brown, C. W., and Kavarnos, G. J. (1982). *Anal. Chem.* **54,** 1479.
Kisner, H. J., Brown, C. W., and Kavarnos, G. J. (1983). *Anal. Chem.* **55,** 1703.
Koenig, J. L. (1964). *Anal. Chem.* **36,** 1045.
Koenig, J. L., and Rodriquez, M. J. M. T. (1981). *Appl. Spectrosc.* **35,** 543.
Krishnan, K. (1984). *ASTM Spec. Tech. Publ.* **850,** 358.
Krishnan, K., and Ferraro, J. R. (1982). *In* "Fourier Transform Infrared Spectroscopy" (J. R. Ferraro and L. J. Basile, eds.) Vol. 3, pp. 149–209. Academic Press, New York.
Krishnan, K., and Mundhe, R. B. (1983). *Proc. Soc. Photo-Opt. Instrum. Eng.* **452,** 71.
Kuehl, D., and Crocombe, R. (1984). *Appl. Spectrosc.* **38,** 907.
Lam, R. B. (1983). *Appl. Spectrosc.* **37,** 567.
Lauer, J. L., Vogel, P., and Seng, G. T. (1985). *Appl. Spectrosc.* **39,** 997.
Leroueille, J. (1982). *Appl. Spectrosc.* **36,** 153.
Lin, L.-T., Mann, C. K., and Vickers, T. J. (1987). *Appl. Spectrosc.* **41,** 422.
Lindberg, W., Persson, J.-A., and Wold, S. (1983). *Anal. Chem.* **55,** 643.
Lorber, A., Wangen, L. E., and Kowalski, B. R. (1987). *J. Chemom.* **1,** 19.
McClure, G. L. (ed.) (1987). *ASTM Spec. Tech. Publ.* **934.**
McClure, G. L. (1988). *In* "Laboratory Methods in Vibrational Spectroscopy" (H. A. Willis, J. H. van der Maas, and R. G. J. Miller, eds.), p. 145. Wiley, New York.
McClure, G. L., Roush, P. B., Williams, J. F., and Lehman, C. A. (1987). *ASTM Spec. Tech. Publ.* **934,** 131.
McCue, M., and Malinowski, E. R. (1981). *Anal. Chim. Acta* **133,** 125.
McKenzie, M. T., and Koenig, J. L. (1985). *Appl. Spectrosc.* **39,** 408.
Maker, P. D., Niki, H., Savage, C. M., and Breitenbach, L. P. (1979). *ACS Symp. Ser.* **1979,** 161.
Malinowski, E. R., and Howery, D. G. (1980). "Factor Analysis in Chemistry." Wiley, New York.

Mann, C. K., Goleniewski, J. R., and Sismanidis, C. A. (1982). *Appl. Spectrosc.* **36,** 223.

Manne, R. (1987). *Chemom. Intell. Lab. Syst.* **2,** 187.

Marcott, C. (1986). *In* "Metals Handbook: Materials Characterization" (R. E. Whan, ed.), Vol, 10, 9th Ed., p. 109. Am. Soc. Met., Metals Park, Ohio.

Maris, M. A., Brown, C. W., and Lavery, D. S. (1983). *Anal. Chem.* **55,** 1694.

Marley, N. A., Mann, C. K., and Vickers, T. J. (1984). *Appl. Spectrosc.* **38,** 540.

Martens, H. A. (1985). "Multivariate Calibration: Quantitative Interpretation of Non-selective Chemical Data. Parts I and II of the 1985 Ph.D. Dissertation," NR Report Nos. 786 and 787. Norwegian Institute of Technology, University of Trondheim, Trondheim, Norway.

Martens, H., and Naes, T. (1984). *Trends Anal. Chem.* **3,** 204.

Massart, D. L., Vandeginste, B. G. M., Deming, S. N., Michotte, Y., and Kaufman, L. (1988). "Chemometrics: A Textbook." Elsevier, Amsterdam.

Meglen, R. R. (1988). *Chemom. Intell. Lab. Syst.* **3,** 17.

Messerschmidt, R. G. (1988). *In* "Infrared Microspectroscopy: Theory and Applications" (R. G. Messerschmidt and M. A. Harthcock, eds.). Dekker, New York.

Miller, J. A., and Obremski, R. J. (1987). *In* "Fourier Transform Infrared Characterization of Polymers" (H. Ishida, ed.), p. 281. Plenum, New York.

Miller, B. E., Danielson, N. D., and Katon, J. E. (1988). *Appl. Spectrosc.* **42,** 401.

Monfre, S. L., and Brown, S. D. (1988). *Anal. Chim. Acta* **200,** 397.

Murthy, R. S. S., and Leyden, D. E. (1986). *Anal. Chem.* **58,** 1228.

Murthy, R. S. S., Blitz, J. P., and Leyden, D. E. (1986). *Anal. Chem.* **58,** 3167.

Nyden, M. R., and Babrauskas, V. (1987). *Chem. Phys. Proc. Combust.* p. 107-1.

Nyden, M. R., Forney, G. P., and Chittur, K. (1988). *Appl. Spectrosc.* **42,** 588.

Osten, D. W. (1988). *J. Chemom.* **2,** 39.

Painter, P. C., Rimmer, S. M., Synder, R. W., and Davis, A. (1981). *Appl. Spectrosc.* **35,** 102.

Ramos, L. S., Beebe, K. R., Carey, W. P., Sanchez, E., Erickson, B. C., Wilson, B. E., Wangen, L. E., and Kowalski, B. R. (1986). *Anal. Chem.* **58,** 294R.

Ramsey, D. A. (1952). *J. Am. Chem. Soc.* **74,** 72.

Rao, G. R., and Zerbi, G. (1984). *Appl. Spectrosc.* **38,** 795.

Reinecke, D., Jansen, A., Fister, F., and Schernau, U. (1987). *Anal. Chem.* **60,** 1221.

Rivord, G. E. (1979). *Sandia National Laboratories Report* SAND79-0968, July 1979.

Reyment, R. A. (1987). *Chemom. Intell. Lab. Syst.* **2,** 79.

Rutan, S. C. (1987). *J. Chemom.* **1,** 7.

Sasaki, K., Kawata, S., and Minami, S. (1986). *Appl. Spectrosc.* **40,** 185.

Scheffe, H. (1958). *J. R. Stat. Soc. B* **20,** 344.

Sharaf, M. A., Illman, D. L., and Kowalski, B. R. (1986). "Chemometrics." Wiley, New York.

Shope, T. B., Vickers, T. J., and Mann, C. K. (1987). *Appl. Spectrosc.* **41,** 908.

Smith, A. L. (1979). "Applied Infrared Spectroscopy." Wiley, New York.

Sperline, R. P., Muralidharan, S., and Freiser, H. (1986). *Appl. Spectrosc.* **40,** 1019.

Sternberg, J. C., Stillo, H. S., and Schwendeman, R. H. (1960). *Anal. Chem.* **32,** 84.

Stone, M. (1974). *J. R. Stat. Soc.* **36,** 111.

Strang, C. R., and Levine, S. P. (1989). *Am. Ind. Hyg. Assoc. J.* **50,** 78.

Strang, C. R., Levine, S. P., and Herget, W. F. (1989). *Am. Ind. Hyg. Assoc. J.* **50,** 70.

Swallow, G. A., and Allen, G. C. (1982). *Oxid. Met.* **17,** 141.

Therrien, M., Laufleur, M., and Pezolet, M. (1985). *Proc. Soc. Photo-Opt. Instrum. Eng.* **553,** 173.

Thompson, M. M., and Palmer, R. A. (1988). *Appl. Spectrosc.* **42,** 945.

Tsuge, A., Uwamino, Y., and Ishizuka, T. (1988). *Appl. Spectrosc.* **42,** 168.

Tyson, L. L., Ling. Y.-C., and Mann, C. K. (1984a). *Appl. Spectrosc.* **38,** 663.

Tyson, L. L., Vickers, T. J., and Mann, C. K. (1984b). *Appl. Spectrosc.* **38,** 697.

Walder, F. T., Vidrine, D. W., and Hansen, G. C. (1984). *Appl. Spectrosc.* **38,** 782.

Ward, K. J., Brinker, C. J., and Haaland, D. M. (1989). Manuscript in preparation.

Wetzel, D. L. (1983). *Anal. Chem.* **55,** 1165A.

Willis, H. A., van der Maas, J. H., and Miller, R. G. J. (eds.) (1987). "Laboratory Methods in Vibrational Spectroscopy." Wiley, New York.

Wold, S. (1978). *Technometrics* **20,** 397.

Wold, S., Esbensen, K., and Geladi, P. (1987). *Chemom. Intell. Lab. Syst.* **2,** 37.

Ying, L., and Levine, S. P. (1989). *Anal. Chem.* **61,** 677.

9

Industrial Applications of GC/FT-IR

Ryujiro Namba
Shiseido Toxicological and Analytical Research Center
Kohoku-ku Yokohama, 223 Japan

Practical Fourier Transform Infrared Spectroscopy

I. INTRODUCTION

The method of infrared analysis has gained wide acceptance for the identification or determination of unknown compounds. Investigations over the past 40 years have made a vast store of data available to scientists interested in using infrared for compound identification. Identification of components in mixtures is hampered by overlapping absorption bands. This problem can be alleviated to a large degree by combining separation techniques with infrared detection of components. Interfacing of chromatographic techniques with dispersive infrared instrumentation has been limited to techniques that save the eluted species for subsequent infrared analysis due to the relatively long analysis times required by dispersive infrared instrumentation and the inherent low sensitivity of the technique. The advent of FT-IR has to a large degree removed these problems, opening the way to real-time infrared chromatographic detection.

FT-IR has risen as an important detector for gas chromatography (GC). GC/IR enables investigators to detect and identify eluted species. The information provided in the GC/IR experiment is complementary to the information provided in a GC/MS experiment. Unique information can be obtained concerning isomeric identification of compounds as well as functional group specific information not available from mass spectroscopy (MS). Despite its lower sensitivity, these advantages, as well as the cost advantage, have brought GC/IR into wide acceptance as a useful technique.

This chapter presents a brief history of GC/IR and describes some of the instrumentation used, the spectral information available, and industrial applications of this technique.

II. DEVELOPMENT OF GC/IR

Initial investigations into GC/IR were conducted with dispersive infrared instrumentation. Low and Freeman (1967) made the first GC/IR investigation with an FT-IR. Using a packed GC column and a TGS detector, they were able to detect common solvents in microgram quantities using a relatively large gas cell as the detector volume. The most important optical improvements in the experiment came with the replacement of the large gas cell by a light pipe. The size of the cell was optimized by Griffiths (1977). The factors of GC separation, absorbance enhancement, high signal-to-noise ratio (S/N), and high throughput were addressed. Further improvement to the light pipe design was made when the inner surface of the cell was coated with gold (Azarraga, 1980). Replacement of the TGS detector with a MCT (mercury-cadmium-telluride) detector played a major role in increasing the sensitivity of the technique.

Originally, slow software hampered data collection. The large amount of data that must be collected in the GC/IR experiment made continuous collection impractical. The main obstacle was the slow speed of the fast Fourier transform (FFT) calculation coupled with the low-speed magnetic tape recording medium that was then available. With these limitations, some method was needed for triggering data collection whenever a GC eluent was about to move through the light pipe. This triggering was accomplished by linking the detector response of a Thermal Conductive Detector (TCD) or Flame Ionization Detector (FID) and collecting FT-IR data only while the signal from the other detector was above a certain level. But this arrangement introduced some other problems. With this method, there was a finite time difference between the chromatogram and the infrared absorption data, so some mismatch between the chromatogram and the spectra might occur. Also, the sensitivity of the GC detector was greater than that of the infrared detector, causing some problems in selection of the threshold signal for infrared data collection. These problems were solved with the timely introduction of higher speed software for the FFT and larger capacity computer memory storage devices.

Several developments have enhanced the technique of GC/IR. Among these are the introduction of spectral search software, which allows for the comparison of unknown spectra to a spectral data base for compound identification by the computer (Hanna *et al.,* 1979a,b). This development has resulted in a very practical system for compound identification. The technique of matrix isolation, in which eluted species are trapped in a matrix of Ar frozen at liquid helium temperatures. This technique resulted in detection in the nanogram range (Reedy *et al.,* 1985; Bourne, 1984). The complementary nature of GC/MS and GC/IR has been routinely exploited, and the use of GC/IR/MS in combination is expected to increase.

III. OUTLINE OF APPARATUS

Figure 1 illustrates a schematic diagram of a GC/FT-IR system. The apparatus consists mainly of a gas chromatograph, FT-IR, light pipe, detector, and computer.

A. FT-IR INTERFEROMETER

A relatively low resolution (8 cm^{-1}) interferometer is sufficient for GC/IR. The absorption bands of the vapor phase components have large half-widths due to contributions from the rotational bands caused by the high measuring temperatures and smaller interaction between the gaseous molecules.

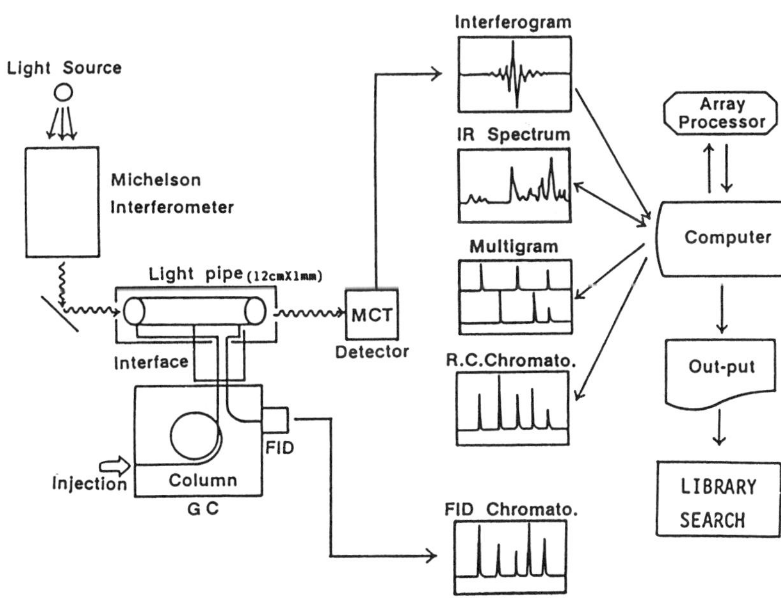

Fig. 1. Schematic diagram of GC/FT-IR system.

The source energy through the light pipe is cut to about 10% as a result of absorptions caused by the light pipe.

To increase the sensitivity of the measurement, a stable infrared source is necessary. A narrow-band MCT detector with high D* is normally used. The bandwidth of the detector may be as narrow as 4000 to 700 cm^{-1}. If the compounds of interest have important absorption bands outside of this range, then a wide-band MCT detector is sometimes used at the cost of sensitivity.

B. Light Pipe

One of the most important components is the light pipe that interfaces the GC with the FT-IR. Of critical importance are the dimensions of the quartz light pipe. It has been found that the light pipe volume should be approximately one half-width of the GC peaks for the best compromise between chromatographic resolution and infrared sensitivity (Griffiths, 1977). For capillary column GC, the light pipe dimensions are typically 1 mm i.d. (inner diameter), 10 cm long for a volume of 100 μl. This is generally used with a flow rate of 2 ml/min, allowing for a peak half-width of 3 sec. For packed columns, light pipes of 2–3 mm i.d., 30–60 cm long, and having a volume of 4 ml are used, allowing one-half peak widths of 10 sec at a flow rate of 30 ml/min.

C. Gas Chromatograph

Most available gas chromatographs can be used. For capillary column GC, an injector capable of split/splitless injection is necessary. The range of the sample quantity that can be determined is called the dynamic range. A wide dynamic range is desirable so that a range of sample concentrations can be measured. The upper limit of the dynamic range can be raised by selection of a column with a higher load capacity. This statement favors the use of packed columns, which have a larger loading capacity than capillary columns. Capillary columns, however, give better chromatographic resolution and efficiency, resulting in narrower peaks for the same amount of sample. This arrangement results in better lower detection limits. The use of capillary columns with relatively thick (0.5–1 μm) film coatings is considered to be the best compromise between column capacity and chromatographic efficiency at present for general use. The thicker the film coating, the greater the retention time, which is good for low boiling point samples. This behavior is undesirable for high boiling point samples, which may be retained on the column too long, thus making the chromatography inefficient. The load quantity is proportional to the film thickness, but it varies with the nature of the compounds and with the combination of the stationary phases (e.g., the load quantity for fatty acids with nonpolar dimethyl silicone is small, but large for the highly polar PEG-20M.)

D. Other GC/IR Methods

The matrix isolation (MI) technique has been previously described (Reedy *et al.*, 1985; Bourne, 1984). A schematic diagram of the GC/MI/FT-IR method is shown in Fig. 2. The carrier gas used is 1–2% argon in helium. The flow eluted from the GC column is blown against a rotating gold disk that is kept at 12 K and held under a reduced pressure (10^{-6} torr). The separated compounds are trapped in a frozen matrix of argon atoms on the disk. Spectra of the compounds are obtained by focusing the source radiation on the frozen sample and collecting the reflection absorption spectrum. The disk has a frozen physical chromatogram since the disk rotates during elution. This frozen chromatogram may then be inter-

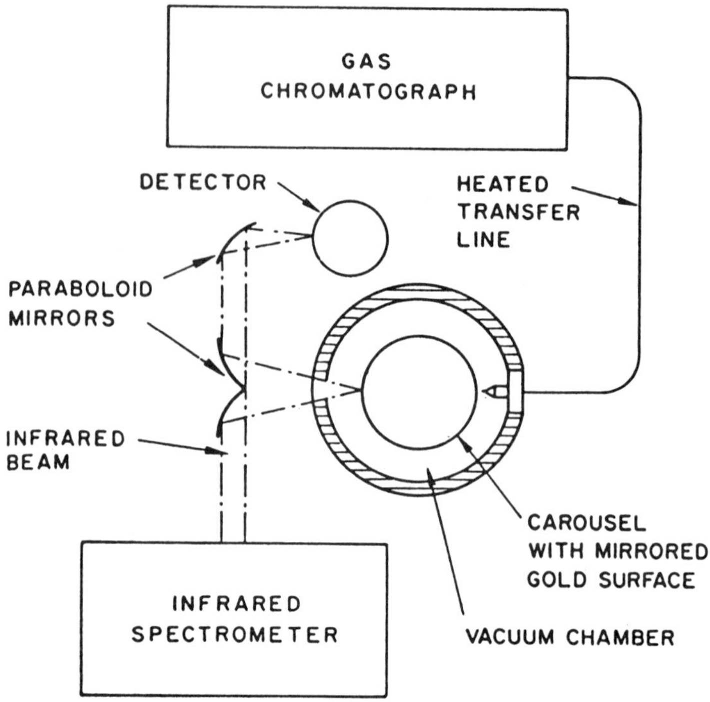

Fig. 2. Schematic diagram of GC/MI/IR system. From Schneider *et al.* (1986). Reproduced from the *Journal of Chromatographic Science* by permission of Preston Publications, a Division of Preston Industries, Inc.

rogated by the infrared beam, and continuous measurement is possible by rotating the disk through the interrogation volume of the infrared beam. Data processing is similar to that of the light pipe method, so it is possible to produce a Gram-Schmidt chromatogram.

The GC/MI/FT-IR method offers several advantages over normal GC/IR. Higher sensitivity and lower detection limits are possible using this method. Subnanogram quantities may be collected. One of the features of the method is that spectral features in the spectra are very sharp. There are two principle reasons for this: first, since the molecules are held in a rigid matrix of surrounding argon atoms, there is no contribution from rotational broadening; second, the analyte molecules are separated from each other by the matrix, so contributions from interactions between neighboring molecules is greatly reduced.

Because of the nature of the technique, it is possible to improve the S/N by increasing the collection time for components of interest, since there are no real time constraints other than those imposed by the investigator. Each component may be studied separately, varying the collection time and the spectral resolution to suit the sample. Sensitivity is increased because it is possible to increase the S/N for minute components. Schneider *et al.* (1986) compared this technique with the conventional light pipe method.

A method in which the carrier gas is blown against an infrared-transparent window (ZnSe) kept at 77 K has been described (Shafer and Griffiths, 1986). In this method, the analyte molecules are frozen to the window while the carrier gas is allowed to dissipate. The chromatogram trapped on the cold window is then analyzed using an infrared microscope in the transmission mode. This method is much simpler than the matrix isolation technique because of the higher measuring temperatures.

E. Data Processing System

The data processing system performs the following tasks: (1) collection of the interferograms in short time intervals; (2) creation of a Gram-Schmidt chromatogram from the interferogram intensities, which reflect the infrared absorption over the entire infrared range; (3) creation of the absorbance spectra of the eluted species by performing the Fourier transformation on the interferograms; (4) creation of functional group chromatograms based on the interferogram intensities over a truncated region of the spectrum; (5) library search of the sample spectra against a stored data base; (6) calculation of chromatogram peak intensities; and, (7) quantitative analysis of the eluted species.

Special software is necessary for the data collection in the GC/IR experiment. As a whole, rapid processing speed is necessary because of the vast amount of data that must be processed. The FFT requires the greatest amount of time among the various operations. The speed of the FFT has been increasing regularly owing to improvement of the software and to hardware advances. Recently, data processing in less than 1 sec has become possible, and real-time observation of the capillary GC/IR spectrum has been achieved.

1. Chromatogram

The Gram-Schmidt chromatogram is presently used by almost all GC/IR instruments. The Gram-Schmidt is a plot of infrared absorbance versus time. In general, the sensitivity of the Gram-Schmidt is a good indicator of the spectral quality of the spectrum represented by the Gram-Schmidt band. In other words, a strong peak will be indicative of a good quality spectrum, and a weak band will be indicative of a marginal spectrum. But in any case, if the signal is strong enough to produce a peak in the chromatogram, then it is strong enough to produce a spectrum with usable information. Figure 3 shows the comparison of relative sensitivities between FID and Gram-Schmidt chromatograms. As can be seen in the figure, similar mono-substituted aromatic species will give similar FID responses. However, the infrared responses may vary greatly with different functional groups, so the appearance of the Gram-Schmidt chromatogram may be markedly different from the FID trace, particularly in intensities.

Another type of chromatogram normally produced in the GC/IR experiment is the functional group chromatogram, also known as a multigram or a chemigram. The intensities of the absorption bands in a specific wavenumber range are used to create the functional group chromatogram. This method bears some resemblance to a mass chromatogram from GC/MS. They differ from each other, however. In the case of a complex mixture, for example, the functional group chromato-

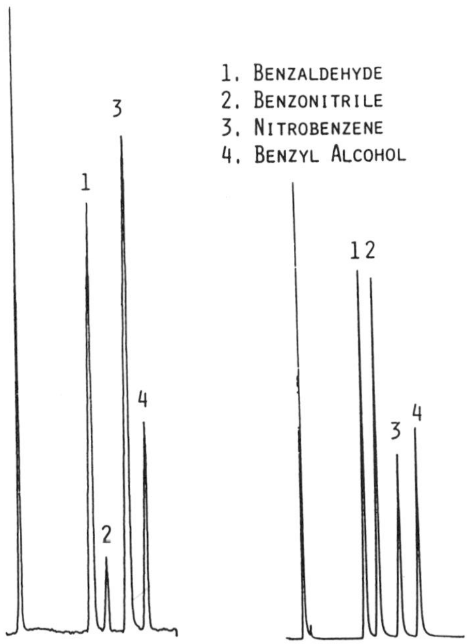

Fig. 3. Comparison of GSC with FID chromatogram on relative sensitivity: (left) Gram-Schmidt chromatogram; (right) FID chromatogram. The concentration ratio of peaks 1, 2, 3, and 4 is 2:2:1:1. Column: 5% FFAP on Chromosorb W, 2 m × 3 mm.

gram will pick out only those compounds with a chosen functional group. The mass chromatogram, however, does not always show the same m/z fragment ion for each functional group (Namba *et al.*, 1981). Figure 4 shows the functional group chromatogram of absolute jasmine oil. In this example the chromatograms based on the —OH, —NH, and carbonyl groups are shown. The carbonyl groups contained in esters, ketones, and aldehydes are indistinguishable from one another because their absorption bands for the carbonyl are very near to one another and so are detected in one of the functional group chromatograms. The carbonyl groups in lactones have a higher wavenumber absorption, so they can be easily distinguished from the other carbonyl-containing compounds

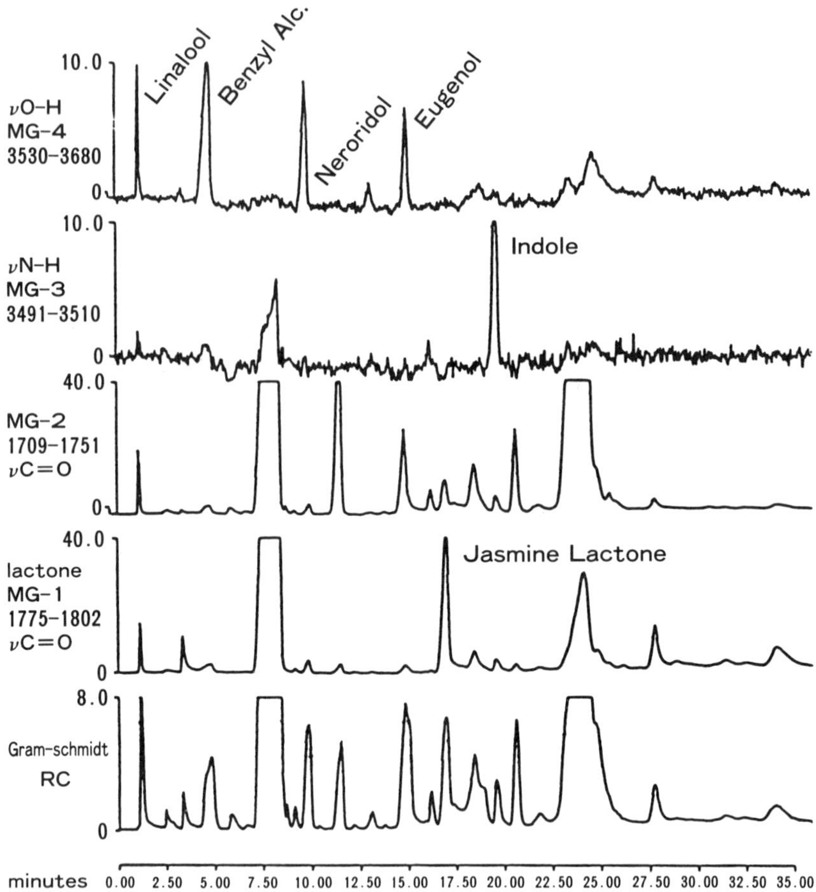

Fig. 4. Functional group chromatograms of jasmine oil. Column: 5% FFAP on Chromosorb W. Column temperature: 80 to 250°C, 10°C/min. Carrier gas: He, 60 ml/min. Light pipe temperature: 250°C Sample: 0.2 μl on column injection of absolute jasmine oil.

by adjusting the range for the functional group chromatogram, as can be seen in the figure.

A method resembling two-dimensional nuclear magnetic resonance has been conceived for visualizing the GC/IR data. On the two-dimensional plane of wavenumber versus retention time (shown in Fig. 5), the absorbance is indicated as

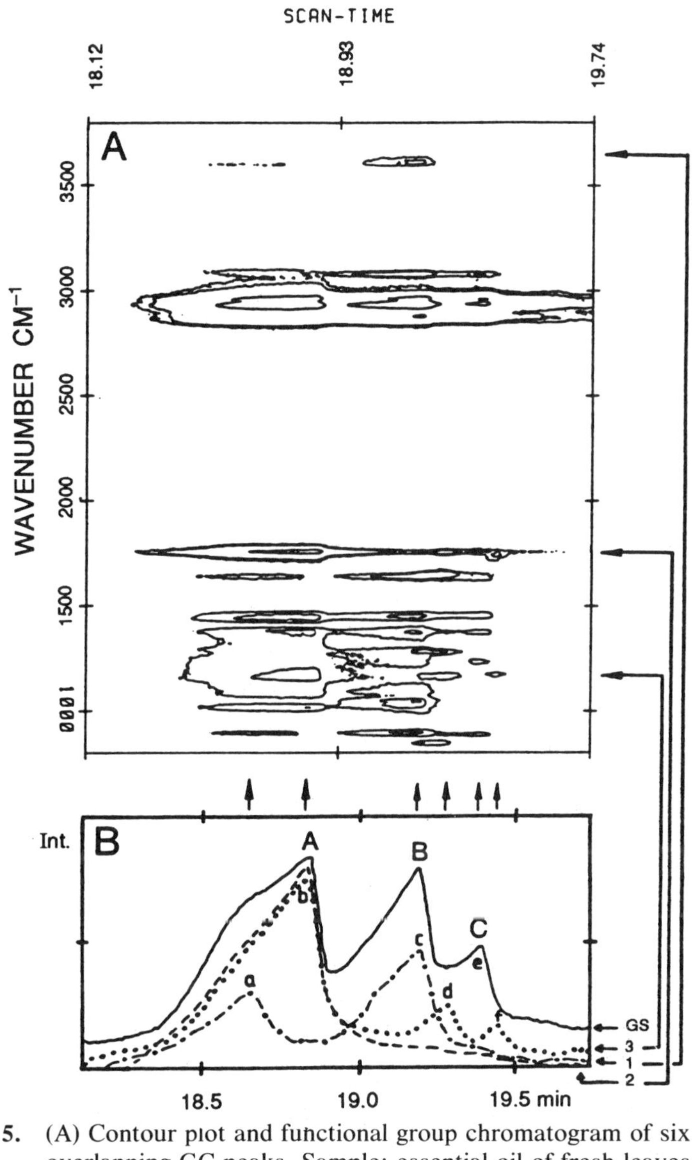

Fig. 5. (A) Contour plot and functional group chromatogram of six overlapping GC peaks. Sample: essential oil of fresh leaves of *Melissa officinalis*. Column: 30 m × 0.25 mm, 0.5 μm DB-WAX (J&W Scientific). Carrier gas: He, 1.5 ml/min. Light pipe: 20 cm × 0.8 mm, 200 °C. Effective time slices: 0.5 sec; Resolution: 8 cm^{-1}. (From Herres *et al.*, 1983.)

contour lines. Spectra of all components can be seen in one glance. This method is effective for getting a "one-look" feel for all of the data or for the analysis of overlapping peaks (Herres *et al.*, 1986).

2. Library Search

A method for comparing spectra obtained in the GC/IR experiment to a large, stored data base of spectra has been developed (Hanna *et al.*, 1979b). The following equation is used to make a calculation of the hit quality index, which is a numerical approximation of how well an unknown spectrum compares to a previously collected and stored spectrum.

$$\mathrm{HQI} = \Sigma\,(U_\mathrm{i} - K_\mathrm{i})^2 \tag{1}$$

In this equation, HQI stands for hit quality index, U_i is the unknown spectrum intensity at i cm^{-1}, and K_i is the known spectrum intensity at i cm^{-1}. The entire spectrum, or region of the spectrum chosen, is compared, and small differences in spectral patterns are detectable. There are now several data bases available containing spectra obtained by GC/IR analysis, including 2300 spectra edited by the Environmental Protection Agency (EPA), and a 9200-spectra collection compiled by Sadtler Research Laboratories.

IV. BASIC FEATURES OF GC/FT-IR SPECTRA

A. Light Pipe Temperature and Sensitivity

It is necessary to keep the light pipe at high temperature in order to keep the eluted components in the vapor phase. The high temperature may lower the sensitivity, however. Table I shows the results measured by Namba *et al.* (1981). When a specific amount of amyl acetate is injected, the peak intensity

Table I Effect of Light Pipe Temperature on Gram-Schmidt Chromatogram Intensity, Interferogram Intensity, and Spectral Absorbance

Light pipe temperature		Temperature ratio (f)	Absorbance (A) (1665–1850 cm^{-1})	A × f	Intensity of the GSC[a]	GSC × f	Intensity of the Interferogram (I)[a]	GSC × f/I
°C	K							
50	323	1.00	100	100	100	100	100	1.00
100	373	1.15	91.4	105	78.4	90	87.3	1.03
150	423	1.31	82.3	108	58.4	77	72.3	1.06
200	473	1.46	73.3	107	41.4	60	60.8	0.99
250	523	1.62	63.9	104	33.2	54	48.8	1.10
300	573	1.79	60.9	108	24.7	44	38.5	1.14
350	623	1.93	55.6	107	20.0	39	31.3	1.23

[a]Normalized. GSC, Gram-Schmidt chromatogram.

of the Gram-Schmidt chromatogram, the absorbance of the 1763-cm^{-1} band of amyl acetate, and the center-burst intensity of the interferogram (without sample) were examined at the same temperatures. The examination revealed that they all markedly decrease with an increase in temperature.

The intensity decrease of the chromatogram was the greatest, and was equivalent to the multiplication of the decreasing ratio of the absorbance with that of the interferogram intensity. The decrease of the absorbance was inversely proportional to the absolute temperature of the light pipe. This result is thought to be caused by the decrease of the sample concentration due to the expansion of the carrier gas.

As to the decrease of the interferogram intensity, various causes have been postulated. Brown *et al.* (1985) conjectured that the emission (nonmodulated light) from the the sides of the light pipe can saturate the MCT detector, thereby resulting in a decrease of the response to the modulated light actually coming through the light pipe. It was confirmed that actual removal of the emission other than the modulated light dramatically increased the intensity of the interferogram.

B. Reproducibility of Spectrum Pattern

When solid samples are analyzed by typical infrared techniques (preparation of KBr pellets, for example), it is often observed that the pattern of the bands varies with the method of preparation of the sample. A tilt in the baseline, fluctuation of peak intensity ratios, and variations in the broadness of bands will all hinder the exact mutual comparison of spectra and further identification by library spectral search methods.

For vapor phase spectra, however, when the S/N ratio is high, it is always possible to get a constant pattern of the spectral bands. Table II presents the changes of the two absorption bands of amyl acetate at different temperatures. Some change

Table II Effect of Light Pipe Temperature on Half-Height Width and Ratio of Absorbances at 1763 cm^{-1} and 1283 cm^{-1}[a]

Light pipe temperature (°C)	Absorbance ratio[b]	Half-height width	
		at 1763 cm^{-1}	at 1283 cm^{-1}
50	1.76	21.7	18.1
100	1.75	23.7	19.1
150	1.74	26.1	19.8
200	1.73	28.3	20.8
250	1.71	30.8	21.7
300	1.70	32.3	22.2
350	1.69	33.1	22.8

[a]Sample: *n*-amyl acetate, 20 μg. Column: 5% FFAP on Chromosorb W, 2 m × 3 mm.

[b]Ratio is peak absorbance at 1160–1340 cm^{-1} divided by peak absorbance at 1665–1850 cm^{-1}.

was recognized when the range of the light pipe temperatures was widened to 50–350°C.

For the half-height of the absorption band and the intensity ratio, the half-height width tends to increase with an increase of temperature. This trend is thought to be due to the influence of the rotation energy. The intensity ratio shows a slight change. This change is very small in the normal measuring temperature range of 150–250°C. It can be said that GC/IR always gives a stable spectrum pattern (Namba, unpublished observations). This characteristic may be one of the greatest advantages of GC/IR over GC/MS.

C. Determination of Alkyl Chain Length by GC/IR Spectra

For aliphatic compounds in general, determination of alkyl chain length is difficult, although homologs of higher fatty acids or higher alcohols can be easily distinguished in infrared spectra, comparing the functional groups with the specific absorption wavenumber.

However, in GC/IR spectra, the reproducibility is high, as discussed earlier. Hence, the relation between the carbon number and the ratio of the absorption band intensities for an alkyl group divided by that for a functional group, as shown in Fig. 6, provides a good linear relationship (Namba *et al.*, 1981). Therefore, in the GC/IR spectra, not only the kind of homologs of aliphatic compounds but also the alkyl chain length can be determined.

D. Identification of Isomers

The capability of identifying isomers is one of the greatest advantages of GC/IR. In GC/MS, when the isomers are both wax esters, for example, of the same molecular weight but different in the combination of fatty acids with alcohols, a clear

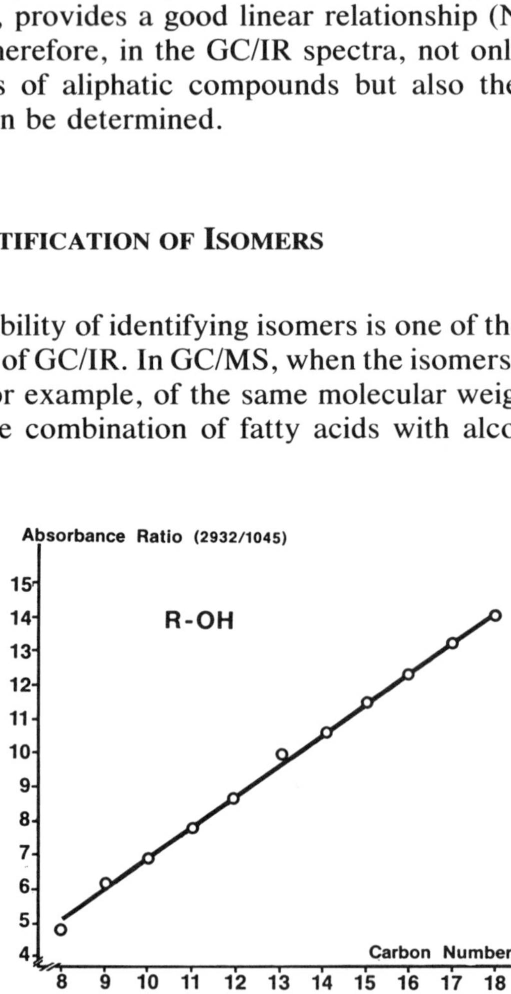

Fig. 6. Relationship between carbon number and absorbance ratio. Sample: alkane-1-ols; carbon number: 8 to 18; weight: 10–20 μg. Light pipe: 2.5 mm × 60 cm, 300°C.

difference will appear in the fragment ions, and the identification can be easily performed. However, this distinction will be difficult to make from the infrared spectrum. But many isomers are identified by GC/IR easily. Identification of positional isomers for substituents of aromatic compounds is one of the strongest points of GC/IR. Identification of ortho, meta, and para isomers may in some cases (aminobenzyl alcohol, for example) be possible even in the mass spectrum, but it is not expected to be successful in all cases. Identification of cis/trans isomers of double-bonded aliphatic compounds is also possible in the GC/IR spectrum. For example, the spectra of methyl *cis*-9-octadecenoate and methyl *trans*-9-octadecenoate show clear differences in the deformation vibration of the double-bonded hydrogen at 943 cm^{-1} for the trans isomer. In GC/MS, these substances show almost the same pattern, so identification is impossible. It is, however, difficult in both GC/IR and GC/MS to know the positions of the double bonds for compounds having a long alkyl chain.

Terpene compounds are one of the groups that have many isomers due to the position of double bonds. The relationship between the structure of these isomers and the spectra has been investigated in the condensed phase since the 1950s (O'Connor and Goldblatt, 1954). In the mass spectrum, fragment ions of the same m/z are apt to form, and only the intensity ratios are likely to differ from one another. In GC/IR, however, all of these substances show specific absorption bands at different wavenumbers. GC/IR is effective because the identification is concretely performed (Kalasinsky and McDonald, 1983) (see Fig. 7).

E. Detection Limit

Detection limit is an important factor in determining the dynamic range of the measurement. It should be noted that GC/IR is a concentration detector, hence the results will change according to the flow rate of the carrier gas, even if the abso-

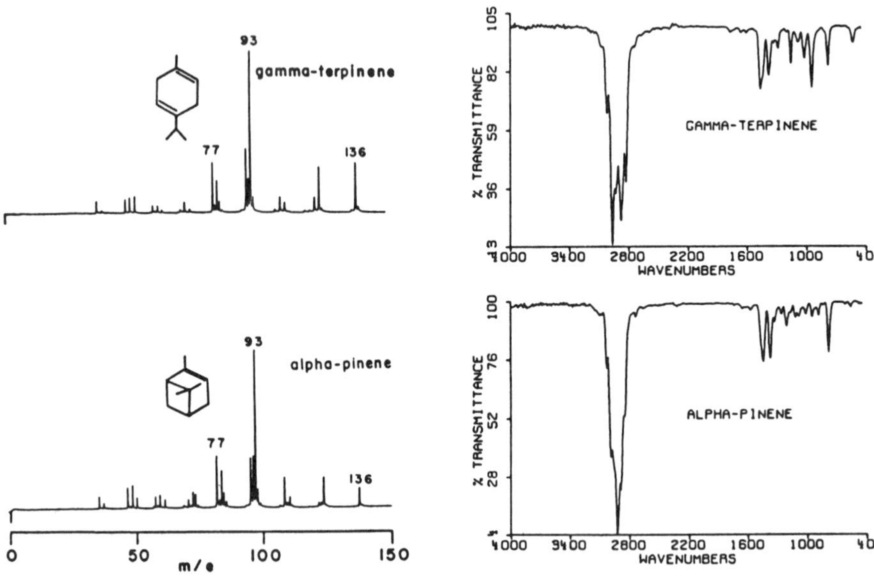

Fig. 7. GC/IR spectra and mass spectra of γ-terpinene and γ-pinene. Light pipe temperature: 250°C. Ionizing voltage: 70 V. (From Kalasinsky & McDonald, 1983. Reproduced from the *Journal of Chromatographic Science* by permission of Preston Publications, a division of Preston Industries, Inc.)

lute amounts of the samples are the same. The greater the number of theoretical plates in the GC column, the higher the concentration of the peak as it moves through the detector volume, and the higher the absorbance peak will be.

Detection sensitivity varies largely with the absorbance of the sample. The sensitivities for detecting a strong absorption band in the spectrum and for measuring clearly the whole spectrum are markedly different, and thus it seems to be difficult to define the sensitivity. Figure 8 shows the spectra for various absolute amounts of isobutylmethacrylate, of measured at a flow rate of 2 ml/min. The entire spectrum is considered to be measured clearly for the sample amount of 550 ng. The spectra were evaluated by library spectral search, and the HQI was used for evaluating the similarity of the spectra. The results revealed that there is a specific relationship between the HQI and the amount of the sample. Major absorption

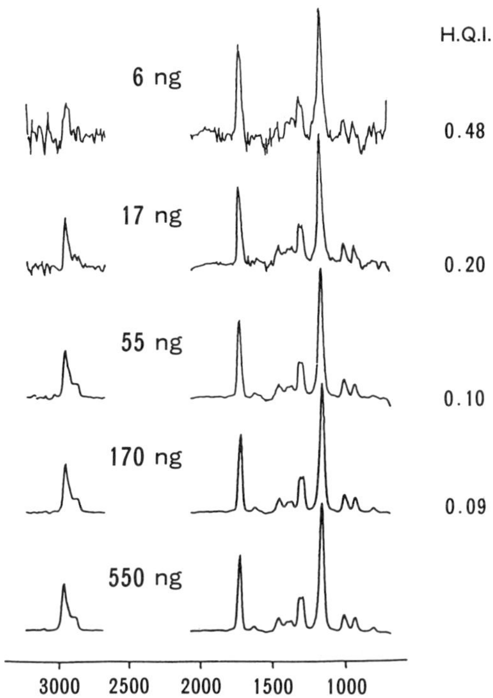

Fig. 8. Spectra of isobutylmethacrylate (6–550 ng) with spectrum of 550 ng used as a reference to calculate HQI.

bands were detected for the sample of 6 ng, but tens of nanograms were thought to be necessary to obtain a well-defined spectrum (Kadowaki and Namba, 1985).

F. Background

Small variations in the purge quality of the FT-IR optical bench and GC interface have a large influence on the quality of the spectrum. Water shows strong absorptions in the ranges of 3800 to 3400 cm^{-1} and 2000 to 1200 cm^{-1} and becomes a problem for microanalysis (see Fig. 9). Problems with purge

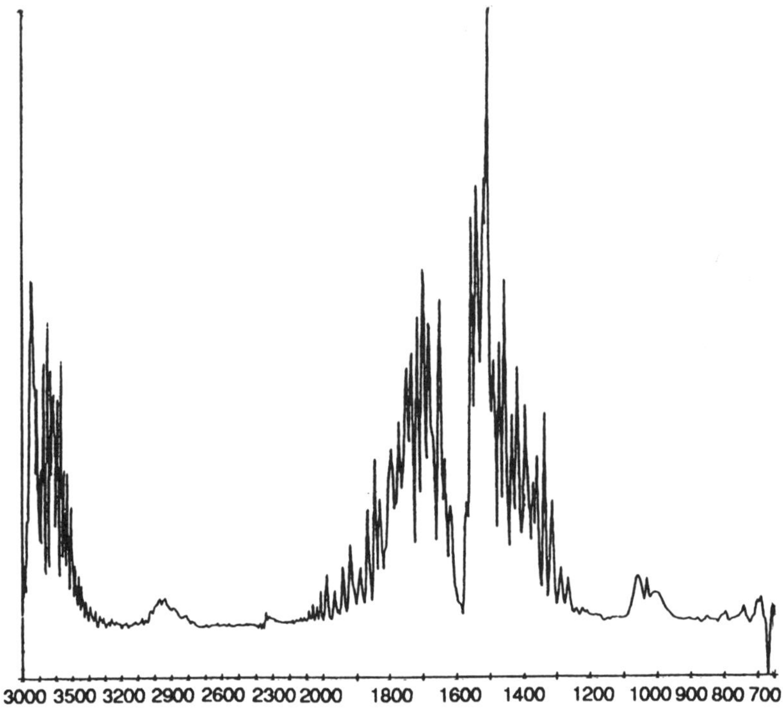

Fig. 9. Vapor phase spectrum of water (background).

variation can be circumvented to a large degree by using as a background the baseline region of the chromatogram immediately before the GC peak. Use of multiple backgrounds in this manner is made simple by software available from FT-IR manufacturers.

V. APPLICATIONS

Applications of GC/FT-IR are increasing. This increase is largely due to the increased identification accuracy deriving from improved library search capabilities, and the capacity of the technique to identify isomers. There are now many laboratories producing satisfactory results using GC/FT-IR along with GC/MS, rather than by using either technique alone. The disadvantage of the lower sensitivity of GC/FT-IR has been partially overcome by investigation of sample concentration methods and by use of GC/MI/FT-IR. For a review of early applications of GC/FT-IR, the reader is directed to an earlier review by Erickson (1979).

A. Foods

The analytical method for detecting ethyl carbamate (EC) in various foods was investigated by Mossoba *et al.* (1988). The carcinogenesis of EC is recognized in animals. Because EC is formed in a fermentation process, the analysis of alcoholic beverages and foods for EC is important. GC/MI/FT-IR is used in the analysis to enhance sensitivity, and good spectra are obtained for sample sizes less than 1 ng. An internal standard method is used for the analysis. The internal standard is EC, isotopically labeled with ^{13}C and ^{15}N (EC_{lab}). The isotopic shift for the asymmetric stretching vibration of C—O—C=O is from 1298 cm^{-1} (EC) to 1326 cm^{-1} (EC_{lab}). Figure 10 shows the spectra of EC, EC_{lab}, and the mixture. The linear dynamic range for the analysis is 0.5 to 9.0 ng. The figure shows the spectrum of 0.6 ng of EC. In the standard addition to whiskey, the recovery percentage was 98–96%. This showed a good correlation with the quantitative value obtained by GC/MS. The application to actual products such as cheese, bread, and yogurt has revealed no detectable amounts of EC (less than 10 parts per billion).

Fehl and Marcott (1986) succeeded in the analysis of a low-concentration sample by injecting a large quantity of the sample using the inject/trap method in order to compensate for low sensitivity. The distillate from stream distillation of 10 g of chocolate was extracted by Freon-11; 50 ml of the Freon

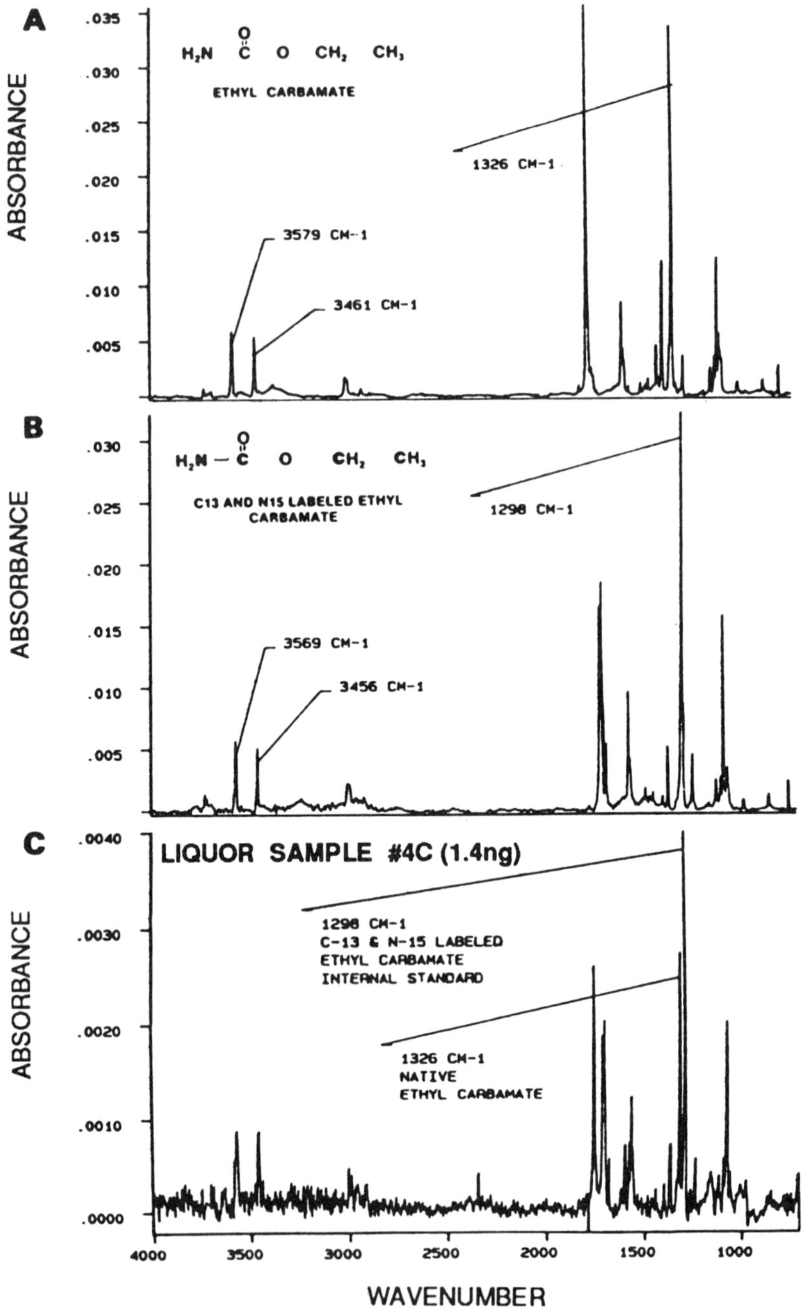

Fig. 10. MI/FT-IR spectra of EC (A), EC_{lab} (B), and 1.4 ng of EC with an internal standard of EC_{lab} (C). (From Mossoba *et al.* Reprinted with permission from *Analytical Chemistry*. Copyright 1988 American Chemical Society.)

solution was concentrated to 0.25 ml; and 10 μl of the concentrated solution was injected for the analysis. The Gram-Schmidt chromatogram and the spectrum of one of the identified components, trimethylpyrazine, are shown in Fig. 11. For this weakly absorbing sample, amounts as small as 300 ng were detected, and the method was considered to be effective for the analysis of minute components in solvents.

Le Quere *et al.* (1987) applied this method to the analysis of wine, Mirabelles plums, and the aromatic components in solvents.

B. Pesticides

Kalasinsky investigated Mirex (dodecachloropentacyclodecane), a pesticide used in the control of fire ants (Kalasinsky,

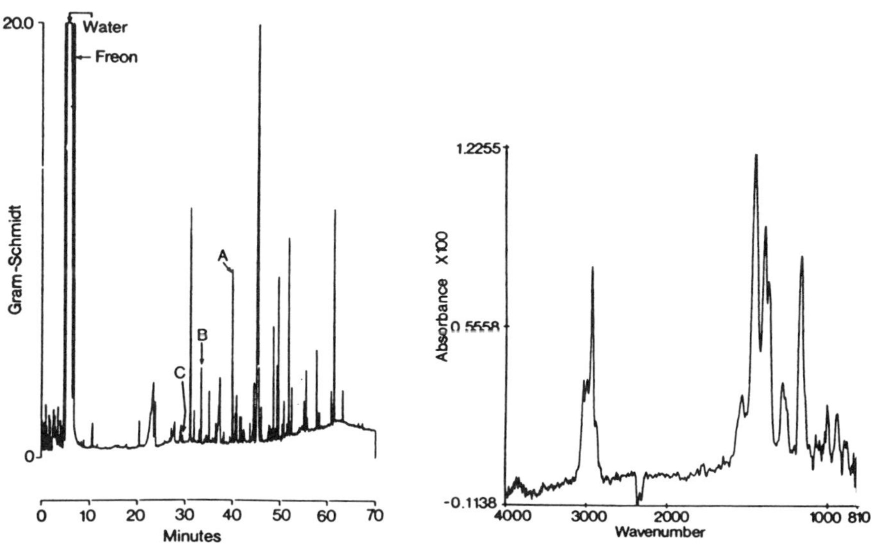

Fig. 11. Gram-Schmidt chromatogram of chocolate extracted volatiles (left) and infrared spectrum of peak C, identified as trimethylopyrazine (right). Column: 80 m × 0.32 mm, DB-1, 1-μm coating. Carrier gas: He, 1.5 ml/min. Light pipe: 12.7 cm × 1 mm, 250°C. (From Fehl and Marcott. Reprinted with permission from *Analytical Chemistry*. Copyright 1986 American Chemical Society.)

1983). Mirex has been analyzed conventionally by electron capture GC or by GC/MS. The identification of Mirex derivatives was difficult because they show quite similar data to one another. Rapid decomposition in the environment is one of the factors required for pesticides. The GC/IR method was applied to the photodecomposition products of Mirex. In the decomposition, some of the chlorine atoms were substituted with hydrogen. The 5,10-dichloro derivative has anti- and syn-isomers. These are difficult to separate by GC, but the spectra of the two isomers have one band that is at 1100 cm^{-1} for one isomer, and 1120 cm^{-1} for the other. There is some overlap of these bands, but the pure spectrum of each can be obtained by calculating the difference between the spectra from the first half and the second half of the GC peak (see Fig. 12). The decomposition of these substances in the environment is quite different from the decomposition in the laboratory. In the environment, the anti-isomer is the major product; in the laboratory, both are produced with a high concentration. GC/IR is effective for the analysis of substances having such stereospecific reactions.

C. Environmental

The application to environmental analysis is one of the fields where GC/FT-IR is most actively used. Gurka *et al.* (1987) have examined the detection limits and the relative sensitivity of various pollutants using the technique. They compared capillary and packed columns, and the sensitivity and identification accuracy of GC/FT-IR with those of GC/MS. Furthermore, they proposed protocols for obtaining reliable analytical results with GC/FT-IR (Gurka, 1985; Gurka *et al.*, 1987; Gurka and Hiatt, 1984).

The still bottom of a herbicide was analyzed by various methods (Gurka *et al.,* 1988). The semivolatile compounds were analyzed using GC/IR/MS and PMR, and the nonvolatile

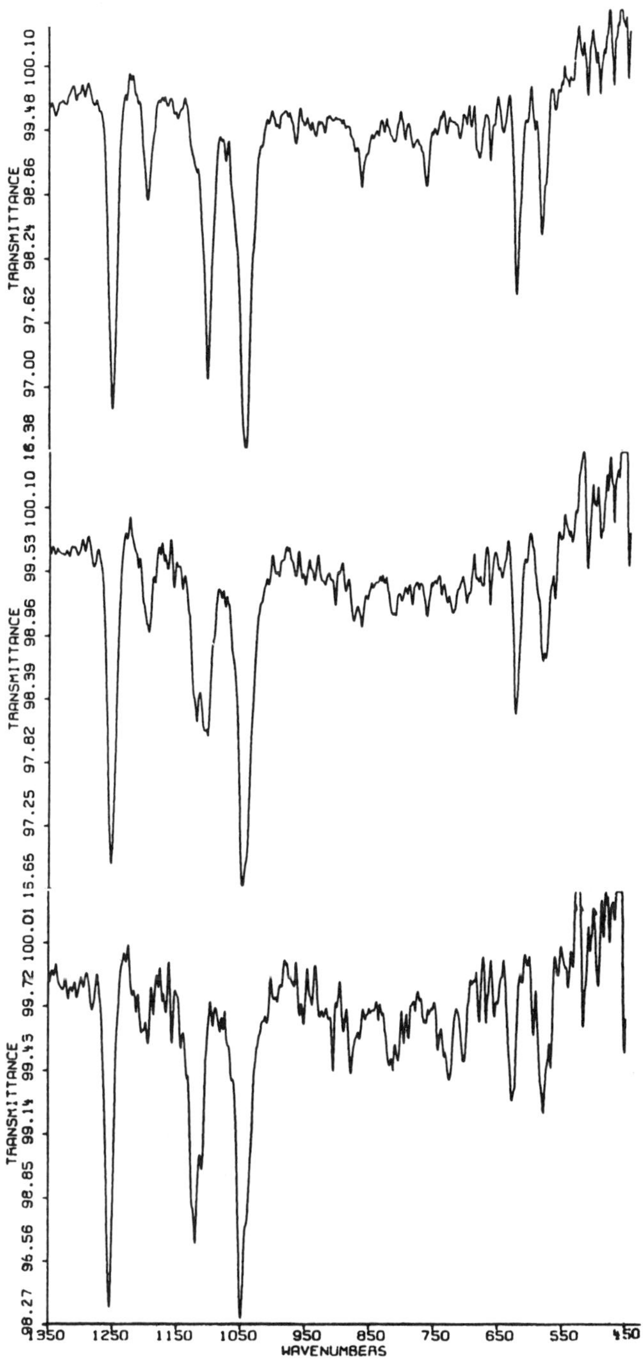

Fig. 12. Spectra of 5,10-dichloro derivatives of Mirex: (top) anti isomer; (middle) anti and syn mixture; (bottom) syn isomer (difference spectrum of B − A). (From Kalasinky, 1983. Reproduced from the *Journal of Chromatographic Science* by permission of Preston Publications, a division of Preston Industries, Inc.)

compounds using diffuse reflectance (DRIFT), thermospray or Fast Atom Bombardment (FAB) MS. In the GC/IR/MS analysis, six major components were recognized, as shown in the Gram-Schmidt chromatogram in Fig. 13. Among the six components, four (retention times of 17.33, 18.82, 21.35, and 27.58 minutes, hereafter referred to as 1 through 4, respectively) were not found in the spectral library data base, so an overall analysis was carried out. The molecular weight and number of chlorine atoms for peak 3 were known from the mass spectrum shown in Fig. 14a, and the existence of trichloroalkenyl ether was indicated by the loss of m/z 161. The absorption of infrared at 766 cm^{-1} suggested the existence of three adjacent hydrogen atoms, and the absorption coincided with that of 2,6-dichlorophenol at 768 cm^{-1}. Thus, the structure shown in Fig. 14a was postulated. However, the position of the Cl atom of the alkenyl group was not determined. For peak 1 in Fig. 13, tetrachlorobenzofuran was suggested by the molecular ion, the isotope cluster, and the indication of the COCl mass loss. Furthermore, the existence of isolated and adjacent hydrogens was suggested by the out-of-plane deformation vibration of the infrared spectra. Thus, the structure shown in Fig. 14b was postulated. This structure has been shown to be carcinogenic.

Polycyclic aromatic compounds (PAC) have a low sensitivity to infrared, but GC/IR is an effective means for the analysis of isomers. Combined with GC/MS, analysis of PAC is quite effective in many cases.

Garg *et al.* (1987) analyzed polycyclic aromatic hydrocarbon (PAH) isomers using GC for quantitative analysis and GC/IR and GC/MS for identification. More than 100 peaks were detected, and GC/IR was proved to be effective for the identification of positional isomers for substituents of aromatic compounds (1-methylnaphthalene, 2-methylnaphthalene, etc.). There were only a few PAH in the spectral library, and identification was chiefly made using the GC/MS data.

Chiu *et al.* (1984) carried out the analysis of PAC using GC/MS and GC/IR complementarily. In the mass spectra of PAC, the molecular ions are strong and the fragmentation is

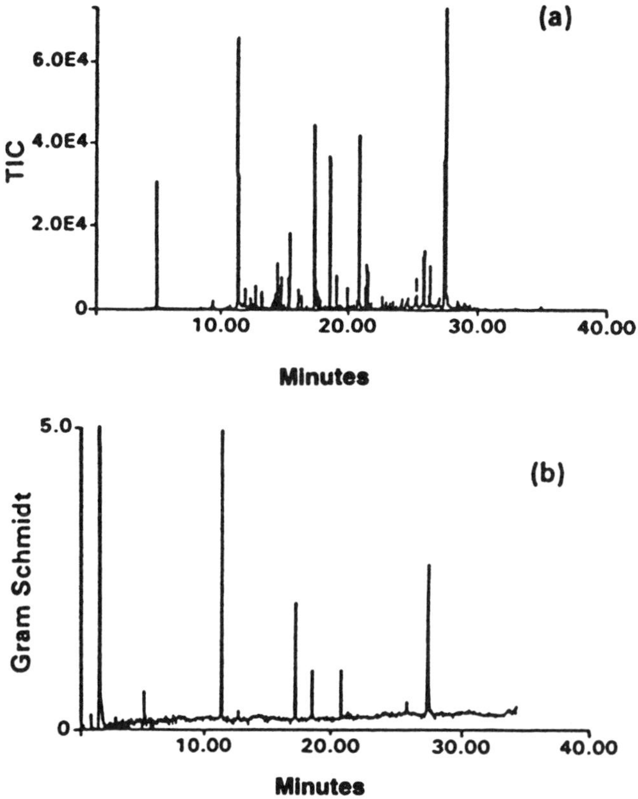

Fig. 13. Total ion chromatogram (a) and Gram-Schmidt chromatogram (b) of herbicide still-bottom extract. Column: 30 m × 0.32 mm, DB-1, 1-μm coating thickness. GC effluent was split 10:1, FT-IR to MS. (From Gurka *et al.*, 1988. Reproduced from the *Journal of Chromatographic Science* by permission of Preston Publications, a division of Preston Industries.)

low, making identification difficult. Structural information was obtained from the GC/IR data. More than 73 peaks were observed in the analysis of coal combustion products.

Among them, for example, 9-fluorenone and 1-phenalenone exhibit similar mass spectra, making identification by mass spectral data difficult. Of the two, only phenalenone is mutagenic, making its identification important. Figure 15

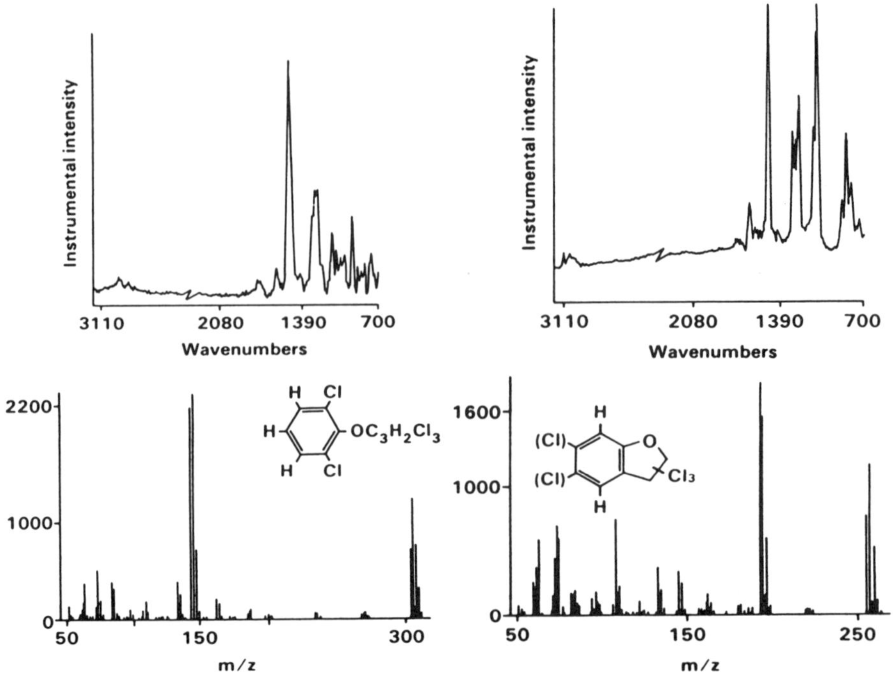

Fig. 14. FT-IR and mass spectra of the components of two peaks of Fig. 13: left, retention time 21.35; right, retention time 17.33 min. (From Gurka *et al.*, 1988. Reproduced from the *Journal of Chromatographic Science* by permission of Preston Publications, a division of Preston Industries, Inc.)

shows the mass and infrared spectra of 9-fluorenone and 1-phenalenone. The mass spectra of these isomers are almost identical. The two can be distinguished by their infrared spectra, however, because the strain of the five-membered ring causes the carbonyl absorption for 9-fluorenone to be shifted to higher energy (1736 cm^{-1}), far from the 1662 cm^{-1} carbonyl absorption of 1-phenalenone.

Another peak could not be identified by mass spectral data. It was not possible to distinguish among indene, phenylpropadiene, phenylpropyne, and isomers of methylphenylacetylene. The absence of acetylenic and asymmetric allenic stretching vibration absorptions at 2100–2250 and at 1950–2000 cm^{-1} was noted in the infrared spectra, pointing to the

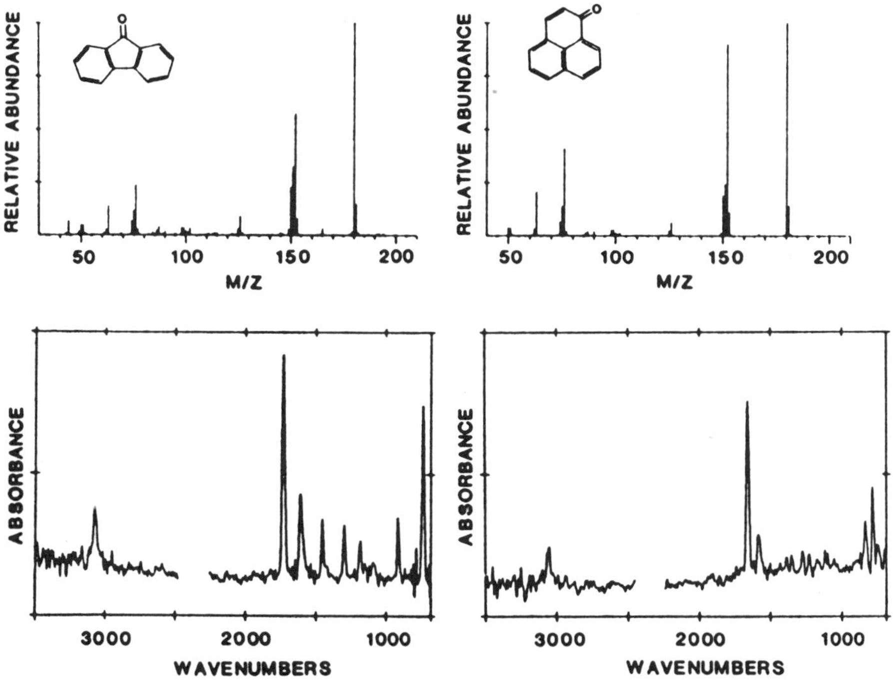

Fig. 15. FT-IR and mass spectra of two peaks of coal combustion products: (left) 9-fluorenone; (right) 1-phenalenone. GC column: 15 m × 0.32 mm, DB-5; film thickness: 0.25 μm (MS), 1 μm (IR). MS ionizing voltage: 70 V. Source temperture: 200°C. (From Chiu *et al.* Reprinted with permission from *Analytical Chemistry*. Copyright 1984 American Chemical Society.)

identification of the peak as indene. This case is an excellent example of the complementary nature of the two techniques.

Polychlorinated biphenyls (PCBs) and dioxins have a large number of isomers, varying in the number and position of Cl substitution. The toxicity of these compounds depends largely on structure, so the identification and quantitative analysis of isomers are important. Schneider *et al.* (1985) applied GC/MI/IR to the analysis of PCB. Detection could be made in the range of 1 to 10 ng with S/N > 10. Various isomers could be identified by absorptions in the fingerprint region. Among 209 theoretically existing isomers of PCB, 40 were stored in the spectral library. Using this data, Aroclor 1221 was analyzed and 98% of the GC-FID peak area was identified. Figure 16 shows the chromatogram and the identification results of peak 5.

Grainger *et al.* (1988) studied 14 isomers of pentachlorodibenzo-*p*-dioxin, mainly analyzing the asymmetrical stretching vibration of C—O—C, and then investigated in detail the relation between the wavenumber of the absorption band and the chemical structure. Wurrey *et al.* (1986) investigated the quantitative analysis of 2,3,7,8-tetrachlorodibenzo-*p*-dioxin using an internal standard labeled with ^{13}C and using GC/MI/FT-IR.

Worley *et al.* (1980) found GC/FT-IR to be a useful method for confirming structures of 4′-chloroacetanilides in water at low levels (parts per billion). They used an extraction/concentration technique starting with 1000 ml of water containing trace amounts of seven acetanilides. Three microli-

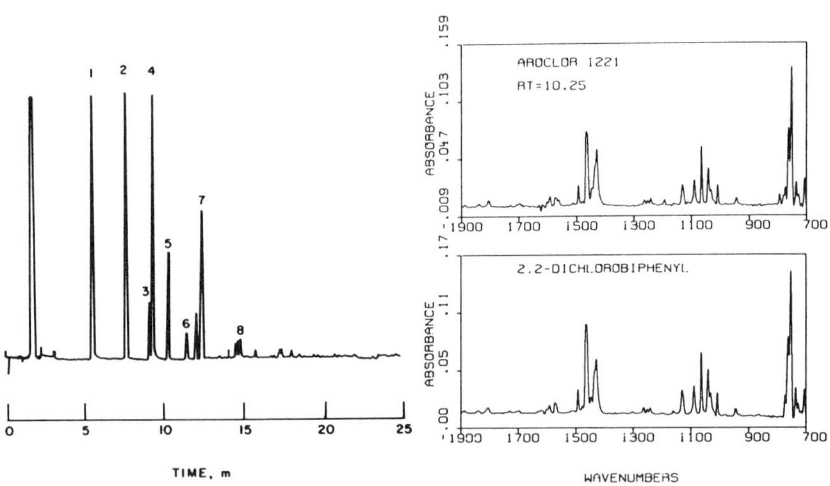

Fig. 16. Analysis of Aroclor 1221 by GC/MI/FT-IR: (left) gas chromatogram by FID; (right) infrared spectra of peak 5 and 2,2-dichlorobiphenyl. Carrier gas: He containing 1.5% Ar. Column: 30 m × 0.25 mm, DB-5, 1-μm coating. Carousel surface temperature: 11 K. (From Schneider *et al.*, 1985. Reproduced from the *Journal of Chromatographic Science* by permission of Preston Publications, a division of Preston Industries, Inc.)

ters of extracted/concentrated solution was injected on a packed column. The effluent was first detected using a nitrogen-selective detector. Application of GC/FT-IR was used to characterize the analyte species. For the GC/FT-IR analysis, a solution containing 4.0 μg of each of the seven acetanilides was injected on to the column. The seven acetanilides were identified by their respective GC/IR spectra, as well as by the functional-group chromatograms. This experiment is a good example of how structural information can be obtained from selected absorbance monitoring of the effluent. Figure 17 shows the five functional-group chromatograms collected during the GC run. The compounds used here are shown in Table

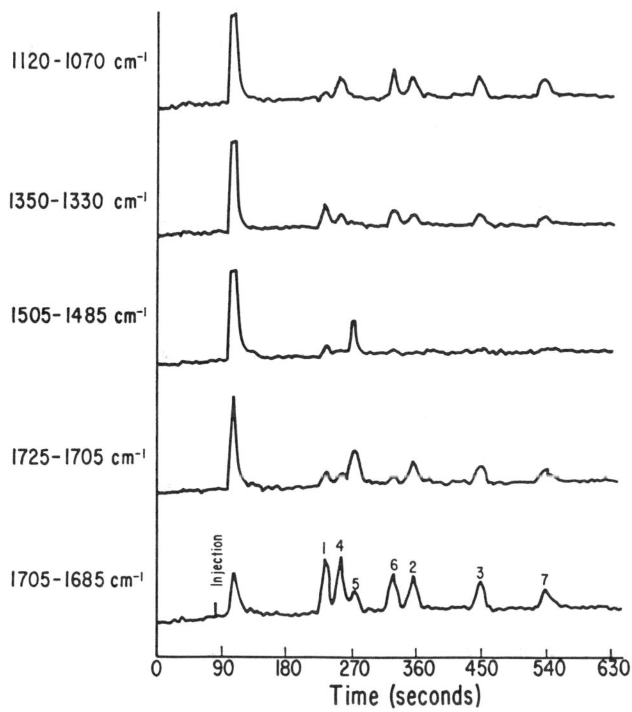

Fig. 17. Set of five functional-group chromatograms for 4.0 μg each of the seven compounds shown in Table III. The first peak after the injection mark is the solvent peak. (From Worley *et al.* Reprinted with permission from *Analytical Chemistry*. Copyright 1980 American Chemical Society.)

III. It can be seen that the molecules are closely related. All seven compounds are evident in the two lower traces of Fig. 17, which are in the region of the amide I carbonyl absorbance. Compound 5 has the highest wavelength absorbance at 1719 cm^{-1}; so it appears only weakly in the lowest trace, but more prominently in the 1725–1705 cm^{-1} window. All of the compounds are tertiary amides except for compound 5, which is a secondary amide. Secondary amides show a strong amide II band around 1495 cm^{-1} in the gas phase. It can be seen in the figure that only compound 5 gives a significant peak in the 1505–1485 cm^{-1} region. Another piece of useful structural information is found in the top trace of Fig. 17. This is the C—O stretching region. Only compounds 1 and 5 do not have the *N*-alkoxymethyl group, and so do not absorb in this region. Based on signal-to-noise data, the authors determined that useful data could be obtained from 1.0 μg or more of an acetanilide in a mixture. From this, it was projected that this method would be useful for confirming acetanilide structures in water at low levels.

Table III Structure of Acetanilides

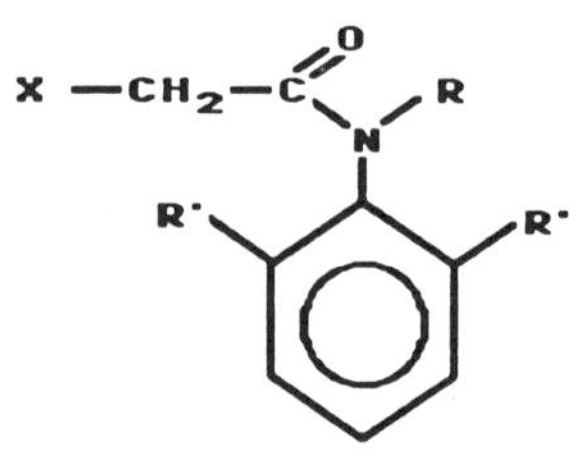

Compound	*X*	*R*	*R'*
1	Cl	*i*-C_3H_7	H
2	Cl	CH_2OCH_3	C_2H_5
3	Cl	$CH_2O_4H_9$	C_2H_5
4	H	CH_2OCH_3	C_2H_5
5	Cl	H	C_2H_5
6	C H	CH_2OCH_3	C_2H_5
7	Cl	$CH_2OC_6H_{13}$	C_2H_5

D. Forensic

In this field, analyses related to the determination of drugs is carried out. Kempert (1988) indicated that the structural difference between methamphetamine and phentermine (the former is a secondary amine, the latter a primary amine) could be easily identified by the absorption around 2800 cm^{-1} in the GC/IR spectrum (Fig. 18). Structural differences had been difficult to elucidate using only MS data. It was also determined that two diastereomers, cocaine and pseudococaine, could be identified by the differences in the spectra in the fingerprint region and for the carbonyl absorption (Fig. 19). Five kinds

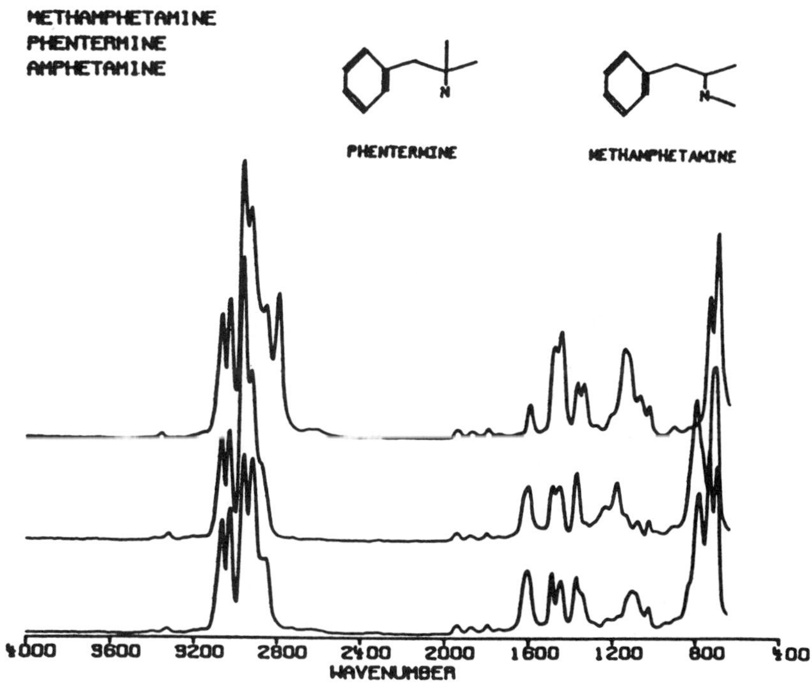

Fig. 18. Spectra of phenethylamine drugs: (from top to bottom) methamphetamine, phentermine, and amphetamine. (From Kempert, 1988.)

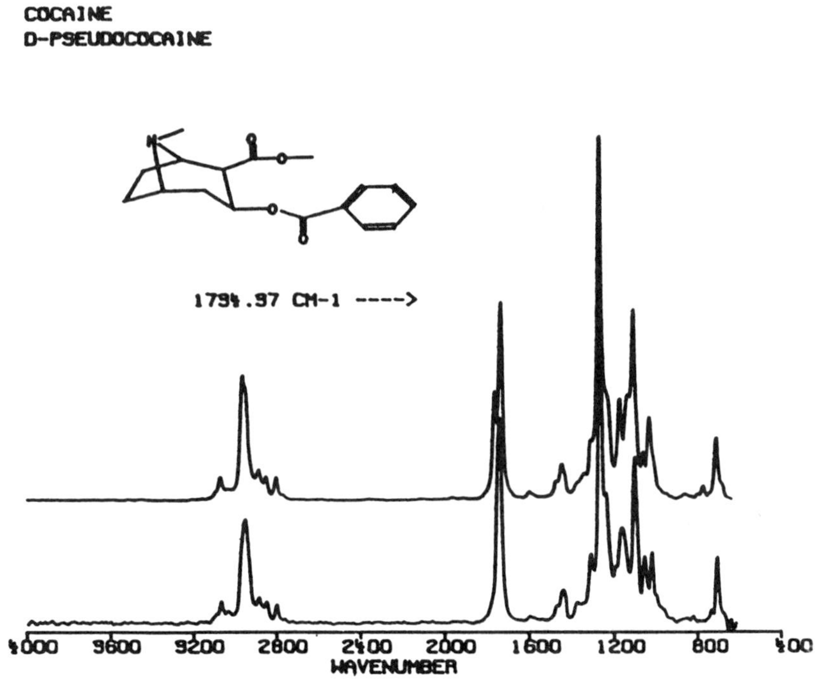

Fig. 19. Spectra of cocaine diastereomers: (top) cocaine (2-exo, 3-exo); (bottom) pseudococaine (2-endo, 3-exo). (From Kempert, 1988.)

of barbiturates were compared. These could be identified by differences in their spectral patterns, except for butabarbital and pentobarbital, which differ only in the length of the alkyl chain. However, these two substances could be identified by the intensity ratio for the stretching vibrations of C—H and C—O. The relative sensitivities of the technique for detecting some drugs was also examined. The results revealed that the sensitivity for pentobarbital was more than 50 times higher than for lysergic acid (LSD).

Deveaux and Huvenne (1985) applied the method to the analysis of components in *Cannabis*. Identifications of cannabinol (CBN), cannabidiol (CBD), cannabichromene (CbCh), Δ8- and Δ9-tetrahydrocannabinol (THC) were made. Figure 20 shows the library search match for Δ-tetrahydrocannabinol.

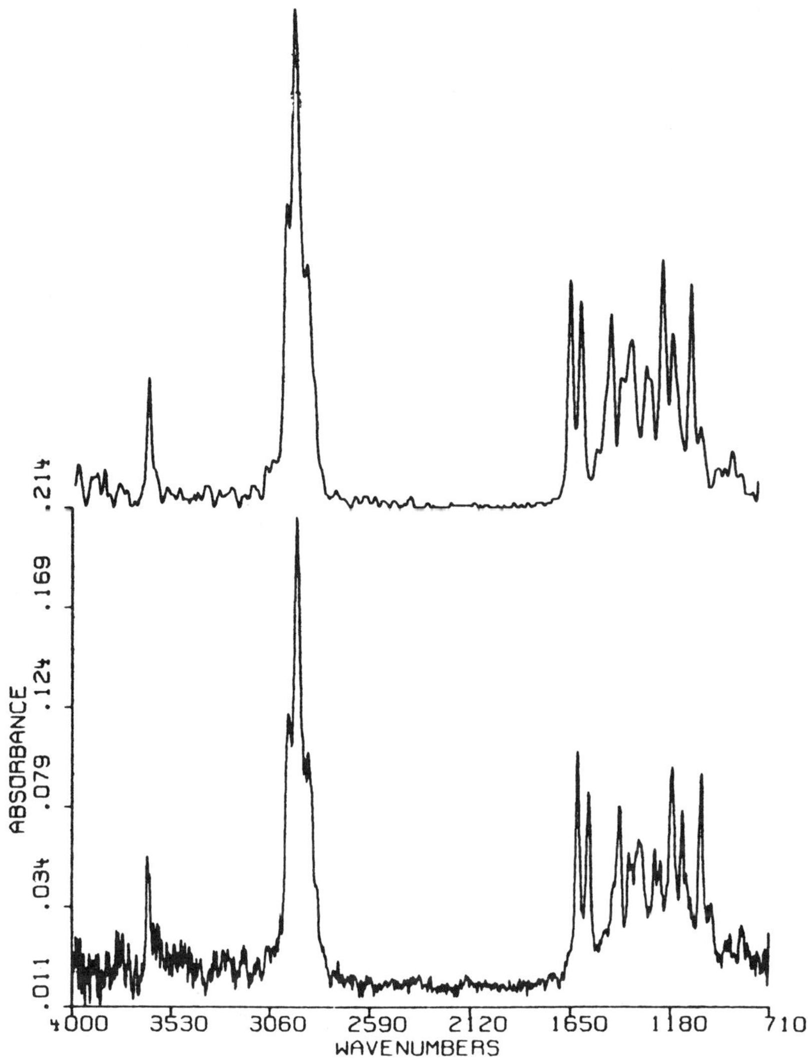

Fig. 20. Identification of △9-tetrahydrocannabinol in a cannabis sample: (top) reference spectrum; (bottom) sample spectrum. (From Deveaux and Havenne, 1985.)

Furthermore, these components were analyzed quantitatively; the ratio of tetrahydrocannabinol plus cannabinol to cannabidiol was found to be related to the producing countries, and the ability to predict the region where the cannabis was grown from the analytical data has been shown.

E. Fragrant compounds

Fragrant compounds are a complex mixture of comparatively low boiling point compounds. They, like terpene and aromatic compounds, contain many isomers. In general, they are good samples for best illustrating the separation and identification capabilities of GC/IR. These methods are used for analyses of perfume and natural fragrant components (Idstein and Schreier 1985; Purcell and Magidman, 1984; Kalasinky and McDonald, 1983). Figure 21 shows the Gram-Schmidt

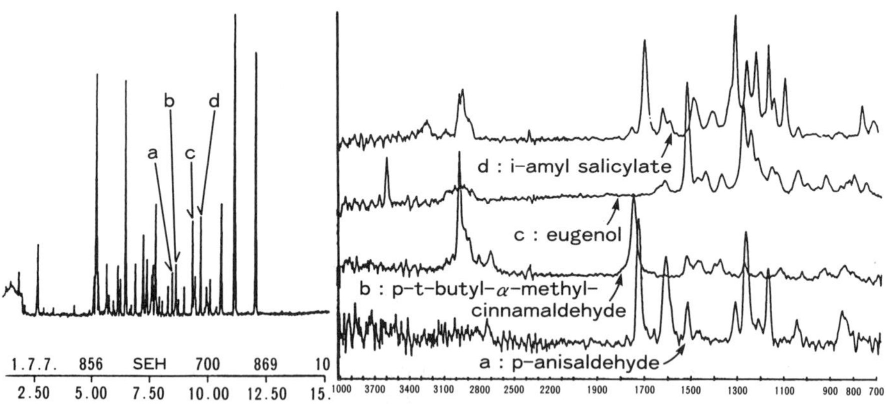

Fig. 21. Gram-Schmidt chromatogram and spectra of perfume. Peaks a–d were identified by library search (Sadtler vapor phase library). Column: DB-WAX (J&W Scientific), 20 m × 0.32 mm, 0.5-μm coating. Carrier gas: He, 2 ml/min. Light pipe: 12 cm × 1 mm, 250°C.

chromatogram of a marketed perfume and the spectra of several of its components. Most of the components contained in these samples could be identified by using a library spectral search (Namba, 1988).

Analyses of natural fragrant components are made by combining head-space sampling and the purge and trap technique of GC/IR. The fragrant components of rose were collected and analyzed to obtain the chromatogram shown in Fig. 22. The indicated peak in the figure is specific for a rose called T-rose. In GC/MS analysis, this component was determined to be a compound having a molecular weight of 152, but the compound could not be identified. In GC/IR analysis, it was

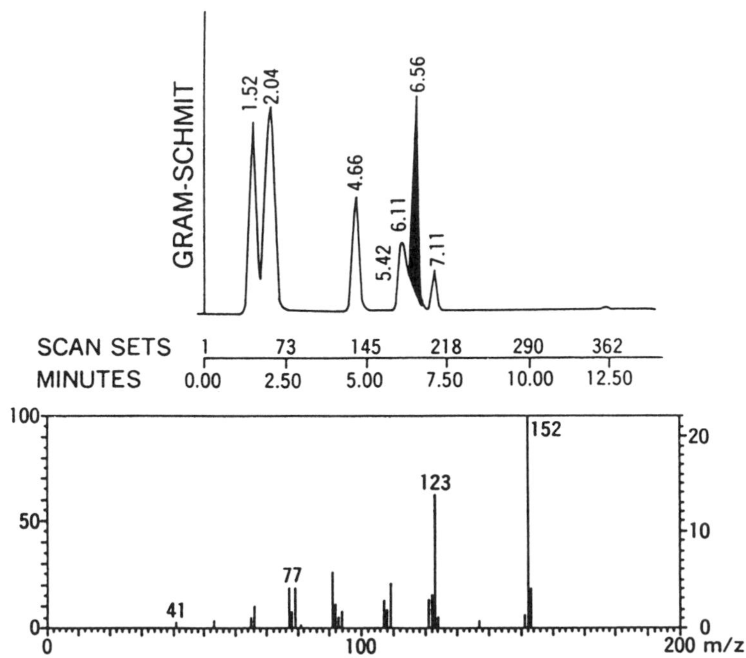

Fig. 22. Purge and trap GC/FT-IR and GC/MS analysis of a rose flower. Column: 5% FFAP on Chromosorb W, 2 m × 3 mm. Carrier: He, 60 ml/min. Light pipe for infrared analysis: 2.5 mm × 60 cm, 250°C; 8-cm^{-1} resolution. Ionizing voltage for MS: 20 V. Ion source temperature for MS: 200°C.

further determined to be the 1,3,5-substituents of aromatic compounds and to have methoxy groups (Fig. 23). Therefore, considering the molecular weight information, the component was determined to be 1,3-dimethoxy-5-methylbenzene. The identification was confirmed by the synthesis of a standard sample.

Gordon *et al.* (1988) made an analysis of flu-cured essential oil. This has a very complex composition, and the separation by capillary GC was thought to be insufficient. Improvement in the separation was made by multidimensional GC.

The megabore chromatogram was divided into 23 divisions, each of these was introduced into the GC capillary column for GC/MS or GC/MI/IR analysis. As a result, 306 compounds were identified, many more than the 211 compounds identified without the megabore preseparation. There were only 98 common components between these two analyses, so the balance were newly identified compounds. Such methods were considered to be very effective for composition analysis of highly complex mixtures.

Kubeczka *et al.* (1985) applied GC/IR to the analysis of

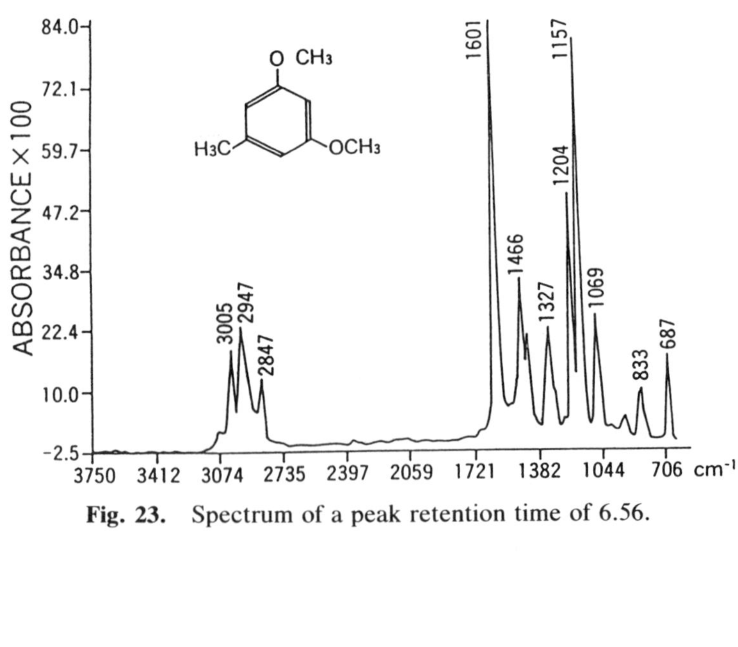

Fig. 23. Spectrum of a peak retention time of 6.56.

essential oils such as geranium oil and identified mainly terpene type compounds. They concluded that further enrichment of the spectral library would result in more effective identification.

Herres *et al.* (1983) analyzed the volatiles of *Cherimoya* fruit. The liquid extract of the pulp sample by high vacuum distillation was subdivided into three parts by silica gel chromatography. Among these, the second and third fractions were used. The components of the second fraction were identified mainly as esters such as butanoate. Lower alcohols and monoterpene alcohol were identified in the third fraction. As a result, 26 components, which included nearly all of the major components, were newly identified.

F. Pyrolysis GC/IR Analysis of Polymer Compounds

Pyrolysis GC (P-GC) has been used for the analysis of polymers, but identification of decomposition components has been difficult. In P-GC/IR, the analyses of the synthetic polymers from which monomers are readily formed are comparatively easy (Liebman and Wampler, 1985; Folster and Herres, 1986; Herres and Folster, 1985; Liebman and Levy, 1983). In the case of natural polymers with high polarity, however, the monomers are hardly obtained; hence, the identification of the decomposition products is difficult. Therefore, comparison of the chromatogram patterns has been used for identification.

The analysis of a molding plastic, NORYL SE1, by pyrolysis GC/FT-IR was reported by Smith *et al.* (1983). The sample was pyrolyzed at 900°C for 20 sec. GC/MS analysis of the same sample previously reported showed only that cresols and dimethylphenols were present. Identification of the exact isomers was made by using spectral search on the computed spectra from the GC/FT-IR analysis. Figure 24 shows the spectra of two of the components identified as *o*-cresol and 2,6-dimethylphenol.

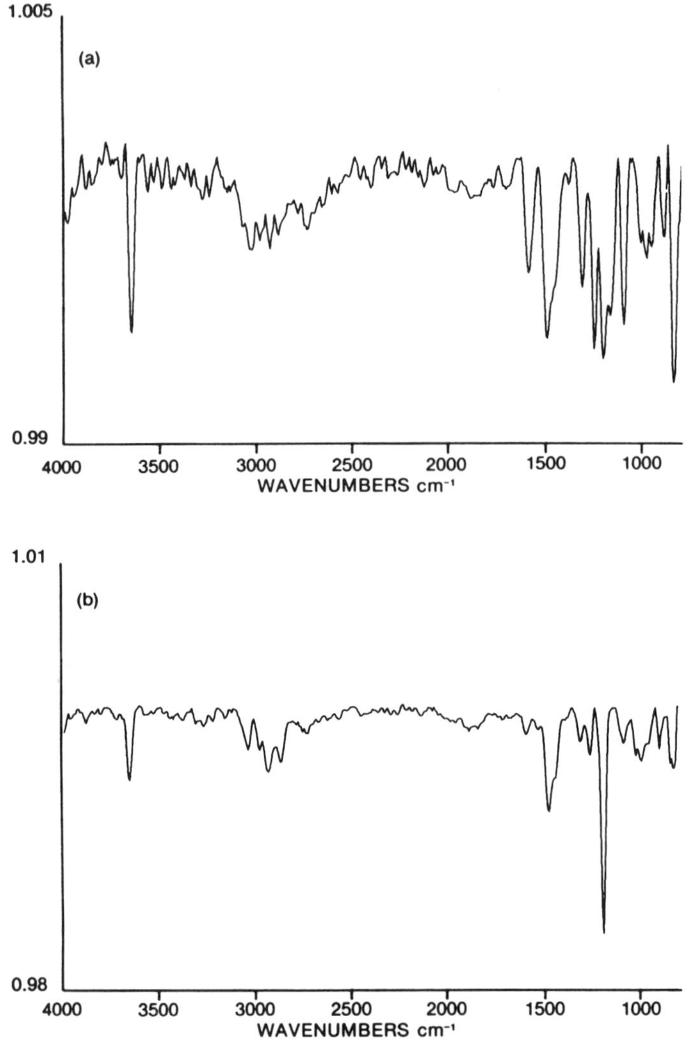

Fig. 24. Transmittance spectra of two of the peaks identified from the pyrolysis GC/IR analysis of NORYL SE1. (a) Identified as *o*-cresol; (b) identified as 2,6-dimethylphenol. (From Smith *et al.*, 1983.)

The author tried to use P-GC/IR measurements of proteins from soybeans. Several components in the complex pyrogram were successfully estimated (Namba, 1988) (see Fig. 25). Phenols derived from the side chain of aromatic amino acids had been identified. The structures of various other decomposition components could be estimated from library spectral search data.

G. Carbohydrates

In GC/IR, highly volatile compounds have been analyzed, but analysis of high boiling point samples having a high polarity has been rare. These samples are analyzed by GC or GC/MS after they are converted into volatile derivatives; such examples, however, are scarce in GC/IR. In GC/MS, it is known that the spectra of the isomers of various aldo-hexoses show almost the same pattern, an outcome making identification difficult. Yoshida *et al.* (1987) analyzed saccharides, which are among the most complex isomers. Monosaccharides are frequently analyzed by converting them to volatile derivatives

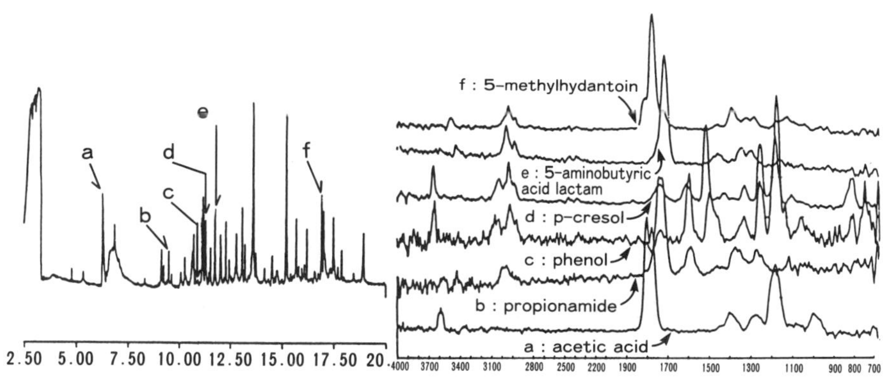

Fig. 25. P-GC/IR analysis of protein (soybean). Pyrolysis: Curier point type, 590°C, 3 sec. Other conditions are the same those described in Fig. 21.

by acetylation. In this study, trimethylsilyl TMS derivatives were investigated. The conversion to TMS derivatives is very convenient, because OH, COOH, NH_2, and SH groups can be simultaneously modified. There is a possibility, however, of hindering the specific absorption bands of the saccharides, because the strong absorptions of the TMS group appear in the regions of 1260, 1100, and 860 cm^{-1}. However, as shown in Fig. 26, all the anomers of glucose show unique spectra.

Furthermore, six aldohexoses also showed unique patterns, as can be seen in Fig. 26. Of the many monosaccharides measured, none showed identical spectral patterns; thus, GC/IR was confirmed as an effective method for the identification of monosaccharides. As an application, the acid hydrolysis decomposition products of pectin, a natural saccharide, were analyzed. The resulting chromatogram and spectra shown in Fig. 27 were obtained as an aid in identifying major components.

Amino acids can be analyzed by the same method. The method is considered to be effective for the simultaneous analyses of monosaccharides and amino acids of proteoglycans, for example.

H. Fuels

Olson *et al.* have conducted an isotope dilution GC/FT-IR study for the analytical determination of carboxylic acids resulting from the oxidation of lignite coals. A set of isotopically ^{18}O-enriched standards were made, and small amounts of these were converted to methyl esters. These were analyzed

Fig. 26. GC/IR spectra of aldohexose trimethylsilyl ethers: (top) glucose anomers [furanoside (A), α-pyranoside (B), β-pyranoside (C)]; (bottom) six aldohexoses; most of the aldohexoses in B displayed two major GC peaks (anomers); the peaks with shorter retention times are shown in this figure. Column: 30 m × 0.32 mm, DB-1, 1-μm coating. Light pipe: 12 cm × 1 mm, 250°C.

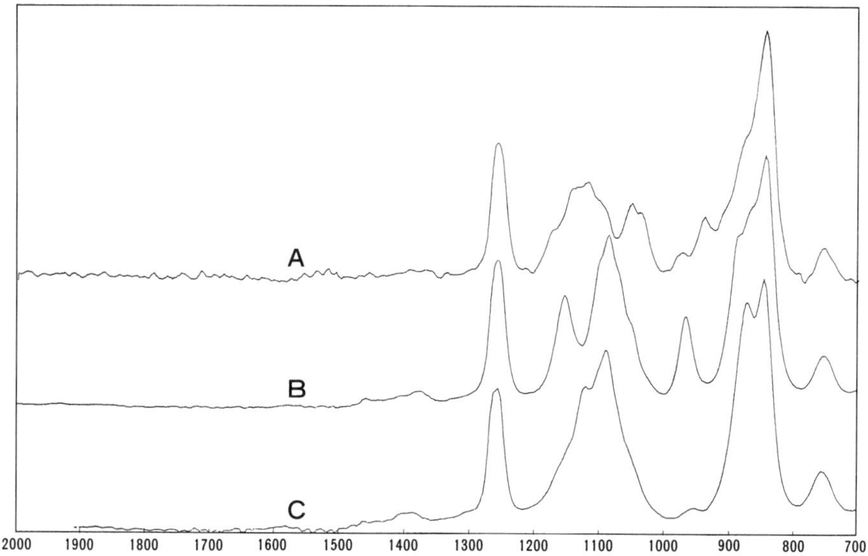

A
B
C
2000
1900
1800
1700
1600
1500
1400
1300
1200
1100
1000
900
800
700

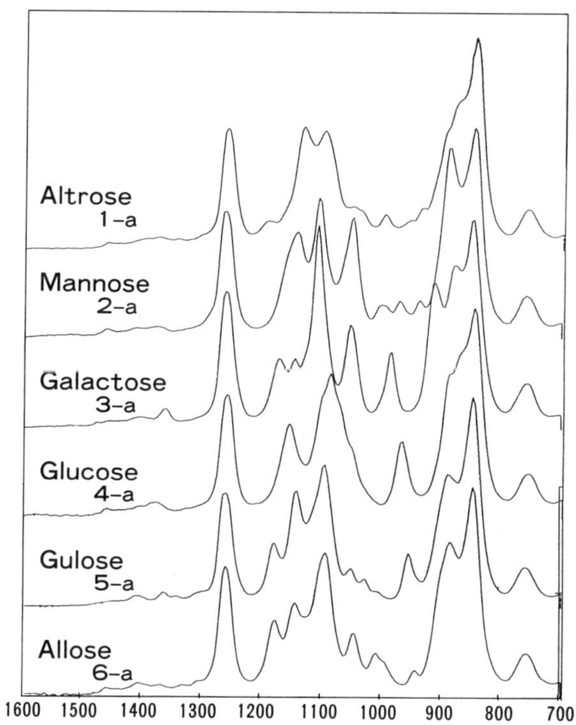

Altrose
1-a
Mannose
2-a
Galactose
3-a
Glucose
4-a
Gulose
5-a
Allose
6-a
1600
1500
1400
1300
1200
1100
1000
900
800
700

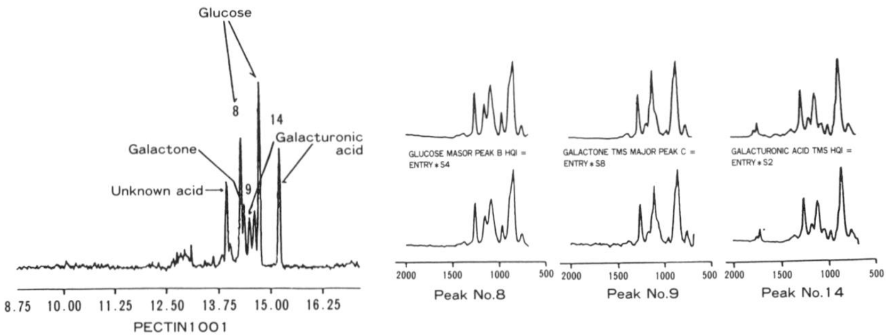

Fig. 27. Analysis of hydrolyzed monosaccharides of pectin: (left) Gram-Schmidt chromatogram. (right) identification results by comparison with reference spectra.

by electron ionization mass spectrometry (EI-MS) for isotopic purity. Calibration curves were prepared using the isotopic samples and analyte acids. Ratios of the absorbances of the carbonyl bands were measured. The carbonyl absorption maxima of the analyte and ^{18}O-enriched standard esters differ by about 33 cm^{-1}. Chromatograms were reconstructed using windows 8 cm^{-1} wide corresponding to the carbonyl stretching bands of the analyte and standard esters.

Two applications of the method were described. The oxidation of a lignite coal with neutral hydrogen peroxide resulted in an aqueous solution of carboxylic acids. This solution was esterified after addition of the ^{18}O-enriched standards and then analyzed by GC/FT-IR.

The second method used RuO_4 oxidation of the same lignite sample, followed by addition of the ^{18}O-enriched standards, conversion to the methyl esters, and isotope dilution ratio measurement with GC/FT-IR. This method yielded results comparable to those determined by GC/MS methods.

Smith *et al.*, (1983) have applied the technique of narrow-bore capillary GC/FT-IR to the analysis of a petroleum distillation fraction. Selectivity was demonstrated by monitoring several wavelength regions, and detection of at least one component representing less than 30 ng of compound was reported.

I. Analytical Methods using GC/FT-IR/MS

Measurements obtained by combining GC/FT-IR and GC/MS have been reported since the development of GC/FT-IR. This GC/FT-IR/MS combination gives the same results as those obtained separately in respective apparatus on the same sample. Hence, essentially no new information will be obtained. It will be convenient, however, particularly for analysis of a complex mixture, if the two kinds of spectra are obtained simultaneously for a GC peak. The analysis of the information obtained from both sources by the same data processing equipment will be effective for the reduction of false positives (Crawford *et al.*, 1982; Gurka and Titus, 1986; Olson and Diehl, 1987; Wang and Isenhour, 1987). Moreover, the improvement of analytical accuracy by adding the molecular weight information by CI and the molecular formula information by high-resolution measurement of the EI mass spectra are now proceeding (Wilkins *et al.*, 1982; Laude *et al.*, 1984, 1985, 1986).

VI. CONCLUSION

As described in this chapter, GC/FT-IR has been used in various areas pertaining to industrial products. Important investigations and applications are proceeding in the environmental field.

In other areas, fewer reports have been forthcoming. This may be because the data in the companies where the techniques are being used are proprietary in nature, or because the number of techniques used in industry is small at present.

GC/FT-IR is recognized as a technique having lower sensitivity than those of GC (FID) and GC/MS, and the features of GC/IR are not fully understood. GC/FT-IR has many features that even GC/MS does not possess. As improvements in the instrumentation and enrichment of the library search methods proceed, GC/FT-IR will become an indispensable analytical method in many industrial laboratories.

ACKNOWLEDGMENT

The author expresses his appreciation to Eiji Kadowaki and Seiichi Yoshida for their cooperation in this work.

REFERENCES

Azarraga, L. V. (1980). *Appl. Spectrosc.* **34,** 224.
Bourne, S. (1984). *Am. Lab. (Fairfield, Conn.)* **16,** 90.
Brown, R. S., Cooper, J. R., and Wilkins, C. L. (1985). *Anal. Chem.* **57,** 2275.
Chiu, K. S., Biemann, K., Krishnan, K., and Hill, S. L. (1984). *Anal. Chem.* **56,** 1610.
Crawford, R. W., Hirschfeld, I., Sanborn, R. H., and Wong, C. M. (1982). *Anal. Chem.* **54,** 817.
de Haseth, J. A., and Isenhour, T. L. (1977). *Anal. Chem.* **49,** 1977.
Deveaux, M., and Huvenne, J. P. (1985). *Trends Anal. Chem.* **4,** 149.
Erickson, M. D. (1979). *Appl. Spectrosc. Rev.* **15,** 261.
Fehl, A. J., and Marcott, C. (1986). *Anal. Chem.* **58,** 2578.
Folster, U., and Herres, W. (1986). *Farbe + Lack* **92,** 13.
Garg, V. N., Bhatt, B. D., Kaushik, V. K., and Murthy, K. R. (1987). *J. Chromatogr. Sci.* **25,** 237.
Gordon, B. M., Uhrig, M. S., Borgerding, M. F., Chung, H. L., Coleman, W. M., Elder, J. F., Jr., Giles, J. A., Moore, D. S., Rix, C. E., and White, E. L. (1988). *J. Chromatogr. Sci.* **26,** 174.
Grainger, J., Reddy, G. T., and Patterson, D. G., Jr. (1988). *Appl. Spectrosc.* **42,** 800.
Griffiths, P. R. (1977). *Appl. Spectrosc.* **31,** 284.
Gurka, D. F. (1985). *Appl. Spectrosc.* **39,** 827.
Gurka, D. F., and Betowski, L. D. (1982). *Anal. Chem.* **54,** 1819.
Gurka, D. F., and Hiatt, M. (1984). *Anal. Chem.* **56,** 1102.
Gurka, D. F., and Laska, P. R. (1982). *J. Chromatogr. Sci.* **20,** 145.
Gurka, D. F., and Titus, R. (1986). *Anal. Chem.* **58,** 2189.

Gurka, D. F., Umana, M., Pellizzari, E. D., Moseley, A., and de Haseth, J. A. (1981). *Appl. Spectrosc.* **39,** 97.

Gurka, D. F., Titus, R., Griffiths, P. R., Henry, D., and Giorgetti, A. (1987). *Anal. Chem.* **59,** 2362.

Gurka, D. F., Betowski, L. D., Jones, T. L., and Pyle, S. M. (1988). *J. Chromatogr. Sci.* **26,** 301.

Hanna, D. A., Hanagac, G., Hohne, B. A., Small, G. W., Wiebolt, R. C., and Isenhour, T. L. (1979a). *J. Chromatogr. Sci.* **17,** 423.

Hanna, D. A, Hanagac, G., Hohne, B. A., Small, G. W., Wiebolt, R. C., and Isenhour, T. L. (1979b). *J. Chromatogr. Sci.* **17,** 434.

Herres, W., and Folster, U. (1985). *Farbe + Lack* **91,** 6.

Herres, W., Idstein, H., and Schreier, P., *J. High Resolut. Chromatogr. Chromatogr. Commun.* **6,** 590 (1983).

Herres, W., Schultze, W., and Kubeczka, K. H. (1986). *J. High Resolut. Chromatogr. Chromatogr. Commun.* **9,** 466.

Idstein, H., and Schreier, P. (1985). *J. Agric. Food Chem.* **33,** 138.

Kadowaki, E., and Namba, R. (1985). *Proc. Annu. Meet. Jpn. Chem. Soc., 50th.*

Kalasinsky, K. S. (1983). *J. Chromatogr. Sci.* **21,** 246.

Kalasinsky, K. S., and McDonald, J. T., Jr. (1983). *J. Chromatogr. Sci.* **21,** 193.

Kempert, K. (1988). *Appl. Spectrosc.* **42,** 845.

Kubeczka, K. H., and Herres, W. (1985). *Top. Flavour Res., Proc. Int. Conf.* p. 109.

Laude, D. A., Jr., Brissey, G. M., Ijames, C. F., Brown, R. S., and Wilkins, C. L. (1984). *Anal. Chem.* **56,** 163.

Laude, D. L., Jr., Johlman, C. L., Cooper, J. R., and Wilkins, C. L. (1985). *Anal. Chem.* **57,** 1044.

Laude, D. L., Jr., Johlman, C. L., Cooper, J. R., and Wilkins, C. L. (1986). *Fresenius' Z. Anal. Chem.* **324,** 839.

Le Quere, J. L., Semon, E., Latrasse, A, and Etrevant, P. (1987). *Sci. Ailments* **7,** 93.

Liebman, S. A., and Levy, E. J. (1983). *J. Chromatogr. Sci.* **21,** 1.

Liebman, S. A., and Wampler, T. P. (1985). *Chromatogr. Sci.* **23,** 53.

Low, M. J. D., and Freeman, S. K. (1967). *Anal. Chem.* **39,** 194.

Mossoba, M. M., Chen, J. T., Brumley, W. C., and Page, S. N. (1988). *Anal. Chem.* **60,** 948.

Namba, R. (1988). *Pap. Annu. Meet. Appl. Spectrosc. Jpn., 24th.*

Namba, R., Ohtsu, Y., Togano, S., and Morikawa, Y. (1981). *Pap. Annu. Meet. Appl. Spectrosc. Jpn., 17th.*

O'Connor, R. T., and Goldblatt, A. (1954). *Anal. Chem.* **28,** 1726.

Ohta, T., Yomogida, K., Omata, A., Morikawa, Y., Nakamura, S., and Suzuki, S. (1984). *Proc. Annu. Meet. TEAC, 28th.*

Olson, E. S., and Diehl, J. W. (1987). *Anal. Chem.* **59,** 443.

Olson, E. S., Diehl, J. W., and Froehlich, M. L. (1988). *Anal. Chem.* **60,** 1920.

Purcell, J. M., and Magidman, P. (1984). *Appl. Spectrosc.* **38,** 181.

Reedy, G. T., Ettinger, D. G., Schneider, J., and Bourne, S. (1985). *Anal. Chem.* **57,** 1602.

Schneider, J. F., Reedy, G. T., and Ettinger, D. G. (1985). *J. Chromatogr. Sci.* **23,** 49.

Schneider, J. F., Demirgian, J. C., and Stickler, J. (1986). *J. Chromatogr. Sci.* **24,** 330.

Shafer, K. H., and Griffiths, P. R. (1986). *Anal. Chem.* **58,** 3254.

Shafer, K. H., Cooke, M., DeRoos, F., Jakobsen, R. J., Rosario, O., and Mulik, J. D. (1981). *Appl. Spectrosc.* **35,** 469.

Shafer, K. H., Hayes, T. L., Brasch, J. W., and Jakobsen, R. J. (1984). *Anal. Chem.* **56,** 237.

Smith, S. L., Garlock, S. E., and Adams, G. E. (1983). *Appl. Spectrosc.* **37,** 192.

Wang, C. P., and Isenhour, T. L. (1987). *Anal. Chem.* **59,** 649.

Wilkins, C. L., Giss, G. N., White, R. L., Brissey, G. M., and Onyiriuka, E. C. (1982). *Anal. Chem.* **54,** 260.

Worley, J. W., Rueppel, M. L., and Rupel, F. L. (1980). *Anal. Chem.* **52,** 1845.

Wurrey, C. J., Bourne, S., and Kleopfer, R. D. (1986). *Anal. Chem.* **58,** 483.

Yoshida, S., Kadowaki, E., Namba, R., and Nakata, O. (1987). *Pap. Annu. Meet. Appl. Spectrosc. Jpn., 23rd.*

Index

A

D

E

F

S

V

W

X

Z